CALCULUS

SECOND EDITION

Student's Solutions Manual
Part II

CALCULUS

SECOND EDITION

Finney/Thomas

Student's Solutions Manual
Part II

Michael B. Schneider
Thomas L. Cochran

Addison-Wesley Publishing Company

Reading, Massachusetts • Menlo Park, California • New York
Don Mills, Ontario • Wokingham, England • Amsterdam • Bonn • Sydney
Singapore • Tokyo • Madrid • San Juan • Milan • Paris

The authors have attempted to make this manual as error free as possible but nobody's "perfeck". If you find errors, we would appreciate knowing them. You are more than welcome to write us at:

Belleville Area College
2500 Carlyle Road
Belleville, Illinois 62221

Reproduced by Addison-Wesley from camera-ready copy supplied by the authors.

ISBN 0-201-53423-1

1 2 3 4 5 6 7 8 9 10 BAH 9897969594

TABLE OF CONTENTS

Note: Student's answers to the critical-thinking exercises (indicated by ◆) may vary. For these exercises, we give one possible answer.

CALCULUS

SECOND EDITION

Student's Solutions Manual
Part II

CHAPTER 9

INFINITE SERIES

9.1 SEQUENCES

1. $a_1 = 0,\ a_2 = -\frac{1}{4},\ a_3 = -\frac{2}{9},\ a_4 = -\frac{3}{16}$

3. $a_1 = 1,\ a_2 = -\frac{1}{3},\ a_3 = \frac{1}{5},\ a_4 = -\frac{1}{7}$

5. $(-1)^2, (-1)^3, (-1)^4, (-1)^5, \ldots \Rightarrow a_n = (-1)^{n+1}$

7. $(-1)^2(1)^2, (-1)^3(2)^2, (-1)^4(3)^2, (-1)^5(4)^2, \ldots \Rightarrow a_n = (-1)^{n+1}(n)^2$

9. $1^2 - 1, 2^2 - 1, 3^2 - 1, 4^2 - 1, \ldots \Rightarrow a_n = n^2 - 1$

11. $4(1) - 3, 4(2) - 3, 4(3) - 3, \ldots \Rightarrow a_n = 4n - 3$

13. $1, \frac{3}{2}, \frac{7}{4}, \frac{15}{8}, \frac{31}{16}, \frac{63}{32}, \frac{127}{64}, \frac{255}{128}, \frac{511}{256}, \frac{1023}{512}$

15. $2, 1, -\frac{1}{2}, -\frac{1}{4}, \frac{1}{8}, \frac{1}{16}, -\frac{1}{32}, -\frac{1}{64}, \frac{1}{128}, \frac{1}{256}$

17. $(1), (1), (1+1), (1+1)+(1), (1+1+1)+(1+1), (1+1+1+1+1)+(1+1+1) \ldots = 1, 1, 2, 3, 5, 8,$ $13, 21, 34, 55$

19. $\lim_{n\to\infty} a_n = \lim_{n\to\infty} \left(2 + (0.1)^n\right) = 2 \Rightarrow$ converges

21. $\lim_{n\to\infty} a_n = \lim_{n\to\infty} \frac{1-2n}{1+2n} = -1 \Rightarrow$ converges

23. $\lim_{n\to\infty} a_n = \lim_{n\to\infty} \frac{1-5n^4}{n^4+8n^3} = -5 \Rightarrow$ converges

25. $\lim_{n\to\infty} a_n = \lim_{n\to\infty} \frac{n^2-2n+1}{n-1} = \lim_{n\to\infty} n - 1 = \infty \Rightarrow$ diverges

27. $\lim_{n\to\infty} a_n = \lim_{n\to\infty} \left(1 + (-1)^n\right)$ does not exist $\Rightarrow$ diverges

29. $\lim_{n\to\infty} a_n = \lim_{n\to\infty} \left(\frac{n+1}{2n}\right)\left(1 - \frac{1}{n}\right) = \frac{1}{2} \Rightarrow$ converges

31. $\lim_{n\to\infty} a_n = \lim_{n\to\infty} \frac{(-1)^{(n+1)}}{2n-1} = 0$

33. $\lim_{n\to\infty} a_n = \lim_{n\to\infty} \frac{\sin n}{n} = 0$, by the Sandwich Theorem for Sequences $\Rightarrow$ converges

35. $\lim_{n\to\infty} a_n = \lim_{n\to\infty} \sqrt{\frac{2n}{n+1}} = \sqrt{\lim_{n\to\infty} \frac{2n}{n+1}} = \sqrt{2} \Rightarrow$ converges

37. $\lim_{n\to\infty} a_n = \lim_{n\to\infty} \tan^{-1} n = \frac{\pi}{2} \Rightarrow$ converges

39. $\lim_{n\to\infty} a_n = \lim_{n\to\infty} \frac{n}{2^n} = \lim_{n\to\infty} \frac{1}{2^n \ln 2} = 0 \Rightarrow$ converges

41. $\lim_{n\to\infty} a_n = \lim_{n\to\infty} \frac{\ln(n+1)}{\sqrt{n}} = 0$, due to the growth rates in section 6.7 $\Rightarrow$ converges

43. $\lim_{n\to\infty} a_n = \lim_{n\to\infty} 8^{1/n} = \lim_{n\to\infty} e^{((\ln 8)/n)} = e^0 = 1 \Rightarrow$ converges

45. $\lim_{n\to\infty} a_n = \lim_{n\to\infty} \left(1+\frac{7}{n}\right)^n = e^7$, due to table 9.1 part 5 $\Rightarrow$ converges

47. $\lim_{n\to\infty} a_n = \lim_{n\to\infty} \frac{1}{(0.9)^n} = \lim_{n\to\infty} \left(\frac{10}{9}\right)^n = \infty \Rightarrow$ diverges

49. $\lim_{n\to\infty} a_n = \lim_{n\to\infty} \sqrt[n]{10n} = \lim_{n\to\infty} \exp\left(\frac{\ln(10n)}{n}\right) = \exp\left(\lim_{n\to\infty} \frac{\ln 10n}{n}\right) = \exp\left(\lim_{n\to\infty} \frac{1}{n}\right) = e^0 = 1 \Rightarrow$ converges;
$\exp[f(n)] = e^{f(n)}$.

51. $\lim_{n\to\infty} a_n = \lim_{n\to\infty} \left(\frac{3}{n}\right)^{1/n} = \frac{\lim_{n\to\infty} \sqrt[n]{3}}{\lim_{n\to\infty} \sqrt[n]{n}} = \frac{1}{1} = 1$, due to table 9.1, parts 2 and 3 $\Rightarrow$ converges

53. $\lim_{n\to\infty} a_n = \lim_{n\to\infty} \frac{\ln n}{n^{1/n}} = \frac{\lim_{n\to\infty} \ln n}{\lim_{n\to\infty} \sqrt[n]{n}} = \frac{\infty}{1} = \infty$, due to table 9.1, part 2 $\Rightarrow$ diverges

55. $\lim_{n\to\infty} \left(\frac{1}{3}\right)^n = 0$, due to table 9.1, part 4 and $\lim_{n\to\infty} \frac{1}{\sqrt{2^n}} = 0$ $\therefore$ $\lim_{n\to\infty} a_n = \lim_{n\to\infty} \left(\left(\frac{1}{3}\right)^n + \frac{1}{\sqrt{s^n}}\right) =$
$\lim_{n\to\infty} \left(\frac{1}{3}\right)^n + \lim_{n\to\infty} \frac{1}{\sqrt{2^n}} = 0 + 0 = 0 \Rightarrow$ converges

57. $\lim_{n\to\infty} a_n = \lim_{n\to\infty} \frac{n!}{n^n} = \lim_{n\to\infty} \frac{1\cdot 2\cdot 3\cdots(n-1)(n)}{n\cdot n\cdot n\cdots n\cdot n} \le \lim_{n\to\infty} \frac{1}{n} = 0$ and $\frac{n!}{n^n} \ge 0 \Rightarrow \lim_{n\to\infty} \frac{n!}{n^n} = 0 \Rightarrow$ converges

59. $\lim_{n\to\infty} a_n = \lim_{n\to\infty} \left(\frac{1}{n}\right)^{1/\ln(n)} = \lim_{n\to\infty} \exp\left(\frac{\ln(1/n)}{\ln(n)}\right) = \exp\left(\lim_{n\to\infty} \frac{\ln(1/n)}{\ln(n)}\right) = \exp(\lim_{n\to\infty}(-1)) = e^{-1} \Rightarrow$ converges;
note, $\exp[f(n)] = e^{f(n)}$.

61. $\lim_{n\to\infty} a_n = \lim_{n\to\infty} \frac{n!}{10^{6n}} = \frac{1}{\lim_{n\to\infty} \frac{(10^6)^n}{n!}} = \infty$, due to table 9.1, part 6 $\Rightarrow$ diverges

63. $\lim_{n\to\infty} a_n = \lim_{n\to\infty} \tanh n = \lim_{n\to\infty} \frac{e^n - e^{-n}}{e^n + e^{-n}} = \lim_{n\to\infty} \frac{e^{2n}-1}{e^{2n}+1} = \lim_{n\to\infty} \frac{2e^{2n}}{2e^{2n}} = \lim_{n\to\infty} 1 = 1 \Rightarrow$ converges

65. $\lim_{n\to\infty} a_n = \lim_{x\to\infty} \ln\left(1+\frac{1}{n}\right)^n = \ln\left(\lim_{n\to\infty}\left(1+\frac{1}{n}\right)^n\right) = \ln e = 1 \Rightarrow$ converges

67. $\lim_{n\to\infty} a_n = \lim_{n\to\infty} \frac{\int_1^n \frac{1}{x}\,dx}{n} = \lim_{n\to\infty} \frac{\frac{1}{n}}{1} = 0 \Rightarrow$ converges

69. $\lim_{n\to\infty} a_n = \lim_{n\to\infty} \frac{n^2 \sin(1/n)}{2n-1} = \lim_{n\to\infty} \frac{\sin\left(\frac{1}{n}\right)}{\frac{2}{n}-\frac{1}{n^2}} = \lim_{n\to\infty} \frac{-\left(\cos\left(\frac{1}{n}\right)\right)\left(\frac{1}{n^2}\right)}{-\frac{2}{n^2}+\frac{2}{n^3}} = \lim_{n\to\infty} \frac{-\cos(1/n)}{-2+2/n} = \frac{1}{2} \Rightarrow$ converges

71. a) True, by definition b) False. The sequence $\{(-1)^n\}$ diverges. c) False. Limits are unique
d) True. If a sequence converges to a limit L, then all subsequences of $\{a_n\}$ also converge to L.
e) False. The subsequence 1, 1, 1, ... of $\{(-1)^n\}$ converges to 1.

73. $\left|\sqrt[n]{0.5} - 1\right| < 10^{-3} \Rightarrow -\frac{1}{1000} < \left(\frac{1}{2}\right)^{1/n} - 1 < \frac{1}{1000} \Rightarrow n > \frac{\ln(1/2)}{\ln(999/1000)} \Rightarrow n > 692.8 \Rightarrow N = 692$

75. $(0.9)^n < 10^{-3} \Rightarrow n\ln(0.9) < -3\ln 10 \Rightarrow n > 65.56 \Rightarrow N = 65$

77. a) No, $\lim_{n \to \infty} A_n = \lim_{n \to \infty} A_o\left(1 + \frac{r}{12}\right)^n = A_o \lim_{n \to \infty}\left(\frac{12+r}{12}\right)^n = \infty$, since $\frac{12+r}{12} > 1$

b) $\{A_n\} = \left\{(\$\,1000)\left(1 + \frac{0.02583^n}{12}\right)\right\} = \left\{(\$\,1000)(1.0021525)^n\right\} \Rightarrow A_1 = \$\,1002.15$, $A_2 = \$\,1004.31$, $A_3 =$ $\$\,1006.47$, $A_4 = \$\,1008.64$, $A_5 = \$\,1010.81$, $A_6 = \$\,1012.98$, $A_7 = \$\,1015.17$, $A_8 = \$\,1017.35$, $A_9 = \$\,1019.54$, $A_{10} = \$\,1021.73$

79. $1, 1, 2, 4, 8, 16, 32, \ldots = 1, 2^0, 2^1, 2^2, 2^3, 2^4, 2^5, \ldots \Rightarrow x_1 = 1$ and $x_n = 2^{n-2}$ for $n \geq 2$

81. a) $f(x) = x^2 - 2$; the sequence converges to $1.414213562 \approx \sqrt{2}$

b) $f(x) = \tan(x) - 1$; the sequence converges to $0.7853981635 \approx \frac{\pi}{4}$

c) $f(x) = e^x$; the sequence $1, 0, -1, -2, -3, -4, -5 \cdots$ diverges

83. $x_1 = 1$, $x_2 = 1 + \cos(1) \approx 1.540302305$, $x_3 = 1 + \cos(1) + \cos(1 + \cos(1)) \approx 1.570791601$,
$x_4 = 1 + \cos(1) + \cos(1 + \cos(1)) + \cos(1 + \cos(1) + \cos(1)) \approx 1.570796327$

9.2 INFINITE SERIES

1. $s_n = \frac{a\left(1 - r^n\right)}{(1 - r)} = \frac{2\left(1 - (1/3)^n\right)}{1 - 1/3} \Rightarrow \lim_{n \to \infty} s_n = \frac{2}{1 - 1/3} = 3$

3. $s_n = \frac{a\left(1 - r^n\right)}{(1 - r)} = \frac{1 - (-1/2)^n}{1 - (-1/2)} \Rightarrow \lim_{n \to \infty} s_n = \frac{1}{3/2} = \frac{2}{3}$

5. $\frac{1}{(n+1)(n+2)} = \frac{1}{n+1} - \frac{1}{n+2} \Rightarrow s_n = \left(\frac{1}{2} - \frac{1}{3}\right) + \left(\frac{1}{3} - \frac{1}{4}\right) + \ldots + \left(\frac{1}{n+1} - \frac{1}{n+2}\right) = \frac{1}{2} - \frac{1}{n+2} \Rightarrow \lim_{n \to \infty} s_n = \frac{1}{2}$

7. $1 - \frac{1}{4} + \frac{1}{16} - \frac{1}{64} + \cdots$, the sum of this geometric series is $\frac{1}{1 + \frac{1}{4}} = \frac{4}{5}$

9. $\frac{7}{4} + \frac{7}{16} + \frac{7}{64} + \ldots$, the sum of this geometric series is $\frac{7/4}{1 - 1/4} = \frac{7}{3}$

11. $(5 + 1) + \left(\frac{5}{2} + \frac{1}{3}\right) + \left(\frac{5}{4} + \frac{1}{9}\right) + \left(\frac{5}{8} + \frac{1}{27}\right) + \ldots$, is the sum of two geometric series; the sum
$\frac{5}{1 - 1/2} + \frac{1}{1 - 1/3} = 10 + \frac{3}{2} = \frac{23}{2}$

13. $(1 + 1) + \left(\frac{1}{2} - \frac{1}{5}\right) + \left(\frac{1}{4} + \frac{1}{25}\right) + \left(\frac{1}{8} - \frac{1}{125}\right) + \ldots$, is the sum of two geometric series; the sum
$\frac{1}{1 - 1/2} + \frac{1}{1 + 1/5} = 2 + \frac{5}{6} = \frac{17}{6}$

15. $\frac{4}{(4n-3)(4n+1)} = \frac{1}{4n-3} - \frac{1}{4n+1} \Rightarrow s_n = \left(1 - \frac{1}{5}\right) + \left(\frac{1}{5} - \frac{1}{9}\right) + \left(\frac{1}{9} - \frac{1}{13}\right) + \ldots + \left(\frac{1}{4n-7} - \frac{1}{4n-3}\right) +$
$\left(\frac{1}{4n-3} - \frac{1}{4n+1}\right) = 1 - \frac{1}{4n+1} \Rightarrow \lim_{n\to\infty} s_n = \lim_{n\to\infty} 1 - \frac{1}{4n+1} = 1$

17. $\frac{40n}{(2n-1)^2(2n+1)^2} = \frac{A}{(2n-1)} + \frac{B}{(2n-1)^2} + \frac{C}{(2n+1)} + \frac{D}{(2n+1)^2} =$
$\frac{A(2n-1)(2n+1)^2 + B(2n+1)^2 + C(2n+1)(2n-1)^2 + D(2n-1)^2}{(2n-1)^2(2n+1)^2} = \Rightarrow$
$A(2n-1)(2n+1)^2 + B(2n+1)^2 + C(2n+1)(2n-1)^2 + D(2n-1)^2 = 40n \Rightarrow A\left(8n^3 + 4n^2 - 2n - 1\right) +$
$B\left(4n^2 + 4n + 1\right) + C\left(8n^3 - 4n^2 - 2n + 1\right) + D\left(4n^2 - 4n + 1\right) = 40n \Rightarrow (8A + 8C)n^3 + (4A + 4B - 4C + 4D)n^2 +$
$(-2A + 4B - 2C - 4D)n + (-A + B + C + D) = 40n \Rightarrow \begin{cases} 8A + 8C = 0 \\ 4A + 4B - 4C + 4D = 0 \\ -2A + 4B - 2C - 4D = 40 \\ -A + B + C + D = 0 \end{cases} \Rightarrow \begin{cases} 8A + 8C = 0 \\ A + B - C + D = 0 \\ -A + 2B - C - 2D = 20 \\ -A + B + C + D = 0 \end{cases} \Rightarrow$
$\begin{cases} B + D = 0 \\ 2B - 2D = 20 \end{cases} \Rightarrow 4B = 20 \Rightarrow B = 5 \text{ and } D = -5 \Rightarrow \begin{cases} A + C = 0 \\ -A + 5 + C - 5 = 0 \end{cases} \Rightarrow C = 0 \text{ and } A = 0.$ Hence,
$\sum_{n=1}^{k} \left(\frac{40n}{(2n-1)^2(2n+1)^2}\right) = 5 \sum_{n=1}^{k} \left(\frac{1}{(2n-1)^2} - \frac{1}{(2n+1)^2}\right) =$
$5\left(\frac{1}{1} - \frac{1}{9} + \frac{1}{9} - \frac{1}{25} + \frac{1}{25} - \ldots - \frac{1}{\left(2(n-1)+1\right)^2} + \frac{1}{(2n-1)^2} - \frac{1}{(2n+1)^2}\right) = 5\left(1 - \frac{1}{(2n+1)^2}\right).$ Therefore, the sum is
$\lim_{n\to\infty} 5\left(1 - \frac{1}{(2n+1)^2}\right) = 5.$

19. A convergent geometric series with a sum of $\frac{1}{1 - 1/\sqrt{2}} = 2 + \sqrt{2}.$

21. A convergent geometric series with a sum of $\frac{3/2}{1 - (-1/2)} = 1.$

23. $\lim_{n\to\infty} \cos(n\pi) = \lim_{n\to\infty} (-1)^n \neq 0 \Rightarrow$ divergence.

25. A convergent geometric series with a sum of $\frac{1}{1 - 1/e^2} = \frac{e^2}{e^2 - 1}.$

27. $\lim_{n\to\infty} a_n = \lim_{n\to\infty} (-1)^{n+1} n \neq 0 \Rightarrow$ divergence.

29. The difference of two convergent geometric series with a sum of $\frac{1}{1 - 2/3} - \frac{1}{1 - 1/3} = 3 - \frac{3}{2} = \frac{3}{2}.$

31. $\lim_{n\to\infty} a_n = \lim_{n\to\infty} \frac{n!}{1000^n} = \frac{1}{\lim_{n\to\infty} \frac{1000^n}{n!}} = \infty \neq 0 \Rightarrow$ divergence. The limit $\lim_{n\to\infty} \frac{1000^n}{n!} = 0$ by table 9.1 part 6

33. $\sum_{n=1}^{\infty} \ln\left(\frac{n}{n+1}\right) = \sum_{n=1}^{\infty} \ln(n) - \ln(n+1) \Rightarrow s_n = \left(\ln(1) - \ln(2)\right) + \left(\ln(2) - \ln(3)\right) + \left(\ln(3) - \ln(4)\right) + \ldots +$
$\left(\ln(n-1) - \ln(n)\right) + \left(\ln(n) - \ln(n+1)\right) = \ln(1) - \ln(n+1) = -\ln(n+1) \Rightarrow \lim_{n\to\infty} s_n = -\infty, \Rightarrow$ the series diverges

35. $s_n = \left(1 - \frac{1}{\sqrt{2}}\right) + \left(\frac{1}{\sqrt{2}} - \frac{1}{\sqrt{3}}\right) + \left(\frac{1}{\sqrt{3}} - \frac{1}{2}\right) + \cdots + \left(\frac{1}{\sqrt{n-1}} - \frac{1}{\sqrt{n}}\right) + \left(\frac{1}{\sqrt{n}} - \frac{1}{\sqrt{n+1}}\right) = 1 - \frac{1}{\sqrt{n+1}} \Rightarrow$
$\lim_{n \to \infty} s_n = 1$. $\therefore$ the series converges to 1.

37. $\sum_{n=1}^{\infty} \left(\frac{1}{\ln(n+2)} - \frac{1}{\ln(n+1)}\right) \Rightarrow s_n = \left(\frac{1}{\ln 3} - \frac{1}{\ln 2}\right) + \left(\frac{1}{\ln 4} - \frac{1}{\ln 3}\right) + \left(\frac{1}{\ln 5} - \frac{1}{\ln 4}\right) + \ldots + \left(\frac{1}{\ln(n+1)} - \frac{1}{\ln(n)}\right) +$
$\left(\frac{1}{\ln(n+2)} - \frac{1}{\ln(n+1)}\right) = -\frac{1}{\ln 2} + \frac{1}{\ln(n+2)} \Rightarrow \lim_{n \to \infty} s_n = \lim_{n \to \infty} \left(-\frac{1}{\ln 2} + \frac{1}{\ln(n+2)}\right) = -\frac{1}{\ln 2} \Rightarrow$ the series converges

39. $a = 1, r = -x$ where $|x| < 1$

41. $a = 3, r = \frac{x-1}{2}$ where $\left|\frac{x-1}{2}\right| < 1 \Rightarrow -1 < x < 3$

43. $0.\overline{23} = \frac{23}{100} + \frac{23}{100}\left(\frac{1}{10^2}\right) + \frac{23}{100}\left(\frac{1}{10^2}\right)^2 + \cdots = \frac{23}{99}$

45. $0.\overline{7} = \frac{7}{10} + \frac{7}{10}\left(\frac{1}{10}\right) + \frac{7}{10}\left(\frac{1}{10}\right)^2 + \cdots = \frac{7}{9}$

47. $0.0\overline{6} = \frac{6}{100} + \frac{6}{100}\left(\frac{1}{10}\right) + \frac{6}{100}\left(\frac{1}{10}\right)^2 + \frac{6}{100}\left(\frac{1}{10}\right)^3 + \cdots = \frac{1}{15}$

49. $1.24\overline{123} = \frac{31}{25} + \frac{123}{10^5} + \frac{123}{10^5}\left(\frac{1}{10^3}\right) + \frac{123}{10^5}\left(\frac{1}{10^3}\right)^2 + \cdots = \frac{31}{25} + \frac{123}{99900} = \frac{41333}{33300}$

51. distance $= 4 + 2\left[(4)\left(\frac{3}{4}\right) + (4)\left(\frac{3}{4}\right)^2 + \ldots\right] = 4 + 2\left[\frac{3}{1 - 3/4}\right] = 28$ m

53. a) $\sum_{n=-2}^{\infty} \frac{1}{(n+4)(n+5)}$ b) $\sum_{n=0}^{\infty} \frac{1}{(n+2)(n+3)}$ c) $\sum_{n=5}^{\infty} \frac{1}{(n-3)(n-2)}$

55. area $= 2^2 + (\sqrt{2})^2 + (1)^2 + \left(\frac{1}{\sqrt{2}}\right)^2 + \ldots = 4 + 2 + 1 + \frac{1}{2} + \ldots = \frac{4}{1 - 1/2} = 8\ \text{m}^2$

57. a) $L_1 = 3, L_2 = 3\left(\frac{4}{3}\right), L_3 = 3\left(\frac{4}{3}\right)^2, \cdots, L_n = 3\left(\frac{4}{3}\right)^{n-1}$. $\therefore \lim_{n \to \infty} L_n = \lim_{n \to \infty} 3\left(\frac{4}{3}\right)^{n-1} = \infty$

b) $A_1 = \frac{1}{2}(1)\left(\frac{\sqrt{3}}{2}\right) = \frac{\sqrt{3}}{4}, A_2 = A_1 + 3\left(\frac{1}{2}\right)\left(\frac{1}{3}\right)\left(\frac{\sqrt{3}}{6}\right) = \frac{\sqrt{3}}{4} + \frac{\sqrt{3}}{12}, A_3 = A_1 + A_2 + 12\left(\frac{1}{2}\right)\left(\frac{1}{9}\right)\left(\frac{\sqrt{3}}{18}\right) =$
$\frac{\sqrt{3}}{4} + \frac{\sqrt{3}}{12} + \frac{\sqrt{3}}{27}, A_4 = A_1 + A_2 + 48\left(\frac{1}{2}\right)\left(\frac{1}{27}\right)\left(\frac{\sqrt{3}}{54}\right), \cdots, A_n = \frac{\sqrt{3}}{4} + \frac{27\sqrt{3}}{64}\left(\frac{4}{9}\right)^2 +$

$\frac{27\sqrt{3}}{64}\left(\frac{4}{9}\right)^3 + \cdots = \frac{\sqrt{3}}{4} + \sum_{n=2}^{\infty} \frac{27\sqrt{3}}{64}\left(\frac{4}{9}\right)^n = \frac{2\sqrt{3}}{5}$

59. Let $a_n = b_n = (1/2)^n$. Then $\sum_{n=1}^{\infty} a_n = \sum_{n=1}^{\infty} b_n = \sum_{n=1}^{\infty} (1/2)^n = 1$, while $\sum_{n=1}^{\infty} (a_n/b_n) = \sum_{n=1}^{\infty} (1)$ diverges.

61. Let $a_n = (1/4)^n$ and $b_n = (1/2)^n$. Then $A = \sum_{n=1}^{\infty} a_n = 1/3$, $B = \sum_{n=1}^{\infty} b_n = 1$ and $\sum_{n=1}^{\infty} (a_n/b_n) =$

$\sum_{n=1}^{\infty} (1/2)^n = 1 \neq A/B$.

63. Yes: $\sum (1/a_n)$ diverges. The reasoning: $\sum a_n$ converges $\Rightarrow a_n \to 0 \Rightarrow 1/a_n \to \infty \Rightarrow \sum (1/a_n)$ diverges

by the nth–Term Test.

9.3 SERIES WITH NONNEGATIVE TERMS–COMPARISON AND INTEGRAL TESTS

1. converges, a geometric series with r = 1/10
3. diverges, by the nth–Term Test for Divergence, since $\lim\limits_{n \to \infty} \frac{n}{n+1} = 1 \neq 0$
5. diverges, by the Limit Comparison Test when compared with $\sum\limits_{n=1}^{\infty} \frac{1}{\sqrt{n}}$ a divergent p–series
7. converges, by the Direct Comparison Test for Convergence, since $\frac{\sin^2 n}{2^n} \leq \frac{1}{2^n}$, the nth term of a convergent geometric series
9. converges, a geometric series with r = 1/8
11. diverges, by the Direct Comparison Test for Convergence, since $\frac{1}{n} < \frac{\ln n}{n}$ for $n \geq 3$
13. converges, a geometric series with r = 2/3
15. diverges, by the Direct Comparison Test for Convergence, since $\frac{1}{n+1} < \frac{1}{1 + \ln n}$ and $\sum\limits_{n=1}^{\infty} \frac{1}{n+1}$ diverges by the Limit Comparison Test when compared with $\sum\limits_{n=1}^{\infty} \frac{1}{n}$
17. diverges, $\lim\limits_{n \to \infty} a_n = \lim\limits_{n \to \infty} \frac{2^n}{n+1} = \lim\limits_{n \to \infty} \frac{2^n \ln 2}{1} = \infty \neq 0$
19. converges, by the Limit Comparison Test when compared with $\sum\limits_{n=1}^{\infty} \frac{1}{n^{3/2}}$
21. diverges, by the Direct Comparison Test for Convergence, since $\frac{1}{n} \leq \frac{\sqrt{n}}{n} \leq \frac{\sqrt{n}}{\ln n}$ for $n \geq 2$
23. diverges, by the Direct Comparison Test since $n > \ln n \Rightarrow \frac{1}{n} < \frac{1}{\ln n}, \frac{1}{\ln n} < \frac{1}{\ln(\ln n)}$ and the harmonic series diverges
25. converges, by the Limit Comparison Test when compared with $\sum\limits_{n=1}^{\infty} \frac{1}{n^2}$, a convergent p–series, for

$$\lim_{n \to \infty} \frac{(\ln n)^2 / n^3}{1/n^2} = \lim_{n \to \infty} \frac{(\ln n)^2}{n} = \lim_{n \to \infty} \frac{2(\ln n)(1/n)}{1} = 2 \lim_{n \to \infty} \frac{\ln n}{n} = 2 \lim_{n \to \infty} \frac{1/n}{1} = 2 \cdot 0 = 0$$

27. diverges: If $f(x) = \sqrt[x]{x} \Rightarrow f'(x) > 0$ when $1 < x < e$ and $f'(x) < 0$ when $x > e$, then $\sqrt[e]{e} > \sqrt[n]{n}$ for all $n \geq 3$.

Also, $3^e > e \Rightarrow 3 > \sqrt[e]{e}$. Consequently, $3n > n\sqrt[n]{n} \Rightarrow \frac{1}{3n} < \frac{1}{n\sqrt[n]{n}} \Rightarrow \sum_{n=1}^{\infty} \frac{1}{n\sqrt[n]{n}}$ diverges, by the Direct Comparison Test for Convergence.

29. diverges, $\lim_{n \to \infty} a_n = \lim_{n \to \infty} \left(1 + \frac{1}{n}\right)^n = e \neq 0$

31. converges, because $\sum_{n=1}^{\infty} \frac{1-n}{n\,2^n} = \sum_{n=1}^{\infty} \frac{1}{n\,2^n} + \sum_{n=1}^{\infty} \frac{-1}{2^n}$ which is the sum of two convergent series, because $\sum_{n=1}^{\infty} \frac{1}{n\,2^n}$ converges, by the Direct Comparison Test for Convergence, since $\frac{1}{n\,2^n} < \frac{1}{2^n}$ and $\sum_{n=1}^{\infty} \frac{-1}{2^n}$ is a convergent geometric series

33. diverges, a geometric series with $r = 1/\ln 2 \approx 1.44 > 1$

35. converges, by the Direct Comparison Test for Convergence, since $\frac{1}{3^{n-1}+1} < \frac{1}{3^{n-1}}$

37. diverges, by the Limit Comparison Test when compared to the divergent harmonic series since $\lim_{n \to \infty} \frac{\sin(1/n)}{1/n} = \lim_{x \to 0} \frac{\sin x}{x} = 1 \neq 0$

39. diverges, by the nth Term Test for divergence for $\lim_{n \to \infty} n \sin(1/n) = \lim_{n \to \infty} \frac{\sin(1/n)}{1/n} = \lim_{x \to 0} \frac{\sin x}{x} = 1 \neq 0$

41. converges, by the Integral Test for $\int_1^{\infty} \frac{\tan^{-1} x}{1+x^2}\,dx = \int_{\pi/4}^{\pi/2} u\,du = u = \left[\frac{u^2}{2}\right]_{\pi/4}^{\pi/2} = \frac{1}{2}\left[(\pi/2)^2 - (\pi/4)^2\right] = \frac{1}{2}\left[\frac{\pi^2}{4} - \frac{\pi^2}{16}\right] = \frac{\pi^2}{8} - \frac{\pi^2}{32} = \frac{4\pi^2 - \pi^2}{32} = \frac{3\pi^2}{32}$ where $u = \tan^{-1} x$, $du = \frac{1}{1+x^2}\,dx$, the lower limit $u = \tan^{-1} 1 = \pi/4$ and the upper limit $u = \lim_{x \to \infty} \tan^{-1} x = \pi/2$

43. converges, by the Limit Comparison Test when compared with $\sum_{n=1}^{\infty} \frac{1}{n^2}$, a convergent p–series, for

$$\lim_{n \to \infty} \frac{(10n+1)/\left(n(n+1)(n+2)\right)}{1/n^2} = \lim_{n \to \infty} \frac{10n^2+n}{(n+1)(n+2)} = 10, \text{ since } \frac{10n^2+n}{(n+1)(n+2)} \text{ is a rational function}$$

45. converges, by the Direct Comparison Test for Convergence, since $\frac{\tan^{-1} n}{n^{1.1}} < \frac{\pi/2}{n^{1.1}}$ and $\sum_{n=1}^{\infty} \frac{\pi/2}{n^{1.1}} = \frac{\pi}{2} \sum_{n=1}^{\infty} \frac{1}{n^{1.1}}$ the product of a convergent p–series and a nonzero constant

47. converges, by the Integral test for $\int_3^{\infty} \frac{1/x}{(\ln x)\sqrt{(\ln x)^2 - 1}}\,dx = \int_{\ln 3}^{\infty} \frac{1}{u\sqrt{u^2-1}}\,du = \lim\limits_{b \to \infty} \left[\sec^{-1}|u|\right]_{\ln 3}^{b} =$

$\lim\limits_{b \to \infty} \left[\sec^{-1}|b| - \sec^{-1}|\ln 3|\right] = \lim\limits_{b \to \infty} \left[\cos^{-1}(1/b) - \sec^{-1}|\ln 3|\right] = \cos^{-1}(0) - \sec^{-1}|\ln 3| = \frac{\pi}{2} - \sec^{-1}|\ln 3| \approx 1.1439,$

where $u = \ln x$, $du = \frac{1}{x}\,dx$, the lower limit is $u = \ln 3$ and the upper limit is $u = \lim\limits_{x \to \infty} \ln x = \infty$

49. converges, by the Limit Comparison Test when compared with $\sum\limits_{n=1}^{\infty} \frac{1}{e^n}$, a convergent geometric series, for

$$\lim_{n \to \infty} \frac{2/\left(1 + e^n\right)}{1/e^n} = \lim_{n \to \infty} \frac{2e^n}{1 + e^n} = \lim_{n \to \infty} \frac{2e^n}{e^n} = \lim_{n \to \infty} 2 = 2$$

51. converges, by the Integral Test, since $\int_1^{\infty} \operatorname{sech} x\,dx = \int_1^{\infty} \frac{2}{e^x + e^{-x}}\,dx = \int_1^{\infty} \frac{2}{e^{-x} + e^x} \cdot \frac{e^x}{e^x}\,dx =$

$$2\lim_{b \to \infty} \int_1^b \frac{e^x}{1 + \left(e^x\right)^2}\,dx = 2\lim_{b \to \infty} \left[\tan^{-1} e^x\right]_1^b = 2\lim_{b \to \infty} \left[\tan^{-1} e^b - \tan^{-1} e\right] = \pi - 2\tan^{-1} e \approx 0{,}71$$

53. converges, by the Limit Comparison Test when compared with $\sum\limits_{n=1}^{\infty} \frac{1}{n^2}$, a convergent p–series, for

$$\lim_{n \to \infty} \frac{(\coth n)/n^2}{1/n^2} = \lim_{n \to \infty} \coth n = \lim_{n \to \infty} \frac{e^n + e^{-n}}{e^n - e^{-n}} = \lim_{n \to \infty} \frac{1 + 1/e^{2n}}{1 - 1/e^{2n}} = 1$$

55. a)

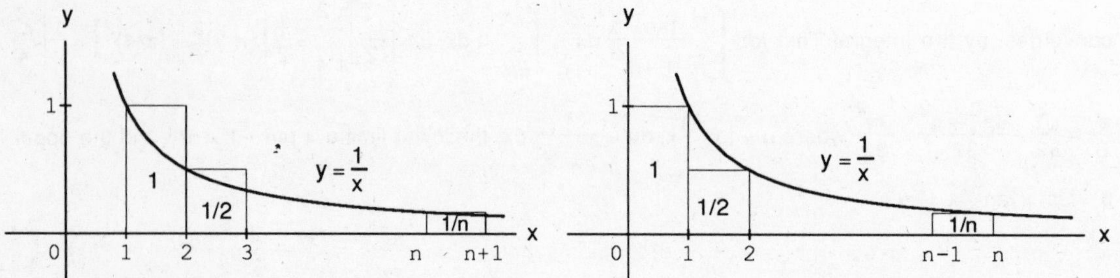

$\int_1^{n+1} \frac{1}{x}\,dx < 1 + \frac{1}{2} + \ldots + \frac{1}{n}$ Graph 9.3.55 $1 + \frac{1}{2} + \ldots + \frac{1}{n} < 1 + \int_1^n \frac{1}{x}\,dx$

b) There are $(13)(365)(24)(60)(60)(10^9)$ seconds in 13 billion years. $s_n \le 1 + \ln n$ where $n =$ $(13)(365)(24)(60)(60)(10^9) \Rightarrow s_n \le 1 + \ln\left((13)(365)(24)(60)(60)(10^9)\right) = 1 + \ln(13) + \ln(365) + \ln(24) +$ $2\ln(60) + 9\ln(10) \approx 41.55$

57. Yes, If $\sum\limits_{n=1}^{\infty} a_n$ is a divergent series of positive numbers, then $(1/2)\sum\limits_{n=1}^{\infty} a_n = \sum\limits_{n=1}^{\infty} \left(a_n/2\right)$ also diverges.

There is no "smallest" divergent series of positive numbers. For any divergent series of positive numbers $\sum_{n=1}^{\infty} a_n$, $\sum_{n=1}^{\infty} \left(a_n/2\right)$ has smaller terms and still diverges.

59. If $\{S_n\}$ is nonincreasing with lower bound M, then $\{-S_n\}$ is a nondecreasing sequence with upper bound – M. By Theorem 7, $\{-S_n\}$ converges, and hence, $\{S_n\}$ converges. If $\{S_n\}$ has no lower bound, then $\{-S_n\}$ has no upper bound and diverges. Hence, $\{S_n\}$ also diverges.

61. a) If $p > 1$, then $\int_2^{\infty} \frac{dx}{x(\ln x)^p} dx = \lim_{b \to \infty} \int_2^b (\ln x)^{-p} \frac{1}{x} dx = \lim_{b \to \infty} \left[\frac{(\ln x)^{1-p}}{1-p}\right]_2^b =$ $\left(\frac{1}{1-p}\right) \lim_{b \to \infty} \left[(\ln b)^{1-p} - (\ln 2)^{1-p}\right] = \frac{(\ln 2)^{1-p}}{p-1}$. If $p = 1$, then $\int_2^{\infty} \frac{dx}{x \ln x} = \lim_{b \to \infty} \int_2^b \frac{1/x}{\ln x} dx =$ $\lim_{b \to \infty} \left[\ln|\ln x|\right]_2^b = \infty$. If $p < 1$, then $\int_2^{\infty} \frac{dx}{x(\ln x)^p} dx = \frac{1}{1-p} \lim_{b \to \infty} \left[(\ln b)^{1-p} - (\ln 2)^{1-p}\right] = \infty$.

b) By the Integral Test, the series would converge if $p > 1$ and diverge if $p \le 1$.

9.4 SERIES WITH NONNEGATIVE TERMS–RATIO AND ROOT TESTS

1. converges, by the Ratio Test for $\lim_{n \to \infty} \frac{a_{n+1}}{a_n} = \lim_{n \to \infty} \left|\frac{(n+1)^2}{2^{n+1}} \frac{2^n}{n^2}\right| = \frac{1}{2} < 1$

3. diverges, by the Ratio Test for $\lim_{n \to \infty} \frac{a_{n+1}}{a_n} = \lim_{n \to \infty} \left|\frac{(n+1)!\, e^{-(n+1)}}{n!\, e^{-n}}\right| = \infty$

5. converges, by the Ratio Test for $\lim_{n \to \infty} \frac{a_{n+1}}{a_n} = \lim_{n \to \infty} \left|\frac{(n+1)^{10}}{10^{n+1}} \frac{10^n}{n^{10}}\right| = \frac{1}{10} < 1$

7. converges, by the Direct Comparison Test for Convergence, since $\frac{2 + (-1)^n}{(1.25)^n} = \left(\frac{4}{5}\right)^n \left[2 + (-1)^n\right] \le \left(\frac{4}{5}\right)^n (3)$

9. diverges, $\lim_{n \to \infty} a_n = \lim_{n \to \infty} \left(1 - \frac{3}{n}\right)^n = \lim_{n \to \infty} \left(1 + \frac{-3}{n}\right)^n = e^{-3} \approx 0.05 \ne 0$ (table 9.1)

11. converges, by the Direct Comparison Test for Convergence, since $\frac{\ln n}{n^3} < \frac{n}{n^3} = \frac{1}{n^2}$ for $n \ge 2$

13. diverges, $\sum_{n=1}^{\infty} \frac{n-1}{n^2} = \sum_{n=1}^{\infty} \frac{1}{n} - \sum_{n=1}^{\infty} \frac{1}{n^2}$, the difference between a divergent and convergent series diverges

15. diverges, by the Direct Comparison Test for Convergence, since $\frac{\ln n}{n} > \frac{1}{n}$ for $n \ge 3$

17. converges, by the Ratio Test for $\lim_{n \to \infty} \frac{a_{n+1}}{a_n} = \lim_{n \to \infty} \left|\frac{(n+2)(n+3)}{(n+1)!} \frac{n!}{(n+1)(n+2)}\right| = 0 < 1$

19. converges, by the Ratio Test for $\lim_{n\to\infty} \frac{a_{n+1}}{a_n} = \lim_{n\to\infty} \left| \frac{(n+4)!}{3!\,(n+1)!\,3^{n+1}} \frac{3!\,n!\,3^n}{(n+3)!} \right| = \frac{1}{3} < 1$

21. converges, by the Ratio Test for $\lim_{n\to\infty} \frac{a_{n+1}}{a_n} = \lim_{n\to\infty} \left| \frac{1}{(2n+3)!} \frac{(2n+1)!}{1} \right| = 0 < 1$

23. converges, by the Root Test for $\lim_{n\to\infty} \sqrt[n]{a_n} = \lim_{n\to\infty} \sqrt[n]{\frac{n}{(\ln n)^n}} = \lim_{n\to\infty} \frac{\sqrt[n]{n}}{\ln n} = \lim_{n\to\infty} \frac{1}{\ln n} = 0 < 1$

25. converges, by the Direct Comparison Test for Convergence, since $\frac{n!\ln n}{n(n+2)!} = \frac{\ln n}{n(n+1)(n+2)} < \frac{1}{(n+1)(n+2)} <$

$\frac{1}{n^2}$ and $\sum_{n=1}^{\infty} \frac{1}{n^2}$ is a convergence p–series

27. converges, by the Ratio Test for $\lim_{n\to\infty} \frac{a_{n+1}}{a_n} = \lim_{n\to\infty} \frac{\left(\frac{1+\sin n}{n}\right) a_n}{a_n} = 0 < 1$

29. diverges, by the Ratio Test for $\lim_{n\to\infty} \frac{a_{n+1}}{a_n} = \lim_{n\to\infty} \left| \frac{\left(\frac{3n-1}{2n+5}\right) a_n}{a_n} \right| = \frac{3}{2} > 1$, since $\frac{3n-1}{2n+5}$ is a rational function

31. converges, by the Ratio Test for $\lim_{n\to\infty} \frac{a_{n+1}}{a_n} = \lim_{n\to\infty} \left| \frac{\left(\frac{2}{n}\right)(a_n)}{a_n} \right| = 0 < 1$

33. converges, by the Ratio Test for $\lim_{n\to\infty} \frac{a_{n+1}}{a_n} = \lim_{n\to\infty} \frac{\frac{\ln n}{n} a_n}{a_n} = \lim_{n\to\infty} \frac{\ln n}{n} = \lim_{n\to\infty} \frac{1/n}{1} = 0 < 1$

35. diverges, by the nth–Term Test for Divergence for the sequence $a_1 = \frac{1}{3}$, $a_2 = \sqrt[2]{\frac{1}{3}}$, $a_3 = \sqrt[3]{\sqrt[2]{\frac{1}{3}}} = \sqrt[6]{\frac{1}{3}}$,

$a_4 = \sqrt[4]{\sqrt[3]{\sqrt[2]{\frac{1}{3}}}} = \sqrt[24]{\frac{1}{3}}$, ... indicates $a_n = \sqrt[n!]{\frac{1}{3}}$ and consequently $\lim_{n\to\infty} a_n = 1$ because $\left\{\sqrt[n!]{\frac{1}{3}}\right\}$ is a

subsequence of $\left\{\sqrt[n]{\frac{1}{3}}\right\}$ whose limit is 1 by table 9.1

37. converges, by the Ratio Test for $\lim_{n\to\infty} \frac{a_{n+1}}{a_n} = \lim_{n\to\infty} \left| \frac{2^{n+1}\,(n+1)!\,(n+1)!}{(2n+2)!} \frac{(2n)!}{2^n\,(n!)(n!)} \right| =$

$\lim_{n\to\infty} \frac{2\,(n+1)^2}{(2n+2)(2n+1)} = \frac{1}{2} < 1$, the limit follows since $\frac{2\,(n+1)^2}{(2n+2)(2n+1)}$ is a rational function

39. diverges, by the Root Test for $\lim_{n\to\infty} \sqrt[n]{a_n} = \lim_{n\to\infty} \sqrt[n]{\frac{(n!)^n}{(n^n)^2}} = \lim_{n\to\infty} \frac{n!}{n^2} = \infty > 1$

41. converges, by the Root Test for $\lim_{n\to\infty} \sqrt[n]{a_n} = \lim_{n\to\infty} \sqrt[n]{\frac{n^n}{2^{n^2}}} = \lim_{n\to\infty} \frac{n}{2^n} = \lim_{n\to\infty} \frac{1}{2^n \ln 2} = 0 < 1$

43. Ratio: $\lim_{n\to\infty} \frac{a_{n+1}}{a_n} = \lim_{n\to\infty} \left|\frac{1}{(n+1)^p} \div \frac{1}{n^p}\right| = \left(\lim_{n\to\infty} \frac{n}{n+1}\right)^p = 1^p = 1 \Rightarrow$ no conclusion

Root: $\lim_{n\to\infty} \sqrt[n]{a_n} = \lim_{n\to\infty} \sqrt[n]{\frac{1}{n^p}} = \lim_{n\to\infty} \frac{1}{\left(\sqrt[n]{n}\right)^p} = \frac{1}{(1)^p} = 1 \Rightarrow$ no conclusion

9.5 ALTERNATING SERIES AND ABSOLUTE CONVERGENCE

1. converges absolutely $\Rightarrow$ converges, by the Absolute Convergence Theorem, since $\sum_{n=1}^{\infty} |a_n|$ is $\sum_{n=1}^{\infty} \frac{1}{n^2}$ a convergent p–series

3. diverges, by the nth–term Test for Divergence since for $n > 10 \Rightarrow \frac{n}{10} > 1 \Rightarrow \lim_{n\to\infty} \left(\frac{n}{10}\right)^n \neq 0 \Rightarrow$ $\sum_{n=10}^{\infty} (-1)^{n+1} \left(\frac{n}{10}\right)^n$ diverges $\Rightarrow \sum_{n=1}^{\infty} (-1)^{n+1} \left(\frac{n}{10}\right)^n$ diverges

5. converges, by the Alternating Series Theorem, since $f(x) = \frac{1}{\ln x} \Rightarrow f'(x) = -\frac{1}{x(\ln x)^2} < 0 \Rightarrow f(x)$ is decreasing and hence $a_n > a_{n+1}$, $a_n > 0$ for $n \geq 1$ and $\lim_{n\to\infty} a_n = \lim_{n\to\infty} \frac{1}{\ln n} = 0$

7. diverges, $\lim_{n\to\infty} a_n = \lim_{n\to\infty} \frac{\ln n}{\ln n^2} = \lim_{n\to\infty} \frac{\ln n}{2 \ln n} = \lim_{n\to\infty} \frac{1}{2} = \frac{1}{2} \neq 0$

9. converges, by the Alternating Series Theorem, since $f(x) = \frac{\sqrt{x} + 1}{x + 1} \Rightarrow f'(x) = \frac{1 - x - 2\sqrt{x}}{2\sqrt{x}\,(x+1)^2} < 0 \Rightarrow f(x)$ is decreasing and hence $a_n > a_{n+1}$, $a_n > 0$ for $n \geq 1$ and $\lim_{n\to\infty} a_n = \lim_{n\to\infty} \frac{\sqrt{n} + 1}{n + 1} = 0$

11. converges absolutely, by the Absolute Convergence Theorem, since $\sum_{n=1}^{\infty} |a_n|$ is $\sum_{n=1}^{\infty} \left(\frac{1}{10}\right)^n$ a convergent geometric series

13. converges conditionally, since $f(x) = x^{-1/2} \Rightarrow f'(x) = \frac{-1}{2\sqrt{x^3}} < 0 \Rightarrow f(x)$ is decreasing and hence $a_n > a_{n+1}$, $a_n > 0$ for $n \geq 1$ and $\lim_{n\to\infty} \frac{1}{\sqrt{n}} = 0 \Rightarrow$ the given series converges, by the Alternating Series Test, but $\sum_{n=1}^{\infty} \frac{1}{\sqrt{n}}$ is a divergent p–series

15. converges absolutely, by the Absolute Convergence Theorem, since $\sum_{n=1}^{\infty} |a_n|$ converges, by the Limit Comparison Test when compared with $\sum_{n=1}^{\infty} \frac{1}{n^2}$ a convergent p–series

17. converges conditionally, since $f(x) = \frac{1}{x+3} \Rightarrow f'(x) = \frac{-1}{(x+3)^2} < 0 \Rightarrow f(x)$ is decreasing and hence $a_n > a_{n+1}$, $a_n > 0$ for $n \geq 1$ and $\lim_{n \to \infty} \frac{1}{n+3} = 0 \Rightarrow$ the given series converges, by the Alternating Series Test, but $\sum_{n=1}^{\infty} \frac{1}{n+3}$ diverges, by the Direct Limit Comparison Test when compared with $\sum_{n=1}^{\infty} \frac{1}{n}$

18. converges absolutely, for the series $\sum_{n=1}^{\infty} \left|\frac{\sin n}{n^2}\right|$ converges by the Direct Comparison Test since $\left|\frac{\sin n}{n^2}\right| \leq \frac{1}{n^2}$

19. diverges, by the nth–Term Test for Divergence for $\lim_{n \to \infty} a_n = \lim_{n \to \infty} \frac{3+n}{5+n} = 1 \neq 0$

20. converges conditionally, for the given seies satisfies all three conditions of the Alternating Series Theorem and therefore converges, but $\sum_{n=1}^{\infty} \frac{1}{\ln n^3}$ diverges because $3 \ln n < 3n \Rightarrow \frac{1}{\ln n^3} > \frac{1}{3}\left(\frac{1}{n}\right)$ and the Direct Comparison Test

21. converges conditionally, since $f(x) = \frac{1}{x^2} + \frac{1}{x} \Rightarrow f'(x) = -\left(\frac{2}{x^3} + \frac{1}{x^2}\right) < 0 \Rightarrow f(x)$ is decreasing and hence $a_n > a_{n+1}$, $a_n > 0$ for $n \geq 1$ and $\lim_{n \to \infty} a_n = \lim_{n \to \infty} \frac{1}{n^2} + \frac{1}{n} = 0 \Rightarrow$ the given series converges, by the Alternating Series Test, but $\sum_{n=1}^{\infty} \frac{1+n}{n^2} = \sum_{n=1}^{\infty} \frac{1}{n^2} + \sum_{n=1}^{\infty} \frac{1}{n}$, the sum of a convergent and divergent series, diverges

22. converges absolutely, by the Absolute Convergence Theorem and the Direct Comparison Test, since $\left|\frac{(-2)^{n+1}}{n+5^n}\right| \leq \frac{2^{n+1}}{n+5^n} < 2\left(\frac{2}{5}\right)^n$, the nth term of a convergent geometric series

23. converges absolutely, by the Absolute Convergence Theorem and the Ratio Test for $\lim_{n \to \infty} \left|\frac{a_{n+1}}{a_n}\right| =$

$$\lim_{n \to \infty} \left| \frac{(n+1)^2 \left(\frac{2}{3}\right)^{n+1}}{n^2 \left(\frac{2}{3}\right)^n} \right| = \frac{2}{3} < 1$$

25. converges absolutely, by the Absolute Convergence Theorem and the Integral Test, since $\int_1^{\infty} \arctan x \left(\frac{1}{1+x^2}\right) dx =$

$$\lim_{b \to \infty} \int_1^b \arctan x \left(\frac{1}{1+x^2}\right) dx = \lim_{b \to \infty} \left[\frac{(\arctan x)^2}{2}\right]_1^b = \frac{3\pi^2}{32}$$

27. diverges, by the nth–Term Test for Divergence since $\lim\limits_{n\to\infty} a_n = \lim\limits_{n\to\infty} \frac{(-1)^n n}{n+1} = \lim\limits_{n\to\infty} (-1)^n \neq 0$

29. converges absolutely, by the Absolute Convergence Theorem and the Ratio Test for $\lim\limits_{n\to\infty} \left|\frac{a_{n+1}}{a_n}\right| =$

$$\lim_{n\to\infty} \left|\frac{(-100)^{n+1}}{(n+1)!} \cdot \frac{n!}{(-100)^n}\right| = \lim_{n\to\infty} \frac{(100)^n \cdot 100 \cdot n!}{(n+1)\cdot(n!)\cdot 100^n} = \lim_{n\to\infty} \frac{100}{n+1} = 0 < 1$$

31. converges absolutely, by the Absolute Convergence Theorem and the Limit Comparison Test when $\sum\limits_{n=1}^{\infty} \frac{1}{n^2+2n+1}$ is compared with $\sum\limits_{n=1}^{\infty} \frac{1}{n^2}$ for $\lim\limits_{n\to\infty} \frac{1/\left(n^2+2n+1\right)}{1/n^2} = \lim\limits_{n\to\infty} \frac{n^2}{n^2+2n+1} = 1$ since $\frac{n^2}{n^2+2n+1}$ is a rational function

33. converges absolutely, by the Absolute Convergence Theorem, since $\left|\frac{\cos n\pi}{n\sqrt{n}}\right| = \left|\frac{(-1)^{n+1}}{n^{3/2}}\right| = \frac{1}{n^{3/2}}$,a convergent p–series

35. converges absolutely, by the Absolute Convergence Theorem and the Root Test for $\lim\limits_{n\to\infty} \sqrt[n]{|a_n|} =$

$$\lim_{n\to\infty} \left(\frac{(n+1)^n}{(2n)^n}\right)^{1/n} = \lim_{n\to\infty} \frac{n+1}{2n} = \frac{1}{2} < 1 \text{ since } \frac{n+1}{2n} \text{ is a rational function}$$

37. diverges, since $\lim\limits_{n\to\infty} |a_n| = \lim\limits_{n\to\infty} \frac{(2n)!}{2^n\, n!\, n} = \lim\limits_{n\to\infty} \frac{(n+1)(n+2)\cdots 2n}{2^n\, n} =$

$$\lim_{n\to\infty} \frac{(n+1)(n+2)\cdots(n+(n-1))}{2^{n-1}} > \lim_{n\to\infty} \left(\frac{n+1}{2}\right)^{n-1} = \infty \neq 0$$

39. converges conditionally, since $\frac{\sqrt{n+1}-\sqrt{n}}{1} \cdot \frac{\sqrt{n+1}+\sqrt{n}}{\sqrt{n+1}+\sqrt{n}} = \frac{1}{\sqrt{n+1}+\sqrt{n}}$ and $\left\{\frac{1}{\sqrt{n+1}+\sqrt{n}}\right\}$ is a decreasing sequence which converges to 0 so $\sum\limits_{n=1}^{\infty} \frac{(-1)^n}{\sqrt{n+1}+\sqrt{n}}$ converges, but for $n > 1/3 \Rightarrow 3n > 1 \Rightarrow 4n > n+1 \Rightarrow$

$2\sqrt{n} > \sqrt{n+1} \Rightarrow 3\sqrt{n} > \sqrt{n+1} + \sqrt{n} \Rightarrow \frac{1}{3\sqrt{n}} < \frac{1}{\sqrt{n+1}+\sqrt{n}} \Rightarrow \sum\limits_{n=1}^{\infty} \frac{1}{\sqrt{n+1}+\sqrt{n}}$ diverges by the Direct Comparison Test

41. converges absolutely, by the Absolute Convergence Test and the Direct Comparison Test, since sech(n) = $\frac{2}{e^n+e^{-n}} = \frac{2e^n}{e^{2n}+1} < \frac{2e^n}{e^{2n}} = \frac{2}{e^n}$, a term from a convergent geometric series where $r = \frac{1}{e} < 1$

43. $|\text{error}| < \left|(-1)^6 \frac{1}{5}\right| = 0.2$

45. $|\text{error}| < \left|(-1)^6 \frac{(0.01)^5}{5}\right| = 2. \times 10^{-11}$

47. $\frac{1}{(2n)!} < \frac{5}{10^6} \Rightarrow (2n)! > \frac{10^6}{5} = 200000 \Rightarrow 2n = 10 \Rightarrow n = 5 \Rightarrow 1 - \frac{1}{2!} + \frac{1}{4!} - \frac{1}{6!} + \frac{1}{8!} \approx$

$0.540302579 \approx 0.54030$

49. a) $a_n \ge a_{n+1}$ fails, since $\frac{1}{3} < \frac{1}{2}$

b) $\left(\frac{1}{3}+\frac{1}{9}+\frac{1}{27}+\dots\right)-\left(\frac{1}{2}+\frac{1}{4}+\frac{1}{8}+\dots\right)=\frac{1/3}{1-1/3}-\frac{1/2}{1-1/2}=\frac{1}{2}-1=-\frac{1}{2}$

51. The unused terms are $\sum_{j=n+1}^{\infty}(-1)^{j+1}a_j=(-1)^{n+1}\left(a_{n+1}-a_{n+2}\right)+(-1)^{n+3}\left(a_{n+3}-a_{n+4}\right)+\dots=$ $(-1)^{n+1}\left[\left(a_{n+1}-a_{n+2}\right)+\left(a_{n+3}-a_{n+4}\right)+\dots\right]$. Each grouped term is positive, hence the remainder has the same sign as $(-1)^{n+1}$, which is the sign of the first unused term.

53. a) True: $\sum_{n=1}^{\infty}(-1)^{n+1}n$ diverges while $\sum_{n=1}^{\infty}(-1)^{n+1}/n$ converges.

b) False: $\sum_{n=1}^{\infty}(-1)^{n+1}/n$ converges conditionally.

c) True, by definition.

d) True. If a series $\sum a_n$ of positive terms converges, it converges absolutely because $\sum |a_n| = \sum a_n$.

9.6 POWER SERIES

1. $\lim_{n\to\infty}\left|\frac{u_{n+1}}{u_n}\right| < 1 \Rightarrow \lim_{n\to\infty}\left|\frac{x^{n+1}}{x^n}\right| < 1 \Rightarrow -1 < x < 1$; when $x=-1$ we have $\sum_{n=1}^{\infty}(-1)^n$, a divergent series; when $x = 1$ we have $\sum_{n=1}^{\infty} 1$, a divergent series.

a) the radius, 1; the interval of convergence, $-1 < x < 1$

b) the interval of absolute convergence, $-1 < x < 1$

c) there are no values for which the series is conditionally convergent

3. $\lim_{n\to\infty}\left|\frac{u_{n+1}}{u_n}\right| < 1 \Rightarrow \lim_{n\to\infty}\left|\frac{(4x+1)^{n+1}}{(4x+1)^n}\right| < 1 \Rightarrow |4x+1| \lim_{n\to\infty} 1 < 1 \Rightarrow |4x+1| < 1 \Rightarrow -1 < 4x+1 < 1 \Rightarrow$ $-1/2 < x < 0$, when $x = -1/2$ we have $\sum_{n=1}^{\infty}(-1)^n(-1)^n = \sum_{n=1}^{\infty}(-1)^{2n} = \sum_{n=1}^{\infty}1^n$ which diverges and when $x = 0$ we have $\sum_{n=1}^{\infty}(-1)^n(1)^n = \sum_{n=1}^{\infty}(-1)^n$ which diverges.

a) the radius, 1/4; the interval of convergence, $-1/2 < x < 0$

b) the interval of absolute convergence, $-1/2 < x < 0$

c) there are no values for which the series is conditionally convergent

5. $\lim_{n\to\infty}\left|\frac{u_{n+1}}{u_n}\right| < 1 \Rightarrow \lim_{n\to\infty}\left|\frac{(x-2)^{n+1}}{10^{n+1}}\;\frac{10^n}{(x-2)^n}\right| < 1 \Rightarrow \frac{|x-2|}{10}\lim_{n\to\infty} 1 < 1 \Rightarrow \frac{|x-2|}{10} < 1 \Rightarrow |x-2| < 10 \Rightarrow$

$-10 < x-2 < 10 \Rightarrow -8 < x < 12$; when $x = -8$ we have $\sum_{n=1}^{\infty}(-1)^n$, a divergent series; when $x = 12$ we have $\sum_{n=1}^{\infty} 1$, a divergent series.

a) the radius, 10; the interval of convergence, $-8 < x < 12$

b) the interval of absolute convergence, $-8 < x < 12$

c) there are no values for which the series is conditionally convergent

7. $\lim_{n\to\infty}\left|\frac{u_{n+1}}{u_n}\right| < 1 \Rightarrow \lim_{n\to\infty}\left|\frac{(n+1)x^{n+1}}{n+3}\;\frac{n+2}{n x^n}\right| < 1 \Rightarrow |x|\lim_{n\to\infty}\left(\frac{(n+1)(n+2)}{(n+3)(n)}\right) < 1 \Rightarrow |x| < 1 \Rightarrow -1 < x < 1$; when $x = -1$ we have $\sum_{n=1}^{\infty}(-1)^n\frac{n}{n+2}$, a divergent series; when $x = 1$ we have $\sum_{n=1}^{\infty}\frac{n}{n+2}$, a divergent series.

a) the radius, 1; the interval of convergence, $-1 < x < 1$

b) the interval of absolute convergence, $-1 < x < 1$

c) there are no values for which the series is conditionally convergent

9. $\lim_{n\to\infty}\left|\frac{u_{n+1}}{u_n}\right| < 1 \Rightarrow \lim_{n\to\infty}\left|\frac{x^{n+1}}{(n+1)\sqrt{n+1}\,3^{n+1}}\cdot\frac{n\sqrt{n}\,3^n}{x^n}\right| < 1 \Rightarrow \frac{|x|}{3}\left(\lim_{n\to\infty}\frac{n}{n+1}\right)\left(\sqrt{\lim_{n\to\infty}\frac{n}{n+1}}\right) < 1 \Rightarrow$

$\frac{|x|}{3}(1)(1) < 1 \Rightarrow |x| < 3 \Rightarrow -3 < x < 3$; when $x = -3$ we have $\sum_{i=1}^{n}\frac{(-1)^n}{n^{3/2}}$ an absolutely convergent series; when $x = 3$ we have $\sum_{i=1}^{n}\frac{1}{n^{3/2}}$ a convergent p–series.

a) the radius, 3; the interval of convergence, $-3 \le x \le 3$

b) the interval of absolute convergence, $-3 \le x \le 3$

c) there are no values for which the series is conditionally convergent

11. $\lim_{n\to\infty}\left|\frac{u_{n+1}}{u_n}\right| < 1 \Rightarrow \lim_{n\to\infty}\left|\frac{x^{n+1}}{(n+1)!}\;\frac{n!}{x^n}\right| < 1 \Rightarrow |x|\lim_{n\to\infty}\frac{1}{n+1} < 1$ for all x.

a) the radius, ∞, this series converges for all x

b) this series absolutely converges for all x

c) there are no values for which the series is conditionally convergent

13. $\lim_{n\to\infty}\left|\frac{u_{n+1}}{u_n}\right| < 1 \Rightarrow \lim_{n\to\infty}\left|\frac{x^{2n+3}}{(n+1)!}\;\frac{n!}{x^{2n+1}}\right| < 1 \Rightarrow x^2\lim_{n\to\infty}\left|\frac{1}{n+1}\right| < 1$ for all x

a) the radius, ∞, this series converges for all x

b) this series absolutely converges for all x

c) there are no values for which the series is conditionally convergent

15. $\lim_{n\to\infty}\left|\frac{u_{n+1}}{u_n}\right| < 1 \Rightarrow \lim_{n\to\infty}\left|\frac{x^{n+1}}{\sqrt{n^2+2n+4}}\cdot\frac{\sqrt{n^2+3}}{x^n}\right| < 1 \Rightarrow |x|\sqrt{\lim_{n\to\infty}\frac{n^2+3}{n^2+2n+4}} < 1 \Rightarrow$

$|x| < 1 \Rightarrow -1 < x < 1$; when $x = -1$ we have $\sum_{n=1}^{\infty}\frac{(-1)^n}{\sqrt{n^2+3}}$, a conditionally convergent series; when $x = 1$ we have $\sum_{n=1}^{\infty}\frac{1}{\sqrt{n^2+3}}$, a divergent series.

a) the radius, 1; the interval of convergence, $-1 \le x < 1$
b) the interval of absolute convergence, $-1 < x < 1$
c) This series is conditionally convergent at $x = -1$.

17. $\lim_{n\to\infty}\left|\frac{u_{n+1}}{u_n}\right| < 1 \Rightarrow \lim_{n\to\infty}\left|\frac{(n+1)(x+3)^{n+1}}{5^{n+1}}\cdot\frac{5^n}{n(x+3)^n}\right| < 1 \Rightarrow \frac{|x+3|}{5}\lim_{n\to\infty}\left(\frac{n+1}{n}\right) < 1 \Rightarrow \frac{|x+3|}{5} < 1 \Rightarrow$

$|x+3| < 5 \Rightarrow -5 < x+3 < 5 \Rightarrow -8 < x < 2$, when $x = -8$ we have $\sum_{n=1}^{\infty}\frac{n(-5)^n}{5^n} = \sum_{n=1}^{\infty}(-1)^n n$ which diverges,

when $x = 2$ we have $\sum_{n=1}^{\infty}\frac{n\,5^n}{5^n} = \sum_{n=1}^{\infty} n$ a divergence series.

a) the radius, 5; the interval of convergence, $-8 < x < 2$
b) the interval of absolute convergence, $-8 < x < 2$
c) there are no values for which the series is conditionally convergent
a) the radius, 4; the interval of convergence, $-4 \le x < 4$
b) the interval of absolute convergence, $-4 < x < 4$
c) This series is conditionally convergent at $x = -4$.

19. $\lim_{n\to\infty}\left|\frac{u_{n+1}}{u_n}\right| < 1 \Rightarrow \lim_{n\to\infty}\left|\frac{\sqrt{n+1}\,x^{n+1}}{3^{n+1}}\cdot\frac{3^n}{\sqrt{n}\,x^n}\right| < 1 \Rightarrow \frac{|x|}{3}\sqrt{\lim_{n\to\infty}\frac{n+1}{n}} < 1 \Rightarrow \frac{|x|}{3}(1) < 1 \Rightarrow |x| < 3 \Rightarrow$

$-3 < x < 3$; both series $\sum_{n=1}^{\infty}\sqrt{n}\,(-1)^n$ and $\sum_{n=1}^{\infty}\sqrt{n}$ diverges, when $x = \pm 3$.

a) the radius, 3; the interval of convergence, $-3 < x < 3$
b) the interval of absolute convergence, $-3 < x < 3$
c) there are no values for which the series is conditionally convergent

21. $\lim_{n\to\infty}\left|\frac{u_{n+1}}{u_n}\right| < 1 \Rightarrow \lim_{n\to\infty}\left|\frac{\left(1+\frac{1}{n+1}\right)^{n+1}x^{n+1}}{\left(1+\frac{1}{n}\right)^n x^n}\right| < 1 \Rightarrow |x|\left(\frac{\lim_{t\to\infty}\left(1+\frac{1}{t}\right)^t}{\lim_{n\to\infty}\left(1+\frac{1}{n}\right)^n}\right) < 1 \Rightarrow |x| < 1$, where

$t = n + 1$ and table 9.1, $\Rightarrow -1 < x < 1$; both series $\sum_{n=1}^{\infty} (-1)^n\left(1+\frac{1}{n}\right)^n$ and $\sum_{n=1}^{\infty}\left(1+\frac{1}{n}\right)^n$ diverge, when $x = \pm 1$ because $\lim_{n\to\infty}\left(1+\frac{1}{n}\right)^n = e \neq 0$.

a) the radius, 1; the interval of convergence, $-1 < x < 1$

b) the interval of absolute convergence, $-1 < x < 1$

c) there are no values for which the series is conditionally convergent

23. $\lim_{n\to\infty}\left|\frac{u_{n+1}}{u_n}\right| < 1 \Rightarrow \lim_{n\to\infty}\left|\frac{(n+1)^{n+1}x^{n+1}}{n^n x^n}\right| < 1 \Rightarrow |x|\left(\lim_{n\to\infty}\left(1+\frac{1}{n}\right)^n\right)\left(\lim_{n\to\infty}(n+1)\right) < 1 \Rightarrow$

$e|x| \lim_{n\to\infty}(n+1) < 1 \Rightarrow$ only $x = 0$ would satisfy this inequality.

a) the radius, 0; this series converges only for $x = 0$

b) this series converges absolutely only for $x = 0$

c) there are no values for which the series is conditionally convergent

25. $\lim_{n\to\infty}\left|\frac{u_{n+1}}{u_n}\right| < 1 \Rightarrow \lim_{n\to\infty}\left|\frac{(x+2)^{n+1}}{(n+1)\,2^{n+1}}\cdot\frac{n\,2^n}{(x+2)^n}\right| < 1 \Rightarrow \frac{|x+2|}{2}\lim_{n\to\infty}\left|\frac{n}{n+1}\right| < 1 \Rightarrow \frac{|x+2|}{2} < 1 \Rightarrow$

$|x+2| < 2 \Rightarrow -2 < x+2 < 2 \Rightarrow -4 < x < 0$, when $x = -4$ we have $\sum_{n=1}^{\infty}\frac{-1}{n}$ a divergent series, when $x = 0$ we have $\sum_{n=1}^{\infty}\frac{(-1)^{n+1}}{n}$ the alternating harmonic series which is conditionally convergent.

a) the radius, 2; the interval of convergence, $-4 < x \leq 0$

b) the interval of absolute convergence, $-4 < x < 0$

c) This series is conditionally convergent at $x = 0$.

27. $\lim_{n\to\infty}\left|\frac{u_{n+1}}{u_n}\right| < 1 \Rightarrow \lim_{n\to\infty}\left|\frac{x^{n+1}}{(n+1)(\ln(n+1))^2}\cdot\frac{n(\ln n)^2}{x^n}\right| < 1 \Rightarrow |x|\left(\lim_{n\to\infty}\frac{n}{n+1}\right)\left(\lim_{n\to\infty}\frac{\ln n}{\ln(n+1)}\right)^2 < 1 \Rightarrow$

$|x|(1)\left(\lim_{n\to\infty}\frac{1/n}{1/(n+1)}\right)^2 < 1 \Rightarrow |x|\left(\lim_{n\to\infty}\frac{n+1}{n}\right)^2 < 1 \Rightarrow |x|(1)^2) < 1 \Rightarrow |x| < 1 \Rightarrow -1 < x < 1$, when $x = -1$ we have $\sum_{n=1}^{\infty}\frac{(-1)^n}{n(\ln n)^2}$ which converges absolutely, where $x = 1$ we have $\sum_{n=1}^{\infty}\frac{1}{n(\ln n)^2}$, which converges.

a) the radius, 1; the interval of convergence, $-1 \leq x \leq 1$

b) the interval of absolute convergence, $-1 \leq x \leq 1$

c) there are no values for which the series is conditionally convergent

29. $\lim_{n\to\infty}\left|\frac{u_{n+1}}{u_n}\right| < 1 \Rightarrow \lim_{n\to\infty}\left|\frac{(4x-5)^{2n+3}}{(n+1)^{3/2}}\cdot\frac{n^{3/2}}{(4x-5)^{2n+1}}\right| < 1 \Rightarrow (4x-5)^2\left(\lim_{n\to\infty}\frac{n}{n+1}\right)^{3/2} < 1 \Rightarrow$

$(4x-5)^2(1) < 1 \Rightarrow (4x-5)^2 < 1 \Rightarrow |4x-5| < 1 \Rightarrow -1 < 4x-5 < 1 \Rightarrow 1 < x < 3/2$, when x = 1 we have

$\sum_{n=1}^{\infty} \frac{(-1)^{2n+1}}{n^{3/2}} = \sum_{n=1}^{\infty} \frac{-1}{n^{3/2}}$ which is absolutely convergent since $\sum_{n=1}^{\infty} \frac{1}{n^{3/2}}$ is a convergent p–series, when

$x = 3/2$ we have $\sum_{n=1}^{\infty} \frac{(1)^{2n+1}}{n^{3/2}}$ which also converges.

a) the radius, 1/4; the interval of convergence, $1 \le x \le 3/2$

b) the interval of absolute convergence, $1 \le x \le 3/2$

c) there are no values for which the series is conditionally convergent

31. $\lim_{n\to\infty} \left|\frac{u_{n+1}}{u_n}\right| < 1 \Rightarrow \lim_{n\to\infty} \left|\frac{(x+\pi)^{n+1}}{\sqrt{n+1}} \cdot \frac{\sqrt{n}}{(x+\pi)^n}\right| < 1 \Rightarrow |x+\pi| \lim_{n\to\infty} \left|\sqrt{\frac{n}{n+1}}\right| < 1 \Rightarrow$

$|x+\pi| \sqrt{\lim_{n\to\infty} \frac{n}{n+1}} < 1 \Rightarrow |x+\pi|\sqrt{1} < 1 \Rightarrow |x+\pi| < 1 \Rightarrow -1 < x+\pi < 1 \Rightarrow -1-\pi < x < 1-\pi$, when

$x = -1-\pi$ we have $\sum_{n=1}^{\infty} \frac{(-1)^n}{\sqrt{n}} = \sum_{n=1}^{\infty} \frac{(-1)^n}{n^{1/2}}$ which converges by the Alternating Series Theorem but

$\sum_{n=1}^{\infty} \left|\frac{(-1)^n}{n^{1/2}}\right| = \sum_{n=1}^{\infty} \frac{1}{n^{1/2}}$ a divergent p–series, when $x = 1-\pi$ we have $\sum_{n=1}^{\infty} \frac{1^n}{\sqrt{n}} = \sum_{n=1}^{\infty} \frac{1}{n^{1/2}}$ a divergent p–series.

a) the radius, 1; the interval of convergence, $(-1-\pi) \le x < (1-\pi)$

b) the interval of absolute convergence, $-1-\pi < x < 1-\pi$

c) This series is conditionally convergent at $x = -1-\pi$.

33. $\lim_{n\to\infty} \left|\frac{u_{n+1}}{u_n}\right| < 1 \Rightarrow \lim_{n\to\infty} \left|\frac{(x-1)^{2n+2}}{4^{n+1}} \cdot \frac{4^n}{(x-1)^{2n}}\right| < 1 \Rightarrow \frac{(x-1)^2}{4} \lim_{n\to\infty} 1 < 1 \Rightarrow (x-1)^2 < 4 \Rightarrow |x-1| < 2 \Rightarrow$

$-2 < x-1 < 2 \Rightarrow -1 < x < 3$, at $x = -1$ we have $\sum_{n=0}^{\infty} \frac{(-2)^{2n}}{4^n} = \sum_{n=0}^{\infty} \frac{4^n}{4^n} = \sum_{n=0}^{\infty} 1$ which diverges; at x = 3

we have $\sum_{n=0}^{\infty} \frac{2^{2n}}{4^n} = \sum_{n=0}^{\infty} \frac{4^n}{4^n} = \sum_{n=0}^{\infty} 1$, which also diverges, therefore the interval of convergence is;

$-1 < x < 3$. The series $\sum_{n=0}^{\infty} \frac{(x-1)^{2n}}{4^n} = \sum_{n=0}^{\infty} \left(\left(\frac{x-1}{2}\right)^2\right)^n$ is a convergent geometric series when $-1 < x < 3$

and the sum is $\frac{1}{1-\left(\frac{x-1}{2}\right)^2} = \frac{1}{\frac{4-(x-1)^2}{4}} = \frac{4}{4-x^2+2x-1} = \frac{4}{3+2x-x^2}$.

35. $\lim_{n\to\infty}\left|\frac{u_{n+1}}{u_n}\right| < 1 \Rightarrow \lim_{n\to\infty}\left|\frac{(\sqrt{x}-2)^{n+1}}{2^{n+1}}\cdot\frac{2^n}{(\sqrt{x}-2)^n}\right| < 1 \Rightarrow |\sqrt{x}-2| < 2 \Rightarrow -2 < \sqrt{x}-2 < 2 \Rightarrow 0 < \sqrt{x} < 4 \Rightarrow$

$0 < x < 16$, when $x = 0$ we have $\sum_{n=0}^{\infty}(-1)^n$ a divergent series, when $x = 16$ we have $\sum_{n=0}^{\infty}(1)^n$ a divergent series,

therefore the interval of convergence is $0 < x < 16$. The series $\sum_{n=0}^{\infty}\left(\frac{\sqrt{x}-2}{2}\right)^n$ is a convergent geometric series

when $0 < x < 16$ and its sum is $= \frac{1}{1-\frac{\sqrt{x}-2}{2}} = \frac{1}{\frac{2-\sqrt{x}+2}{2}} = \frac{1}{\frac{4-\sqrt{x}}{2}} = \frac{2}{4-\sqrt{x}}$

37. $\lim_{n\to\infty}\left|\frac{u_{n+1}}{u_n}\right| < 1 \Rightarrow \lim_{n\to\infty}\left|\left(\frac{x^2+1}{3}\right)^{n+1}\cdot\left(\frac{3}{x^2+1}\right)^n\right| < 1 \Rightarrow \frac{(x^2+1)}{3}\left(\lim_{n\to\infty} 1\right) < 1 \Rightarrow \frac{x^2+1}{3} < 1 \Rightarrow x^2 < 2 \Rightarrow$

$|x| < \sqrt{2} \Rightarrow -\sqrt{2} < x < \sqrt{2}$, at $x = \pm\sqrt{2}$ we have $\sum_{n=0}^{\infty}(1)^n$ which diverges, therefore the interval of convergence

is $-\sqrt{2} < x < \sqrt{2}$. The series $\sum_{n=0}^{\infty}\left(\frac{x^2+1}{3}\right)^n$ is a convergence geometric series when $-\sqrt{2} < x < \sqrt{2}$ and its sum

is $\frac{1}{1-\left(\frac{x^2+1}{3}\right)} = \frac{1}{\frac{3-x^2-1}{3}} = \frac{3}{2-x^2}$

39. $\lim_{n\to\infty}\left|\frac{(x-3)^{n+1}}{2^{n+1}}\frac{2^n}{(x-3)^n}\right| < 1 \Rightarrow |x-3| < 2 \Rightarrow 1 < x < 5$; both series: $\sum_{n=1}^{\infty}(1)$, when $x = 1$ and $\sum_{n=1}^{\infty}(-1)^n$,

when $x = 5$ diverge and therefore the interval of convergence is $1 < x < 5$. The sum of this convergent geometric series is $\frac{1}{1+\frac{x-3}{2}} = \frac{2}{x-1}$. $f(x) = 1 - \frac{1}{2}(x-3) + \frac{1}{4}(x-3)^2 + \ldots + \left(-\frac{1}{2}\right)^n(x-3)^n + \ldots = \frac{2}{x-1} \Rightarrow f'(x) = -\frac{1}{2} +$ $\frac{1}{2}(x-3) + \ldots + \left(-\frac{1}{2}\right)^n n(x-3)^{n-1} + \ldots$ is convergent when $1 < x < 5$, and diverges when $x = 1$ or 5. The sum is $\frac{-2}{(x-1)^2}$ the derivative of $\frac{2}{x-1}$.

41. a) $\ln|\sec x| + C = \int \tan x\,dx = \int x + \frac{x^3}{3} + \frac{2x^5}{15} + \frac{17x^7}{315} + \frac{62x^9}{2835} + \ldots dx = \frac{x^2}{2} + \frac{x^4}{12} + \frac{x^6}{45} + \frac{17x^8}{2520} +$ $\frac{31x^{10}}{14175} + \ldots + C$, but $x = 0 \Rightarrow C = 0 \Rightarrow \ln|\sec x| = \frac{x^2}{2} + \frac{x^4}{12} + \frac{x^6}{45} + \frac{17x^8}{2520} + \frac{31x^{10}}{14175} + \ldots$, when $-\frac{\pi}{2} < x < \frac{\pi}{2}$

b) $\sec^2 x = \frac{d(\tan x)}{dx} = \frac{d\left(x + \frac{x^3}{3} + \frac{2x^5}{15} + \frac{17x^7}{315} + \frac{62x^9}{2835} + \ldots\right)}{dx} = 1 + x^2 + \frac{2x^4}{3} + \frac{17x^6}{45} + \frac{62x^8}{315} + \ldots$, when $-\frac{\pi}{2} < x < \frac{\pi}{2}$

c) $\sec^2 x = (\sec x)(\sec x) = \left(1 + \frac{x^2}{2} + \frac{5x^4}{24} + \frac{61x^6}{720} + \ldots\right)\left(1 + \frac{x^2}{2} + \frac{5x^4}{24} + \frac{61x^6}{720} + \ldots\right) = 1 + \left(\frac{1}{2} + \frac{1}{2}\right)x^2 +$ $\left(\frac{5}{24} + \frac{1}{4} + \frac{5}{24}\right)x^4 + \left(\frac{61}{720} + \frac{5}{48} + \frac{5}{48} + \frac{61}{720}\right)x^6 + \ldots = 1 + x^2 + \frac{2x^4}{3} + \frac{17x^6}{45} + \ldots$

43. a) If $f(x) = \sum_{n=0}^{\infty} a_n x^n$, then $f^{(k)}(x) = \sum_{n=k}^{\infty} n(n-1)(n-2)\cdots(n-(k-1))\, a_n x^{n-k}$ and $f^{(k)}(0) = k!\, a_k \Rightarrow$

$a_k = \frac{f^{(k)}(0)}{k!}$ and likewise if $f(x) = \sum_{n=0}^{\infty} b_n x^n$, then $b_k = \frac{f^{(k)}(0)}{k!} \Rightarrow a_k = b_k$ for every nonnegative integer k.

b) If $f(x) = \sum_{n=0}^{\infty} a_n x^n = 0$ for all x, then $f^{(k)}(x) = 0$ for all x and from part a) $a_k = 0$ for every nonnegative integer k.

45. The series $\sum_{n=1}^{\infty} \frac{x^n}{n}$ converges conditionally at the left–hand endpoint of its interval of convergence $[-1,1)$. The series $\sum_{n=1}^{\infty} \frac{x^n}{(n^2)}$ converges absolutely at the left–hand endpoint of its interval of convergence $[-1,1]$.

47. The given series converges for $-\infty < x < \infty$, so its derivative must converge for $-\infty < x < \infty$ by Theorem 12. The derivative is identical with the original series, so $f'(x) = f(x)$. This means that $f(x) = C e^x$ for some constant C. But $f(0) = 1$, so $C e^0 = C \cdot 1$ and $C = 1$. Hence $f(x) = e^x$.

9.7 TAYLOR AND MACLAURIN SERIES

1. $f(x) = \ln x$, $f'(x) = \frac{1}{x}$, $f''(x) = -\frac{1}{x^2}$, $f'''(x) = \frac{2}{x^3}$, $f(1) = \ln 1 = 0$, $f'(1) = 1$, $f''(1) = -1$, $f'''(1) = 2 \Rightarrow P_0(x) = 0$,
$P_1(x) = x - 1$, $P_2(x) = (x-1) - \frac{1}{2}(x-1)^2$, $P_3(x) = (x-1) - \frac{1}{2}(x-1)^2 + \frac{1}{3}(x-1)^3$

3. $f(x) = \frac{1}{x} = x^{-1}$, $f'(x) = -x^{-2}$, $f''(x) = 2x^{-3}$, $f'''(x) = -6x^{-4}$, $f(2) = \frac{1}{2}$, $f'(2) = -\frac{1}{4}$, $f''(2) = \frac{1}{4}$, $f'''(2) = -\frac{3}{8} \Rightarrow$
$P_0(x) = \frac{1}{2}$, $P_1(x) = \frac{1}{2} - \frac{1}{4}(x-2)$, $P_2(x) = \frac{1}{2} - \frac{1}{4}(x-2) + \frac{1}{8}(x-2)^2$, $P_3(x) = \frac{1}{2} - \frac{1}{4}(x-2) + \frac{1}{8}(x-2)^2 - \frac{1}{16}(x-2)^3$

5. $f(x) = \sin x \Rightarrow f'(x) = \cos x \Rightarrow f''(x) = -\sin x \Rightarrow f'''(x) = -\cos x \Rightarrow f(\pi/4) = \sin \pi/4 = \frac{\sqrt{2}}{2}$, $f'(\pi/4) = \cos \pi/4 = \frac{\sqrt{2}}{2}$,
$f''(\pi/4) = -\sin \pi/4 = -\frac{\sqrt{2}}{2}$, $f'''(\pi/4) = -\cos \pi/4 = -\frac{\sqrt{2}}{2} \Rightarrow P_0 = \frac{\sqrt{2}}{2}$, $P_1(x) = \frac{\sqrt{2}}{2} + \frac{\sqrt{2}}{2}\left(x - \frac{\pi}{4}\right)$,
$P_2(x) = \frac{\sqrt{2}}{2} + \frac{\sqrt{2}}{2}\left(x - \frac{\pi}{4}\right) - \frac{\sqrt{2}}{4}\left(x - \frac{\pi}{4}\right)^2$, $P_3(x) = \frac{\sqrt{2}}{2} + \frac{\sqrt{2}}{2}\left(x - \frac{\pi}{4}\right) - \frac{\sqrt{2}}{4}\left(x - \frac{\pi}{4}\right)^2 - \frac{\sqrt{2}}{12}\left(x - \frac{\pi}{4}\right)^3$

7. $f(x) = \sqrt{x} = x^{1/2} \Rightarrow f'(x) = (1/2)x^{-1/2} \Rightarrow f''(x) = (-1/4)x^{-3/2} \Rightarrow f'''(x) = (3/8)x^{-5/2} \Rightarrow$
$f(4) = \sqrt{4} = 4^{1/2} = 2$, $f'(4) = (1/2)4^{-1/2} = \frac{1}{4}$, $f''(4) = (-1/4)4^{-3/2} = -\frac{1}{32}$, $f'''(4) = (3/8)4^{-5/2} = \frac{3}{256} \Rightarrow$
$P_0(x) = 2$, $P_1(x) = 2 + \frac{1}{4}(x-4)$, $P_2(x) = 2 + \frac{1}{4}(x-4) - \frac{1}{64}(x-4)^2$,
$P_3(x) = 2 + \frac{1}{4}(x-4) - \frac{1}{64}(x-4)^2 + \frac{1}{512}(x-4)^3$

9. $e^x = \sum_{n=0}^{\infty} \frac{x^n}{n!} \Rightarrow e^{-5x} = \sum_{n=0}^{\infty} \frac{(-5x)^n}{n!} = 1 - 5x + \frac{25x^2}{2!} - \frac{125x^3}{3!} + \frac{625x^4}{4!} - \cdots$

11. $\sin x = \sum_{n=0}^{\infty} \frac{(-1)^n x^{2n+1}}{(2n+1)!} \Rightarrow \sin 3x = \sum_{n=0}^{\infty} \frac{(-1)^n (3x)^{2n+1}}{(2n+1)!} = \sum_{n=0}^{\infty} \frac{(-1)^n (3)^{2n+1} (x)^{2n+1}}{(2n+1)!} =$
$3x - \frac{(3x)^3}{3!} + \frac{(3x)^5}{5!} - \ldots$

13. $7\cos(-x) = 7\cos(x) = 7 \sum_{n=0}^{\infty} \frac{(-1)^n x^{2n}}{(2n)!} = 7 - \frac{7x^2}{2!} + \frac{7x^4}{4!} - \frac{7x^6}{6!} + \ldots$, since cosine is an even function

15. $\cos x = \sum_{n=0}^{\infty} \frac{(-1)^n x^{2n}}{(2n)!} \Rightarrow \cos\sqrt{x} = \sum_{n=0}^{\infty} \frac{(-1)^n \left(x^{1/2}\right)^{2n}}{(2n)!} = \sum_{n=0}^{\infty} \frac{(-1)^n x^n}{(2n)!} = 1 - \frac{x}{2!} + \frac{x^2}{4!} - \frac{x^3}{6!} + \ldots$

17. $e^x = \sum_{n=0}^{\infty} \frac{x^n}{n!} \Rightarrow x\,e^x = x\left(\sum_{n=0}^{\infty} \frac{x^n}{n!}\right) = \sum_{n=0}^{\infty} \frac{x^{n+1}}{n!} = x + x^2 + \frac{x^3}{2!} + \frac{x^4}{3!} + \frac{x^5}{4!} + \ldots$

19. $\sin x = \sum_{n=0}^{\infty} \frac{(-1)^n x^{2n+1}}{(2n+1)!} \Rightarrow \frac{\sin x}{x} = \frac{1}{x}\left(\sum_{n=0}^{\infty} \frac{(-1)^n x^{2n+1}}{(2n+1)!}\right) = \sum_{n=0}^{\infty} \frac{(-1)^n x^{2n}}{(2n+1)!} = 1 - \frac{x^2}{3!} + \frac{x^4}{5!} - \frac{x^6}{7!} + \ldots$

21. $\cos x = \sum_{n=0}^{\infty} \frac{(-1)^n x^{2n}}{(2n)!} \Rightarrow \frac{x^2}{2} - 1 + \cos x = \frac{x^2}{2} - 1 + \sum_{n=0}^{\infty} \frac{(-1)^n x^{2n}}{(2n)!} = \frac{x^2}{2} - 1 + 1 - \frac{x^2}{2} + \frac{x^4}{4!} - \frac{x^6}{6!} + \frac{x^8}{8!} - \frac{x^{10}}{10!} + \ldots =$
$\frac{x^4}{4!} - \frac{x^6}{6!} + \frac{x^8}{8!} - \frac{x^{10}}{10!} + \ldots$

23. $\cos^2 x = \frac{1}{2} + \frac{\cos 2x}{2} = \frac{1}{2} + \frac{1}{2} \sum_{n=0}^{\infty} \frac{(-1)^n (2x)^{2n}}{(2n)!} = \frac{1}{2} + \frac{1}{2}\left[1 - \frac{(2x)^2}{2!} + \frac{(2x)^4}{4!} - \frac{(2x)^6}{6!} + \frac{(2x)^8}{8!} - \ldots\right] =$
$1 - \frac{(2x)^2}{2 \cdot 2!} + \frac{(2x)^4}{2 \cdot 4!} - \frac{(2x)^6}{2 \cdot 6!} + \frac{(2x)^8}{2 \cdot 8!} - \ldots$

25. $\cosh x = \frac{e^x + e^{-x}}{2} = \frac{1}{2}\left[\left(1 + x + \frac{x^2}{2!} + \frac{x^3}{3!} + \frac{x^4}{4!} + \ldots\right) + \left(1 - x + \frac{x^2}{2!} - \frac{x^3}{3!} + \frac{x^4}{4!} - \ldots\right)\right] = 1 + \frac{x^2}{2!} + \frac{x^4}{4!} + \frac{x^6}{6!} + \ldots$

27. $f(x) = e^{\tan x} \Rightarrow f'(x) = e^{\tan x} \sec^2 x \Rightarrow f''(x) = e^{\tan x} \sec^4 x + 2\,e^{\tan x} \sec^2 x \tan x = e^{\tan x} \sec^2 x\left(\sec^2 x + 2\tan x\right) \Rightarrow$
$f'''(x) = e^{\tan x} \sec^2 x\left(\sec^2 x + 2\tan x\right)^2 + e^{\tan x} \sec^2 x\left(2\sec^2 x \tan x + 2\sec^2 x\right) =$
$e^{\tan x} \sec^2 x\left[\left(\sec^2 x + 2\tan x\right)^2 + \left(2\sec^2 x \tan x + 2\sec^2 x\right)\right]$. $\therefore\ e^{\tan x} = 1 + x + \frac{x^2}{2} + \frac{f'''(c)}{3!} x^3$
where c is between 0 and x.

29. $f(x) = \ln(\cos x) \Rightarrow f'(x) = -\frac{\sin x}{\cos x} = -\tan x \Rightarrow f''(x) = -\sec^2 x \Rightarrow f'''(x) = -2\sec^2 x \tan x$. $\therefore \ln(\cos x) =$
$-\frac{x^2}{2} + \frac{f'''(c)}{3!} x^3$ where c is between 0 and x.

31. $f(x) = \sinh x \Rightarrow f'(x) = \cosh x \Rightarrow f''(x) = \sinh x \Rightarrow f'''(x) = \cosh x$. $\therefore \cosh x = x + \frac{f'''(c)}{3!} x^3$
where c is between 0 and x.

33. If $e^x = \sum_{n=0}^{\infty} \frac{f^{(n)}(a)}{n!}(x-a)^n$ and $f(x) = e^x$, we have $f^{(n)}(a) = e^a$ for all n = 0, 1, 2, 3,...;

$e^x = e^a\left[\frac{(x-a)^0}{0!} + \frac{(x-a)^1}{1!} + \frac{(x-a)^2}{2!} + \ldots\right] = e^a\left[1 + (x-a) + \frac{(x-a)^2}{2!} + \ldots\right]$, at x = a.

35. $\left|R_3(x)\right| = \left|\frac{-\cos c}{5!}(x-0)^5\right| < \left|\frac{x^5}{5!}\right| < 5 \times 10^{-4} \Rightarrow -5 \times 10^{-4} < \frac{x^5}{5!} < 5 \times 10^{-4} \Rightarrow$

$-\sqrt[5]{5!\left(5 \times 10^{-4}\right)} < x < \sqrt[5]{5!\left(5 \times 10^{-4}\right)} \Rightarrow -0.569679052 < x < 0.569679052$

37. $\sin x = x + R_1(x)$, when $|x| < 10^{-3} \Rightarrow \left|R_1(x)\right| = \left|\frac{-\cos c}{3!}x^3\right| < \left|\frac{(1)\,x^3}{3!}\right| < \frac{\left(10^{-3}\right)^3}{3!} = 1.67 \times 10^{-10}$

From exercise 51 in section 9.5, $R_1(x)$ has the same sign as $-\frac{x^3}{3!}$. $x < \sin x \Rightarrow 0 < \sin x - x =$ $R_1(x)$, which has the same sign as $-\frac{x^3}{3!} \Rightarrow x < 0 \Rightarrow -10^{-3} < x < 0$.

39. $\left|R_2(x)\right| = \left|\frac{e^c x^3}{3!}\right| < \frac{3^{(0.1)}(0.1)^3}{3!} = 0.00018602$, where c is between 0 and x.

41. sin x, when x = 0.1; the sum is $\sin(0.1) \approx 0.099833416$

43. $\sin x = x - \frac{x^3}{3!} + \frac{x^5}{5!} - \frac{x^7}{7!} + \ldots$; $\frac{d(\sin x)}{dx} = \frac{d\left(x - \frac{x^3}{3!} + \frac{x^5}{5!} - \frac{x^7}{7!} + \ldots\right)}{dx} = 1 - \frac{x^2}{2!} + \frac{x^4}{4!} - \frac{x^6}{6!} + \ldots = \cos x$;

$\frac{d(\cos x)}{dx} = \frac{d\left(1 - \frac{x^2}{2!} + \frac{x^4}{4!} - \frac{x^6}{6!} + \ldots\right)}{dx} = -x + \frac{x^3}{3!} - \frac{x^5}{5!} + \ldots = -\sin x$; $\frac{d\left(e^x\right)}{dx} =$

$\frac{d\left(1 + x + \frac{x^2}{2!} + \frac{x^3}{3!} + \frac{x^4}{4!} + \ldots\right)}{dx} = 1 + x + \frac{x^2}{2!} + \frac{x^3}{3!} + \frac{x^4}{4!} + \ldots = e^x$

45. $2\,[\cos x]\,[\sin x] = 2\left[1 - \frac{x^2}{2!} + \frac{x^4}{4!} - \frac{x^6}{6!} + \ldots\right]\left[x - \frac{x^3}{3!} + \frac{x^5}{5!} - \frac{x^7}{7!} + \ldots\right] =$

$2\left[x - \frac{4x^3}{3!} + \frac{16x^5}{5!} - \frac{64x^7}{7!} + \frac{256x^9}{9!} - \ldots\right] = 2x - \frac{8x^3}{3!} + \frac{32x^5}{5!} - \frac{128\,x^7}{7!} + \frac{512\,x^9}{9!} - \cdots = \sin 2x$

47. a) $\sin x = x - \frac{x^3}{3!} + \frac{x^5}{5!} - \frac{x^7}{7!} + \ldots \Rightarrow \frac{\sin x}{x} = 1 - \frac{x^2}{3!} + \frac{x^4}{5!} - \frac{x^6}{7!} + \ldots$, $s_1 = 1$ and $s_2 = 1 - \frac{x^2}{6}$; If L is the sum of the series representing $\frac{\sin x}{x}$, then $L - s_1 = \frac{\sin x}{x} - 1 < 0$ and $L - s_2 = \frac{\sin x}{x} - \left(1 - \frac{x^2}{6}\right) > 0$, by the Alternating Series Estimation Theorem. $\therefore 1 - \frac{x^2}{6} < \frac{\sin x}{x} < 1$.

b) The graph of $y = \frac{\sin x}{x}$, $x \neq 0$, is bounded below by the graph of $y = 1 - \frac{x^2}{6}$ and above by the graph of y = 1 as indicated in part a.

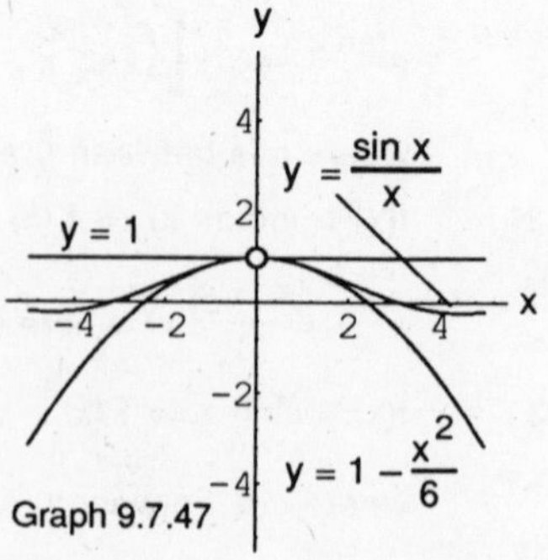

Graph 9.7.47

49. $\sin^2 x = \left(\frac{1-\cos 2x}{2}\right) = \frac{1}{2} - \frac{1}{2}\cos 2x = \frac{1}{2} - \frac{1}{2}\left(1 - \frac{(2x)^2}{2!} + \frac{(2x)^4}{4!} - \frac{(2x)^6}{6!} + \cdots\right) = \frac{2x^2}{2!} - \frac{2^3x^4}{4!} + \frac{2^5x^6}{6!} - \cdots,$

$\frac{d}{dx}(\sin^2 x) = \frac{d}{dx}\left(\frac{2x^2}{2!} - \frac{2^3x^4}{4!} + \frac{2^5x^6}{6!} - \cdots\right) = 2x - \frac{(2x)^3}{3!} + \frac{(2x)^5}{5!} - \frac{(2x)^7}{7!} + \cdots = \sin 2x$

51. a) $e^{-i\pi} = \cos(-\pi) + i\sin(-\pi) = -1 + i(0) = -1$ b) $e^{i\pi/4} = \cos\left(\frac{\pi}{4}\right) + i\sin\left(\frac{\pi}{4}\right) = \frac{1}{\sqrt{2}} + \frac{i}{\sqrt{2}} = \left(\frac{1}{\sqrt{2}}\right)(1+i)$

c) $e^{-i\pi/2} = \cos\left(-\frac{\pi}{2}\right) + i\sin\left(-\frac{\pi}{2}\right) = 0 + i(-1) = -i$

53. $e^x = 1 + x + \frac{x^2}{2!} + \frac{x^3}{3!} + \frac{x^4}{4!} + \ldots \Rightarrow e^{i\theta} = 1 + i\theta + \frac{(i\theta)^2}{2!} + \frac{(i\theta)^3}{3!} + \frac{(i\theta)^4}{4!} + \ldots$ and $e^{-i\theta} =$

$1 - i\theta + \frac{(-i\theta)^2}{2!} + \frac{(-i\theta)^3}{3!} + \frac{(-i\theta)^4}{4!} + \ldots = 1 - i\theta + \frac{(i\theta)^2}{2!} - \frac{(i\theta)^3}{3!} + \frac{(i\theta)^4}{4!} - \ldots, \frac{e^{i\theta} + e^{-i\theta}}{2} =$

$$\frac{\left(1 + i\theta + \frac{(i\theta)^2}{2!} + \frac{(i\theta)^3}{3!} + \frac{(i\theta)^4}{4!} + \ldots\right) + \left(1 - i\theta + \frac{(i\theta)^2}{2!} - \frac{(i\theta)^3}{3!} + \frac{(i\theta)^4}{4!} - \ldots\right)}{2} = 1 - \frac{\theta^2}{2!} + \frac{\theta^4}{4!} - \frac{\theta^6}{6!} + \ldots = \cos\theta;$$

$$\frac{e^{i\theta} - e^{-i\theta}}{2} = \frac{\left(1 + i\theta + \frac{(i\theta)^2}{2!} + \frac{(i\theta)^3}{3!} + \frac{(i\theta)^4}{4!} + \ldots\right) - \left(1 - i\theta + \frac{(i\theta)^2}{2!} - \frac{(i\theta)^3}{3!} + \frac{(i\theta)^4}{4!} - \ldots\right)}{2i} = \theta - \frac{\theta^3}{3!} + \frac{\theta^5}{5!} - \frac{\theta^7}{7!} + \ldots = \sin\theta$$

55. $e^x \sin x = \left(1 + x + \frac{x^2}{2!} + \frac{x^3}{3!} + \frac{x^4}{4!} + \ldots\right)\left(x - \frac{x^3}{3!} + \frac{x^5}{5!} - \frac{x^7}{7!} + \ldots\right) = (1)\,x + (1)\,x^2 + \left(-\frac{1}{6} + \frac{1}{2}\right)x^3 + \left(-\frac{1}{6} + \frac{1}{6}\right)x^4 +$
$\left(\frac{1}{120} - \frac{1}{12} + \frac{1}{24}\right)x^5 + \ldots = x + x^2 + \frac{1}{3}x^3 - \frac{1}{30}x^5 \ldots; e^x \cdot e^{ix} = e^{(1+ix)} = e^x(\cos x + i\sin x) = e^x\cos x +$
$i\left(e^x \sin x\right) \Rightarrow e^x \sin x$ is the series of the imaginary part of $e^{(1+ix)}$; $e^x \sin x$ will converge for all x

57. If $f(x) = \sum_{n=0}^{\infty} a_n x^n$, then $f^{(k)}(x) = \sum_{n=k}^{\infty} n(n-)(n-2)\cdots(n)(k-1)\,a_k\,x^{n-k}$ and $f^{(k)}(0) = k!\,a_k \Rightarrow a_k = \frac{f^{(k)}(0)}{k!}$ for k a nonnegative integer. ∴ the coefficients of f(x) are identical with the corresponding coefficients in the Maclaurin series of f(x) and the statements follow.

9.8 CALCULATIONS WITH TAYLOR SERIES

1. $(1+x)^{1/2} = 1^{1/2} + \frac{(1/2)(1)^{-1/2}x}{1!} + \frac{(1/2)(-1/2)(1)^{-3/2}x^2}{2!} + \frac{(1/2)(-1/2)(-3/2)(1)^{-5/2}x^3}{3!} + \ldots.$ The first four terms are $1, \frac{x}{2}, -\frac{x^2}{8}, \frac{x^3}{16}$.

3. $(1+3x)^{1/3} = 1^{1/3} + \frac{(1/3)(1)^{-2/3}(3x)}{1!} + \frac{(1/3)(-2/3)(1)^{-5/3}(3x)^2}{2!} + \frac{(1/3)(-2/3)(-5/3)(1)^{-8/3}(3x)^3}{3!} + \ldots.$ The first four terms are $1, x, -x^2, \frac{5x^3}{3}$.

5. $(1+x/2)^{-2} = 1^{-2} + \frac{(-2)(1)^{-3}(x/2)}{1!} + \frac{(-2)(-3)(1)^{-4}(x/2)^2}{2!} + \frac{(-2)(-3)(-4)(1)^{-5}(x/2)^3}{3!} + \ldots.$ The first four terms are $1, -x, \frac{3x^2}{4}, -\frac{x^3}{2}$.

7. $\left(1+x^3\right)^{-1/2} = (1)^{-1/2} + \frac{(-1/2)(1)^{-3/2}\left(x^3\right)^1}{1!} + \frac{(-1/2)(-3/2)(1)^{-5/2}\left(x^3\right)^2}{2!} + \frac{(-1/2)(-3/2)(-5/2)(1)^{-7/2}\left(x^3\right)^3}{3!} + \ldots$

The first four terms are 1, $-\frac{x^3}{2}, \frac{3x^6}{8}, -\frac{5x^9}{16}$.

9. $(1+1/x)^{1/2} = 1^{1/2} + \frac{(1/2)(1)^{-1/2}(1/x)}{1!} + \frac{(1/2)(-1/2)(1)^{-3/2}(1/x)^2}{2!} + \frac{(1/2)(-1/2)(-3/2)(1)^{-5/2}(1/x)^3}{3!} + \ldots$ The first four terms are 1, $\frac{1}{2x}, -\frac{1}{8x^2}, \frac{1}{16x^3}$.

11. $(1+x)^4 = (1)^4 + \frac{(4)(1)^3(x)}{1!} + \frac{(4)(3)(1)^2(x)^2}{2!} + \frac{(4)(3)(2)(1)^1(x)^3}{3!} + \frac{(4)(3)(2)(1)(1)^0(x)^4}{4!} = 1 + 4x + 6x^2 + 4x^3 + x^4$.

13. $(1-2x)^3 = \left(1+(-2x)\right)^3 = (1)^3 + \frac{(3)(1)^2(-2x)^1}{1!} + \frac{(3)(2)(1)^1(-2x)^2}{2!} + \frac{(3)(2)(1)(1)^0(-2x)^3}{3!} = 1 - 6x + 12x^2 - 8x^3$.

15. $\int_0^{0.2} \sin x^2\, dx = \int_0^{0.2}\left(x^2 - \frac{x^6}{3!} + \frac{x^{10}}{5!} - \cdots\right)dx = \left[\frac{x^3}{3} - \frac{x^7}{7\cdot 3!} + \cdots\right]_0^{0.2} \approx \left[\frac{x^3}{3}\right]_0^{0.2} \approx 0.00267$ with error $|E| \le \frac{(.2)^7}{7\cdot 3!} \approx 0.0000003$

17. $\int_0^{0.1} x^2 e^{-x^2}\, dx = \int_0^{0.1} x^2\left(1 - x^2 + \frac{x^4}{2!} - \frac{x^6}{3!} + \cdots\right)dx = \int_0^{0.1}\left(x^2 - x^4 + \frac{x^6}{2!} - \cdots\right)dx =$ $\left[\frac{x^3}{3} - \frac{x^5}{5} + \cdots\right]_0^{0.1} \approx \left[\frac{x^3}{3}\right]_0^{0.1} \approx 0.00033$ with error $|E| \le \frac{(0.1)^5}{5} \approx 0.000002$

19. $\int_0^{0.4} \frac{1-e^{-x}}{x}\, dx = \int_0^{0.4} \frac{1}{x}\left(1 - \left(1 - x + \frac{x^2}{2} - \frac{x^3}{3!} + \cdots\right)\right)dx = \int_0^{0.4}\left(1 - \frac{x}{2!} + \frac{x^2}{3!} - \frac{x^3}{4!} + \cdots\right)dx =$ $\left[x - \frac{x^2}{2\cdot 2!} + \frac{x^3}{3\cdot 3!} - \frac{x^4}{4\cdot 4!} + \cdots\right]_0^{0.4} \approx \left[x - \frac{x^2}{2\cdot 2!} + \frac{x^3}{3\cdot 3!}\right]_0^{0.4} \approx 0.36356$ with error $|E| \le \frac{(0.4)^4}{4\cdot 4!} \approx 0.000266$

21. $\int_0^{0.1} \frac{1}{\sqrt{1+x^4}}\, dx = \int_0^{0.1}\left(1 - \frac{x^4}{2} + \frac{3x^8}{8} - \cdots\right)dx = \left[x - \frac{x^5}{10} + \cdots\right]_0^{0.1} \approx [x]_0^{0.1} \approx 0.1$ with error $|E| \le \frac{(0.1)^5}{10} = 0.000001$

23. $\int_0^{0.1} \frac{\sin x}{x}\, dx = \int_0^{0.1} 1 - \frac{x^2}{3!} + \frac{x^4}{5!} - \frac{x^6}{7!} + \ldots dx = \left[x - \frac{x^3}{3\cdot 3!} + \frac{x^5}{5\cdot 5!} - \frac{x^7}{7\cdot 7!} + \cdots\right]_0^{0.1} \approx$ $\left[x - \frac{x^3}{3\cdot 3!} + \frac{x^5}{5\cdot 5!}\right]_0^{0.1} \approx 0.099944461$

25. $\left(1+x^4\right)^{1/2} = (1)^{1/2} + \frac{1/2}{1}(1)^{-1/2}\left(x^4\right) + \frac{(1/2)(-1/2)}{2!}(1)^{-3/2}\left(x^4\right)^2 + \frac{(1/2)(-1/2)(-3/2)}{3!}(1)^{-5/2}\left(x^4\right)^3 +$ $\frac{(1/2)(-1/2)(-3/2)(-5/2)}{4!}(1)^{-7/2}\left(x^4\right)^4 + \ldots = 1 + \frac{x^4}{2} - \frac{x^8}{8} + \frac{x^{12}}{16} - \frac{5x^{16}}{128} + \ldots;$

$\int_0^{0.1} 1 + \frac{x^4}{2} - \frac{x^8}{8} + \frac{x^{12}}{16} - \frac{5x^{16}}{128} + \ldots dx = \left[x + \frac{x^5}{10} - \frac{x^9}{72} + \frac{x^{13}}{208} - \frac{5x^{17}}{2176} + \cdots\right]_0^{0.1} \approx 0.100001$

27. $\ln\left(\frac{1+x}{1-x}\right) = \ln(1+x) - \ln(1-x) = \left(x - \frac{x^2}{2} + \frac{x^3}{3} - \frac{x^4}{4} + \ldots\right) - \left(-x - \frac{x^2}{2} - \frac{x^3}{3} - \frac{x^4}{4} - \ldots\right) = 2\left(x + \frac{x^3}{3} + \frac{x^5}{5} + \ldots\right)$

29. $\tan^{-1}x = x - \frac{x^3}{3} + \frac{x^5}{5} - \frac{x^7}{7} + \frac{x^9}{9} - \ldots + \frac{(-1)^{n-1}x^{2n-1}}{2n-1} + \ldots$ and the $|\text{error}| = \left|\frac{(-1)^{n-1}x^{2n-1}}{2n-1}\right| = \frac{1}{2n-1}$, when $x = 1$; $\frac{1}{2n-1} < \frac{1}{10^3} \Rightarrow n > \frac{1001}{2} = 500.5 \Rightarrow$ the first term not used is $501^{st} \Rightarrow$ we must use 500 terms

31. $\tan^{-1}x = x - \frac{x^3}{3} + \frac{x^5}{5} - \frac{x^7}{7} + \frac{x^9}{9} - \cdots + \frac{(-1)^{n-1}x^{2n-1}}{2n-1} + \cdots$; when the series representing $48\tan^{-1}\left(\frac{1}{18}\right)$ has an error of magnitude less than 10^{-6}, then the series representing $48\tan^{-1}\left(\frac{1}{18}\right) + 32\tan^{-1}\left(\frac{1}{57}\right) - 20\tan^{-1}\left(\frac{1}{239}\right)$ will also have an error of magnitude less than 10^{-6}; $\frac{\left(\frac{1}{18}\right)^{2n-1}}{2n-1} < \frac{1}{10^6} \Rightarrow n > 3 \Rightarrow 3$ terms

33. a) $\left(1-x^2\right)^{-1/2} \approx 1 + \frac{x^2}{2} + \frac{3x^4}{8} + \frac{5x^6}{16} \Rightarrow \sin^{-1}x \approx x + \frac{x^3}{6} + \frac{3x^5}{40} + \frac{5x^7}{112}$;

$\lim_{n\to\infty}\left|\frac{1\cdot3\cdot5\cdots(2n-1)(2n+1)\,x^{2n+3}}{2\cdot4\cdot6\cdots(2n)(2n+2)(2n+3)}\cdot\frac{2\cdot4\cdot6\cdots(2n)(2n+1)}{1\cdot3\cdot5\cdots(2n-1)\,x^{2n+1}}\right| < 1 \Rightarrow x^2\lim_{n\to\infty}\left|\frac{(2n+1)(2n+1)}{(2n+2)(2n+3)}\right| < 1 \Rightarrow$

$|x| < 1 \Rightarrow$ the radius of convergence is 1

b) since $\frac{d}{dx}\left(\cos^{-1}x\right) = -\left(1-x^2\right)^{-1/2} \Rightarrow \cos^{-1}x = \frac{\pi}{2} - \sin^{-1}x \approx \frac{\pi}{2} - \left(x + \frac{x^3}{6} + \frac{3x^5}{40} + \frac{5x^7}{112}\right) \approx \frac{\pi}{2} - x - \frac{x^3}{6} - \frac{3x^5}{40} - \frac{5x^7}{112}$

35. $\frac{d}{dx}\left(\frac{-1}{1+x}\right) = \frac{d}{dx}\left(\frac{-1}{1-(-x)}\right) = \frac{d}{dx}\left((-1) + (-1)(-x) + (-1)(-x)^2 + (-1)(-x)^3 + (-1)(-x)^4 + \cdots\right) =$
$1 - 2x + 3x^2 - 4x^3 + \cdots = \frac{1}{(1+x)^2}$

37. Assume that the solution has the form $y = a_0 + a_1x + a_2x^2 + \cdots + a_{n-1}x^{n-1} + a_nx^n + \cdots$. $\frac{dy}{dx} = a_1 + 2a_2x + \cdots + na_nx^{n-1} + \cdots$. Now $\frac{dy}{dx} + y = \left(a_1 + a_0\right) + \left(2a_2 + a_1\right)x + \left(3a_3 + a_2\right)x^2 + \cdots + \left(na_n + a_{n-1}\right)x^{n-1} + \cdots = 0 \Rightarrow$ $a_1 + a_0 = 0$, $2a_2 + a_1 = 0$, $3a_3 + a_2 = 0$ and in general $na_n + a_{n-1} = 0$. Since $y = 1$ when $x = 0$ we have $a_0 = 1$.

Therefore $a_1 = -1$, $a_2 = \frac{-a_1}{2\cdot1} = \frac{1}{2}$, $a_3 = \frac{-a_2}{3} = -\frac{1}{3\cdot2}$, $\cdots$, $a_n = \frac{-a_{n-1}}{n} = \frac{(-1)^n}{n!} \Rightarrow y = 1 - x + \frac{1}{2}x^2 - \frac{1}{3\cdot2}x^3 + \cdots + \frac{(-1)^n}{n!}x^n + \cdots = \sum_{n=0}^{\infty}\frac{(-1)^n x^n}{n!} = e^{-x}$.

39. Assume that the solution has the form $y = a_0 + a_1x + a_2x^2 + \cdots + a_{n-1}x^{n-1} + a_nx^n + \cdots$. $\frac{dy}{dx} = a_1 + 2a_2x + \cdots + na_nx^{n-1} + \cdots$. Now $\frac{dy}{dx} - y = \left(a_1 - a_0\right) + \left(2a_2 - a_1\right)x + \left(3a_3 - a_2\right)x^2 + \cdots + \left(na_n - a_{n-1}\right)x^{n-1} + \cdots = 1 \Rightarrow$ $a_1 - a_0 = 1$, $2a_2 - a_1 = 0$, $3a_3 - a_2 = 0$ and in general $na_n - a_{n-1} = 0$. Since $y = 0$ when $x = 0$ we have $a_0 = 0$,

$a_1 = 1$, $a_2 = \frac{a_1}{2} = \frac{1}{2}$, $a_3 = \frac{a_2}{3} = \frac{1/2}{3} = \frac{1}{3\cdot2}$, $a_4 = \frac{a_3}{4} = \frac{1/(3\cdot2)}{4} = \frac{1}{4\cdot3\cdot2}$, $\cdots$, $a_n = \frac{a_{n-1}}{n} = \frac{1}{n!} \Rightarrow y = 0 + 1x + \frac{1}{2}x^2 + \frac{1}{3\cdot2}x^3 + \frac{1}{4\cdot3\cdot2}x^4 + \cdots + \frac{1}{n!}x^n + \cdots = \left[1 + 1x + \frac{1}{2}x^2 + \frac{1}{3\cdot2}x^3 + \frac{1}{4\cdot3\cdot2}x^4 + \cdots + \frac{1}{n!}x^n + \cdots\right] - 1 = \left[\sum_{n=0}^{\infty}\frac{x^n}{n!}\right] - 1 = e^x - 1$.

41. Assume that the solution has the form $y = a_0 + a_1x + a_2x^2 + \cdots + a_{n-1}x^{n-1} + a_nx^n + \cdots$. $\frac{dy}{dx} = a_1 + 2a_2x + \cdots + na_nx^{n-1} + \cdots$. Now $\frac{dy}{dx} - 2y = \left(a_1 - 2a_0\right) + \left(2a_2 - 2a_1\right)x + \left(3a_3 - 2a_2\right)x^2 + \cdots + \left(na_n - 2a_{n-1}\right)x^{n-1} + \cdots = 0$.

Since $y = 1$ when $x = 0$ we have $a_0 = 1$. Therefore $a_1 = 2a_0 = 2(1) = 2$, $a_2 = \frac{2}{2}a_1 = \frac{2}{2}(2) = \frac{2^2}{2}$, $a_3 = \frac{2}{3}a_2 = \frac{2}{3}\left(\frac{2^2}{2}\right) =$

$\frac{2^3}{3 \cdot 2}, \cdots, a_n = \frac{2}{n} a_{n-1} = \frac{2}{n}\left(2^{n-1}\right) = \frac{2^n}{n!} \Rightarrow y = 1 + 2x + \frac{2^2}{2}x^2 + \frac{2^3}{3 \cdot 2}x^3 + \cdots + \frac{2^n}{n!}x^n + \cdots = 1 + (2x) + \frac{(2x)^2}{2!} +$

$\frac{(2x)^3}{3!} + \cdots + \frac{(2x)^n}{n!} + \cdots = \sum_{n=0}^{\infty} \frac{(2x)^n}{n!} = e^{2x}.$

43. Assume that the solution has the form $y = a_0 + a_1x + a_2x^2 + \cdots + a_{n-1}x^{n-1} + a_nx^n + \cdots$. $\frac{dy}{dx} = a_1 + 2a_2x + \cdots + na_nx^{n-1} + \cdots$. Now $\frac{dy}{dx} - y = (a_1 - a_0) + (2a_2 - a_1)x + (3a_3 - a_2)x^2 + \cdots + (na_n - a_{n-1})x^{n-1} + \cdots = x \Rightarrow$ $a_1 - a_0 = 0, 2a_2 - a_1 = 1, 3a_3 - a_2 = 0$ and in general $na_n - a_{n-1} = 0$. Since $y = 0$ when $x = 0$ we have $a_0 = 0$, $a_1 = 0, a_2 = \frac{1 + a_1}{2} = \frac{1}{2}, a_3 = \frac{a_2}{3} = \frac{1/2}{3} = \frac{1}{3 \cdot 2}, a_4 = \frac{a_3}{4} = \frac{1/(3 \cdot 2)}{4} = \frac{1}{4 \cdot 3 \cdot 2}, \cdots, a_n = \frac{a_{n-1}}{n} = \frac{1}{n!} \Rightarrow y = 0 + 0x + \frac{1}{2}x^2 + \frac{1}{3 \cdot 2}x^3 + \frac{1}{4 \cdot 3 \cdot 2}x^4 + \cdots + \frac{1}{n!}x^n + \cdots = \left[1 + 1x + \frac{1}{2}x^2 + \frac{1}{3 \cdot 2}x^3 + \frac{1}{4 \cdot 3 \cdot 2}x^4 + \cdots + \frac{1}{n!}x^n + \cdots\right] - 1 - x =$ $\left[\sum_{n=0}^{\infty} \frac{x^n}{n!}\right] - 1 - x = e^x - 1 - x = e^x - x - 1.$

45. Assume that the solution has the form $y = a_0 + a_1x + a_2x^2 + \cdots + a_{n-1}x^{n-1} + a_nx^n + \cdots$. $\frac{dy}{dx} = a_1 + 2a_2x + \cdots + na_nx^{n-1} + \cdots$. Now $\frac{dy}{dx} + y = (a_1 + a_0) + (2a_2 + a_1)x + (3a_3 + a_2)x^2 + \cdots + (na_n + a_{n-1})x^{n-1} + \cdots = x \Rightarrow$ $a_1 + a_0 = 0, 2a_2 + a_1 = 1, 3a_3 + a_2 = 0$ and in general $na_n + a_{n-1} = 0$. Since $y = 0$ when $x = 0$ we have $a_0 = 0$. Therefore $a_1 = 0, a_2 = \frac{1 - a_1}{2} = \frac{1}{2}, a_3 = \frac{-a_2}{3} = -\frac{1}{3 \cdot 2}, \cdots, a_n = \frac{-a_{n-1}}{n} = \frac{(-1)^n}{n!} \Rightarrow y = 0 - 0x + \frac{1}{2}x^2 - \frac{1}{3 \cdot 2}x^3 + \cdots + \frac{(-1)^n}{n!}x^n + \cdots = \left[1 - 1x + \frac{1}{2}x^2 - \frac{1}{3 \cdot 2}x^3 + \cdots + \frac{(-1)^n}{n!}x^n + \cdots\right] - 1 + x = \left[\sum_{n=0}^{\infty} \frac{(-1)^n x^n}{n!}\right] - 1 + x =$ $e^{-x} - 1 + x = e^{-x} + x - 1.$

9.P PRACTICE EXERCISES

1. converges to 1, since $\lim_{n \to \infty} a_n = \lim_{n \to \infty}\left(1 + \frac{(-1)^n}{n}\right) = 1$

3. converges to -1, since $\lim_{n \to \infty} a_n = \lim_{n \to \infty}\left(\frac{1 - 2^n}{2^n}\right) = \lim_{n \to \infty}\left(\frac{1}{2^n} - 1\right) = -1$

5. diverges, since $\left\{\sin \frac{n\pi}{2}\right\} = 0, 1, 0, -1, 0, 1, \ldots$

7. converges to 0, since $\lim_{n \to \infty} a_n = \lim_{n \to \infty} \frac{\ln n^2}{n} = 2 \lim_{n \to \infty} \frac{1/n}{1} = 0$

9. converges to 1, since $\lim_{n \to \infty} a_n = \lim_{n \to \infty}\left(\frac{n + \ln n}{n}\right) = \lim_{n \to \infty} \frac{1 + 1/n}{1} = 1$

11. converges to e^5, since $\lim_{n\to\infty} a_n = \lim_{n\to\infty} \left(\frac{n+5}{n}\right)^n = \lim_{n\to\infty} \left(1+\frac{5}{n}\right)^n = e^5$ by table 9.1

13. converges to 3, since $\lim_{n\to\infty} a_n = \lim_{n\to\infty} \left(\frac{3^n}{n}\right)^{1/n} = \frac{3}{\lim_{n\to\infty} n^{1/n}} = \frac{3}{1} = 3$ by table 9.1

15. diverges, since $\lim_{n\to\infty} a_n = \lim_{n\to\infty} \frac{(n+1)!}{n!} = \lim_{n\to\infty} n+1 = \infty$

17. Rewrite $\frac{1}{(2n-3)(2n-1)}$ as $\frac{A}{2n-3} + \frac{B}{2n-1} \Rightarrow \frac{A(2n-1)+B(2n-3)}{(2n-3)(2n-1)} = \frac{1}{(2n-3)(2n-1)} \Rightarrow A(2n-1) + B(2n-3) = 1 \Rightarrow (2A+2B)n + (-A-3B) = 1 \Rightarrow \begin{cases} A+B=0 \\ -A-3B=1 \end{cases} \Rightarrow -2B = 1 \Rightarrow B = -1/2$ and $A = 1/2$.

$s_n = \sum_{k=3}^{n} \frac{1}{(2k-3)(2k-1)} = \sum_{k=3}^{n} \left(\frac{1/2}{2k-3} + \frac{-1/2}{2k-1}\right) = \frac{1}{2} \sum_{k=3}^{n} \left(\frac{1}{2k-3} - \frac{1}{2k-1}\right) =$
$\frac{1}{2}\left[\frac{1}{3} - \frac{1}{5} + \frac{1}{5} - \frac{1}{7} + \frac{1}{7} - \frac{1}{9} + \cdots + \frac{1}{2n-5} - \frac{1}{2n-3} + \frac{1}{2n-3} - \frac{1}{2n-1}\right] = \frac{1}{2}\left[\frac{1}{3} - \frac{1}{2n-1}\right]$. The sum is $\lim_{n\to\infty} s_n =$
$\lim_{n\to\infty} \frac{1}{2}\left[\frac{1}{3} - \frac{1}{2n-1}\right] = \frac{1}{6}$.

19. Rewrite $\frac{9}{(3n-1)(3n+2)}$ as $\frac{A}{3n-1} + \frac{B}{3n+2} \Rightarrow \frac{A(3n+2)+B(3n-1)}{(3n-1)(3n+2)} = \frac{9}{(3n-1)(3n+2)} \Rightarrow (3A+3B)n + (2A-B) = 9 \Rightarrow \begin{cases} A+B=0 \\ 2A-B=9 \end{cases} \Rightarrow 3A = 9 \Rightarrow A = 3$ and $B = -3$. $s_n = \sum_{k=1}^{n} \frac{9}{(3k-1)(3k+2)} =$

$\sum_{k=1}^{n} \left(\frac{3}{3k-1} + \frac{-3}{3k+2}\right) = 3 \sum_{k=1}^{n} \left(\frac{1}{3k-1} - \frac{1}{3k+2}\right) =$
$3\left(\frac{1}{2} - \frac{1}{5} + \frac{1}{5} - \frac{1}{8} + \frac{1}{8} - \frac{1}{11} + \cdots + \frac{1}{3n-4} - \frac{1}{3n-1} + \frac{1}{3n-1} - \frac{1}{3n+2}\right) = 3\left(\frac{1}{2} - \frac{1}{3n+2}\right)$. The sum is $\lim_{n\to\infty} s_n =$
$\lim_{n\to\infty} 3\left(\frac{1}{2} - \frac{1}{3n+2}\right) = \frac{3}{2}$.

21. $\sum_{n=0}^{\infty} e^{-n} = \sum_{n=0}^{\infty} \frac{1}{e^n}$ a convergent geometric series with $r = \frac{1}{e}$ and $a = 1$. The sum is $\frac{1}{1-\frac{1}{e}} = \frac{1}{\frac{e-1}{e}} = \frac{e}{e-1}$.

23. diverges, a p–series where $p = \frac{1}{2}$

25. Since, $f(x) = \frac{1}{x^{1/2}} \Rightarrow f'(x) = -\frac{1}{2x^{3/2}} < 0 \Rightarrow f(x)$ is decreasing $\Rightarrow a_{n+1} < a_n$ and $\lim_{n\to\infty} a_n = \lim_{n\to\infty} \frac{(-1)^n}{\sqrt{n}} = 0$,

the $\sum_{n=1}^{\infty} \frac{(-1)^n}{\sqrt{n}}$ converges, by the Alternating Series Theorem. The series $\sum_{n=1}^{\infty} \frac{1}{\sqrt{n}}$ diverges $\Rightarrow$ the given series converges conditionally.

27. The given series does not converge absolutely by the Direct Comparison Test since, $\frac{1}{\ln(n+1)} > \frac{1}{n+1}$, the nth term of a divergent series. Since $f(x) = \frac{1}{\ln(x+1)} \Rightarrow f'(x) = -\frac{1}{(\ln(x+1))^2(x+1)} < 0 \Rightarrow$ f(x) is decreasing $\Rightarrow$ $a_{n+1} < a_n$ and $\lim_{n\to\infty} a_n = \lim_{n\to\infty} \frac{1}{\ln(n+1)} = 0$, the given series converges conditionally, by the Alternating Series Test.

29. converges absolutely by the Direct Comparison Test for $\frac{\ln n}{n^3} < \frac{n}{n^3} = \frac{1}{n^2}$ the nth term of a convergent p–series

31. $\lim_{n\to\infty} \frac{\frac{1}{n\sqrt{n^2+1}}}{\frac{1}{n^2}} = \sqrt{\lim_{n\to\infty} \frac{n^2}{n^2+1}} = \sqrt{1} = 1 \Rightarrow$ converges absolutely, by the Limit Comparison Test

33. converges absolutely, by the Ratio Test, since $\lim_{n\to\infty} \left|\frac{n+2}{(n+1)!} \cdot \frac{n!}{n+1}\right| = \lim_{n\to\infty} \frac{n+2}{(n+1)^2} = 0 < 1$

35. converges absolutely, by the Ratio Test, since $\lim_{n\to\infty} \left|\frac{3^{n+1}}{(n+1)!} \cdot \frac{n!}{3^n}\right| = \lim_{n\to\infty} \left|\frac{3}{n+1}\right| = 0 < 1$

37. converges absolutely, since $\lim_{n\to\infty} \frac{\frac{1}{n^{3/2}}}{\frac{1}{\sqrt{n(n+1)(n+2)}}} = \sqrt{\lim_{n\to\infty} \frac{n(n+1)(n+2)}{n^3}} = 1$, by the Limit Comparison Test

39. $\lim_{n\to\infty} \left|\frac{u_{n+1}}{u_n}\right| < 1 \Rightarrow \lim_{n\to\infty} \left|\frac{(x+4)^{n+1}}{(n+1)\,3^{n+1}} \cdot \frac{n\,3^n}{(x+4)^n}\right| < 1 \Rightarrow \frac{|x+4|}{3} \lim_{n\to\infty} \left(\frac{n}{n+1}\right) < 1 \Rightarrow \frac{|x+4|}{3} < 1 \Rightarrow |x+4| < 3 \Rightarrow$

$-3 < x + 4 < 3 \Rightarrow -7 < x < -1$, at $x = -7$ we have $\sum_{n=1}^{\infty} \frac{(-1)^n 3^n}{n\,3^n} = \sum_{n=1}^{\infty} \frac{(-1)^n}{n}$, the alternating harmonic series

which is conditionally convergent, at $x = -1$ we have $\sum_{n=1}^{\infty} \frac{3^n}{n\,3^n} = \sum_{n=1}^{\infty} \frac{1}{n}$, the divergent harmonic series.

a) the radius, 3; the interval of convergence, $-7 \le x < -1$

b) the interval of absolute convergence, $-7 < x < -1$

c) This series is conditionally convegence at $x = -7$.

41. $\lim_{n\to\infty} \left|\frac{u_{n+1}}{u_n}\right| < 1 \Rightarrow \lim_{n\to\infty} \left|\frac{(3x-1)^{n+1}}{(n+1)^2} \cdot \frac{n^2}{(3x-1)^n}\right| < 1 \Rightarrow |3x-1| \lim_{n\to\infty} \frac{n^2}{(n+1)^2} < 1 \Rightarrow |3x-1|\,(1) < 1 \Rightarrow$

$|3x-1| < 1 \Rightarrow -1 < 3x - 1 < 1 \Rightarrow 0 < 3x < 2 \Rightarrow 0 < x < 2/3$, at $x = 0$ we have $\sum_{n=1}^{\infty} \frac{(-1)^{n-1}(-1)^n}{n^2} =$

$\sum_{n=1}^{\infty} \frac{(-1)^{2n-1}}{n^2} = -\sum_{n=1}^{\infty} \frac{1}{n^2}$, the product of a nonzero constant and a convergent p–series is a convergent series which is also absolutely convergent; at x = 2/3 we have $\sum_{n=1}^{\infty} \frac{(-1)^{n-1}(1)^n}{n^2} = \sum_{n=1}^{\infty} \frac{(-1)^{n-1}}{n^2}$ an absolutely convergent series.

a) the radius, 1/3; the interval of convergence, $0 \le x \le 2/3$

b) the interval of absolute convergence, $0 \le x \le 2/3$

c) there are no values for which the series is conditionally converent

43. $\lim_{n \to \infty} \left|\frac{u_{n+1}}{u_n}\right| < 1 \Rightarrow \lim_{n \to \infty} \left|\frac{x^{n+1}}{(n+1)^{n+1}} \cdot \frac{n^n}{x^n}\right| < 1 \Rightarrow |x| \lim_{n \to \infty} \left|\left(\frac{n}{n+1}\right)^n \left(\frac{1}{n+1}\right)\right| < 1 \Rightarrow \frac{|x|}{e} \lim_{n \to \infty} \left(\frac{1}{n+1}\right) < 1 \Rightarrow$ $\frac{|x|}{e} \cdot 0 < 1.$

a) the radius, ∞; the interval of convergence is all x

b) the interval of absolute convergence is all x

c) there are no values for which the series is conditionally convergent

45. $\lim_{n \to \infty} \left|\frac{u_{n+1}}{u_n}\right| < 1 \Rightarrow \lim_{n \to \infty} \left|\frac{(n+2)\, x^{2n+1}}{3^{n+1}} \cdot \frac{3^n}{(n+1)\, x^{2n-1}}\right| < 1 \Rightarrow \frac{x^2}{3} \lim_{n \to \infty} \left|\frac{n+2}{n+1}\right| < 1 \Rightarrow -\sqrt{3} < x < \sqrt{3}$; when $x = \pm\sqrt{3}$ both series $\sum_{n=1}^{\infty} -\frac{n+1}{\sqrt{3}}$ and $\sum_{n=1}^{\infty} \frac{n+1}{\sqrt{3}}$ diverge.

a) the radius, $\sqrt{3}$, the interval of convergence, $-\sqrt{3} < x < \sqrt{3}$

b) the interval of absolute convergence, $-\sqrt{3} < x < \sqrt{3}$

c) there are no values for which the series is conditionally convergent

47. a) $\sum_{n=1}^{\infty} \left[\sin\left(\frac{1}{2n}\right) - \sin\left(\frac{1}{2n+1}\right)\right] = \sum_{n=2}^{\infty} (-1)^n \sin\left(\frac{1}{n}\right)$, converges by the Alternating Series Theorem for f(x) = $\sin\left(x^{-1}\right) \Rightarrow f'(x) = -\left(\cos\left(x^{-1}\right)\right)\left(x^{-2}\right) = \frac{-\cos\left(\frac{1}{x}\right)}{x^2} < 0$ and $\lim_{n \to \infty} \sin\left(\frac{1}{n}\right) = 0$

b) By using the original series we have $\sin\left(\frac{1}{42}\right) - \sin\left(\frac{1}{43}\right) \approx 0.00055$, the estimate is too small, by the Alternating Series Estimation Theorem. When using $\sum_{n=2}^{\infty} (-1)^n \sin\left(\frac{1}{n}\right)$ we have $\sin\frac{1}{41} \approx 0.2439$ which implies the same conclusion because the decision is based on the sign not the magnitude.

49. The given series is in the form $1 - x + x^2 - x^3 + \ldots + (-x)^n + \ldots = \frac{1}{1+x}$, where $x = \frac{1}{4}$. The sum is $\frac{1}{1 + 1/4} = \frac{4}{5}$.

51. The given series is in the form $x - \frac{x^3}{3!} + \frac{x^5}{5!} - \ldots + (-1)^n \frac{x^{2n+1}}{(2n+1)!} + \ldots = \sin x$, where $x = \pi$. The sum is $\sin \pi = 0$.

53. The given series is in the form $1 + x + \frac{x^2}{2!} + \frac{x^3}{3!} + \ldots + \frac{x^n}{n!} + \ldots = e^x$, where $x = \ln 2$. The sum is $e^{\ln(2)} = 2$.

55. $f(x) = \sqrt{3+x^2} = \left(3+x^2\right)^{1/2} \Rightarrow f(-1) = 2,\ f'(-1) = -\frac{1}{2},\ f''(-1) = \frac{3}{8},\ f'''(-1) = \frac{9}{32};\ \sqrt{3+x^2} = 2 - \frac{(x+1)}{2\cdot 1!} + \frac{3(x+1)^2}{2^3\cdot 2!} + \frac{9(x+1)^3}{2^5\cdot 3!} + \ldots$

57. Consider $\frac{1}{1-2x}$ as the sum of a convergent geometric series with $a = 1$ and $r = 2x \Rightarrow \frac{1}{1-2x} = 1 + (2x) + (2x)^2 + (2x)^3 + \ldots = \sum_{n=0}^{\infty} (2x)^n$ where $|2x| < 1 \Rightarrow |x| < \frac{1}{2}$.

59. $\sin x = \sum_{n=0}^{\infty} \frac{(-1)^n x^{2n+1}}{(2n+1)!} \Rightarrow \sin \pi x = \sum_{n=0}^{\infty} \frac{(-1)^n (\pi x)^{2n+1}}{(2n+1)!} = \sum_{n=0}^{\infty} \frac{(-1)^n \pi^{2n+1} x^{2n+1}}{(2n+1)!}$

61. $\cos x = \sum_{n=0}^{\infty} \frac{(-1)^n x^{2n}}{(2n)!} \Rightarrow \cos\left(x^{5/2}\right) = \sum_{n=0}^{\infty} \frac{(-1)^n \left(x^{5/2}\right)^{2n}}{(2n)!} = \sum_{n=0}^{\infty} \frac{(-1)^n x^{5n}}{(2n)!}$

63. $e^x = \sum_{n=0}^{\infty} \frac{x^n}{n!} \Rightarrow e^{(\pi x/2)} = \sum_{n=0}^{\infty} \frac{(\pi x/2)^n}{n!} = \sum_{n=0}^{\infty} \frac{\pi^n x^n}{2^n n!}$

65. $\int_0^{1/2} \exp\left(-x^3\right) dx = \int_0^{1/2} 1 - x^3 + \frac{x^6}{2!} - \frac{x^9}{3!} + \frac{x^{12}}{4!} + \ldots dx = \left[x - \frac{x^4}{4} + \frac{x^7}{7\cdot 2!} - \frac{x^{10}}{10\cdot 3!} + \frac{x^{13}}{13\cdot 4!} - \ldots\right]_0^{1/2} \approx \frac{1}{2} - \frac{1}{2^4\cdot 4} + \frac{1}{2^7\cdot 7\cdot 2!} - \frac{1}{2^{10}\cdot 10\cdot 3!} \approx 0.484917151$; note $\exp[f(x)] = e^{f(x)}$

67. Diverges. The nth partial sum $\ln\left[\frac{1}{2}\right] + \ln\left[\frac{2}{3}\right] + \cdots + \ln\left[\frac{n-1}{n}\right] + \ln\left[\frac{n}{n+1}\right] = \ln\left[\frac{1}{2}\cdot\frac{2}{3}\cdot \cdots \cdot \frac{n-1}{n}\cdot\frac{n}{n+1}\right] = \ln\left[\frac{1}{n+1}\right]$ approaches $-\infty$ as $n \to \infty$.

69. Diverges because the nth term, $\left(1 - (1/n)\right)^n$ approaches $1/e$ as $n \to \infty$. To see why, let $y = \left(1 - (1/n)\right)^n$. Then $\ln y = n \ln\left[\frac{n-1}{n}\right] = \frac{\ln(n-1) - \ln n}{1/n}$ and L'Hôpital's rule gives $\lim_{n\to\infty} \ln y = \lim_{n\to\infty} \frac{\left(1/(n-1)\right) - (1/n)}{-\left(1/n^2\right)} = \lim_{n\to\infty} -\frac{n^2}{n^2 - n} = -1$. Hence $\lim_{n\to\infty} y = e^{-1} = 1/e$.

71. Yes. $\sum_{n=1}^{\infty} a_n b_n$ converges. Since $\sum_{n=1}^{\infty} a_n$ converges, $a_n \to 0$, and from some index N on we have $a_n < 1$.

Therefore $a_n b_n$ for $n \geq N$ and $\sum_{n=1}^{\infty} a_n b_n$ converges by the Direct Comparison Test.

73. Assume that the solution has the form $y = a_0 + a_1 x + a_2 x^2 + \cdots + a_{n-1}x^{n-1} + a_n x^n + \cdots$. $\frac{dy}{dx} = a_1 + 2a_2 x + \cdots + na_n x^{n-1} + \cdots$. Now $\frac{dy}{dx} + y = \left(a_1 + a_0\right) + \left(2a_2 + a_1\right)x + \left(3a_3 + a_2\right)x^2 + \cdots + \left(na_n + a_{n-1}\right)x^{n-1} + \cdots = 0 \Rightarrow a_1 + a_0 = 0,\ 2a_2 + a_1 = 0,\ 3a_3 + a_2 = 0$ and in general $na_n + a_{n-1} = 0$. Since $y = -1$ when $x = 0$ we have $a_0 = -1$. Therefore $a_1 = 1,\ a_2 = \frac{-a_1}{2\cdot 1} = -\frac{1}{2},\ a_3 = \frac{-a_2}{3} = \frac{1}{3\cdot 2},\ a_4 = \frac{-a_3}{4} = -\frac{1}{4\cdot 3\cdot 2},\ \ldots,\ a_n = \frac{-a_{n-1}}{n} = \frac{-1}{n}\frac{(-1)^n}{(n-1)!} =$

$\frac{(-1)^{n+1}}{n!} \Rightarrow y = -1 + x - \frac{1}{2}x^2 + \frac{1}{3\cdot 2}x^3 - \cdots + \frac{(-1)^{n+1}}{n!}x^n + \cdots =$

$-1\left[1 - x + \frac{1}{2}x^2 - \frac{1}{3\cdot 2}x^3 - \cdots + \frac{(-1)^n}{n!}x^n + \cdots\right] = -1\left[\sum_{n=0}^{\infty} \frac{(-1)^n x^n}{n!}\right] = -e^{-x}$.

75. Assume that the solution has the form $y = a_0 + a_1x + a_2x^2 + \cdots + a_{n-1}x^{n-1} + a_nx^n + \cdots$. $\frac{dy}{dx} = a_1 + 2a_2x + \cdots + na_nx^{n-1} + \cdots$. Now $\frac{dy}{dx} - y = (a_1 - a_0) + (2a_2 - a_1)x + (3a_3 - a_2)x^2 + \cdots + (na_n - a_{n-1})x^{n-1} + \cdots = x \Rightarrow$ $a_1 - a_0 = 0$, $2a_2 - a_1 = 1$, $3a_3 - a_2 = 0$ and in general $na_n - a_{n-1} = 0$. Since $y = 1$ when $x = 0$ we have $a_0 = 1$, $a_1 = 1$, $a_2 = \frac{1 + a_1}{2} = \frac{2}{2}$, $a_3 = \frac{a_2}{3} = \frac{2/2}{3} = \frac{2}{3\cdot 2}$, $a_4 = \frac{a_3}{4} = \frac{2/(3\cdot 2)}{4} = \frac{2}{4\cdot 3\cdot 2}$, $\ldots$, $a_n = \frac{a_{n-1}}{n} = \frac{2}{n!} \Rightarrow y = 1 + 1x + \frac{2}{2}x^2 + \frac{2}{3\cdot 2}x^3 + \frac{1}{4\cdot 2\cdot 2}x^4 + \cdots + \frac{2}{n!}x^n + \cdots = 2\left[1 + 1x + \frac{1}{2}x^2 + \frac{1}{3\cdot 2}x^3 + \frac{1}{4\cdot 3\cdot 2}x^4 + \cdots + \frac{1}{n!}x^n + \cdots\right] - 1 - x = 2\left[\sum_{n=0}^{\infty} \frac{x^n}{n!}\right] - 1 - x = 2e^x - 1 - x = 2e^x - x - 1$.

Assume that the solution has the form [illegible]

[illegible] $(\bar{y}_{\cdot} - a)^2 - (\bar{y}_{\cdot} - a)^2$ [illegible]

[illegible] and [illegible]

CHAPTER 10

CONIC SECTIONS, PARAMETRIZED CURVES, AND POLAR COORDINATES

10.1 CONIC SECTIONS AND QUADRATIC EQUATIONS

1. $x = \frac{y^2}{8} \Rightarrow 4p = 8 \Rightarrow p = 2$. $\therefore$ Focus is (2,0), directrix is $x = -2$.

3. $y = -\frac{x^2}{6} \Rightarrow 4p = 6 \Rightarrow p = \frac{3}{2}$. $\therefore$ Focus is $(0,-\frac{3}{2})$, directrix is $y = \frac{3}{2}$.

5. $\frac{x^2}{4} - \frac{y^2}{9} = 1 \Rightarrow c = \sqrt{4+9} = \sqrt{13} \Rightarrow$ Foci are $\left(\pm\sqrt{13},0\right)$. $e = \frac{c}{a} = \frac{\sqrt{13}}{2} \Rightarrow \frac{a}{e} = \frac{2}{\frac{\sqrt{13}}{2}} = \frac{4}{\sqrt{13}} \Rightarrow$ Directrices are $x = \pm\frac{4}{\sqrt{13}}$.

   Asymptotes are $y = \pm\frac{3}{2}x$.

7. $\frac{x^2}{2} + y^2 = 1 \Rightarrow c = \sqrt{2-1} = 1 \Rightarrow$ Foci are $(\pm1,0)$. $e = \frac{c}{a} = \frac{1}{\sqrt{2}} \Rightarrow$ Directrices are $x = \pm\frac{\sqrt{2}}{\frac{1}{\sqrt{2}}} = \pm2$.

9.

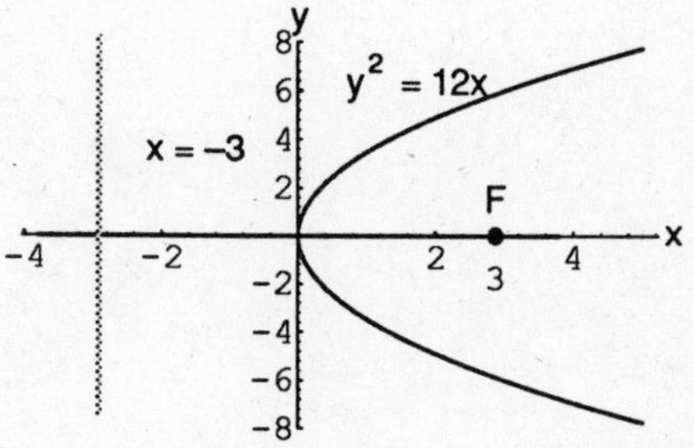

Graph 10.1.9

$y^2 = 12x \Rightarrow \frac{y^2}{12} = x \Rightarrow 4p = 12 \Rightarrow p = 3$

$\therefore$ focus is (3,0), directrix is $x = -3$

11.

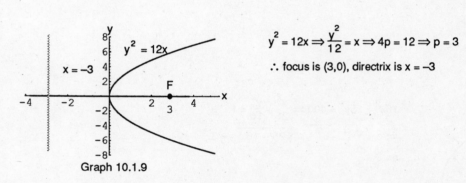

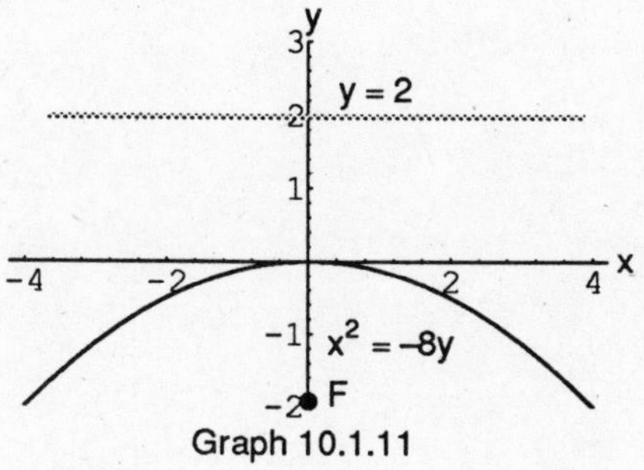

Graph 10.1.11

$x^2 = -8y \Rightarrow \frac{x^2}{-8} = y \Rightarrow 4p = 8 \Rightarrow p = 2$

$\therefore$ focus is (0,–2), directrix is $y = 2$.

13.

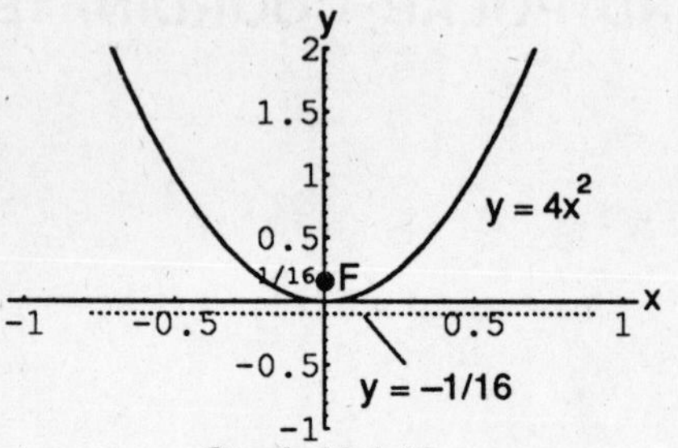

Graph 10.1.13

$y = 4x^2 \Rightarrow y = \frac{x^2}{1/4} \Rightarrow 4p = \frac{1}{4} \Rightarrow p = \frac{1}{16}$.

$\therefore$ focus is $\left(0, \frac{1}{16}\right)$, directrix is $y = -\frac{1}{16}$

15.

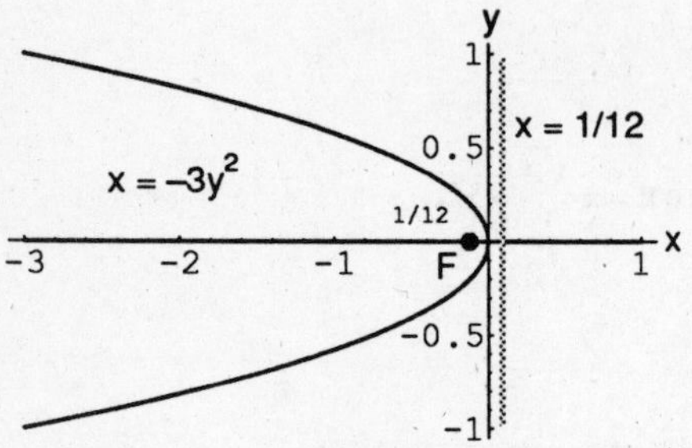

Graph 10.1.15

$x = -3y^2 \Rightarrow x = -\frac{y^2}{1/3} \Rightarrow 4p = \frac{1}{3} \Rightarrow p = \frac{1}{12}$

$\therefore$ focus is $\left(-\frac{1}{12}, 0\right)$, directrix is $x = \frac{1}{12}$

17.

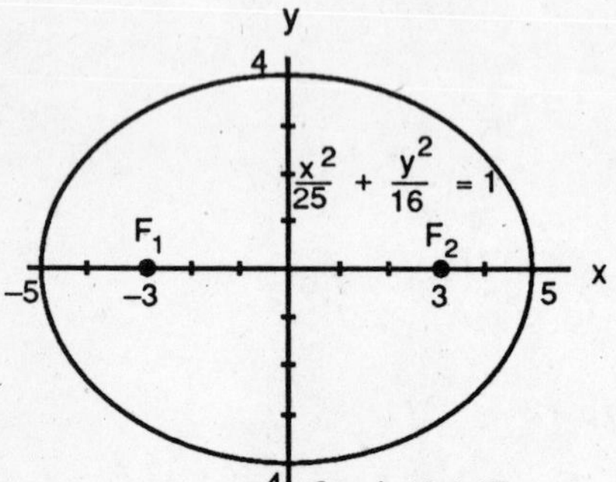

Graph 10.1.17

$16x^2 + 25y^2 = 400 \Rightarrow \frac{x^2}{25} + \frac{y^2}{16} = 1$

$\Rightarrow c = \sqrt{a^2 - b^2} = \sqrt{25 - 16} = 3.$

$e = \frac{c}{a} = \frac{3}{5}$

19.

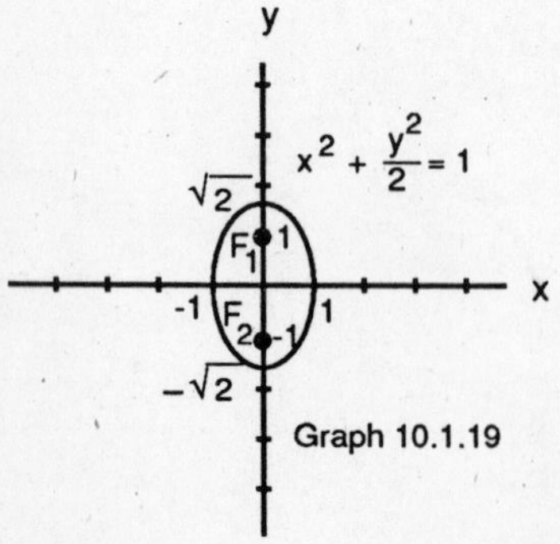

Graph 10.1.19

$2x^2 + y^2 = 2 \Rightarrow x^2 + \frac{y^2}{2} = 1$

$\Rightarrow c = \sqrt{a^2 - b^2} = \sqrt{2 - 1} = 1$

$e = \frac{c}{a} = \frac{1}{\sqrt{2}}$

21.

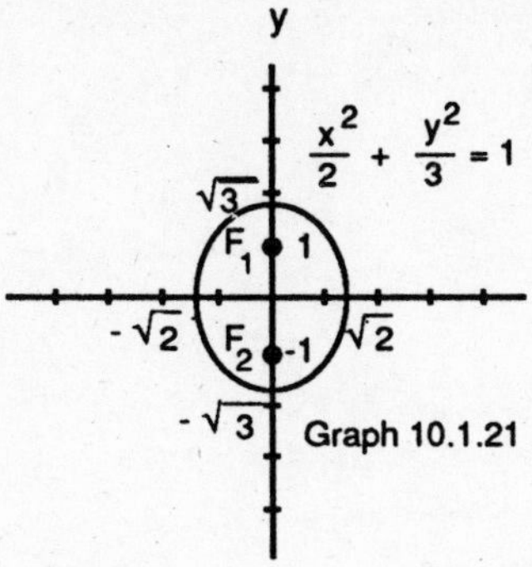

Graph 10.1.21

$3x^2 + 2y^2 = 6 \Rightarrow \frac{x^2}{2} + \frac{y^2}{3} = 1$

$\Rightarrow c = \sqrt{a^2 - b^2} = \sqrt{3-2} = 1$

$e = \frac{c}{a} = \frac{1}{\sqrt{3}}$

23.

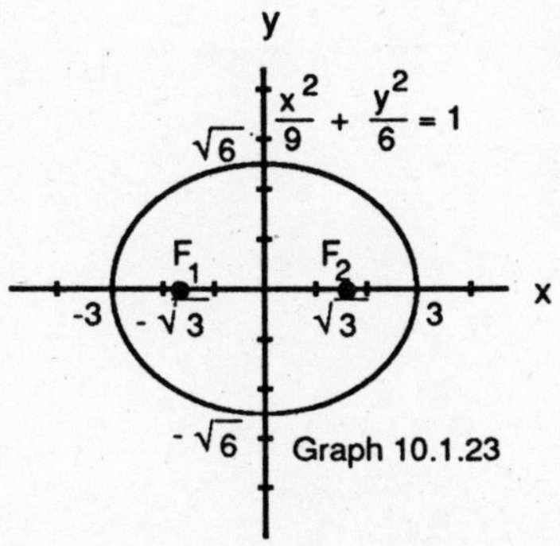

Graph 10.1.23

$6x^2 + 9y^2 = 54 \Rightarrow \frac{x^2}{9} + \frac{y^2}{6} = 1$

$c = \sqrt{a^2 - b^2} = \sqrt{9-6} = \sqrt{3}$

$e = \frac{c}{a} = \frac{\sqrt{3}}{3}$

25.

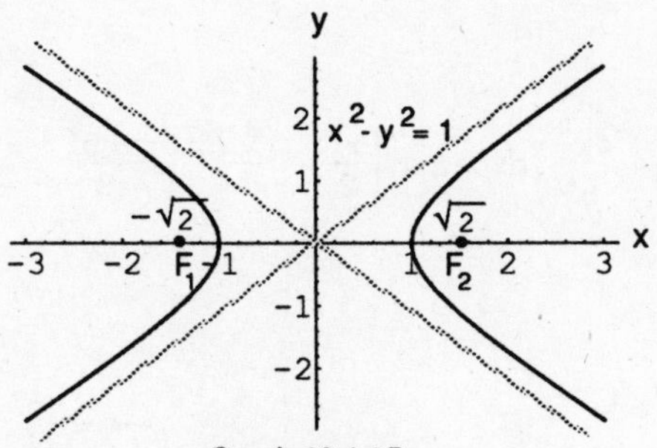

Graph 10.1.25

$x^2 - y^2 = 1 \Rightarrow c = \sqrt{a^2 + b^2} = \sqrt{1+1} = \sqrt{2}$

$e = \frac{c}{a} = \frac{\sqrt{2}}{1} = \sqrt{2}$

Asymptotes are $y = \pm x$

27.

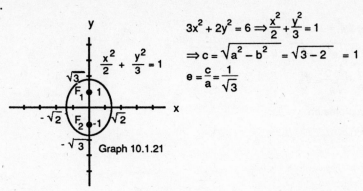

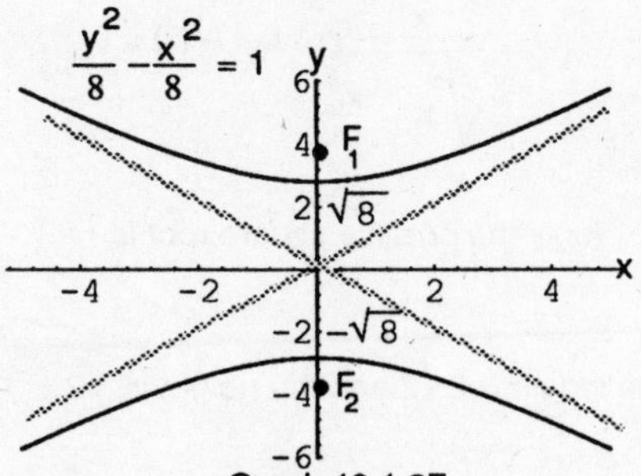

Graph 10.1.27

$y^2 - x^2 = 8 \Rightarrow \frac{y^2}{8} - \frac{x^2}{8} = 1$

$\Rightarrow c = \sqrt{a^2 + b^2} = \sqrt{8+8} = 4$

$e = \frac{c}{a} = \frac{4}{\sqrt{8}} = \sqrt{2}$ Asymptotes are $y = \pm x$.

29.

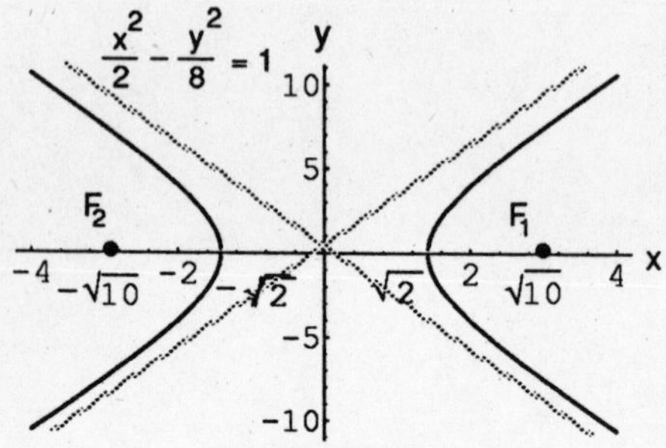

Graph 10.1.29

$8x^2 - 2y^2 = 16 \Rightarrow \frac{x^2}{2} - \frac{y^2}{8} = 1$

$\Rightarrow c = \sqrt{a^2 + b^2} = \sqrt{2+8} = \sqrt{10}$

$e = \frac{c}{a} = \frac{\sqrt{10}}{\sqrt{2}} = \sqrt{5}$

Asymptotes are $y = \pm 2x$.

31.

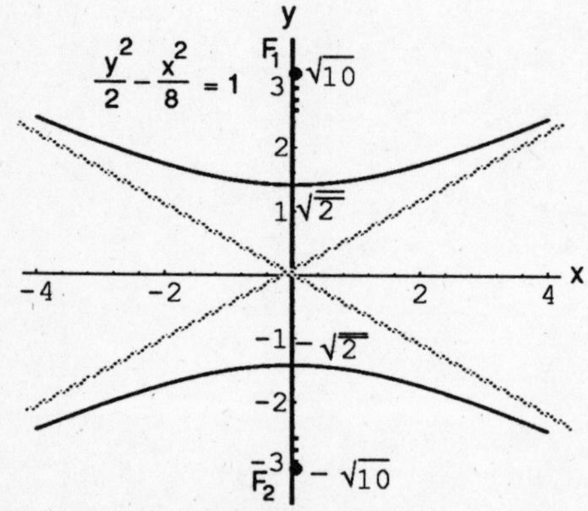

Graph 10.1.31

$8y^2 - 2x^2 = 16 \Rightarrow \frac{y^2}{2} - \frac{x^2}{8} = 1$

$\Rightarrow c = \sqrt{a^2 + b^2} = \sqrt{2+8} = \sqrt{10}$

$e = \frac{c}{a} = \frac{\sqrt{10}}{\sqrt{2}} = \sqrt{5}$ Asymptotes are $y = \pm\frac{x}{2}$.

33. Volume of the Parabolic Solid: $V_1 = \int_0^{b/2} 2\pi x\left(h - \frac{4h}{b^2}x^2\right)dx = 2\pi h \int_0^{b/2}\left(x - \frac{4x^3}{b^2}\right)dx = 2\pi h\left[\frac{x^2}{2} - \frac{x^4}{b^2}\right]_0^{b/2} = \frac{\pi h b^2}{8}$

Volume of the Cone: $V_2 = \frac{1}{3}\pi\left(\frac{b}{2}\right)^2 h = \frac{1}{3}\pi\left(\frac{b^2}{4}\right)h = \frac{\pi h b^2}{12}$ $\therefore V_1 = \frac{3}{2}V_2$

35. Let $y = \sqrt{1 - \frac{x^2}{4}}$ on the interval $0 \le x \le 2$. Then the area of the inscribed rectangle is given by $A(x) = 2x\left(2\sqrt{1 - \frac{x^2}{4}}\right)$ $= 4x\sqrt{1 - \frac{x^2}{4}}$ since the length is $2x$ and the height is $2y$. Then $A'(x) = 4\sqrt{1 - \frac{x^2}{4}} - \frac{x^2}{\sqrt{1 - \frac{x^2}{4}}}$. Let $A'(x) = 0$

$\Rightarrow 4\sqrt{1 - \frac{x^2}{4}} - \frac{x^2}{\sqrt{1 - \frac{x^2}{4}}} = 0 \Rightarrow 4\left(1 - \frac{x^2}{4}\right) - x^2 = 0 \Rightarrow x^2 = 2 \Rightarrow x = \sqrt{2}$ (only the positive square root is in the interval). Since $A(0) = 0$, $A(2) = 0$, $A(\sqrt{2}) = 4$ is the maximum area when the length is $2\sqrt{2}$ and the height is $\sqrt{2}$.

37. $9x^2 - 4y^2 = 36$, $x = 4 \Rightarrow y^2 = \frac{9x^2 - 36}{4} \Rightarrow y = \frac{3}{2}\sqrt{x^2 - 4}$

$$V = \int_2^4 \pi\left(\frac{3}{2}\sqrt{x^2 - 4}\right)^2 dx = \frac{9\pi}{4}\int_2^4 \left(x^2 - 4\right) dx = \frac{9\pi}{4}\left[\frac{x^3}{3} - 4x\right]_2^4 = 24\pi$$

39.

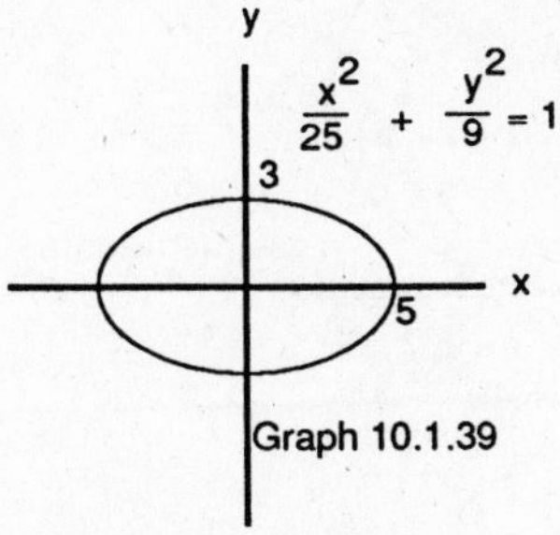

Graph 10.1.39

$e = \frac{4}{5} \Rightarrow$ Let $c = 4$ and $a = 5$. Then $c^2 = a^2 - b^2 \Rightarrow 16 = 25 - b^2 \Rightarrow b^2 = 9 \Rightarrow b = 3$. $\therefore \frac{x^2}{25} + \frac{y^2}{9} = 1$

41. $\frac{dr_A}{dt} = \frac{dr_B}{dt} \Rightarrow \int \frac{dr_A}{dt} = \int \frac{dr_B}{dt} \Rightarrow r_A + C_1 = r_B + C_2 \Rightarrow r_A - r_B = C$, a constant $\Rightarrow$ The points, P(t), lie on a hyperbola with foci A and B.

43. Volume of the solid created by rotating region A: $V_1 = \int_0^b \pi\left(\frac{y^2}{k}\right)^2 dy = \left[\frac{\pi y^5}{5k^2}\right]_0^b = \frac{\pi b^5}{5k^2}$ where $y = b$ is the upper boundary of the rectangle. The volume created by rotating the rectangle formed by the lines and the coordinate axes: $V_2 = \pi r^2 h = \pi\left(\frac{b^2}{k}\right)^2 b = \frac{\pi b^5}{k^2}$. $\therefore$ the volume created by rotating region B is $V_3 = V_2 - V_1 = \frac{\pi b^5}{k^2} - \frac{\pi b^5}{5k^2} = \frac{4\pi b^5}{5k^2}$.
$\therefore$ the ratio is $\frac{4\pi b^5/5k^2}{\pi b^5/5k^2} = \frac{4}{1}$ or 4 to 1.

45. PF will always equal PB because the string has constant length AB.

47. a) $y^2 = 8x \Rightarrow 4p = 8 \Rightarrow p = 2 \Rightarrow$ directrix is $x = -2$, focus is (2,0). Vertex is (0,0). $\therefore$ the new directrix is $x = -1$, new focus is (3,−2), and the new vertex is (1,−2).

b)

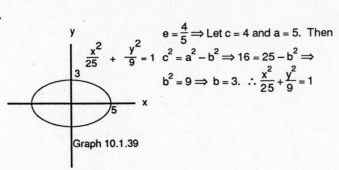

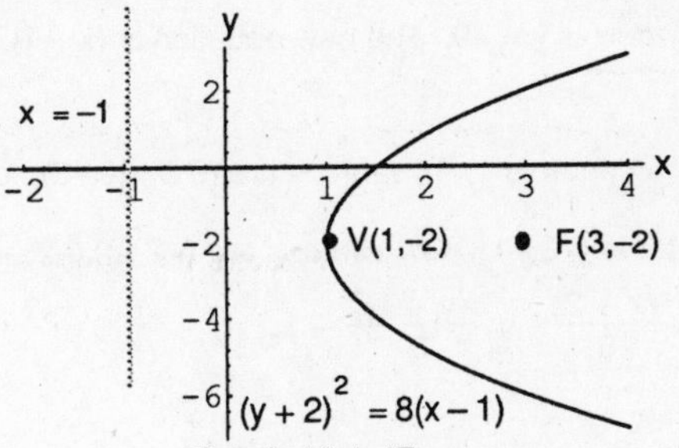

Graph 10.1.47

49. a) $\frac{x^2}{16}+\frac{y^2}{9}=1 \Rightarrow$ center is (0,0), vertices are (–4,0), (4,0). b)
$c=\sqrt{a^2-b^2}=\sqrt{7} \Rightarrow$ foci are $\left(\sqrt{7},\ 0\right), \left(-\sqrt{7},\ 0\right)$.
$\therefore$ new center is (4,3), new vertices are (0,3), (8,3), and new foci are $\left(-\sqrt{7}+4,\ 3\right), \left(\sqrt{7}+4,\ 3\right)$.

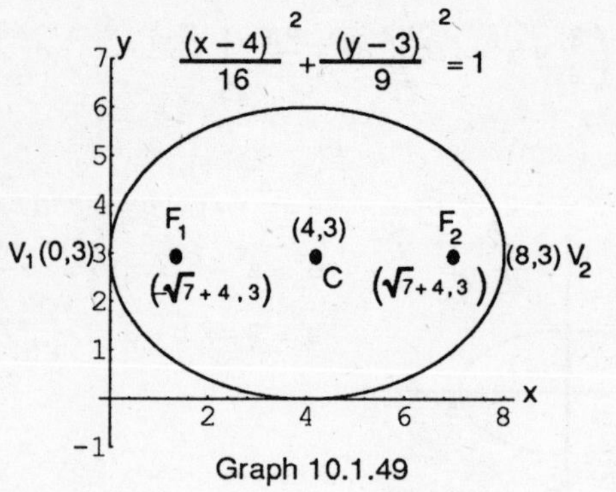

Graph 10.1.49

51. a) $\frac{x^2}{16}-\frac{y^2}{9}=1 \Rightarrow$ center is (0,0), vertices are (–4,0), (4,0). b)
Asymptotes are $\frac{x}{4}=\pm\frac{y}{3}$ or $y=\pm\frac{3x}{4}$. $c=\sqrt{a^2+b^2}=\sqrt{25}$ $=5 \Rightarrow$ foci are (–5,0), (5,0). $\therefore$ the new center is (2,0), new vertices are (–2,0), (6,0), new foci are (–3,0), (7,0), and new asymptotes are $y=\pm\frac{3(x-2)}{4}$.

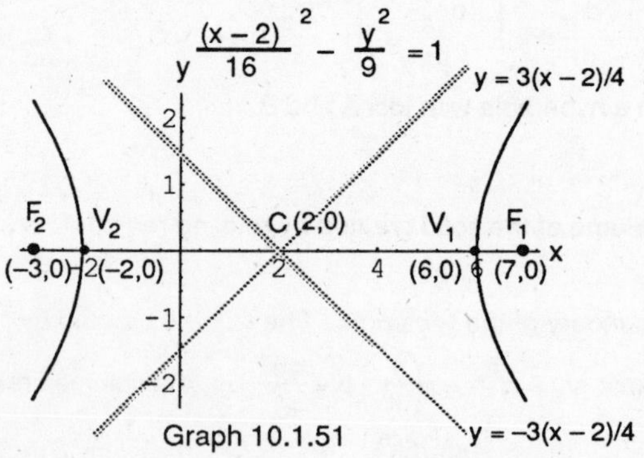

Graph 10.1.51

53. $y^2=4x \Rightarrow 4p=4 \Rightarrow p=1 \Rightarrow$ focus is (1,0), directrix is $x=-1$. Vertex is (0,0). $\therefore$ new vertex is (–2,–3), new focus is (–1,–3), and new directrix is $x=-3$. The new equation is $(y+3)^2=4(x+2)$.

55. $x^2=8y \Rightarrow 4p=8 \Rightarrow p=2 \Rightarrow$ focus is (0,2), directrix is $y=-2$. Vertex is (0,0). $\therefore$ new vertex is (1,–7), new focus is (1,–5), and new directrix is $y=-9$. The new equation is $(x-1)^2=8(y+7)$.

57. $\frac{x^2}{6}+\frac{y^2}{9}=1 \Rightarrow$ center is (0,0), vertices are (0,3), (0,–3). $c=\sqrt{a^2-b^2}=\sqrt{9-6}=\sqrt{3} \Rightarrow$ foci are $\left(0,\sqrt{3}\right), \left(0,-\sqrt{3}\right)$.
$\therefore$ new center is (–2,–1), new vertices are (–2,2), (–2,–4), new foci are $\left(-2,-1+\sqrt{3}\right), \left(-2,-1-\sqrt{3}\right)$. The new equation is $\frac{(x+2)^2}{6}+\frac{(y+1)^2}{9}=1$.

59. $\frac{x^2}{3}+\frac{y^2}{2}=1 \Rightarrow$ center is (0,0), vertices are $\left(\sqrt{3}\ ,0\right), \left(-\sqrt{3}\ ,0\right)$. $c=\sqrt{a^2-b^2}=\sqrt{3-2}=1 \Rightarrow$ foci are(–1,0), (1,0).
$\therefore$ new center is (2,3), new vertices are $\left(2+\sqrt{3}\ ,3\right), \left(2-\sqrt{3}\ ,\ 3\right)$, new foci are (1,3), (3,3). The new equation is $\frac{(x-2)^2}{3}+\frac{(y-3)^2}{2}=1$.

61. $\frac{x^2}{4} - \frac{y^2}{5} = 1 \Rightarrow$ center is (0,0), vertices are (2,0), (−2,0). $c = \sqrt{a^2 + b^2} = \sqrt{4+5} = 3 \Rightarrow$ foci are (3,0), (−3,0).

Asymptotes are $\pm\frac{x}{2} = \frac{y}{\sqrt{5}} \Rightarrow y = \pm\frac{x\sqrt{5}}{2}$. ∴ new center is (2,2), new vertices are (4,2), (0,2), new foci are (5,2), (−1,2).

The new asymptotes are $y - 2 = \pm\frac{(x-2)\sqrt{5}}{2}$. The new equation is $\frac{(x-2)^2}{4} - \frac{(y-2)^2}{5} = 1$.

63. $y^2 - x^2 = 1 \Rightarrow$ center is (0,0), vertices are (0,1), (0,−1). $c = \sqrt{a^2 + b^2} = \sqrt{1+1} = \sqrt{2} \Rightarrow$ foci are $(0,\sqrt{2})$, $(0,-\sqrt{2})$.
Asymptotes are $y = \pm x$. ∴ new center is (−1,−1), new vertices are (−1,0), (−1,−2), new foci are $(-1,-1+\sqrt{2})$, $(-1,-1-\sqrt{2})$. The new asymptotes are $y + 1 = \pm(x+1)$. The new equation is $(y+1)^2 - (x+1)^2 = 1$.

65. $x^2 + 5y^2 + 4x = 1 \Rightarrow x^2 + 4x + 4 + 5y^2 = 5 \Rightarrow (x+2)^2 + 5y^2 = 5 \Rightarrow \frac{(x+2)^2}{5} + y^2 = 1$, which is the equation of an ellipse. The center is (−2,0), the vertices are $(-2+\sqrt{5},0)$, $(-2-\sqrt{5},0)$. $c = \sqrt{a^2 - b^2} = \sqrt{5-1} = 2 \Rightarrow$ the foci are (−4,0), (0,0).

67. $x^2 + 2y^2 - 2x - 4y = -1 \Rightarrow x^2 - 2x + 1 + 2(y^2 - 2y + 1) = 2 \Rightarrow (x-1)^2 + 2(y-1)^2 = 2 \Rightarrow \frac{(x-1)^2}{2} + (y-1)^2 = 1$, which is the equation of an ellipse. The center is (1,1), the vertices are $(1+\sqrt{2},1)$, $(1-\sqrt{2},1)$. $c = \sqrt{a^2 - b^2} = \sqrt{2-1} = 1$ $\Rightarrow$ the foci are (2,1), (0,1).

69. $x^2 - y^2 - 2x + 4y = 4 \Rightarrow x^2 - 2x + 1 - (y^2 - 4y + 4) = 1 \Rightarrow (x-1)^2 - (y-2)^2 = 1$, which is the equation of a hyperbola. The center is (1,2), the vertices are (2,2), (0,2). $c = \sqrt{a^2 + b^2} = \sqrt{1+1} = \sqrt{2} \Rightarrow$ the foci are $(1+\sqrt{2},2)$, $(1-\sqrt{2},2)$. The asymptotes are $y - 2 = \pm(x-1)$.

71. $2x^2 - y^2 + 6y = 3 \Rightarrow 2x^2 - (y^2 - 6y + 9) = -6 \Rightarrow \frac{(y-3)^2}{6} - \frac{x^2}{3} = 1$, which is the equation of a hyperbola. The center is (0,3), the vertices are $(0,3+\sqrt{6})$, $(0,3-\sqrt{6})$. $c = \sqrt{a^2 + b^2} = \sqrt{6+3} = 3 \Rightarrow$ the foci are (0,6), (0,0).
The asymptotes are $\frac{y-3}{\sqrt{6}} = \pm\frac{x}{\sqrt{3}} \Rightarrow y = \pm x\sqrt{2} + 3$.

10.2 QUADRATIC EQUATIONS IN *X* AND *Y*

1. $x^2 - 3xy + y^2 - x = 0 \Rightarrow B^2 - 4AC = (-3)^2 - 4(1)(1) = 5 > 0 \Rightarrow$ Hyperbola

3. $3x^2 - 7xy + \sqrt{17}\,y^2 = 1 \Rightarrow B^2 - 4AC = (-7)^2 - 4(3)\sqrt{17} = -0.477 < 0 \Rightarrow$ Ellipse

5. $x^2 + 2xy + y^2 + 2x - y + 2 = 0 \Rightarrow B^2 - 4AC = (2)^2 - 4(1)(1) = 0 \Rightarrow$ Parabola

7. $x^2 + 4xy + 4y^2 - 3x = 6 \Rightarrow B^2 - 4AC = 4^2 - 4(1)(4) = 0 \Rightarrow$ Parabola

9. $xy + y^2 - 3x = 5 \Rightarrow B^2 - 4AC = 1^2 - 4(0)(1) = 1 > 0 \Rightarrow$ Hyperbola

11. $3x^2 - 5xy + 2y^2 - 7x - 14y = -1 \Rightarrow B^2 - 4AC = (-5)^2 - 4(3)(2) = 1 > 0 \Rightarrow$ Hyperbola

13. $x^2 - 3xy + 3y^2 + 6y = 7 \Rightarrow B^2 - 4AC = (-3)^2 - 4(1)(3) = -3 < 0 \Rightarrow$ Ellipse

15. $6x^2 + 3xy + 2y^2 + 17y + 2 = 0 \Rightarrow B^2 - 4AC = 3^2 - 4(6)(2) = -39 < 0 \Rightarrow$ Ellipse

17. $\cot 2\alpha = \frac{A - C}{B} = \frac{0}{1} = 0 \Rightarrow 2\alpha = \frac{\pi}{2} \Rightarrow \alpha = \frac{\pi}{4}$. $\therefore x = x'\cos\alpha - y'\sin\alpha,\ y = x'\sin\alpha + y'\cos\alpha \Rightarrow$
$x = x'\frac{\sqrt{2}}{2} - y'\frac{\sqrt{2}}{2},\ y = x'\frac{\sqrt{2}}{2} + y'\frac{\sqrt{2}}{2} \Rightarrow \left(\frac{\sqrt{2}}{2}x' - \frac{\sqrt{2}}{2}y'\right)\left(\frac{\sqrt{2}}{2}x' + \frac{\sqrt{2}}{2}y'\right) = 2 \Rightarrow \frac{1}{2}x'^2 - \frac{1}{2}y'^2 = 2 \Rightarrow x'^2 - y'^2 = 4$
$\Rightarrow$ Hyperbola

19. $\cot 2\alpha = \frac{A - C}{B} = \frac{3 - 1}{2\sqrt{3}} = \frac{1}{\sqrt{3}} \Rightarrow 2\alpha = \frac{\pi}{3} \Rightarrow \alpha = \frac{\pi}{6}$. $\therefore x = x'\cos\alpha - y'\sin\alpha,\ y = x'\sin\alpha + y'\cos\alpha \Rightarrow$
$x = \frac{\sqrt{3}}{2}x' - \frac{1}{2}y',\ y = \frac{1}{2}x' + \frac{\sqrt{3}}{2}y' \Rightarrow 3\left(\frac{\sqrt{3}}{2}x' - \frac{1}{2}y'\right)^2 + 2\sqrt{3}\left(\frac{\sqrt{3}}{2}x' - \frac{1}{2}y'\right)\left(\frac{1}{2}x' + \frac{\sqrt{3}}{2}y'\right) + \left(\frac{1}{2}x' + \frac{\sqrt{3}}{2}y'\right)^2 -$
$8\left(\frac{\sqrt{3}}{2}x' - \frac{1}{2}y'\right) + 8\sqrt{3}\left(\frac{1}{2}x' + \frac{\sqrt{3}}{2}y'\right) = 0 \Rightarrow 4x'^2 + 16y' = 0$, Parabola

21. $\cot 2\alpha = \frac{A - C}{B} = \frac{1 - 1}{-2} = 0 \Rightarrow 2\alpha = \frac{\pi}{2} \Rightarrow \alpha = \frac{\pi}{4}$. $\therefore x = x'\cos\alpha - y'\sin\alpha,\ y = x'\sin\alpha + y'\cos\alpha \Rightarrow$
$x = x'\frac{\sqrt{2}}{2} - y'\frac{\sqrt{2}}{2},\ y = x'\frac{\sqrt{2}}{2} + y'\frac{\sqrt{2}}{2} \Rightarrow \left(\frac{\sqrt{2}}{2}x' - \frac{\sqrt{2}}{2}y'\right)^2 - 2\left(\frac{\sqrt{2}}{2}x' - \frac{\sqrt{2}}{2}y'\right)\left(\frac{\sqrt{2}}{2}x' + \frac{\sqrt{2}}{2}y'\right) + \left(\frac{\sqrt{2}}{2}x' + \frac{\sqrt{2}}{2}y'\right)^2 = 2$
$\Rightarrow y'^2 = 1$, Parallel Horizontal Lines.

23. $\cot 2\alpha = \frac{A - C}{B} = \frac{\sqrt{2} - \sqrt{2}}{2\sqrt{2}} = 0 \Rightarrow 2\alpha = \frac{\pi}{2} \Rightarrow \alpha = \frac{\pi}{4}$. $\therefore x = x'\cos\alpha - y'\sin\alpha,\ y = x'\sin\alpha + y'\cos\alpha \Rightarrow$
$x = x'\frac{\sqrt{2}}{2} - y'\frac{\sqrt{2}}{2},\ y = x'\frac{\sqrt{2}}{2} + y'\frac{\sqrt{2}}{2} \Rightarrow \sqrt{2}\left(\frac{\sqrt{2}}{2}x' - \frac{\sqrt{2}}{2}y'\right)^2 + 2\sqrt{2}\left(\frac{\sqrt{2}}{2}x' - \frac{\sqrt{2}}{2}y'\right)\left(\frac{\sqrt{2}}{2}x' + \frac{\sqrt{2}}{2}y'\right) +$
$\sqrt{2}\left(\frac{\sqrt{2}}{2}x' + \frac{\sqrt{2}}{2}y'\right)^2 - 8\left(\frac{\sqrt{2}}{2}x' - \frac{\sqrt{2}}{2}y'\right) + 8\left(\frac{\sqrt{2}}{2}x' + \frac{\sqrt{2}}{2}y'\right) = 0 \Rightarrow 2\sqrt{2}\,x'^2 + 8\sqrt{2}\,y' = 0$, Parabola

25. $\cot 2\alpha = \frac{A - C}{B} = \frac{3-3}{2} = 0 \Rightarrow 2\alpha = \frac{\pi}{2} \Rightarrow \alpha = \frac{\pi}{4}$. $\therefore x = x'\cos\alpha - y'\sin\alpha,\ y = x'\sin\alpha + y'\cos\alpha \Rightarrow$

$x = x'\frac{\sqrt{2}}{2} - y'\frac{\sqrt{2}}{2},\ y = x'\frac{\sqrt{2}}{2} + y'\frac{\sqrt{2}}{2} \Rightarrow 3\left(\frac{\sqrt{2}}{2}x' - \frac{\sqrt{2}}{2}y'\right)^2 + 2\left(\frac{\sqrt{2}}{2}x' - \frac{\sqrt{2}}{2}y'\right)\left(\frac{\sqrt{2}}{2}x' + \frac{\sqrt{2}}{2}y'\right) + 3\left(\frac{\sqrt{2}}{2}x' + \frac{\sqrt{2}}{2}y'\right)^2$

$= 19 \Rightarrow 4x'^2 + 2y'^2 = 19$, Ellipse.

27. $\tan 2\alpha = \frac{-1}{1-3} = \frac{1}{2} \Rightarrow 2\alpha \approx 26.57° \Rightarrow \alpha \approx 13.28° \Rightarrow \sin\alpha \approx 0.23,\ \cos\alpha \approx 0.97$. Then $A' \approx 0.88,\ B' \approx 0.00,\ C' \approx 3.12$,

$D' \approx 0.74,\ E' \approx -1.20$, and $F' = -3 \Rightarrow 0.88x'^2 + 3.12y'^2 + 0.74x' - 1.20y' - 3 = 0$, an ellipse.

29. $\tan 2\alpha = \frac{-4}{1-4} = \frac{4}{3} \Rightarrow 2\alpha \approx 53.13° \Rightarrow \alpha \approx 26.56° \Rightarrow \sin\alpha \approx 0.45,\ \cos\alpha \approx 0.89$. Then $A' \approx 0.00,\ B' \approx 0.00,\ C' \approx 5.00$,

$D' \approx 0,\ E' \approx 0$, and $F' = -5 \Rightarrow 5.00y'^2 - 5 = 0$ or $y' = \pm 1.00$, parallel lines.

31. $\tan 2\alpha = \frac{5}{3-2} = 5 \Rightarrow 2\alpha \approx 78.69° \Rightarrow \alpha \approx 39.34° \Rightarrow \sin\alpha \approx 0.63,\ \cos\alpha \approx 0.77$. Then $A' \approx 5.05,\ B' \approx 0.00,\ C' \approx -0.05$,

$D' \approx -5.07,\ E' \approx -6.19$, and $F' = -1 \Rightarrow 5.05x'^2 - 0.05y'^2 - 5.07x' - 6.19y' - 1 = 0$, a hyperbola.

33. a) $\frac{x'^2}{b^2} + \frac{y'^2}{a^2} = 1$

b) $\frac{y'^2}{a^2} - \frac{x'^2}{b^2} = 1$

c) $x'^2 + y'^2 = a^2$

d) $y = mx \Rightarrow y - mx = 0 \Rightarrow D = -m,\ E = 1.\ \alpha = 90°$

$\Rightarrow D' = 1, E' = m \Rightarrow my' + x' = 0 \Rightarrow y' = -\frac{1}{m}x'$

e) $y = mx + b \Rightarrow y - mx - b = 0 \Rightarrow D = -m,\ E = 1 \Rightarrow$

$D' = 1,\ E' = m$ (See part d above.) Also $F' = -b$

$\Rightarrow my' + x' - b = 0 \Rightarrow y' = -\frac{1}{m}x' + \frac{b}{m}$.

35. a) $A' = \cos 45° \sin 45° = \left(\frac{\sqrt{2}}{2}\right)\left(\frac{\sqrt{2}}{2}\right) = \frac{1}{2},\ B' = 0,\ C' = -\cos 45° \sin 45° = -\frac{1}{2},\ F' = -1 \Rightarrow$

$\frac{1}{2}x'^2 - \frac{1}{2}y'^2 = 1$ or $x'^2 - y'^2 = 2$.

b) $A' = \frac{1}{2},\ C' = -\frac{1}{2}$ (See part a above.) $D' = E' = B' = 0,\ F' = -a \Rightarrow \frac{1}{2}x'^2 - \frac{1}{2}y'^2 = a$ or $x'^2 - y'^2 = 2a$

37. Yes, the graph is a hyperbola. With $AC < 0$ we have $-4AC > 0$ and $B^2 - 4AC > 0$

39. Let α be any angle. Then $A' = \cos^2\alpha + \sin^2\alpha = 1,\ B' = 0,\ C' = \sin^2\alpha + \cos^2\alpha = 1,\ D' = E' = 0,\ F' = -a^2$

$\Rightarrow x'^2 + y'^2 = a^2$.

10.3 PARAMETRIZATIONS OF CURVES

1.

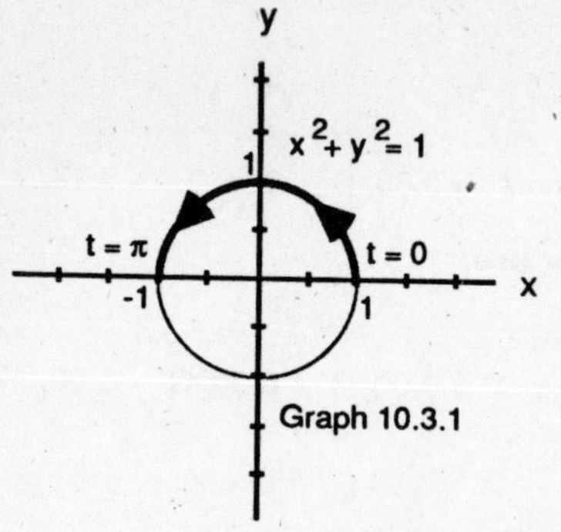

Graph 10.3.1

$x = \cos t,\ y = \sin t,\ 0 \le t \le \pi \Rightarrow$
$\cos^2 t + \sin^2 t = 1 \Rightarrow x^2 + y^2 = 1$

3.

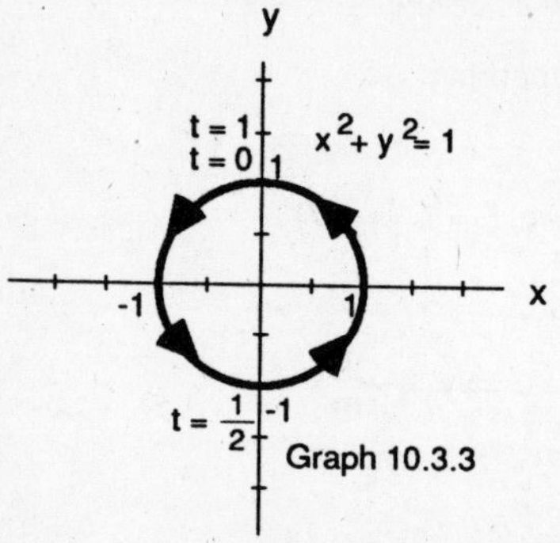

Graph 10.3.3

$x = \sin[2\pi(1 - t)],\ y = \cos[2\pi(1 - t)],\ 0 \le t \le 1 \Rightarrow$
$\sin^2[2\pi(1 - t)] + \cos^2[2\pi(1 - t)] = 1 \Rightarrow x^2 + y^2 = 1$

5.

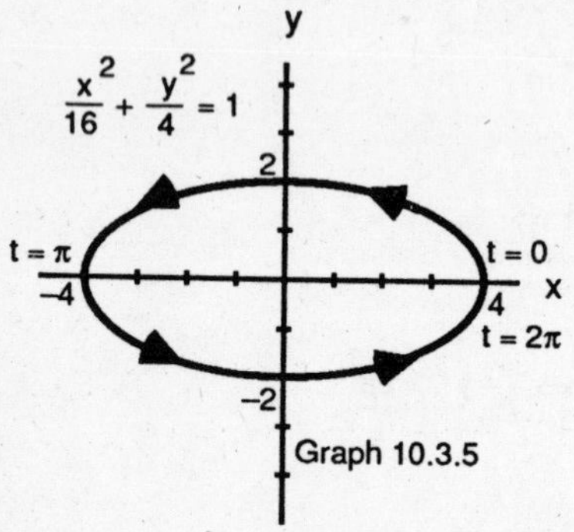

Graph 10.3.5

$x = 4\cos t,\ y = 2\sin t,\ 0 \le t \le 2\pi \Rightarrow$
$\frac{16\cos^2 t}{16} + \frac{4\sin^2 t}{4} = 1 \Rightarrow \frac{x^2}{16} + \frac{y^2}{4} = 1$

7.

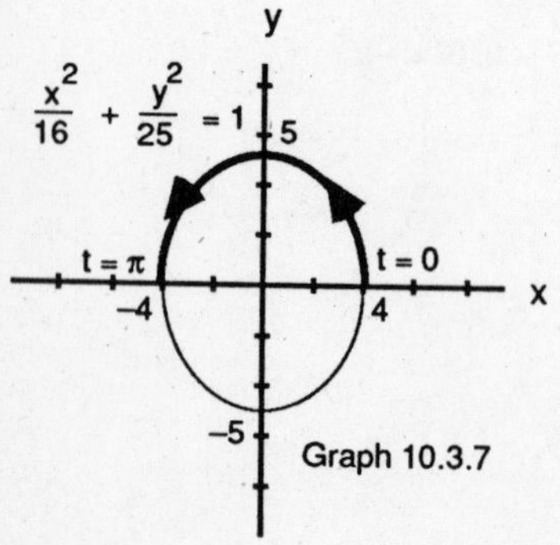

Graph 10.3.7

$x = 4\cos t,\ y = 5\sin t,\ 0 \le t \le \pi \Rightarrow$
$\frac{16\cos^2 t}{16} + \frac{25\sin^2 t}{25} = 1 \Rightarrow \frac{x^2}{16} + \frac{y^2}{25} = 1$

9.

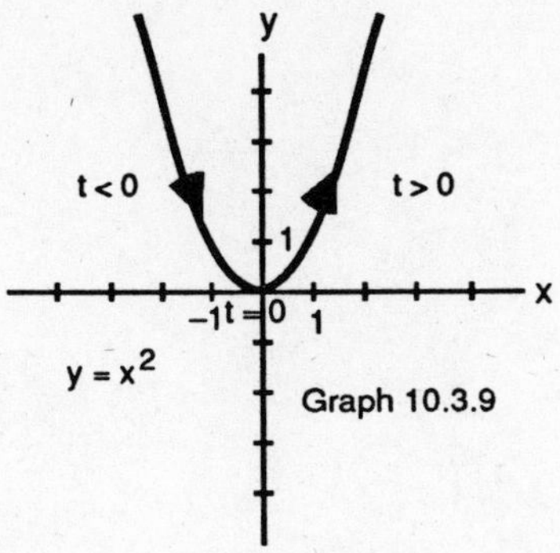

Graph 10.3.9

$x = 3t,\ y = 9t^2,\ -\infty < t < \infty \Rightarrow$
$y = x^2$

11.

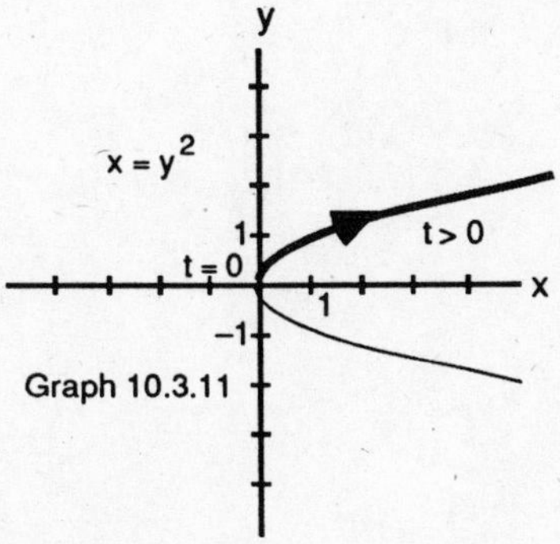

Graph 10.3.11

$x = t,\ y = \sqrt{t},\ t \geq 0$
$x = y^2$

13.

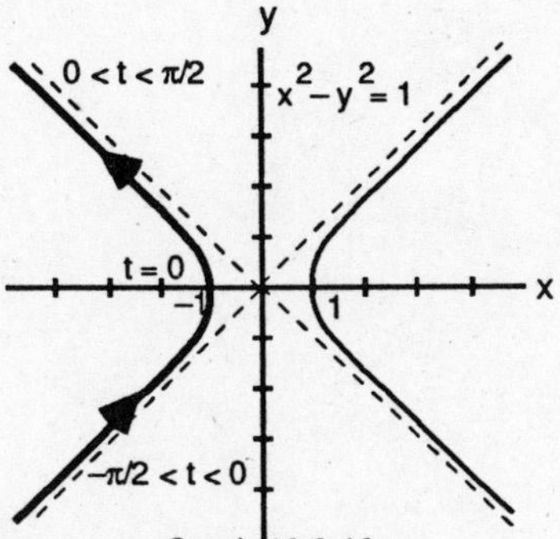

Graph 10.3.13

$x = -\sec t,\ y = \tan t,\ -\frac{\pi}{2} < t < \frac{\pi}{2} \Rightarrow$
$\sec^2 t - \tan^2 t = 1 \Rightarrow x^2 - y^2 = 1$

15.

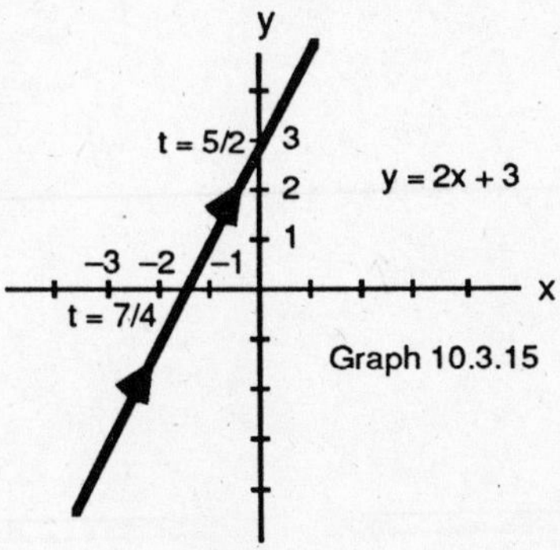

Graph 10.3.15

$x = 2t - 5,\ y = 4t - 7,\ -\infty < t < \infty \Rightarrow$
$x + 5 = 2t \Rightarrow 2(x + 5) = 4t \Rightarrow$
$y = 2(x + 5) - 7 \Rightarrow y = 2x + 3$

17.

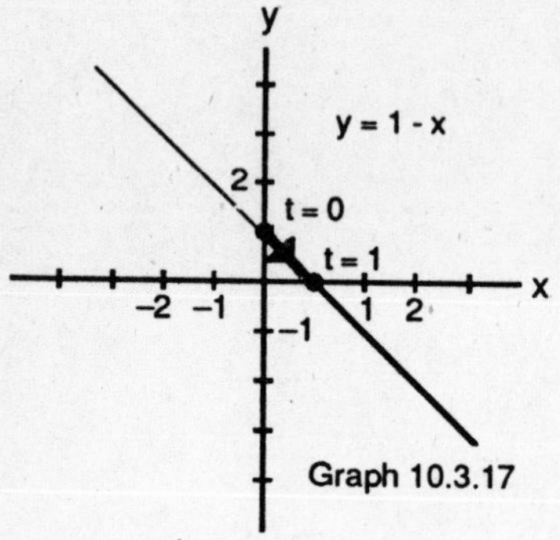

Graph 10.3.17

$x = t, y = 1 - t, 0 \le t \le 1 \Rightarrow$
$y = 1 - x$

19.

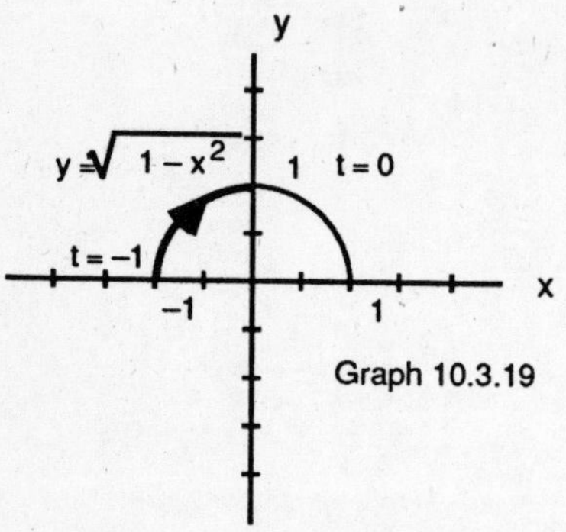

Graph 10.3.19

$x = t, y = \sqrt{1 - t^2}, -1 \le t \le 0 \Rightarrow$
$y = \sqrt{1 - x^2}$

21.

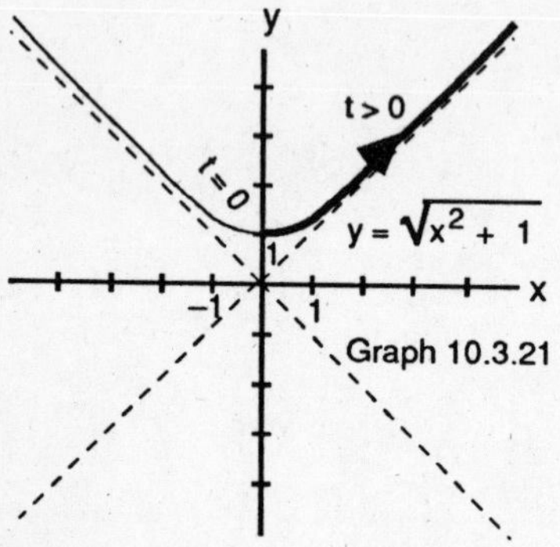

Graph 10.3.21

$x = t^2, y = \sqrt{t^4 + 1}, t \ge 0 \Rightarrow$
$y = \sqrt{x^2 + 1}, x \ge 0$

23.

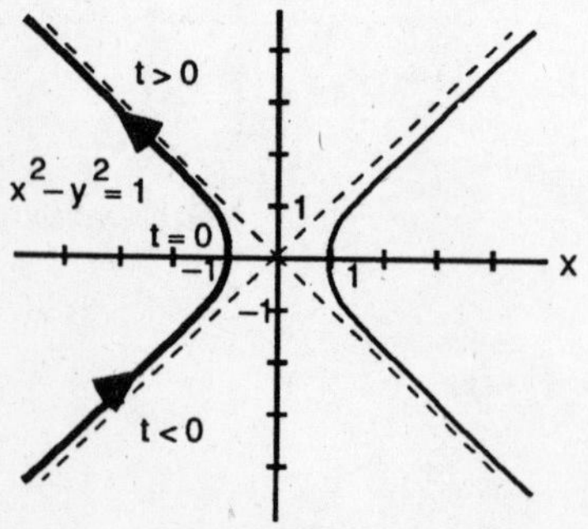

Graph 10.3.23

$x = -\cosh t, y = \sinh t, -\infty < t < \infty \Rightarrow$
$\cosh^2 t - \sinh^2 t = 1 \Rightarrow x^2 - y^2 = 1$

25. a) $x = a\cos t,\ y = -a\sin t,\ 0 \le t \le 2\pi$

b) $x = a\cos t,\ y = a\sin t,\ 0 \le t \le 2\pi$

c) $x = a\cos t,\ y = -a\sin t,\ 0 \le t \le 4\pi$

d) $x = a\cos t,\ y = a\sin t,\ 0 \le t \le 4\pi$

27.

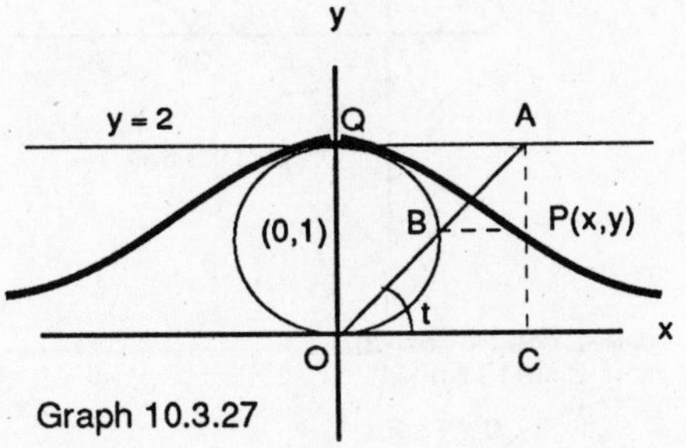

Graph 10.3.27

Extend the vertical line through A to the x–axis, let C be the point of intersection. Then OC = AQ = x and $\tan t = \frac{2}{OC} = \frac{2}{x} \Rightarrow x = \frac{2}{\tan t} = 2\cot t$. $\sin t = \frac{2}{OA} \Rightarrow$ $OA = \frac{2}{\sin t}$. $(AB)(OA) = (AQ)^2 \Rightarrow AB\left(\frac{2}{\sin t}\right) = x^2 \Rightarrow$ $AB\left(\frac{2}{\sin t}\right) = \left(\frac{2}{\tan t}\right)^2 \Rightarrow AB = \frac{2\sin t}{\tan^2 t}$.

$y = 2 - AB\sin t \Rightarrow y = 2 - \left(\frac{2\sin t}{\tan^2 t}\right)\sin t = 2 - \frac{2\sin^2 t}{\tan^2 t} =$ $2 - 2\cos^2 t = 2\sin^2 t$. $\therefore$ let $x = 2\cot t,\ y = 2\sin^2 t,\ 0 < t < \pi$

29. $d = \sqrt{(x-2)^2 + \left(y - \frac{1}{2}\right)^2} \Rightarrow d^2 = (x-2)^2 + \left(y - \frac{1}{2}\right)^2 = (t-2)^2 + \left(t^2 - \frac{1}{2}\right)^2 \Rightarrow d^2 = t^4 - 4t + \frac{17}{4}$

$\frac{d(d^2)}{dt} = 4t^3 - 4$. Let $4t^3 - 4 = 0 \Rightarrow t = 1$ The second derivative is always positive for $t \ne 0 \Rightarrow$ $t = 1$ gives a local minimum which is an absolute minimum since it is the only extremum. $\therefore$ the closest point on the parabola is (1,1).

31. a) $x = x_0 + (x_1 - x_0)t,\ y = y_0 + (y_1 - y_0)t \Rightarrow t = \frac{x - x_0}{x_1 - x_0} \Rightarrow y = y_0 + (y_1 - y_0)\left(\frac{x - x_0}{x_1 - x_0}\right) \Rightarrow$

$y - y_0 = \left(\frac{y_1 - y_0}{x_1 - x_0}\right)(x - x_0)$ which is the equation of the line through the points.

b) $x = x_1 t,\ y = y_1 t$ (Answer not unique)

c) $x = -1 + t,\ y = t$ or $x = -t,\ y = 1 - t$

33. a)

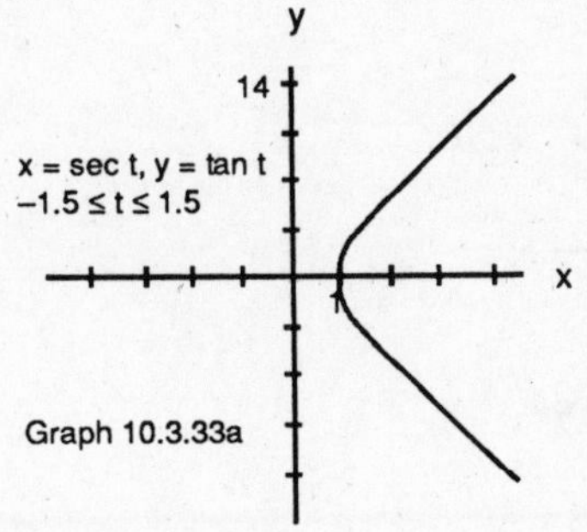

Graph 10.3.33a

b)

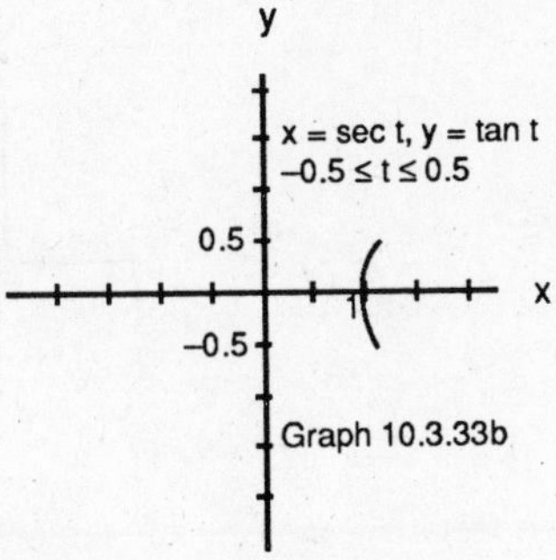

Graph 10.3.33b

c)

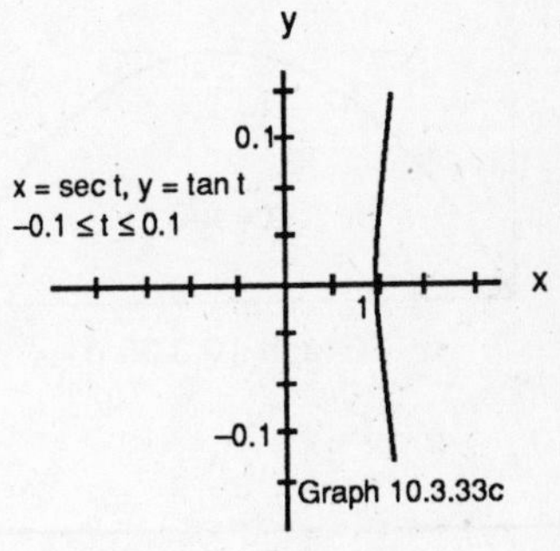

Graph 10.3.33c

35. a)

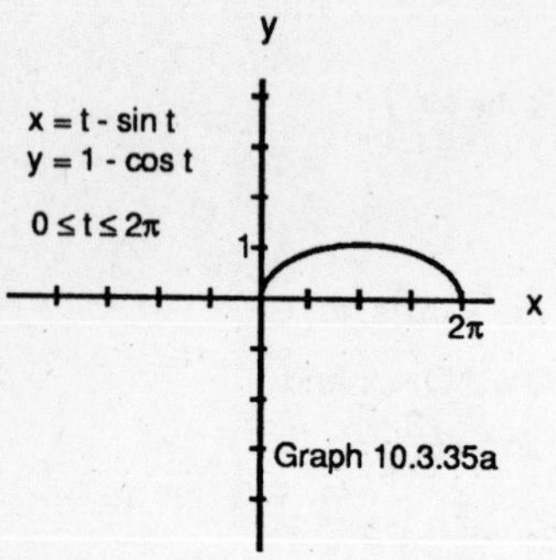

x = t - sin t
y = 1 - cos t

$0 \le t \le 2\pi$

Graph 10.3.35a

b)

x = t - sin t
y = 1 - cos t
$0 \le t \le 4\pi$

Graph 10.3.35b

c)

x = t - sin t
y = 1 - cos t
$\pi \le t \le 3\pi$

Graph 10.3.35c

37. a)

x = 2 cos t + cos 2t

y = 2 sin t - sin 2t

$0 \le t \le 2\pi$

Graph 10.3.37a

b)

x = -2 cos t + cos(-2t)
y = -2 sin t - sin(-2t)

$0 \le t \le 2\pi$

Graph 10.3.37b

39. a)

$\alpha = \pi/4$

Graph 10.3.39 a

b)

$\alpha = \pi/6$

Graph 10.3.39 b

c)

$\alpha = \pi/3$

Graph 10.3.39 c

d)

$\alpha = \pi/2$

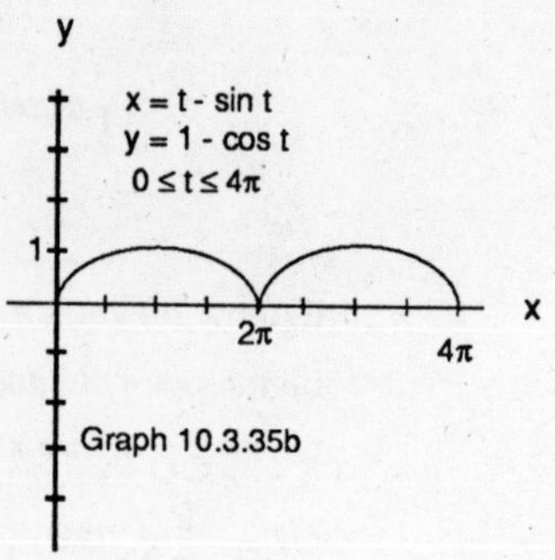

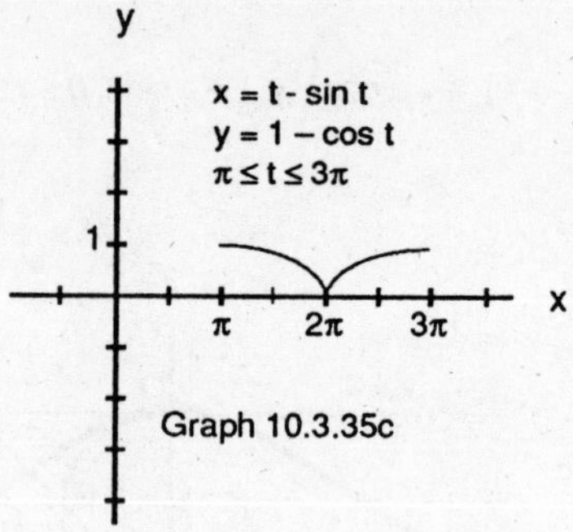

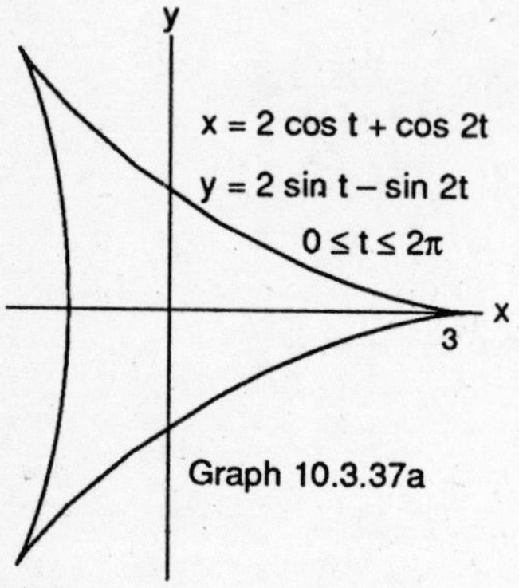

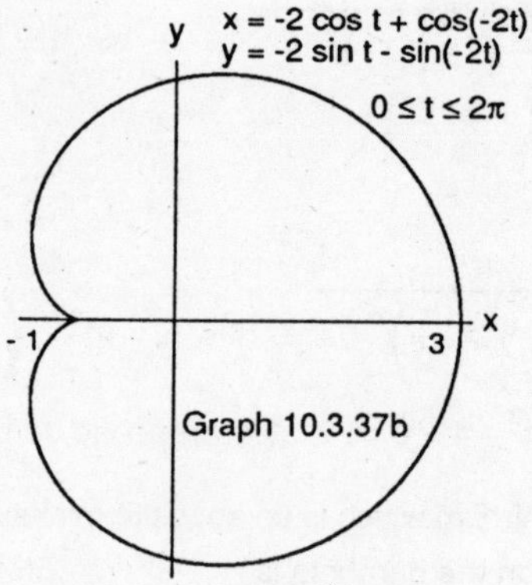

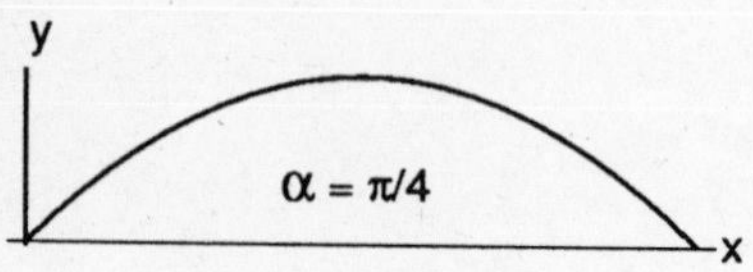

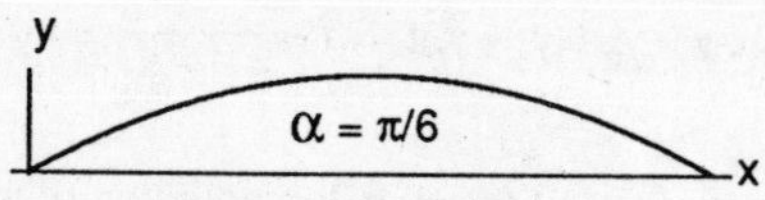

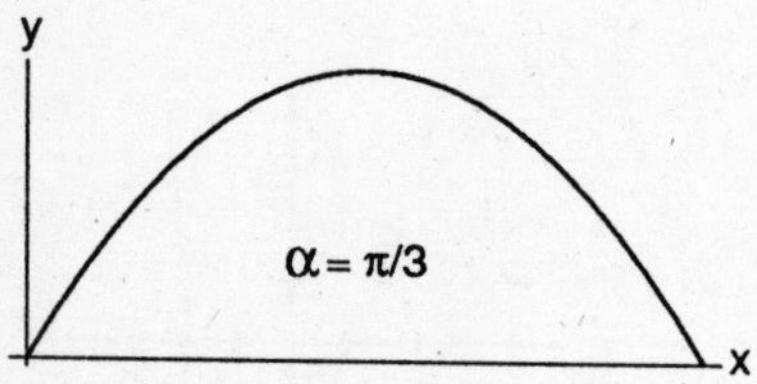

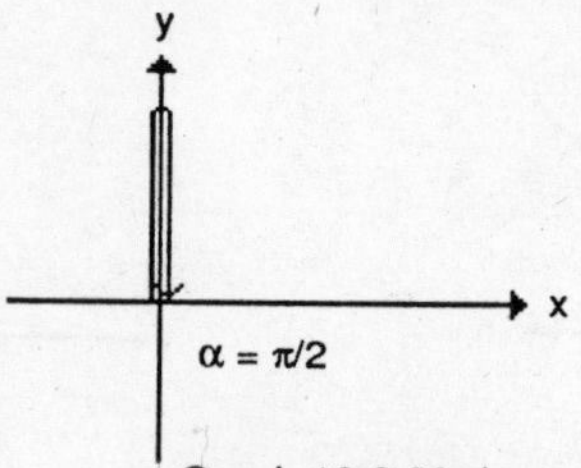

Graph 10.3.39 d

10.4 CALCULUS WITH PARAMETRIZED CURVES

1. $t=\frac{\pi}{4} \Rightarrow x = 2\cos\frac{\pi}{4} = \sqrt{2}$, $y = 2\sin\frac{\pi}{4} = \sqrt{2}$. $\frac{dx}{dt} = -2\sin t$, $\frac{dy}{dt} = 2\cos t \Rightarrow \frac{dy}{dx} = \frac{2\cos t}{-2\sin t} = -\cot t$. $\therefore \frac{dy}{dx}\left(\frac{\pi}{4}\right) = -\cot\frac{\pi}{4} = -1$.
$\therefore$ Tangent line is $y - \sqrt{2} = -1\left(x - \sqrt{2}\right) \Rightarrow y = -x + 2\sqrt{2}$.
$\frac{dy'}{dt} = \csc^2 t \Rightarrow \frac{d^2y}{dx^2} = \frac{\csc^2 t}{-2\sin t} = -\frac{1}{2\sin^3 t} \Rightarrow \frac{d^2y}{dx^2}\left(\frac{\pi}{4}\right) = -\sqrt{2}$

3. $t=\frac{\pi}{4} \Rightarrow x = 4\sin\frac{\pi}{4} = 2\sqrt{2}$, $y = 2\cos\frac{\pi}{4} = \sqrt{2}$. $\frac{dx}{dt} = 4\cos t$, $\frac{dy}{dt} = -2\sin t \Rightarrow \frac{dy}{dx} = \frac{-2\sin t}{4\cos t} = -\frac{1}{2}\tan t \Rightarrow$
$\frac{dy}{dx}\left(\frac{\pi}{4}\right) = -\frac{1}{2}\tan\frac{\pi}{4} = -\frac{1}{2}$. $\therefore$ Tangent line is $y - \sqrt{2} = -\frac{1}{2}\left(x - 2\sqrt{2}\right) \Rightarrow y = -\frac{1}{2}x + 2\sqrt{2}$.
$\frac{dy'}{dt} = -\frac{1}{2}\sec^2 t \Rightarrow \frac{d^2y}{dx^2} = \frac{-\frac{1}{2}\sec^2 t}{4\cos t} = -\frac{1}{8\cos^3 t} \Rightarrow \frac{d^2y}{dx^2}\left(\frac{\pi}{4}\right) = -\frac{\sqrt{2}}{4}$

5. $t=\frac{1}{4} \Rightarrow x = \frac{1}{4}$, $y = \frac{1}{2}$. $\frac{dx}{dt} = 1$, $\frac{dy}{dt} = \frac{1}{2\sqrt{t}} \Rightarrow \frac{dy}{dx} = \frac{1}{2\sqrt{t}} \Rightarrow \frac{dy}{dx}\left(\frac{1}{4}\right) = \frac{1}{2\sqrt{\frac{1}{4}}} = 1$.
$\therefore$ Tangent line is $y - \frac{1}{2} = 1\left(x - \frac{1}{4}\right) \Rightarrow y = x + \frac{1}{4}$. $\frac{dy'}{dt} = -\frac{1}{4}t^{-3/2} \Rightarrow \frac{d^2y}{dx^2} = -\frac{1}{4}t^{-3/2} \Rightarrow \frac{d^2y}{dx^2}\left(\frac{1}{4}\right) = -2$.

7. $t=\frac{\pi}{6} \Rightarrow x = \sec\frac{\pi}{6} = \frac{2}{\sqrt{3}}$, $y = \tan\frac{\pi}{6} = \frac{1}{\sqrt{3}}$. $\frac{dx}{dt} = \sec t\tan t$, $\frac{dy}{dt} = \sec^2 t \Rightarrow \frac{dy}{dx} = \frac{\sec^2 t}{\sec t\tan t} = \csc t \Rightarrow$
$\frac{dy}{dx}\left(\frac{\pi}{6}\right) = \csc\frac{\pi}{6} = 2$. $\therefore$ Tangent line is $y - \frac{1}{\sqrt{3}} = 2\left(x - \frac{2}{\sqrt{3}}\right) \Rightarrow y = 2x - \sqrt{3}$. $\frac{dy'}{dt} = -\csc t\cot t \Rightarrow$
$\frac{d^2y}{dx^2} = \frac{-\csc t\cot t}{\sec t\tan t} = -\cot^3 t \Rightarrow \frac{d^2y}{dx^2}\left(\frac{\pi}{6}\right) = -3\sqrt{3}$

9. $t = -1 \Rightarrow x = 5$, $y = 1$. $\frac{dx}{dt} = 4t$, $\frac{dy}{dt} = 4t^3 \Rightarrow \frac{dy}{dx} = \frac{4t^3}{4t} = t^2 \Rightarrow \frac{dy}{dx}(-1) = (-1)^2 = 1$. $\therefore$ Tangent line is $y - 1 = 1(x - 5)$
$\Rightarrow y = x - 4$. $\frac{dy'}{dt} = 2t \Rightarrow \frac{d^2y}{dx^2} = \frac{2t}{4t} = \frac{1}{2} \Rightarrow \frac{d^2y}{dx^2}(-1) = \frac{1}{2}$

11. $t=\frac{\pi}{3} \Rightarrow x = \frac{\pi}{3} - \sin\frac{\pi}{3} = \frac{\pi}{3} - \frac{\sqrt{3}}{2}$, $y = 1 - \cos\frac{\pi}{3} = 1 - \frac{1}{2} = \frac{1}{2}$. $\frac{dx}{dt} = 1 - \cos t$, $\frac{dy}{dt} = \sin t \Rightarrow \frac{dy}{dx} = \frac{\sin t}{1 - \cos t}$
$\Rightarrow \frac{dy}{dx}\left(\frac{\pi}{3}\right) = \frac{\sin\frac{\pi}{3}}{1 - \cos\frac{\pi}{3}} = \frac{\frac{\sqrt{3}}{2}}{\frac{1}{2}} = \sqrt{3}$. $\therefore$ Tangent line is $y - \frac{1}{2} = \sqrt{3}\left(x - \frac{\pi}{3} + \frac{\sqrt{3}}{2}\right) \Rightarrow y = \sqrt{3}x - \frac{\pi\sqrt{3}}{3} + 2$.
$\frac{dy'}{dt} = \frac{(1 - \cos t)\cos t - \sin t(\sin t)}{(1 - \cos t)^2} = \frac{-1}{1 - \cos t} \Rightarrow \frac{d^2y}{dx^2} = \frac{\frac{-1}{1 - \cos t}}{1 - \cos t} = \frac{-1}{(1 - \cos t)^2} \Rightarrow \frac{d^2y}{dx^2}\left(\frac{\pi}{3}\right) = -4$

13. $x^2 - 2tx + 2t^2 = 4 \Rightarrow 2x\frac{dx}{dt} - 2x - 2t\frac{dx}{dt} + 4t = 0 \Rightarrow (2x - 2t)\frac{dx}{dt} = 2x - 4t \Rightarrow \frac{dx}{dt} = \frac{2x - 4t}{2x - 2t} = \frac{x - 2t}{x - t}$. $2y^3 - 3t^2 = 4 \Rightarrow$
$6y^2\frac{dy}{dt} - 6t = 0 \Rightarrow \frac{dy}{dt} = \frac{6t}{6y^2} = \frac{t}{y^2}$. $\therefore \frac{dy}{dx} = \frac{dy/dt}{dx/dt} = \frac{t/y^2}{(x - 2t)/(x - t)} = \frac{t(x - t)}{y^2(x - 2t)}$. $t = 2 \Rightarrow x^2 - 2(2)x + 2(2)^2 = 4$
$\Rightarrow x^2 - 4x + 4 = 0 \Rightarrow (x - 2)^2 = 0 \Rightarrow x = 2$. $t = 2 \Rightarrow 2y^3 - 3(2)^2 = 4 \Rightarrow 2y^3 = 16 \Rightarrow y^3 = 8 \Rightarrow y = 2$.
$\therefore \frac{dy}{dx}\Big|_{t=2} = \frac{2(2 - 2)}{(2)^2(2 - 2(2))} = 0$.

15. $x + 2x^{3/2} = t^2 + t \Rightarrow \frac{dx}{dt} + 3x^{1/2}\frac{dx}{dt} = 2t + 1 \Rightarrow \frac{dx}{dt}\left(1 + 3x^{1/2}\right) = 2t + 1 \Rightarrow \frac{dx}{dt} = \frac{2t+1}{1+3x^{1/2}}$. $y\sqrt{t+1} + 2t\sqrt{y} = 4 \Rightarrow$
$\frac{dy}{dt}\sqrt{t+1} + y\left(\frac{1}{2}\right)(t+1)^{-1/2} + 2\sqrt{y} + 2t\left(\frac{1}{2}\right)y^{-1/2}\frac{dy}{dt} = 0 \Rightarrow \frac{dy}{dt}\sqrt{t+1} + \frac{y}{2\sqrt{t+1}} + 2\sqrt{y} + \frac{t}{\sqrt{y}}\frac{dy}{dt} = 0 \Rightarrow$
$\Rightarrow \frac{dy}{dt}\left(\sqrt{t+1} + \frac{t}{\sqrt{y}}\right) = \frac{-y}{2\sqrt{t+1}} - 2\sqrt{y} \Rightarrow \frac{dy}{dt} = \frac{\left(-y/\left(2\sqrt{t+1}\right)\right) - 2\sqrt{y}}{\sqrt{t+1} + t/\sqrt{y}} = \frac{-y\sqrt{y} - 4y\sqrt{t+1}}{2\sqrt{y}\,(t+1) + 2t\sqrt{t+1}}$
$\therefore \frac{dy}{dx} = \frac{\left(-y\sqrt{y} - 4y\sqrt{t+1}\right)/\left(2\sqrt{y}\,(t+1) + 2t\sqrt{t+1}\right)}{(2t+1)/\left(1+3x^{1/2}\right)}$. $t = 0 \Rightarrow x + 2x^{3/2} = 0 \Rightarrow x\left(1 + 2x^{1/2}\right) = 0 \Rightarrow x = 0$. $t = 0$
$\Rightarrow y\sqrt{0+1} + 2(0)\sqrt{y} = 4 \Rightarrow y = 4$. $\therefore \left.\frac{dy}{dx}\right|_{t=0} = \frac{\left(-4\sqrt{4} - 4(4)\sqrt{0+1}\right)/\left(2\sqrt{4}\,(0+1) + 2(0)\sqrt{0+1}\right)}{(2(0)+1)/\left(1+3(0)^{1/2}\right)} = -6.$

17. $\frac{dx}{dt} = -\sin t, \frac{dy}{dt} = 1 + \cos t \Rightarrow \sqrt{\left(\frac{dx}{dt}\right)^2 + \left(\frac{dy}{dt}\right)^2} = \sqrt{(-\sin t)^2 + (1+\cos t)^2} = \sqrt{2 + 2\cos t}$.

$\therefore$ Length $= \int_0^{\pi} \sqrt{2+2\cos t}\, dt = \sqrt{2}\int_0^{\pi}\sqrt{\frac{1-\cos t}{1-\cos t}(1+\cos t)}\, dt = \sqrt{2}\int_0^{\pi}\sqrt{\frac{\sin^2 t}{1-\cos t}}\, dt =$

$\sqrt{2}\int_0^{\pi}\frac{\sin t}{\sqrt{1-\cos t}}\, dt$ (since $\sin t \ge 0$ on $[0,\pi]$) $= \sqrt{2}\int_0^{2} u^{-1/2}\, du = \sqrt{2}\left[2u^{1/2}\right]_0^2 = 4.$

(Let $u = 1 - \cos t \Rightarrow du = \sin t\, dt$; $t = 0 \Rightarrow u = 0$, $t = \pi \Rightarrow u = 2$)

19. $\frac{dx}{dt} = t, \frac{dy}{dt} = (2t+1)^{1/2} \Rightarrow \sqrt{\left(\frac{dx}{dt}\right)^2 + \left(\frac{dy}{dt}\right)^2} = \sqrt{t^2 + \left((2t+1)^{1/2}\right)^2} = |t+1| = t + 1$ since $0 \le t \le 4$.

$\therefore$ Length $= \int_0^4 (t+1)\, dt = \left[\frac{t^2}{2} + t\right]_0^4 = 12.$

21. $\frac{dx}{dt} = 8t\cos t, \frac{dy}{dt} = 8t\sin t \Rightarrow \sqrt{\left(\frac{dx}{dt}\right)^2 + \left(\frac{dy}{dt}\right)^2} = \sqrt{(8t\cos t)^2 + (8t\sin t)^2} = \sqrt{64t^2\cos^2 t + 64t^2\sin^2 t} = |8t|$

$= 8t$ since $0 \le t \le \frac{\pi}{2}$. $\therefore$ Length $= \int_0^{\pi/2} 8t\, dt = \left[4t^2\right]_0^{\pi/2} = \pi^2.$

23. $\frac{dx}{dt} = -\sin t, \frac{dy}{dt} = \cos t \Rightarrow \sqrt{\left(\frac{dx}{dt}\right)^2 + \left(\frac{dy}{dt}\right)^2} = \sqrt{(-\sin t)^2 + (\cos t)^2} = 1.$

$\therefore$ Area $= \int_0^{2\pi} 2\pi(2 + \sin t)1\, dt = 2\pi[2t - \cos t]_0^{2\pi} = 8\pi^2$

25. $\frac{dx}{dt} = 1, \frac{dy}{dt} = t + \sqrt{2} \Rightarrow \sqrt{\left(\frac{dx}{dt}\right)^2 + \left(\frac{dy}{dt}\right)^2} = \sqrt{1^2 + (t+\sqrt{2})^2} = \sqrt{t^2 + 2\sqrt{2}\,t + 3}.$

$\therefore$ Area $= \int_{-\sqrt{2}}^{\sqrt{2}} 2\pi(t+\sqrt{2})\sqrt{t^2 + 2\sqrt{2}\,t + 3}\, dt = \int_1^9 \pi\sqrt{u}\, du = \left[\frac{2}{3}\pi u^{3/2}\right]_1^9 = \frac{52}{3}\pi$

(Let $u = t^2 + 2\sqrt{2}\,t + 3 \Rightarrow du = 2t + 2\sqrt{2}$; $t = -\sqrt{2} \Rightarrow u = 1$, $t = \sqrt{2} \Rightarrow u = 9$)

27. $\frac{dx}{dt} = 2, \frac{dy}{dt} = 1 \Rightarrow \sqrt{\left(\frac{dx}{dt}\right)^2 + \left(\frac{dy}{dt}\right)^2} = \sqrt{2^2 + 1^2} = \sqrt{5}$. $\therefore$ Area $= \int_0^1 2\pi(t+1)\sqrt{5}\,dt =$

$2\pi\sqrt{5}\left[\frac{t^2}{2} + t\right]_0^1 = 3\pi\sqrt{5}$. The slant height is $\sqrt{5} \Rightarrow$ Area $= \pi(1+2)\sqrt{5} = 3\pi\sqrt{5}$.

29. a) $\frac{dx}{dt} = -2\sin 2t, \frac{dy}{dt} = 2\cos 2t \Rightarrow \sqrt{\left(\frac{dx}{dt}\right)^2 + \left(\frac{dy}{dt}\right)^2} = \sqrt{(-2\sin 2t)^2 + (2\cos 2t)^2} = 2$.

$\therefore$ Length $= \int_0^{\pi/2} 2\,dt = [2t]_0^{\pi/2} = \pi$.

b) $\frac{dx}{dt} = \pi\cos \pi t, \frac{dy}{dt} = -\pi\sin \pi t \Rightarrow \sqrt{\left(\frac{dx}{dt}\right)^2 + \left(\frac{dy}{dt}\right)^2} = \sqrt{(\pi\cos \pi t)^2 + (-\pi\sin \pi t)^2} = \pi$.

$\therefore$ Length $= \int_{-1/2}^{1/2} \pi\,dt = [\pi t]_{-1/2}^{1/2} = \pi$.

31. $x = x \Rightarrow \frac{dx}{dx} = 1$. $y = f(x) \Rightarrow \frac{dy}{dx} = f'(x)$. Then $L = \int_a^b \sqrt{\left(\frac{dx}{dt}\right)^2 + \left(\frac{dy}{dt}\right)^2}\,dt = \int_a^b \sqrt{\left(\frac{dx}{dx}\right)^2 + \left(\frac{dy}{dx}\right)^2}\,dx =$

$\int_a^b \sqrt{1 + \left(\frac{dy}{dx}\right)^2}\,dx$.

33. $\frac{dx}{dt} = \cos t, \frac{dy}{dt} = 2\cos 2t \Rightarrow \frac{dy}{dx} = \frac{2\cos 2t}{\cos t} = \frac{2(2\cos^2 t - 1)}{\cos t}$. Let $\frac{2(2\cos^2 t - 1)}{\cos t} = 0 \Rightarrow 2\cos^2 t - 1 = 0$

$\Rightarrow \cos t = \pm\frac{1}{\sqrt{2}} \Rightarrow t = \frac{\pi}{4}, \frac{3\pi}{4}, \frac{5\pi}{4}, \frac{7\pi}{4}$. In the 1st quadrant, $t = \frac{\pi}{4} \Rightarrow x = \sin\frac{\pi}{4} = \frac{\sqrt{2}}{2}$, $y = \sin 2\left(\frac{\pi}{4}\right) = 1$

$\Rightarrow \left(\frac{\sqrt{2}}{2}, 1\right)$ is the point in the 1st quadrant where the tangent line is horizontal.

$x = 0, y = 0 \Rightarrow \sin t = 0 \Rightarrow t = 0$ or $t = \pi$ and $\sin 2t = 0 \Rightarrow t = 0, \frac{\pi}{2}, \pi, \frac{3\pi}{2}$. $\therefore$ $t = 0$ and $t = \pi$ give the tangent lines at the origin. $\frac{dy}{dx}(0) = 2 \Rightarrow y = 2x$ and $\frac{dy}{dx}(\pi) = -2 \Rightarrow y = -2x$.

$\sin 3t = 0 \Rightarrow t = 0, \frac{\pi}{3}, \frac{2\pi}{3}, \pi, \frac{4\pi}{3}, \frac{5\pi}{3}$. $\therefore$ $t = 0$ and $t = \pi$ give the tangent lines at the origin.

$\frac{dy}{dx}(0) = \frac{3\cos 3(0)}{2\cos 2(0)} = \frac{3}{2} \Rightarrow y = \frac{3}{2}x$. $\frac{dy}{dx}(\pi) = \frac{3\cos 3\pi}{2\cos 2\pi} = -\frac{3}{2} \Rightarrow y = -\frac{3}{2}x$.

35.

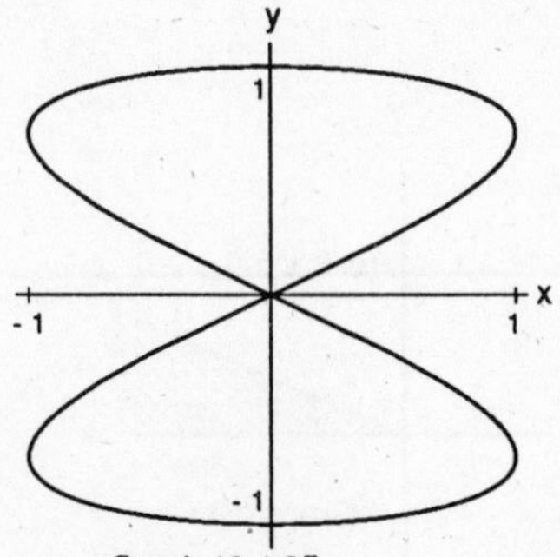

Graph 10.4.35

37.

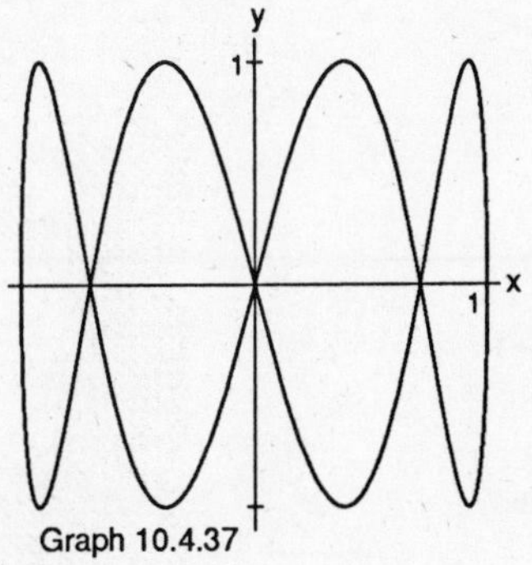

Graph 10.4.37

39.

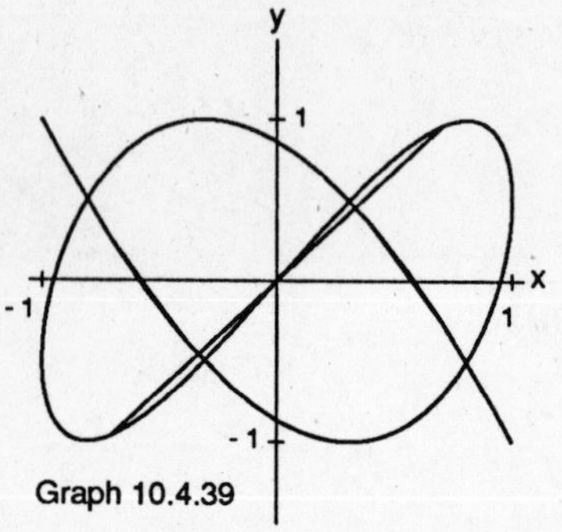

Graph 10.4.39

41.

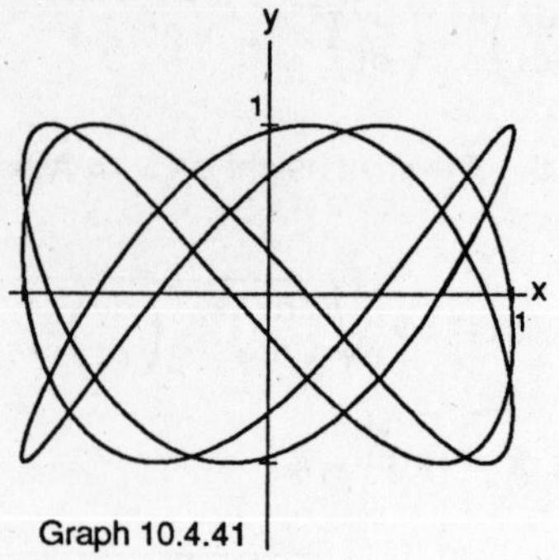

Graph 10.4.41

10.5 POLAR COORDINATES

1. a, c; b, d; e, k; g, j; h, f; i, l; m, o; n, p

3. a) $\left(2, \frac{\pi}{2} + 2n\pi\right)$ and $\left(-2, \frac{\pi}{2} + (2n+1)\pi\right)$, n an integer

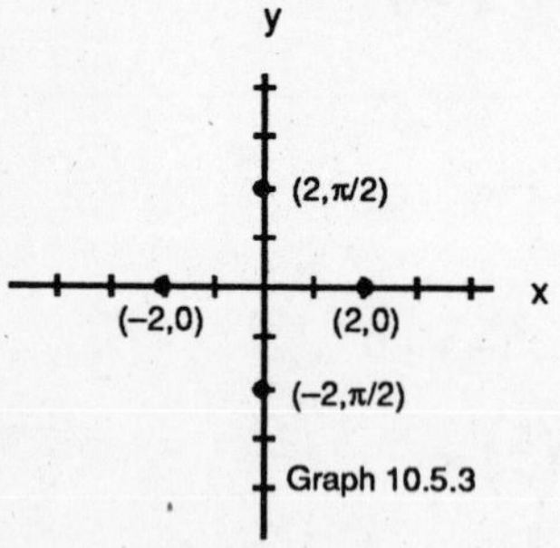

Graph 10.5.3

b) $(2, 2n\pi)$ and $(-2, (2n+1)\pi)$, n an integer

c) $\left(2, \frac{3\pi}{2} + 2n\pi\right)$ and $\left(-2, \frac{3\pi}{2} + (2n+1)\pi\right)$, n an integer

d) $(2, (2n+1)\pi)$ and $(-2, 2n\pi)$, n an integer

5. a) $x = \sqrt{2}\cos\frac{\pi}{4} = 1,\ y = \sqrt{2}\sin\frac{\pi}{4} = 1 \Rightarrow (1,1)$

b) $x = 1\cos 0 = 1,\ y = 1\sin 0 = 1 \Rightarrow (1,0)$

c) $x = (0)\cos\frac{\pi}{2} = 0,\ y = 0\sin\frac{\pi}{2} = 0 \Rightarrow (0,0)$

d) $x = -\sqrt{2}\cos\frac{\pi}{4} = -1,\ y = -\sqrt{2}\sin\frac{\pi}{2} = -1 \Rightarrow (-1,-1)$

e) $x = -3\cos\frac{5\pi}{6} = \frac{3\sqrt{3}}{2},\ y = -3\sin\frac{5\pi}{6} = -\frac{3}{2}$
$\Rightarrow \left(\frac{3\sqrt{3}}{2}, -\frac{3}{2}\right)$

f) $x = 5\cos(\tan^{-1}\frac{4}{3}) = 3,\ y = 5\sin(\tan^{-1}\frac{4}{3}) = 4 \Rightarrow (3,4)$

g) $x = -1\cos 7\pi = 1,\ y = -1\sin 7\pi = 0 \Rightarrow (1,0)$

h) $x = 2\sqrt{3}\cos\frac{2\pi}{3} = -\sqrt{3},\ y = 2\sqrt{3}\sin\frac{2\pi}{3} = 3 \Rightarrow (-\sqrt{3}, 3)$

7.

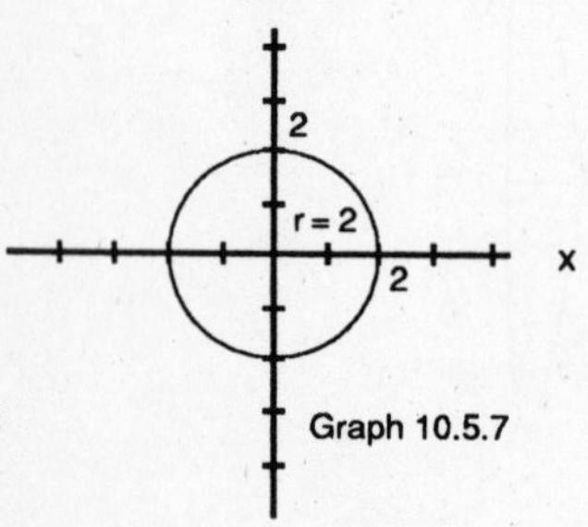

Graph 10.5.7

9.

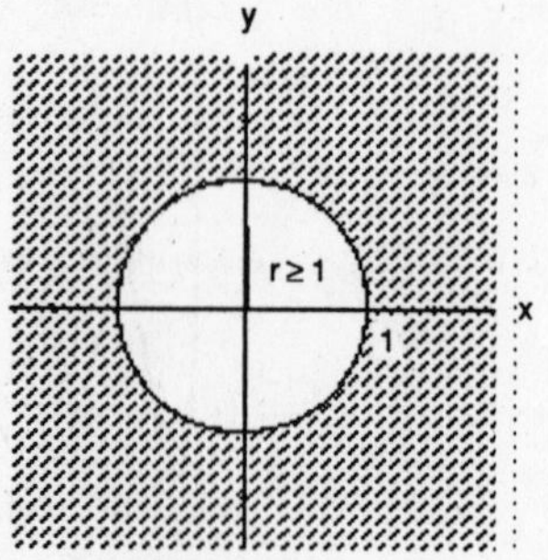

Graph 10.5.9

11.

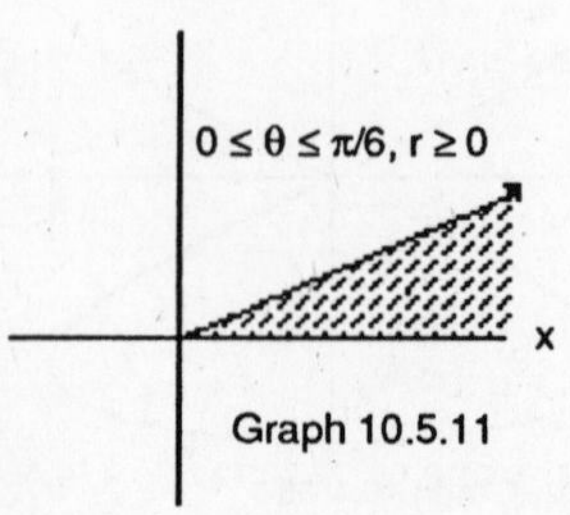

Graph 10.5.11

13.

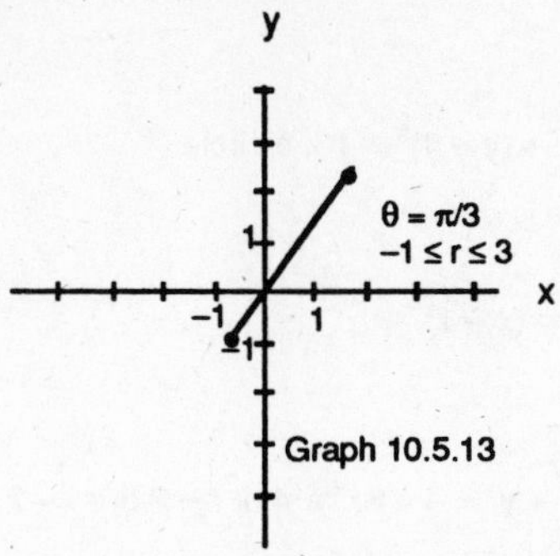

15.

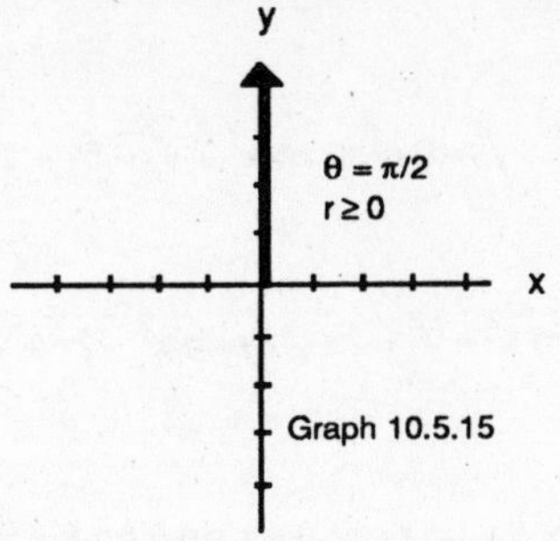

17.

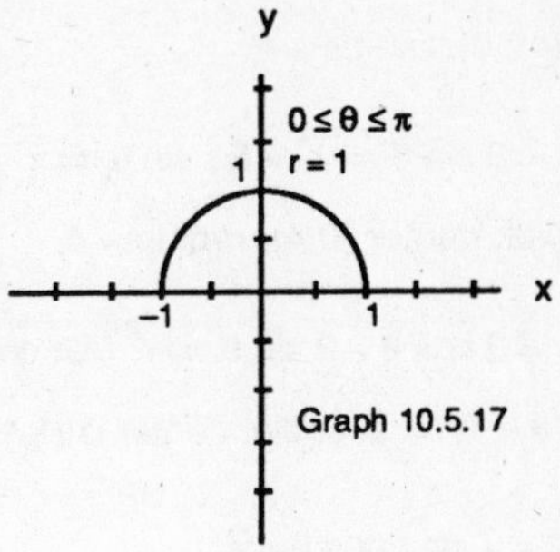

19.

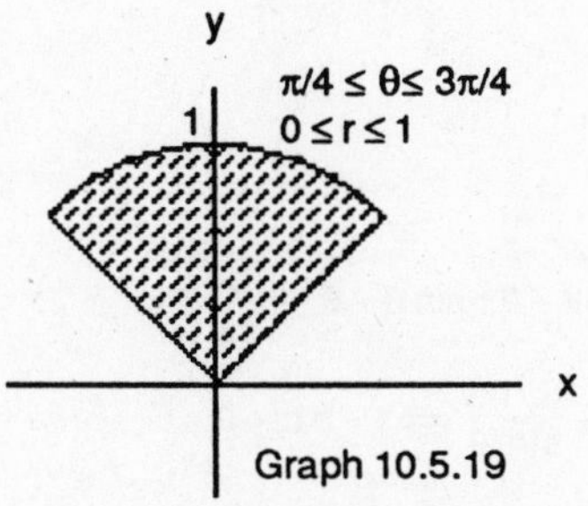

21.

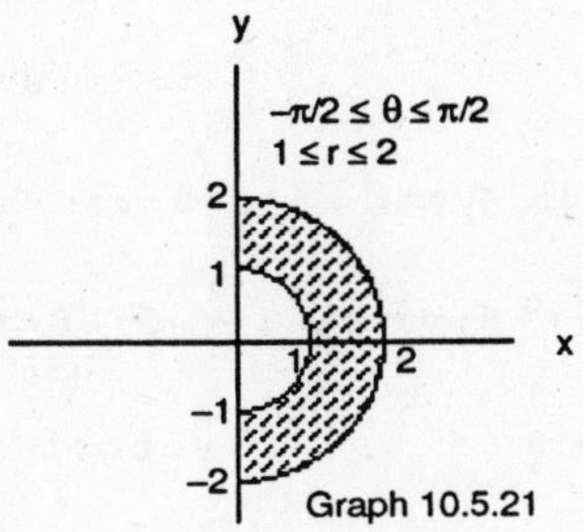

23. $r \cos \theta = 2 \Rightarrow x = 2$, vertical line through (2,0).

25. $r \sin \theta = 0 \Rightarrow y = 0$, the x–axis.

27. $r = 4 \csc \theta \Rightarrow r = \dfrac{4}{\sin \theta} \Rightarrow r \sin \theta = 4 \Rightarrow y = 4$, a horizontal line through (0,4).

29. $r \cos \theta + r \sin \theta = 1 \Rightarrow x + y = 1$, line, $m = -1$, $b = 1$

31. $r^2 = 1 \Rightarrow x^2 + y^2 = 1$, circle, C = (0,0), r = 1

33. $r = \dfrac{5}{\sin \theta - 2\cos \theta} \Rightarrow r \sin \theta - 2r \cos \theta = 5 \Rightarrow$

$y - 2x = 5$, line, $m = 2$, $b = 5$

35. $r = \cot \theta \csc\theta = \dfrac{\cos \theta}{\sin \theta}\left(\dfrac{1}{\sin \theta}\right) \Rightarrow r \sin^2\theta = \cos \theta \Rightarrow r^2\sin^2\theta = r \cos \theta \Rightarrow y^2 = x$, parabola, vertex is (0,0), opens right.

37. $r = \csc \theta\, e^{r \cos \theta} \Rightarrow r \sin \theta = e^{r \cos \theta} \Rightarrow y = e^x$, the natural exponential function

39. $r^2 + 2r^2\cos \theta \sin \theta = 1 \Rightarrow x^2 + y^2 + 2xy = 1 \Rightarrow x^2 + 2xy + y^2 = 1 \Rightarrow (x + y)^2 = 1 \Rightarrow x + y = \pm 1$, two straight lines, slope of each is -1, y–intercepts are ± 1.

41. $r^2 = -4r\cos\theta \Rightarrow x^2 + y^2 = -4x \Rightarrow x^2 + 4x + y^2 = 0 \Rightarrow x^2 + 4x + 4 + y^2 = 4 \Rightarrow (x+2)^2 + y^2 = 4$, a circle with center $(-2,0)$, radius $= 2$.

43. $r = 8\sin\theta \Rightarrow r^2 = 8r\sin\theta \Rightarrow x^2 + y^2 = 8y \Rightarrow x^2 + y^2 - 8y = 0 \Rightarrow x^2 + y^2 - 8y + 16 = 16 \Rightarrow x^2 + (y-4)^2 = 16$, a circle with center $(0,4)$, radius $= 4$.

45. $r = 2\cos\theta + 2\sin\theta \Rightarrow r^2 = 2r\cos\theta + 2r\sin\theta \Rightarrow x^2 + y^2 = 2x + 2y \Rightarrow x^2 - 2x + y^2 - 2y = 0 \Rightarrow (x-1)^2 + (y-1)^2 = 2$, circle, center is $(1,1)$, $r = \sqrt{2}$.

47. $x = 7 \Rightarrow r\cos\theta = 7$

49. $x = y \Rightarrow r\cos\theta = r\sin\theta \Rightarrow \theta = \frac{\pi}{4}$

51. $x^2 + y^2 = 4 \Rightarrow r^2 = 4$ or $r = 2$ or $r = -2$

53. $\frac{x^2}{9} + \frac{y^2}{4} = 1 \Rightarrow 4x^2 + 9y^2 = 36 \Rightarrow$
$4r^2\cos^2\theta + 9r^2\sin^2\theta = 36$

55. $y^2 = 4x \Rightarrow r^2\sin^2\theta = 4r\cos\theta$
$\Rightarrow r\sin^2\theta = 4\cos\theta$

57. $x^2 + (y-2)^2 = 4 \Rightarrow x^2 + y^2 - 4y + 4 = 4 \Rightarrow x^2 + y^2 = 4y \Rightarrow r^2 = 4r\sin\theta \Rightarrow r = 4\sin\theta$

59. $(x-3)^2 + (y+1)^2 = 4 \Rightarrow x^2 - 6x + 9 + y^2 + 2y + 1 = 4 \Rightarrow x^2 + y^2 = 6x - 2y - 6 \Rightarrow r^2 = 6r\cos\theta - 2r\sin\theta - 6$

61. a) $x = a \Rightarrow r\cos\theta = a \Rightarrow r = \frac{a}{\cos\theta} \Rightarrow r = a\sec\theta$ b) $y = b \Rightarrow r\sin\theta = b \Rightarrow r = \frac{b}{\sin\theta} \Rightarrow r = b\csc\theta$

10.6 POLAR GRAPHS

1.

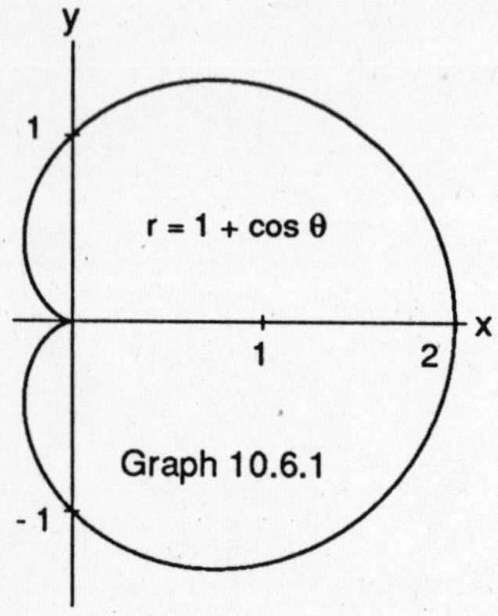

Graph 10.6.1

$1 + \cos(-\theta) = 1 + \cos\theta = r \Rightarrow$ symmetric about the x-axis. $1 + \cos(-\theta) \neq -r$ and $1 + \cos(\pi - \theta) = 1 - \cos\theta \neq r \Rightarrow$ not symmetric about the y-axis. $\therefore$ not symmetric about the origin.

3.

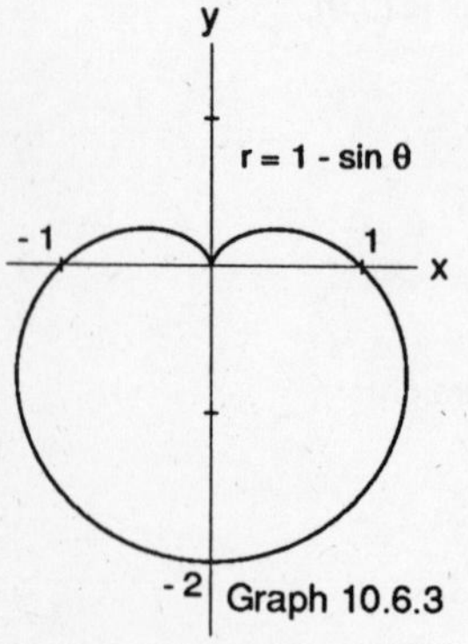

Graph 10.6.3

$1 - \sin(-\theta) = 1 + \sin\theta \neq r$ and $1 - \sin(\pi - \theta) = 1 - \sin\theta \neq -r \Rightarrow$ not symmetric about the x-axis. $1 - \sin(\pi - \theta) = 1 - \sin\theta = r \Rightarrow$ symmetric about the y-axis. $\therefore$ not symmetric about the origin.

5.

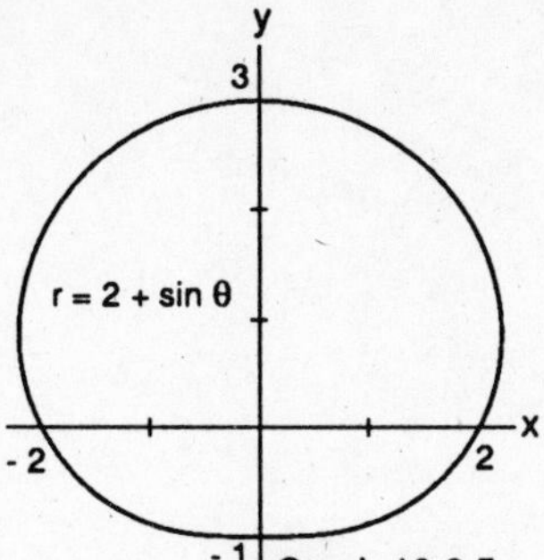

Graph 10.6.5

$2 + \sin(-\theta) = 2 - \sin\theta \neq r$ and $2 + \sin(\pi - \theta) = 2 + \sin\theta \neq -r \Rightarrow$ not symmetric about the x-axis. $2 + \sin(\pi - \theta) = 2 + \sin\theta = r \Rightarrow$ symmetric about the y-axis. ∴ not symmetric about the origin.

7.

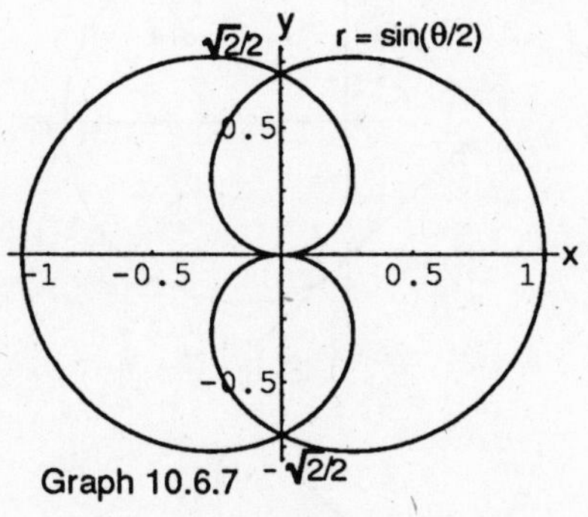

Graph 10.6.7

$\sin(-\theta/2) = -\sin(\theta/2) = -r \neq r$ and $\sin\left(\frac{\pi - \theta}{2}\right) = \sin\left(\frac{\pi}{2} - \frac{\theta}{2}\right)$ $= \cos(\theta/2) \neq -r \Rightarrow$ not symmetric about the x-axis. $\sin(-\theta/2) = -\sin(\theta/2) = -r \Rightarrow$ symmetric about the y-axis. ∴ not symmetric about the origin.

9.

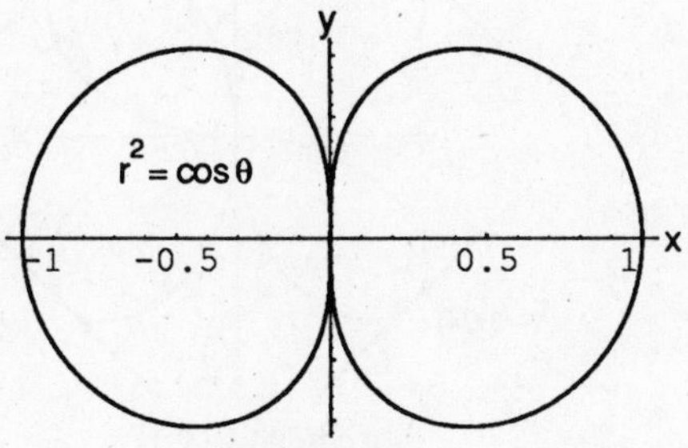

Graph 10.6.9

$\cos(-\theta) = \cos\theta = r^2 \Rightarrow (r,-\theta)$ and $(-r,-\theta)$ are on the graph when (r,θ) is on the graph $\Rightarrow$ symmetric about the x-axis and the y-axis. ∴ symmetric about the origin.

11.

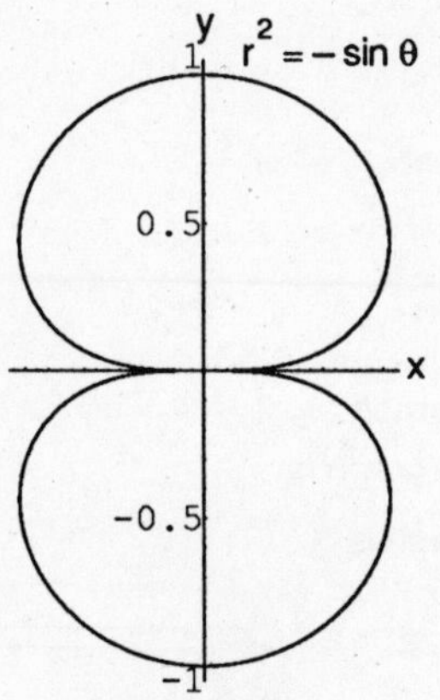

Graph 10.6.11

$-\sin(\pi - \theta) = -\sin\theta = r^2 \Rightarrow (r,\pi - \theta)$ and $(-r,\pi - \theta)$ are on the graph when (r,θ) is on the graph $\Rightarrow$ symmetric about the y-axis and the x-axis. ∴ symmetric about the origin.

13.

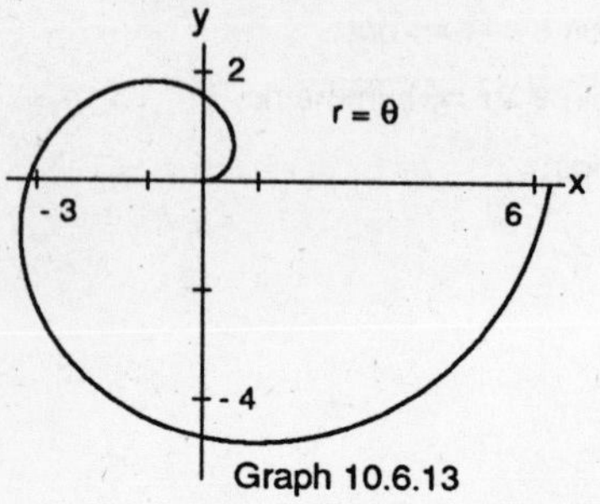

Graph 10.6.13

15. $\theta = \frac{\pi}{2} \Rightarrow r = -1 \Rightarrow \left(-1, \frac{\pi}{2}\right)$, $\theta = -\frac{\pi}{2} \Rightarrow r = -1 \Rightarrow \left(-1, -\frac{\pi}{2}\right)$

$r' = \frac{dr}{d\theta} = -\sin\theta$. Slope $= \frac{r'\sin\theta + r\cos\theta}{r'\cos\theta - r\sin\theta} = \frac{-\sin^2\theta + r\cos\theta}{-\sin\theta\cos\theta - r\sin\theta}$

$\Rightarrow$ Slope at $\left(-1, \frac{\pi}{2}\right) = \frac{-\sin^2\left(\frac{\pi}{2}\right) + (-1)\cos\frac{\pi}{2}}{-\sin\frac{\pi}{2}\cos\frac{\pi}{2} - (-1)\sin\frac{\pi}{2}} = -1$. Slope at $\left(-1, -\frac{\pi}{2}\right)$

$$= \frac{-\sin^2\left(-\frac{\pi}{2}\right) + (-1)\cos\left(-\frac{\pi}{2}\right)}{-\sin\left(-\frac{\pi}{2}\right)\cos\left(-\frac{\pi}{2}\right) - (-1)\sin\left(-\frac{\pi}{2}\right)} = 1$$

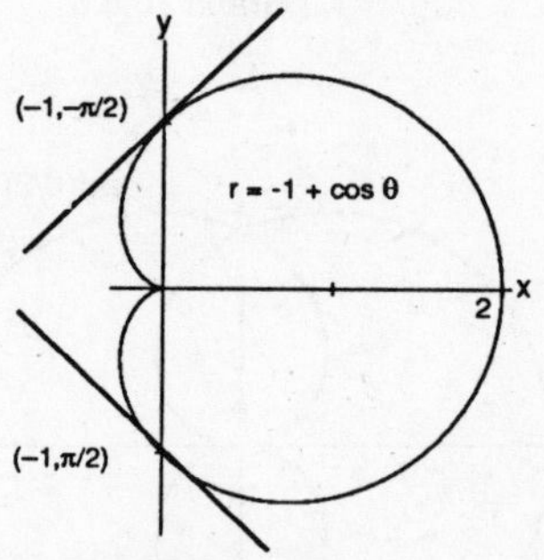

Graph 10.6.15

17. $\theta = \frac{\pi}{4} \Rightarrow r = 1 \Rightarrow \left(1, \frac{\pi}{4}\right)$. $\theta = -\frac{\pi}{4} \Rightarrow r = -1 \Rightarrow \left(-1, -\frac{\pi}{4}\right)$

$\theta = \frac{3\pi}{4} \Rightarrow r = -1 \Rightarrow \left(-1, \frac{3\pi}{4}\right)$. $\theta = -\frac{3\pi}{4} \Rightarrow r = 1 \Rightarrow \left(1, -\frac{3\pi}{4}\right)$

$r' = \frac{dr}{d\theta} = 2\cos 2\theta \Rightarrow$

Slope $= \frac{r'\sin\theta + r\cos\theta}{r'\cos\theta - r\sin\theta} = \frac{2\cos 2\theta\sin\theta + r\cos\theta}{2\cos 2\theta\cos\theta - r\sin\theta} \Rightarrow$

Slope at $\left(1, \frac{\pi}{4}\right) = \frac{2\cos\left(\frac{\pi}{2}\right)\sin\frac{\pi}{4} + (1)\cos\frac{\pi}{4}}{2\cos\left(\frac{\pi}{2}\right)\cos\frac{\pi}{4} - (1)\sin\frac{\pi}{4}} = -1$

Slope at $\left(-1, -\frac{\pi}{4}\right) = \frac{2\cos\left(-\frac{\pi}{2}\right)\sin\left(-\frac{\pi}{4}\right) + (-1)\cos\left(-\frac{\pi}{4}\right)}{2\cos\left(-\frac{\pi}{2}\right)\cos\left(-\frac{\pi}{4}\right) - (-1)\sin\left(-\frac{\pi}{4}\right)} = 1$

Slope at $\left(-1, \frac{3\pi}{4}\right) = \frac{2\cos\left(\frac{3\pi}{2}\right)\sin\left(\frac{3\pi}{4}\right) + (-1)\cos\left(\frac{3\pi}{4}\right)}{2\cos\left(\frac{3\pi}{2}\right)\cos\left(\frac{3\pi}{4}\right) - (-1)\sin\left(\frac{3\pi}{4}\right)} = 1$

Slope at $\left(1, -\frac{3\pi}{4}\right) = \frac{2\cos\left(-\frac{3\pi}{2}\right)\sin\left(-\frac{3\pi}{4}\right) + (1)\cos\left(-\frac{3\pi}{4}\right)}{2\cos\left(-\frac{3\pi}{2}\right)\cos\left(-\frac{3\pi}{4}\right) - (1)\sin\left(-\frac{3\pi}{4}\right)} = -1$

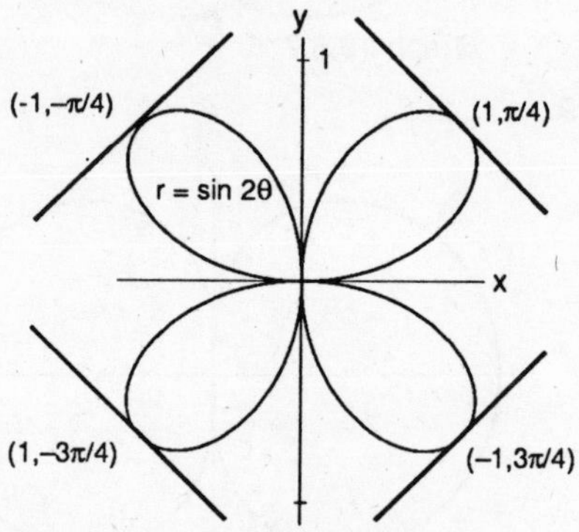

Graph 10.6.17

19.

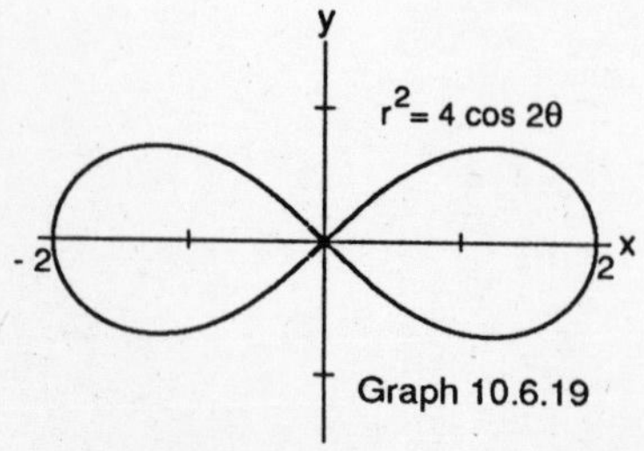

Graph 10.6.19

Since $(\pm r, -\theta)$ are on the graph when (r, θ) is on the graph $\left((\pm r)^2 = 4\cos 2(-\theta) \Rightarrow r^2 = 4\cos 2\theta\right)$, the graph is symmetric about the x-axis and the y-axis $\Rightarrow$ the graph is symmetric about the origin.

21.

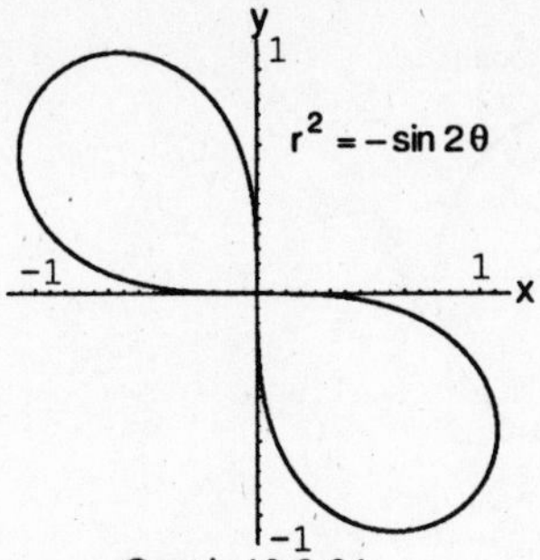

Graph 10.6.21

Since (r,θ) on the graph $\Rightarrow (-r,\theta)$ is on the graph $\left((\pm r)^2 = -\sin 2\theta \Rightarrow r^2 = -\sin 2\theta\right)$, the graph is symmetric about the origin. But $-\sin 2(-\theta) = -(-\sin 2\theta) = \sin 2\theta \neq r^2$ and $-\sin 2(\pi - \theta) = -\sin(2\pi - 2\theta) = -\sin(-2\theta) = -(-\sin 2\theta) = \sin 2\theta \neq r^2 \Rightarrow$ the graph is not symmetric about the x-axis. $\therefore$ the graph is not symmetric about the y-axis.

23. a)

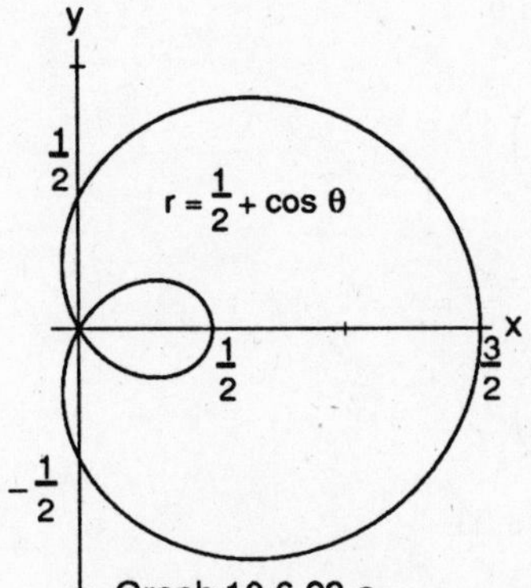

Graph 10.6.23 a

23. b)

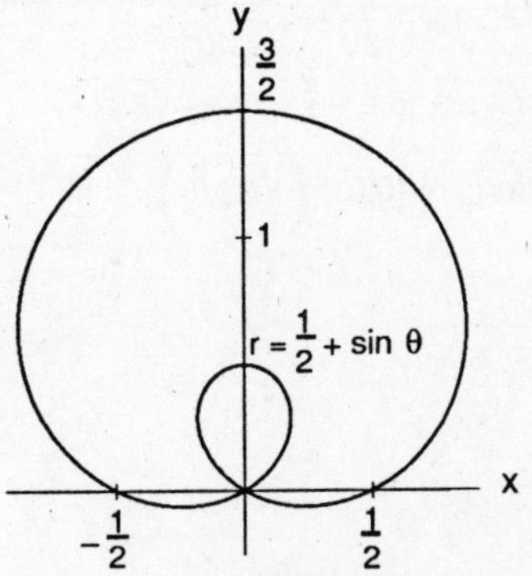

Graph 10.6.23 b

25. a)

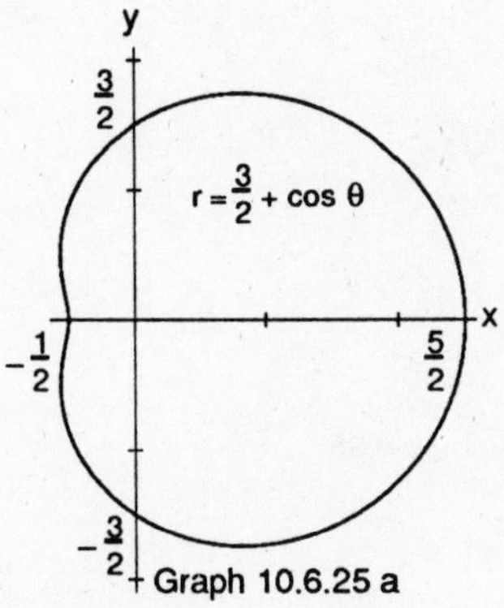

Graph 10.6.25 a

25. b)

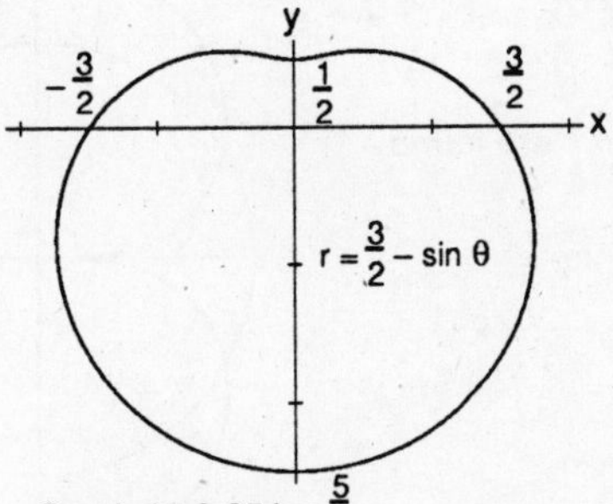

Graph 10.6.25 b $-\frac{5}{2}$

27.

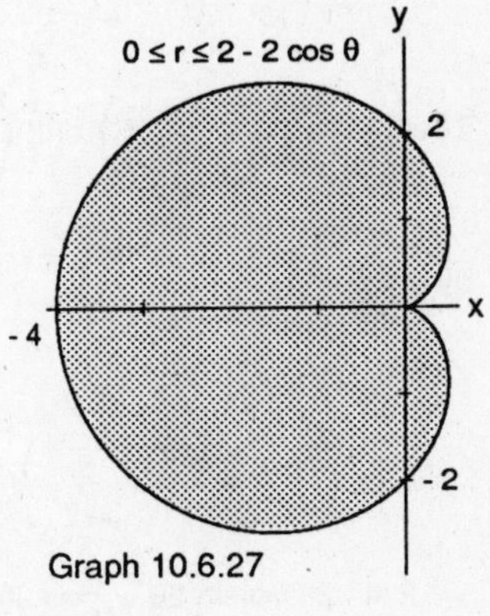

Graph 10.6.27

29. $\left(2, \frac{3\pi}{4}\right)$ is the same point as $\left(-2, -\frac{\pi}{4}\right)$. $r = 2\sin 2\left(-\frac{\pi}{4}\right) = 2\sin\left(-\frac{\pi}{2}\right) = -2 \Rightarrow \left(-2, -\frac{\pi}{4}\right)$ is on the graph $\Rightarrow \left(2, \frac{3\pi}{4}\right)$ is on the graph.

31. $1 + \cos\theta = 1 - \cos\theta \Rightarrow \cos\theta = 0 \Rightarrow$

$\theta = \frac{\pi}{2}, \frac{3\pi}{2} \Rightarrow r = 1$. Points of intersection are

$\left(1, \frac{\pi}{2}\right)$ and $\left(1, \frac{3\pi}{2}\right)$. The point of

intersection, (0,0), is found by graphing.

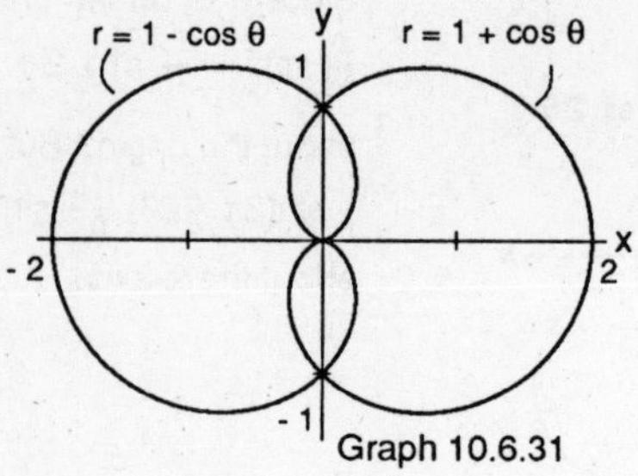

Graph 10.6.31

33. $2\sin\theta = 2\sin 2\theta \Rightarrow \sin\theta = \sin 2\theta \Rightarrow \sin\theta = 2\sin\theta\cos\theta$
$\Rightarrow \sin\theta - 2\sin\theta\cos\theta = 0 \Rightarrow \sin\theta(1 - 2\cos\theta) = 0 \Rightarrow$
$\sin\theta = 0$ or $\cos\theta = 1/2 \Rightarrow \theta = 0, \frac{\pi}{3}$, or $-\frac{\pi}{3}$.
$\theta = 0 \Rightarrow r = 0, \theta = \frac{\pi}{3} \Rightarrow r = \sqrt{3}, \theta = -\frac{\pi}{3} \Rightarrow r = -\sqrt{3}$.
The points of intersection are (0,0), $\left(\sqrt{3}, \frac{\pi}{3}\right)$,
$\left(-\sqrt{3}, -\frac{\pi}{3}\right)$.

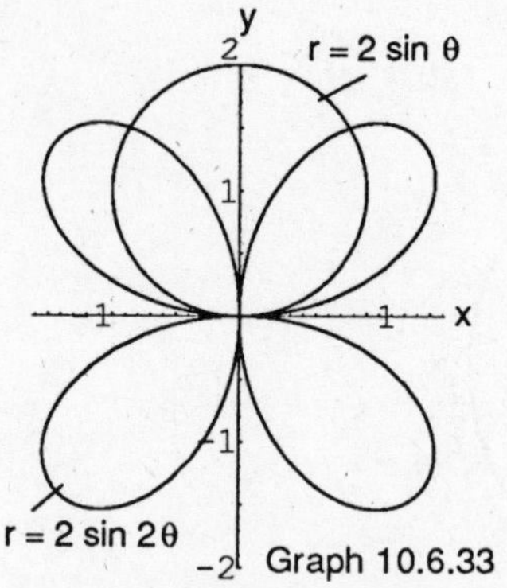

Graph 10.6.33

35. $\left(\sqrt{2}\right)^2 = 4\sin\theta \Rightarrow \frac{1}{2} = \sin\theta \Rightarrow \theta = \frac{\pi}{6}, \frac{5\pi}{6}$.

The points of intersection are $\left(\sqrt{2}, \frac{\pi}{6}\right), \left(\sqrt{2}, \frac{5\pi}{6}\right)$.

The points $\left(\sqrt{2}, -\frac{\pi}{6}\right)$ and $\left(\sqrt{2}, -\frac{5\pi}{6}\right)$ are found

by graphing.

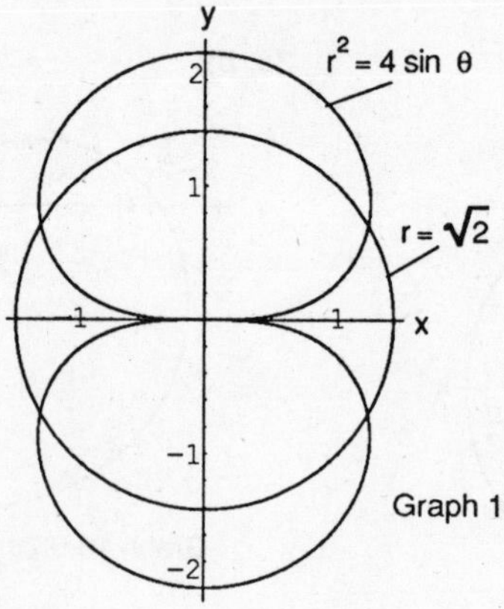

Graph 10.6.35

37. $1 = 2\sin 2\theta \Rightarrow \sin 2\theta = \frac{1}{2} \Rightarrow 2\theta = \frac{\pi}{6}, \frac{5\pi}{6}, \frac{13\pi}{6}, \frac{17\pi}{6} \Rightarrow$
$\theta = \frac{\pi}{12}, \frac{5\pi}{12}, \frac{13\pi}{12}, \frac{17\pi}{12}$. The points of intersection are
$\left(1, \frac{\pi}{12}\right), \left(1, \frac{5\pi}{12}\right), \left(1, \frac{13\pi}{12}\right), \left(1, \frac{17\pi}{12}\right)$.
No other points are found by graphing.

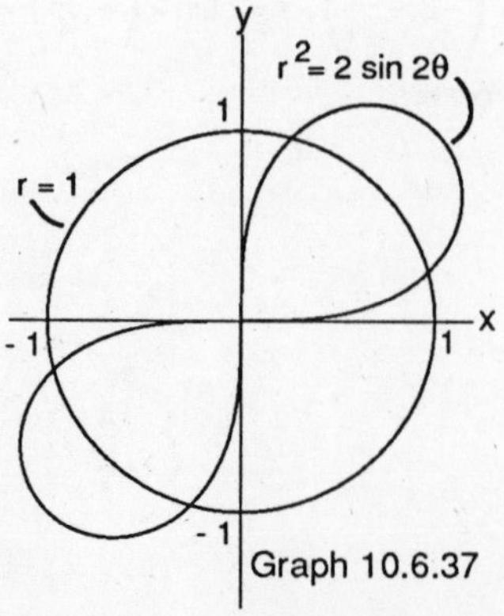

Graph 10.6.37

39. a) $r^2 = -4\cos\theta \Rightarrow \cos\theta = -\frac{r^2}{4}$. $r = 1 - \cos\theta \Rightarrow r = 1 - \left(-\frac{r^2}{4}\right) \Rightarrow 0 = r^2 - 4r + 4 \Rightarrow (r - 2)^2 = 0 \Rightarrow$

$r = 2$. $\therefore \cos\theta = -\frac{2^2}{4} = -1 \Rightarrow \theta = \pi$ $\therefore (2,\pi)$ is a point of intersection.

b) $r = 0 \Rightarrow 0^2 = 4\cos\theta \Rightarrow \cos\theta = 0 \Rightarrow \theta = \frac{\pi}{2}, \frac{3\pi}{2} \Rightarrow \left(0,\frac{\pi}{2}\right)$ or $\left(0,\frac{3\pi}{2}\right)$ is on the graph.

$r = 0 \Rightarrow 0 = 1 - \cos\theta \Rightarrow \cos\theta = 1 \Rightarrow \theta = 0 \Rightarrow (0,0)$ is on the graph. Since $(0,0) = \left(0,\frac{\pi}{2}\right)$,

the graphs intersect at the origin.

41. $r^2 = \sin 2\theta$ and $r^2 = \cos 2\theta$ are generated completely for $0 \le \theta \le \frac{\pi}{2}$. Then $\sin 2\theta = \cos 2\theta$ yields $2\theta = \frac{\pi}{4}$ as the only solution on that interval $\Rightarrow$

$\theta = \frac{\pi}{8} \Rightarrow r^2 = \sin 2\left(\frac{\pi}{8}\right) = \frac{1}{\sqrt{2}} \Rightarrow r = \pm\frac{1}{\sqrt[4]{2}}$.

$\therefore$ Points of intersection are $\left(\pm\frac{1}{\sqrt[4]{2}}, \frac{\pi}{8}\right)$

The point of intersection (0,0) is found by graphing.

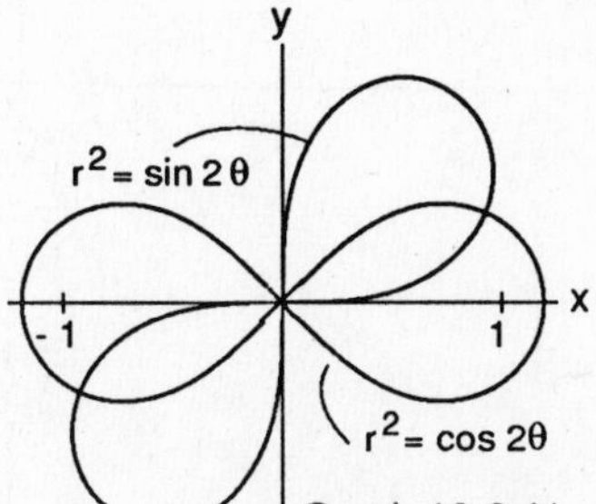

Graph 10.6.41

43. $1 = 2\sin 2\theta \Rightarrow \sin 2\theta = \frac{1}{2} \Rightarrow 2\theta = \frac{\pi}{6}, \frac{5\pi}{6}, \frac{13\pi}{6}, \frac{17\pi}{6}$

$\Rightarrow \theta = \frac{\pi}{12}, \frac{5\pi}{12}, \frac{13\pi}{12}, \frac{17\pi}{12}$. Points of intersection are

$\left(1,\frac{\pi}{12}\right), \left(1,\frac{5\pi}{12}\right), \left(1,\frac{13\pi}{12}\right)$, and $\left(1,\frac{17\pi}{12}\right)$.

Points of intersection $\left(1,\frac{7\pi}{12}\right), \left(1,\frac{11\pi}{12}\right), \left(1,\frac{19\pi}{12}\right)$, and $\left(1,\frac{23\pi}{12}\right)$ found by graphing and symmetry.

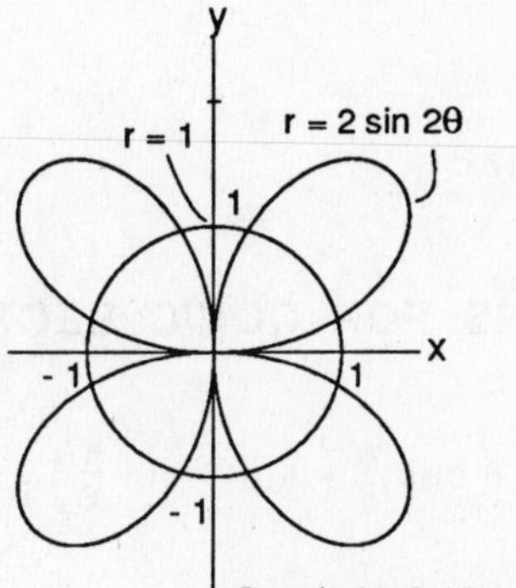

Graph 10.6.43

45.

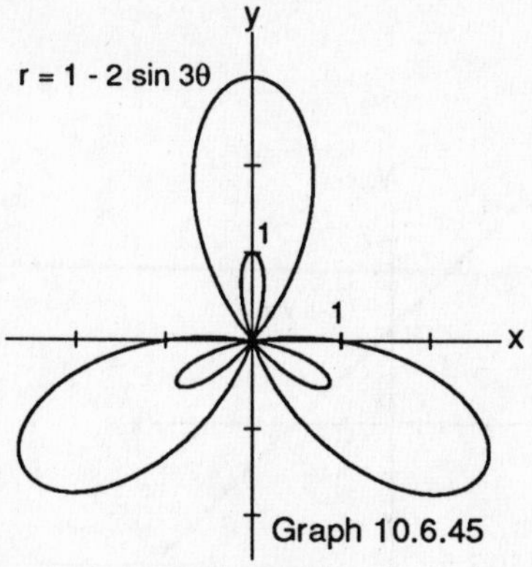

Graph 10.6.45

47. a)

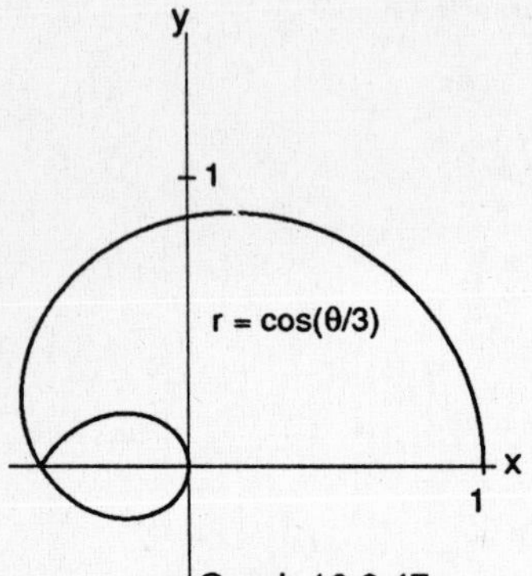

Graph 10.6.47 a

47. b)

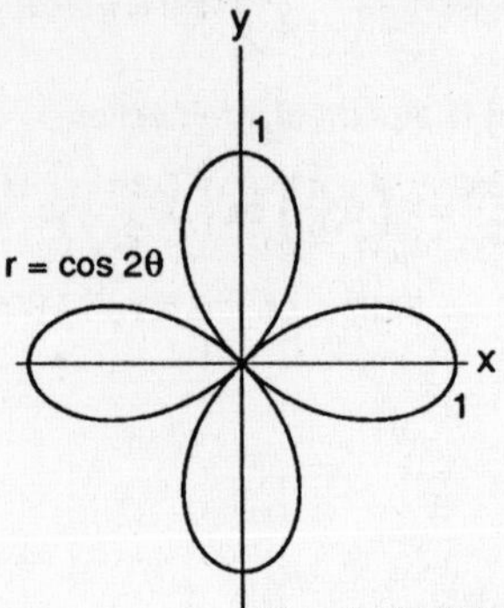

Graph 10.6.47 b

47. c)

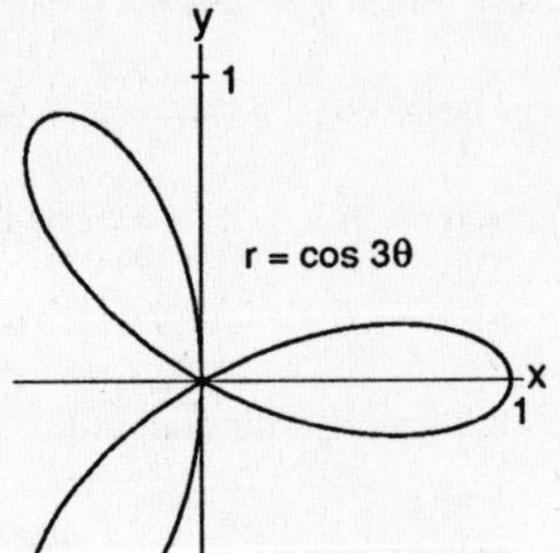

Graph 10.6.47 c

47. d)

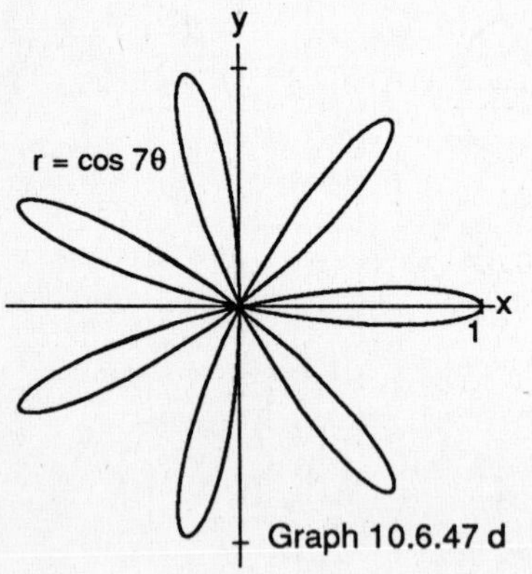

Graph 10.6.47 d

10.7 POLAR EQUATIONS FOR CONIC SECTIONS

1. $r\cos\left(\theta-\frac{\pi}{6}\right)=5 \Rightarrow r\left(\cos\theta\cos\frac{\pi}{6}+\sin\theta\sin\frac{\pi}{6}\right)=5 \Rightarrow \frac{\sqrt{3}}{2}r\cos\theta+\frac{1}{2}r\sin\theta=5 \Rightarrow$
$\frac{\sqrt{3}}{2}x+\frac{1}{2}y=5 \Rightarrow \sqrt{3}x+y=10.$

3. $r\cos\left(\theta-\frac{4\pi}{3}\right)=3 \Rightarrow r\left(\cos\theta\cos\frac{4\pi}{3}+\sin\theta\sin\frac{4\pi}{3}\right)=3 \Rightarrow -\frac{1}{2}r\cos\theta-\frac{\sqrt{3}}{2}r\sin\theta=3 \Rightarrow$
$-\frac{1}{2}x-\frac{\sqrt{3}}{2}y=3 \Rightarrow x+\sqrt{3}y=-6$

5. $r\cos\left(\theta-\frac{\pi}{4}\right)=\sqrt{2} \Rightarrow r\left(\cos\theta\cos\frac{\pi}{4}+\sin\theta\sin\frac{\pi}{4}\right)=$
$\sqrt{2} \Rightarrow \frac{1}{\sqrt{2}}r\cos\theta+\frac{1}{\sqrt{2}}r\sin\theta=\sqrt{2} \Rightarrow$
$\frac{1}{\sqrt{2}}x+\frac{1}{\sqrt{2}}y=\sqrt{2} \Rightarrow x+y=2.$

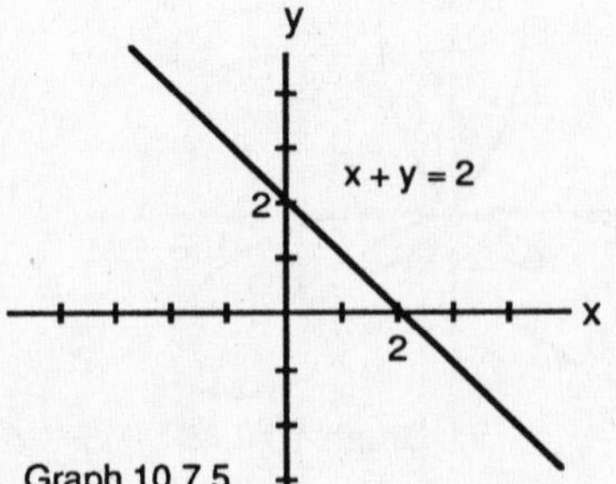

Graph 10.7.5

7. $r\cos\left(\theta - \frac{2\pi}{3}\right) = 3 \Rightarrow r\left(\cos\theta\cos\frac{2\pi}{3} + \sin\theta\sin\frac{2\pi}{3}\right) = 3 \Rightarrow$
$-\frac{1}{2}r\cos\theta + \frac{\sqrt{3}}{2}r\sin\theta = 3 \Rightarrow -\frac{1}{2}x + \frac{\sqrt{3}}{2}y = 3 \Rightarrow -x + \sqrt{3}\,y = 6 \Rightarrow$
$y = \frac{\sqrt{3}}{3}x + 2\sqrt{3}.$

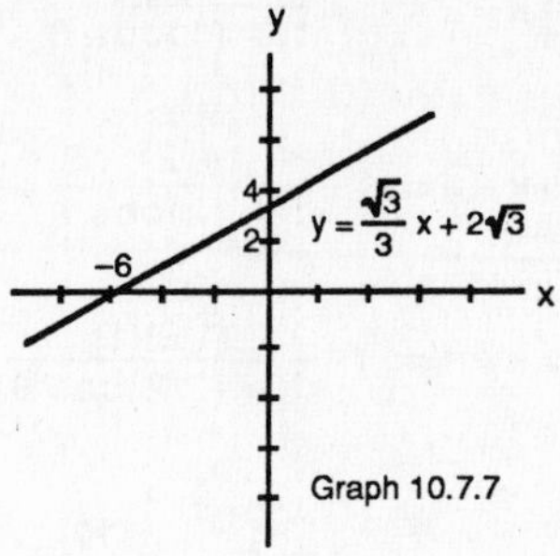

Graph 10.7.7

9. $r = 2(4)\cos\theta = 8\cos\theta$

11. $r = 2\sqrt{2}\sin\theta$

13.

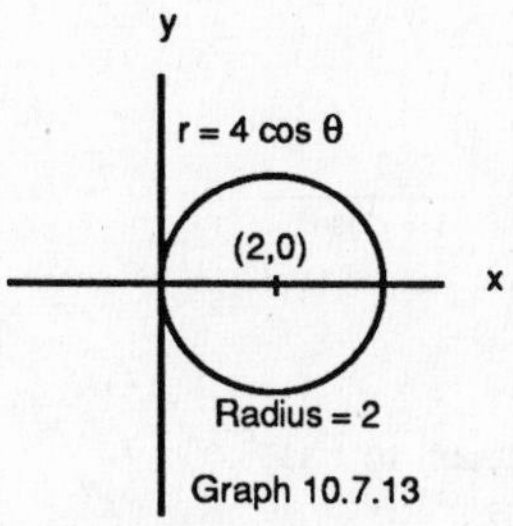

Graph 10.7.13

15.

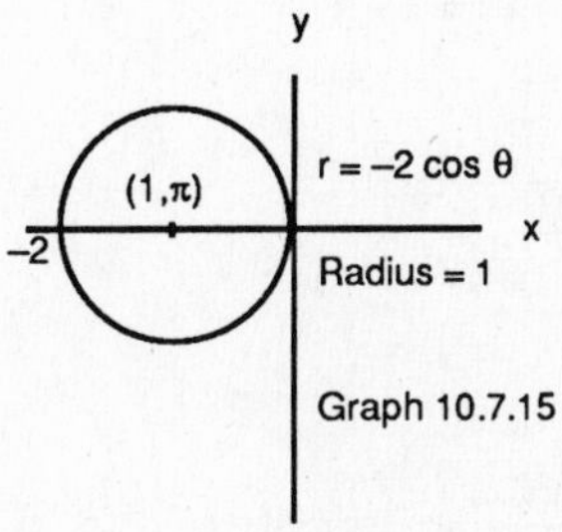

Graph 10.7.15

17. $(x-6)^2 + y^2 = 36 \Rightarrow C = (6,0),\ a = 6 \Rightarrow r = 12\cos\theta$ is the polar equation.

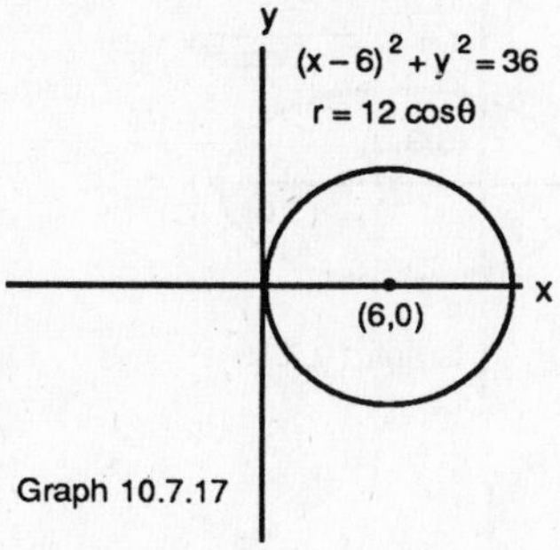

Graph 10.7.17

19. $x^2 + (y-5)^2 = 25 \Rightarrow C = (0,5),\ a = 5 \Rightarrow r = 10\sin\theta$ is the polar equation.

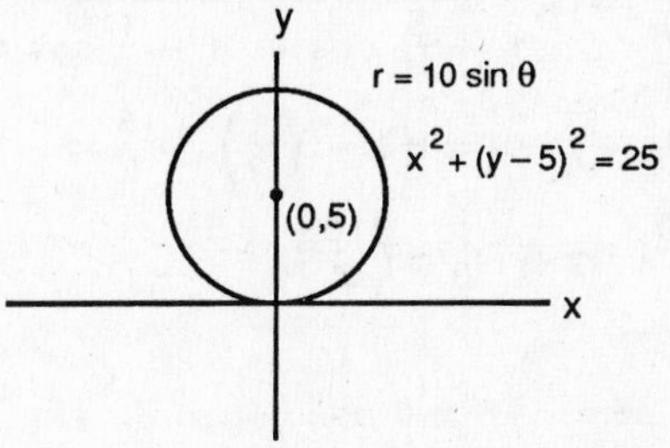

Graph 10.7.19

21. $x^2 + 2x + y^2 = 0 \Rightarrow (x+1)^2 + y^2 = 1 \Rightarrow C = (-1,0),\ a = 1$
$\Rightarrow r = -2\cos\theta$ is the polar equation.

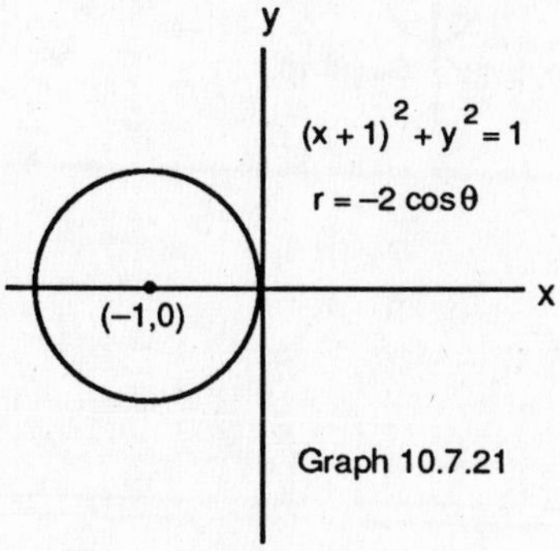

Graph 10.7.21

23. $x^2 + y^2 + y = 0 \Rightarrow x^2 + \left(y + \frac{1}{2}\right)^2 = \frac{1}{4} \Rightarrow C = \left(0, -\frac{1}{2}\right),$
$a = \frac{1}{2} \Rightarrow r = -\sin\theta$ is the polar equation.

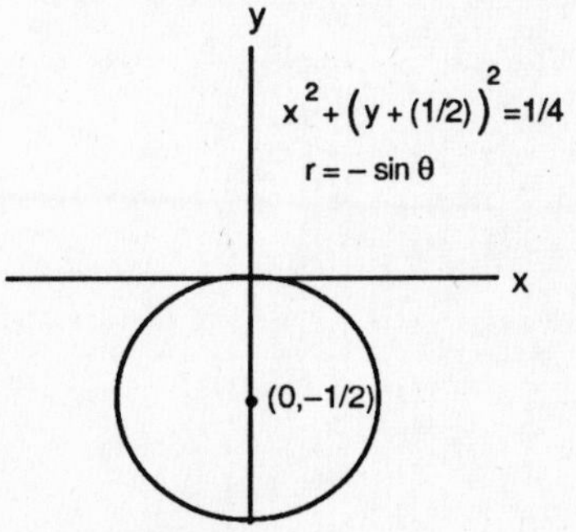

Graph 10.7.23

25. $e = 1, x = 2 \Rightarrow k = 2 \Rightarrow r = \dfrac{2(1)}{1 + (1)\cos\theta} = \dfrac{2}{1 + \cos\theta}$

27. $e = 2, x = 4 \Rightarrow k = 4 \Rightarrow r = \dfrac{4(2)}{1 + (2)\cos\theta} = \dfrac{8}{1 + 2\cos\theta}$

29. $e = \frac{1}{2}, x = 1 \Rightarrow k = 1 \Rightarrow r = \dfrac{(1/2)(1)}{1 + (1/2)\cos\theta} = \dfrac{1}{2 + \cos\theta}$

31. $e = \frac{1}{5}, y = -10 \Rightarrow k = 10 \Rightarrow r = \dfrac{\frac{1}{5}(10)}{1 - \frac{1}{5}\sin\theta} = \dfrac{2}{1 - \frac{1}{5}\sin\theta} = \dfrac{10}{5 - \sin\theta}$

33. $r = \dfrac{1}{1 + \cos\theta} \Rightarrow e = 1, k = 1 \Rightarrow x = 1$

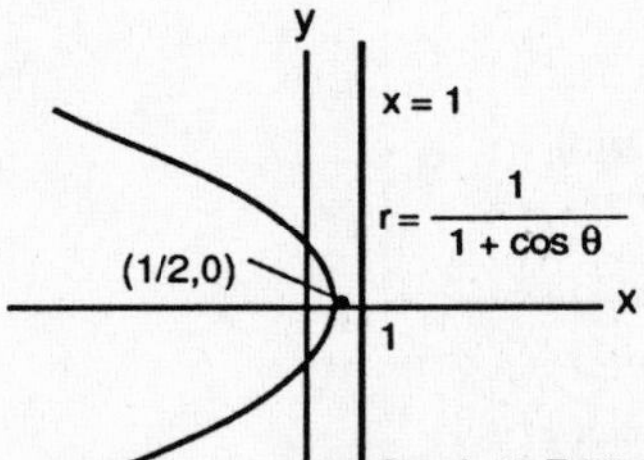

Graph 10.7.33

35. $r = \dfrac{25}{10 - 5\cos\theta} \Rightarrow r = \dfrac{\frac{25}{10}}{1 - \frac{5}{10}\cos\theta} = \dfrac{\frac{5}{2}}{1 - \frac{1}{2}\cos\theta} \Rightarrow$

$e = \frac{1}{2}, k = 5.\ a(1 - e^2) = ke \Rightarrow a\left(1 - \left(\frac{1}{2}\right)^2\right) = \frac{5}{2} \Rightarrow$

$\frac{3}{4}a = \frac{5}{2} \Rightarrow a = \frac{10}{3}.\ a - ae = \frac{10}{3} - \left(\frac{10}{3}\right)\frac{1}{2} = \frac{5}{3}$

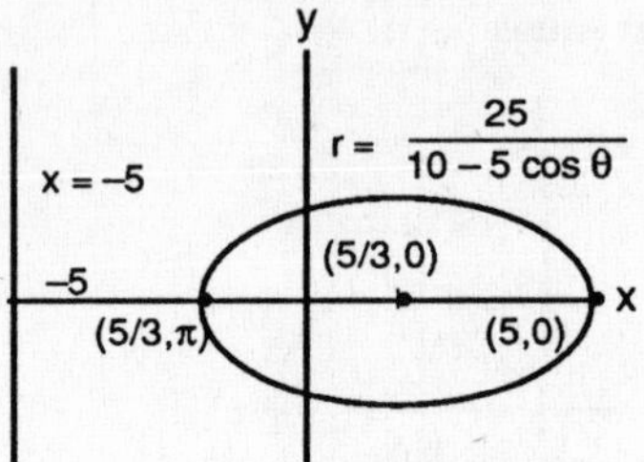

Graph 10.7.35

37. $r = \dfrac{400}{16 + 8\sin\theta} \Rightarrow r = \dfrac{\frac{400}{16}}{1 + \frac{8}{16}\sin\theta} \Rightarrow r = \dfrac{25}{1 + \frac{1}{2}\sin\theta} \Rightarrow$

$e = \frac{1}{2}, k = 50.\ a(1 - e^2) = ke \Rightarrow a\left(1 - \left(\frac{1}{2}\right)^2\right) = 25 \Rightarrow$

$\frac{3}{4}a = 25 \Rightarrow a = \frac{100}{3}.\ a - ae = \frac{100}{3} - \frac{100}{3}\left(\frac{1}{2}\right) = \frac{50}{3}.$

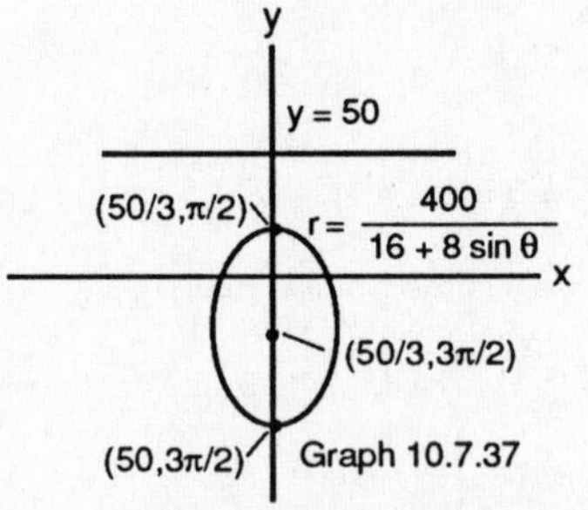

Graph 10.7.37

39. $r = \frac{8}{2 - 2\sin\theta} \Rightarrow r = \frac{4}{1 - \sin\theta} \Rightarrow e = 1, k = 4$

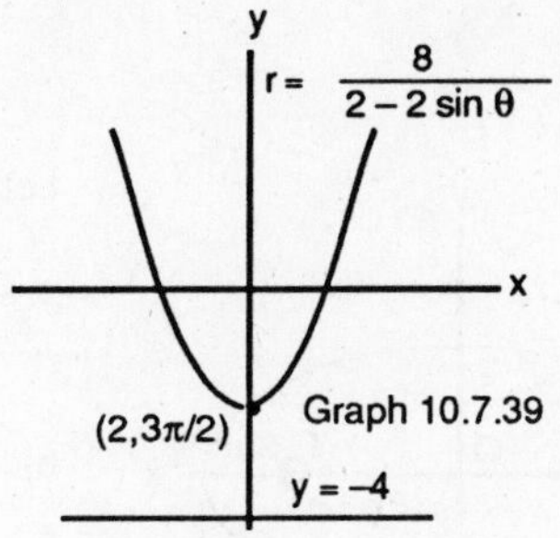

Graph 10.7.39

41.

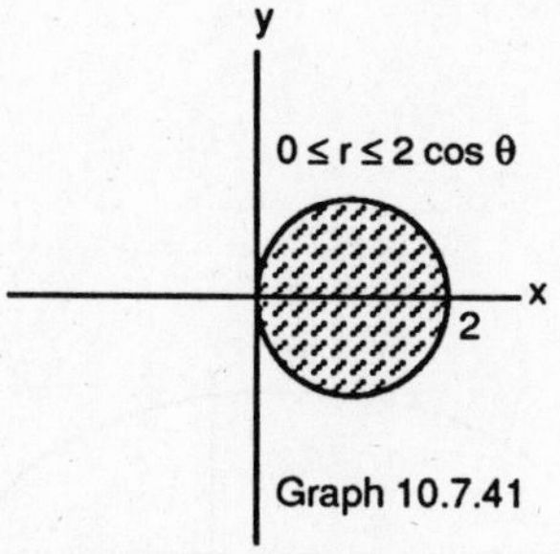

Graph 10.7.41

43. a) Perihelion $= a - ae = a(1 - e)$

Aphelion $= ea + a = a(1 + e)$

b)

Planet	Perihelion	Aphelion
Mercury	0.3075 AU	0.4667 AU
Venus	0.7184 AU	0.7282 AU
Earth	0.9833 AU	1.0167 AU
Mars	1.3817 AU	1.6663 AU
Jupiter	4.9512 AU	5.4548 AU
Saturn	9.0210 AU	10.0570 AU
Uranus	18.2977 AU	20.0623 AU
Neptune	29.8135 AU	30.3065 AU
Pluto	29.6549 AU	49.2251 AU

45. a) $r = 4\sin\theta \Rightarrow r^2 = 4r\sin\theta \Rightarrow x^2 + y^2 = 4y$ and $r = \sqrt{3}\sec\theta$
$\Rightarrow r = \frac{\sqrt{3}}{\cos\theta} \Rightarrow r\cos\theta = \sqrt{3} \Rightarrow x = \sqrt{3}$. $x = \sqrt{3} \Rightarrow$
$\left(\sqrt{3}\right)^2 + y^2 = 4y \Rightarrow y^2 - 4y + 3 = 0 \Rightarrow (y - 3)(y - 1) = 0 \Rightarrow$
$y = 3$ or $y = 1$. $\therefore$ in Cartesian coordinates, the points are
$\left(\sqrt{3}, 3\right), \left(\sqrt{3}, 1\right)$. In polar coordinates, $4\sin\theta = \sqrt{3}\sec\theta$
$\Rightarrow 4\sin\theta\cos\theta = \sqrt{3} \Rightarrow 2\sin\theta\cos\theta = \frac{\sqrt{3}}{2} \Rightarrow \sin 2\theta = \frac{\sqrt{3}}{2} \Rightarrow$
$2\theta = \frac{\pi}{3}$ or $\frac{2\pi}{3} \Rightarrow \theta = \frac{\pi}{6}$ or $\frac{\pi}{3}$. $\theta = \frac{\pi}{6} \Rightarrow r = 2$, $\theta = \frac{\pi}{3} \Rightarrow r = 2\sqrt{3}$
$\Rightarrow \left(2, \frac{\pi}{6}\right)$ and $\left(2\sqrt{3}, \frac{\pi}{3}\right)$ are the points in polar coordinates.

b)

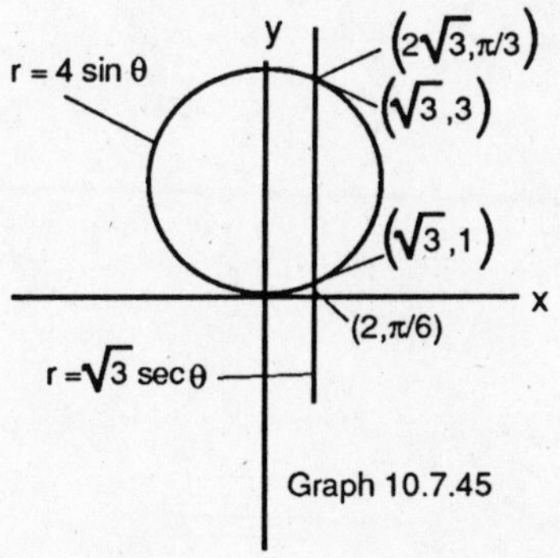

Graph 10.7.45

47. $r\cos\theta = 4 \Rightarrow x = 4 \Rightarrow k = 4$. Parabola $\Rightarrow e = 1$. $\therefore r = \dfrac{4}{1 + \cos\theta}$

49. a)

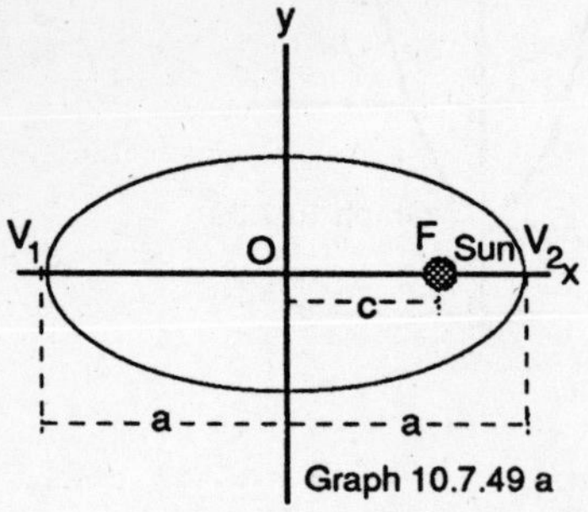

Graph 10.7.49 a

Let the ellipse be the orbit, with the Sun at one focus. Then $r_{max} = a + c$, $r_{min} = a - c \Rightarrow \dfrac{r_{max} - r_{min}}{r_{max} + r_{min}} = \dfrac{(a + c) - (a - c)}{(a + c) + (a - c)} = \dfrac{2c}{2a} = \dfrac{c}{a} = e$.

b) Let F_1, F_2 be the foci. Then $PF_1 + PF_2 = 10$ where P is any point on the ellipse. If P is a vertex, then $PF_1 = a + c$, $PF_2 = a - c \Rightarrow (a + c) + (a - c) = 10 \Rightarrow$ $2a = 10 \Rightarrow a = 5$. Since $e = \dfrac{c}{a}$, $0.2 = \dfrac{c}{5} \Rightarrow c = 1.0 \Rightarrow$ the pins are 2 inches apart.

51.

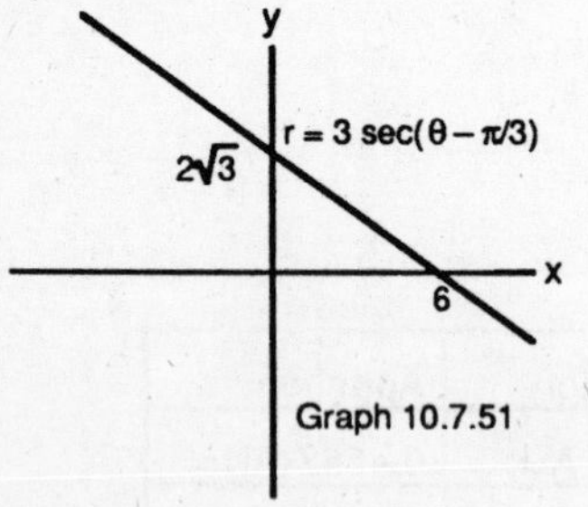

Graph 10.7.51

53.

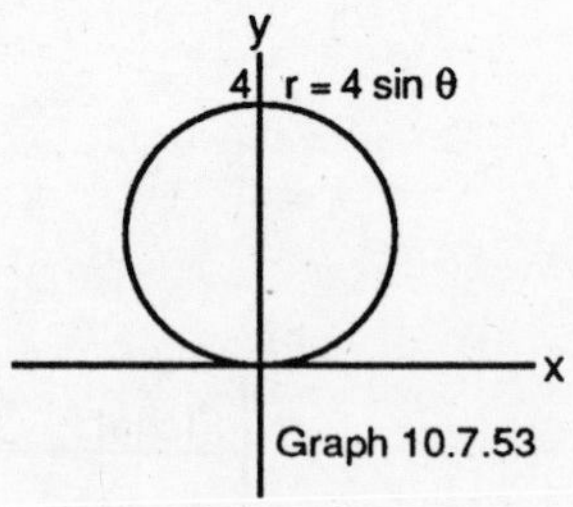

Graph 10.7.53

55.

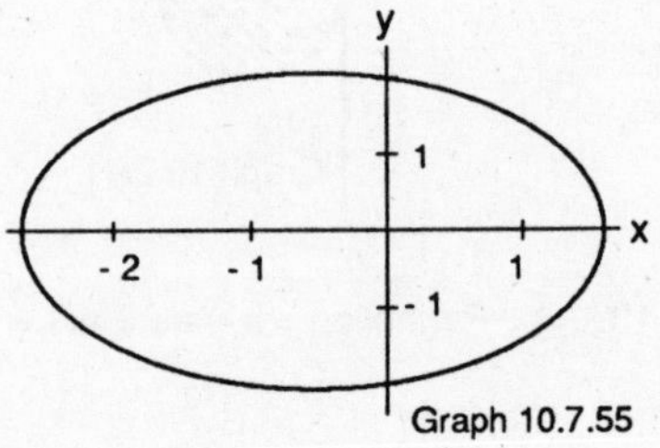

Graph 10.7.55

57.

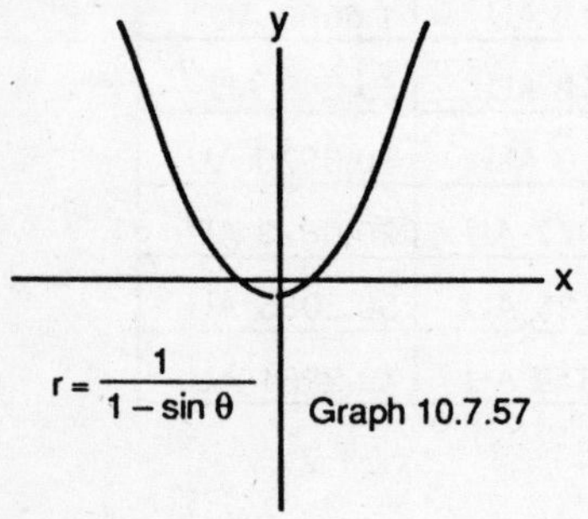

Graph 10.7.57

59.

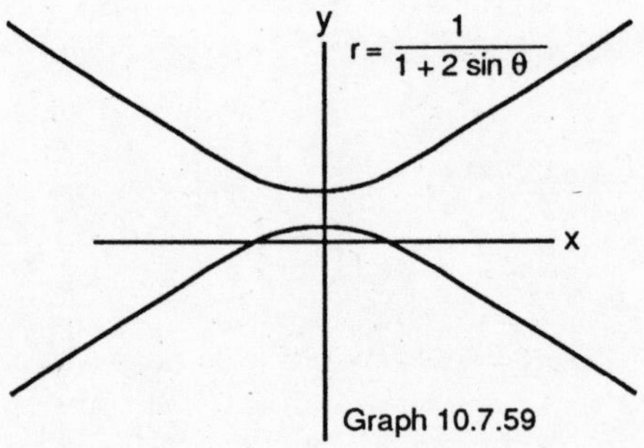

Graph 10.7.59

10.8 INTEGRATION IN POLAR COORDINATES

1. $A = \int_0^{\pi/4} \frac{1}{2}\left(2\sqrt{\cos\theta}\right)^2 d\theta = \int_0^{\pi/4} 2\cos\theta\, d\theta = [2\sin\theta]_0^{\pi/4} = 2\sin\frac{\pi}{4} - 2\sin 0 = \sqrt{2}$

3. $A = \int_0^{\pi} \frac{1}{2}\left(1 + e^{\theta/\pi}\right)^2 d\theta \quad \int_0^{\pi} \frac{1}{2}\left(1 + 2e^{\theta/\pi} + e^{2\theta/\pi}\right) d\theta = \int_0^{\pi} \left(\frac{1}{2} + e^{\theta/\pi} + \frac{1}{2}e^{2\theta/\pi}\right) d\theta =$

$\left[\frac{1}{2}\theta + \pi e^{\theta/\pi} + \frac{\pi}{4}e^{2\theta/\pi}\right]_0^{\pi} = \left(\frac{1}{2}\pi + \pi e^{\pi/\pi} + \frac{\pi}{4}e^{2\pi/\pi}\right) - \left(0 + \pi e^0 + \frac{\pi}{4}e^0\right) = -\frac{3\pi}{4} + \pi e + \frac{\pi}{4}e^2$

5. $A = \int_0^{2\pi} \frac{1}{2}(4 + 2\cos\theta)^2 d\theta = \int_0^{2\pi} \frac{1}{2}(16 + 16\cos\theta + 4\cos^2\theta)\, d\theta = \int_0^{2\pi} \left(8 + 8\cos\theta + 2\left(\frac{1 + \cos 2\theta}{2}\right)\right) d\theta$

$= \int_0^{2\pi} (9 + 8\cos\theta + \cos 2\theta)\, d\theta = \left[9\theta + 8\sin\theta + \frac{1}{2}\sin 2\theta\right]_0^{2\pi} = 18\pi$

7. $A = 2\int_0^{\pi/4} \frac{1}{2}\cos^2 2\theta\, d\theta = \int_0^{\pi/4} \frac{1 + \cos 4\theta}{2} d\theta = \frac{1}{2}\left[\theta + \frac{\sin 4\theta}{4}\right]_0^{\pi/4} = \frac{\pi}{8}$

9. $A = \int_0^{\pi/2} \frac{1}{2}(4\sin 2\theta)\, d\theta = \int_0^{\pi/2} 2\sin 2\theta\, d\theta = [-\cos 2\theta]_0^{\pi/2} = 2$

11. $r = 2\cos\theta,\ r = 2\sin\theta \Rightarrow 2\cos\theta = 2\sin\theta \Rightarrow \cos\theta = \sin\theta \Rightarrow \theta = \frac{\pi}{4} \quad \therefore A = 2\int_0^{\pi/4} \frac{1}{2}(2\sin\theta)^2 d\theta = \int_0^{\pi/4} 4\sin^2\theta\, d\theta$

$= \int_0^{\pi/4} 4\left(\frac{1 - \cos 2\theta}{2}\right) d\theta = \int_0^{\pi/4} (2 - 2\cos 2\theta)\, d\theta = [2\theta - \sin 2\theta]_0^{\pi/4} = \frac{\pi}{2} - 1$

13. $r = 2,\ r = 2(1 - \cos\theta) \Rightarrow 2 = 2(1 - \cos\theta) \Rightarrow \cos\theta = 0 \Rightarrow \theta = \pm\frac{\pi}{2}$. Sketch a graph to see the region.

$A = 2\int_0^{\pi/2} \frac{1}{2}(2(1 - \cos\theta))^2 d\theta + \frac{1}{2}$ of the area of the circle $= \int_0^{\pi/2} 4(1 - 2\cos\theta + \cos^2\theta)\, d\theta + \frac{1}{2}\pi(2)^2$

$= \int_0^{\pi/2} \left(4 - 8\cos\theta + 4\left(\frac{1 + \cos 2\theta}{2}\right)\right) d\theta + 2\pi = \int_0^{\pi/2} (6 - 8\cos\theta + 2\cos 2\theta)\, d\theta + 2\pi =$

$[6\theta - 8\sin\theta + \sin 2\theta]_0^{\pi/2} + 2\pi = 5\pi - 8$

15. $r = \sqrt{3}, r^2 = 6\cos 2\theta \Rightarrow 3 = 6\cos 2\theta \Rightarrow \cos 2\theta = \frac{1}{2} \Rightarrow \theta = \frac{\pi}{6}$ in the 1st quadrant. Use symmetry to find the area.

$$\therefore A = 4\int_0^{\pi/6}\left(\frac{1}{2}(6\cos 2\theta) - \frac{1}{2}(\sqrt{3})^2\right)d\theta = 2\int_0^{\pi/6}(6\cos 2\theta - 3)\,d\theta = 2[3\sin 2\theta - 3\theta]_0^{\pi/6} = 3\sqrt{3} - \pi$$

17. $r = 1, r = -2\cos\theta \Rightarrow 1 = -2\cos\theta \Rightarrow \cos\theta = -\frac{1}{2} \Rightarrow \theta = \frac{2\pi}{3}$ in quadrant II.

$$A = 2\int_{2\pi/3}^{\pi}\frac{1}{2}\left((-2\cos\theta)^2 - 1^2\right)d\theta = \int_{2\pi/3}^{\pi}\left(4\cos^2\theta - 1\right)d\theta = \int_{2\pi/3}^{\pi}(2(1+\cos 2\theta) - 1)\,d\theta$$

$$= \int_{2\pi/3}^{\pi}(1 + 2\cos 2\theta)\,d\theta = [\theta + \sin 2\theta]_{2\pi/3}^{\pi} = \frac{\pi}{3} + \frac{\sqrt{3}}{2}$$

19. $r = 6.\ r = 3\csc\theta \Rightarrow r\sin\theta = 3\ \therefore\ 6\sin\theta = 3 \Rightarrow \sin\theta = \frac{1}{2} \Rightarrow \theta = \frac{\pi}{6}, \frac{5\pi}{6}$.

$$A = \int_{\pi/6}^{5\pi/6}\frac{1}{2}\left(6^2 - \left(\frac{3}{\sin\theta}\right)^2\right)d\theta = \int_{\pi/6}^{5\pi/6}\left(18 - \frac{9}{2}\csc^2\theta\right)d\theta = \left[18\theta + \frac{9}{2}\cot\theta\right]_{\pi/6}^{5\pi/6} = 12\pi - 9\sqrt{3}.$$

21. a)

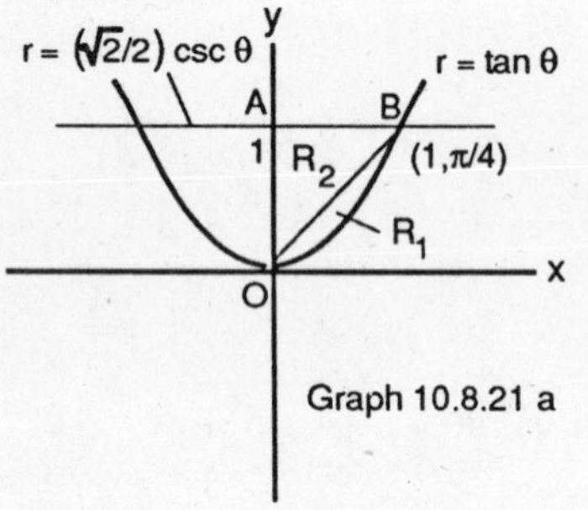

Graph 10.8.21 a

$r = \tan\theta, r = \left(\sqrt{2}/2\right)\csc\theta \Rightarrow \tan\theta = \left(\sqrt{2}/2\right)\csc\theta \Rightarrow \sin^2\theta = \left(\sqrt{2}/2\right)\cos\theta \Rightarrow$ $1 - \cos^2\theta = \left(\sqrt{2}/2\right)\cos\theta \Rightarrow \cos^2\theta + \left(\sqrt{2}/2\right)\cos\theta - 1 = 0 \Rightarrow \cos\theta = -\sqrt{2}$ or $\frac{\sqrt{2}}{2}$ (Use the quadratic formula.) $\Rightarrow \theta = \frac{\pi}{4}$ (the solution in the first quadrant).

$\therefore$ The area of R_1 is $A_1 = \int_0^{\pi/4}\frac{1}{2}\tan^2\theta\,d\theta = \frac{1}{2}\int_0^{\pi/4}(\sec^2\theta - 1)\,d\theta =$

$\frac{1}{2}[\tan\theta - \theta]_0^{\pi/4} = \left(\frac{1}{2}\left(\tan\frac{\pi}{4} - \frac{\pi}{4}\right)\right) = \frac{1}{2} - \frac{\pi}{8}$.

$AO = \left(\sqrt{2}/2\right)\csc\frac{\pi}{2} = \frac{\sqrt{2}}{2}$, $OB = \left(\sqrt{2}/2\right)\csc\frac{\pi}{4} = 1$

$\Rightarrow AB = \sqrt{1^2 - \left(\sqrt{2}/2\right)^2} = \sqrt{2}/2 \Rightarrow$ the area of R_2 is $A_2 = \frac{1}{2}\left(\sqrt{2}/2\right)\left(\sqrt{2}/2\right) = 1/4$. $\therefore$ the area of the region shaded in the text is $2\left(\frac{1}{2} - \frac{\pi}{8} + \frac{1}{4}\right) = \frac{3}{2} - \frac{\pi}{4}$. Note: the area must be found this way since no common interval generates the region. For example, the interval $0 \le \theta \le \pi/4$ generates the arc OB of $r = \tan\theta$ but does not generate the segment AB of the line $r = \csc\theta$. Instead the interval generates the half–line from B to $+\infty$ on the line $r = \csc\theta$.

b) $\lim_{\theta \to (\pi/2)^-}\tan\theta = +\infty$. The line $x = 1$ is $r = \sec\theta$ in polar coordinates. Then $\lim_{\theta \to (\pi/2)^-}(\tan\theta - \sec\theta)$

$= \lim_{\theta \to (\pi/2)^-}\left(\frac{\sin\theta}{\cos\theta} - \frac{1}{\cos\theta}\right) = \lim_{\theta \to (\pi/2)^-}\left(\frac{\sin\theta - 1}{\cos\theta}\right) = \lim_{\theta \to (\pi/2)^-}\left(\frac{\cos\theta}{-\sin\theta}\right) = 0 \Rightarrow r = \tan\theta \to r =$ $\sec\theta$ as $\theta \to \frac{\pi}{2}^- \Rightarrow r = \sec\theta$ $(x = 1)$ is a vertical asymptote of $r = \tan\theta$. Similarly, $r = -\sec\theta$ is the polar equation of $x = -1$ and $\lim_{\theta \to (-\pi/2)^+}(\tan\theta - (-\sec\theta)) = 0 \Rightarrow r = -\sec\theta$ $(x = -1)$ is a vertical asymptote of $r = \tan\theta$.

23. $r = \theta^2, 0 \le \theta \le \sqrt{5} \Rightarrow \frac{dr}{d\theta} = 2\theta$. $\therefore$ Length $= \int_0^{\sqrt{5}} \sqrt{(\theta^2)^2 + (2\theta)^2}\, d\theta = \int_0^{\sqrt{5}} \sqrt{\theta^4 + 4\theta^2}\, d\theta =$

$\int_0^{\sqrt{5}} |\theta|\sqrt{\theta^2 + 4}\, d\theta =$ (since $\theta \ge 0$) $\int_0^{\sqrt{5}} \theta\sqrt{\theta^2 + 4}\, d\theta = \int_4^9 \frac{1}{2}\sqrt{u}\, du = \frac{1}{2}\left[\frac{2}{3}u^{3/2}\right]_4^9 = \frac{19}{3}$

Let $u = \theta^2 + 4 \Rightarrow \frac{1}{2}du = \theta d\theta$; $\theta = 0 \Rightarrow u = 4$, $\theta = \sqrt{5} \Rightarrow u = 9$

25. $r = 1 + \cos\theta \Rightarrow \frac{dr}{d\theta} = -\sin\theta$. $\therefore$ Length $= \int_0^{2\pi} \sqrt{(1 + \cos\theta)^2 + (-\sin\theta)^2}\, d\theta =$

$2\int_0^{\pi} \sqrt{1 + 2\cos\theta + \cos^2\theta + \sin^2\theta}\, d\theta = 2\int_0^{\pi} \sqrt{2 + 2\cos\theta}\, d\theta =$

$2\int_0^{\pi} \sqrt{\frac{4(1 + \cos\theta)}{2}}\, d\theta = 4\int_0^{\pi} \sqrt{\frac{1 + \cos\theta}{2}}\, d\theta = 4\int_0^{\pi} \cos\frac{1}{2}\theta\, d\theta = 4\left[2\sin\frac{1}{2}\theta\right]_0^{\pi} = 8$

27. $r = \cos^3\frac{\theta}{3} \Rightarrow \frac{dr}{d\theta} = -\sin\frac{\theta}{3}\cos^2\frac{\theta}{3}$. $\therefore$ Length $= \int_0^{\pi/4} \sqrt{\left(\cos^3\frac{\theta}{3}\right)^2 + \left(-\sin\frac{\theta}{3}\cos^2\frac{\theta}{3}\right)^2}\, d\theta =$

$\int_0^{\pi/4} \sqrt{\cos^6\frac{\theta}{3} + \sin^2\frac{\theta}{3}\cos^4\frac{\theta}{3}}\, d\theta = \int_0^{\pi/4} \cos^2\frac{\theta}{3}\sqrt{\cos^2\frac{\theta}{3} + \sin^2\frac{\theta}{3}}\, d\theta = \int_0^{\pi/4} \cos^2\frac{\theta}{3}\, d\theta =$

$\int_0^{\pi/4} \frac{1 + \cos(2\theta/3)}{2}\, d\theta = \left[\frac{\theta + \frac{3}{2}\sin\frac{2\theta}{3}}{2}\right]_0^{\pi/4} = \frac{\pi}{8} + \frac{3}{8}$

29. $r = \sqrt{1 + \cos 2\theta} \Rightarrow \frac{dr}{d\theta} = \frac{1}{2}(1 + \cos 2\theta)^{-1/2}(-2\sin 2\theta) = \frac{-\sin 2\theta}{\sqrt{1 + \cos 2\theta}} \Rightarrow \left(\frac{dr}{d\theta}\right)^2 = \frac{\sin^2 2\theta}{1 + \cos 2\theta}$.

$r^2 = 1 + \cos 2\theta \Rightarrow r^2 + \left(\frac{dr}{d\theta}\right)^2 = 1 + \cos 2\theta + \frac{\sin^2 2\theta}{1 + \cos 2\theta} = 1 + \cos 2\theta + \left(\frac{\sin^2 2\theta}{1 + \cos 2\theta}\right)\left(\frac{1 - \cos 2\theta}{1 - \cos 2\theta}\right) =$

$1 + \cos 2\theta + \frac{\sin^2 2\theta\,(1 - \cos 2\theta)}{1 - \cos^2 2\theta} = 1 + \cos 2\theta + \frac{\sin^2 2\theta\,(1 - \cos 2\theta)}{\sin^2 2\theta} = 1 + \cos 2\theta + 1 - \cos 2\theta = 2.$

$\therefore L = \int_0^{\pi\sqrt{2}} \sqrt{2}\, d\theta = \left[\sqrt{2}\,\theta\right]_0^{\pi\sqrt{2}} = 2\pi.$

31. $r = \sqrt{\cos 2\theta}, 0 \le \theta \le \frac{\pi}{4} \Rightarrow \frac{dr}{d\theta} = \frac{1}{2}(\cos 2\theta)^{-1/2}(-\sin 2\theta)(2) = \frac{-\sin 2\theta}{\sqrt{\cos 2\theta}}$.

$\therefore$ Surface Area $= \int_0^{\pi/4} 2\pi r\cos\theta\sqrt{\left(\sqrt{\cos 2\theta}\right)^2 + \left(\frac{-\sin 2\theta}{\sqrt{\cos 2\theta}}\right)^2}\, d\theta =$

$\int_0^{\pi/4} 2\pi\sqrt{\cos 2\theta}\cos\theta\sqrt{\cos 2\theta + \frac{\sin^2 2\theta}{\cos 2\theta}}\, d\theta = \int_0^{\pi/4} 2\pi\sqrt{\cos 2\theta}\cos\theta\sqrt{\frac{1}{\cos 2\theta}}\, d\theta =$

31. (Continued)

$$\int_0^{\pi/4} 2\pi \cos\theta \, d\theta = \left[2\pi \sin\theta\right]_0^{\pi/4} = \pi\sqrt{2}$$

33. $r^2 = \cos 2\theta \Rightarrow r = \pm\sqrt{\cos 2\theta}$. Use $r = \sqrt{\cos 2\theta}$ on $\left[0, \frac{\pi}{4}\right]$. Then $\frac{dr}{d\theta} = \frac{1}{2}(\cos 2\theta)^{-1/2}(-\sin 2\theta)(2) =$

$\frac{-\sin 2\theta}{\sqrt{\cos 2\theta}}$. $\therefore$ Surface Area $= \int_0^{\pi/4} 2\pi\sqrt{\cos 2\theta}\sin\theta\sqrt{\left(\sqrt{\cos 2\theta}\right)^2 + \left(\frac{-\sin 2\theta}{\sqrt{\cos 2\theta}}\right)^2}\, d\theta =$

$$\int_0^{\pi/4} 2\pi\sqrt{\cos 2\theta}\sin\theta\sqrt{\cos 2\theta + \frac{\sin^2 2\theta}{\cos 2\theta}}\, d\theta = \int_0^{\pi/4} 2\pi\sqrt{\cos 2\theta}\sin\theta\sqrt{\frac{1}{\cos 2\theta}}\, d\theta =$$

$$\int_0^{\pi/4} 2\pi \sin\theta \, d\theta = \left[-2\pi \cos\theta\right]_0^{\pi/4} = \pi\left(2 - \sqrt{2}\right)$$

35. Let $r = f(\theta)$. Then $x = f(\theta)\cos\theta \Rightarrow \frac{dx}{d\theta} = f'(\theta)\cos\theta - f(\theta)\sin\theta \Rightarrow \left(\frac{dx}{d\theta}\right)^2 = (f'(\theta)\cos\theta - f(\theta)\sin\theta)^2 = (f'(\theta))^2\cos^2\theta -$ $2f'(\theta)f(\theta)\sin\theta\cos\theta + (f(\theta))^2\sin^2\theta$. $y = f(\theta)\sin\theta \Rightarrow \frac{dy}{d\theta} = f'(\theta)\sin\theta + f(\theta)\cos\theta \Rightarrow \left(\frac{dy}{d\theta}\right)^2 = (f'(\theta)\sin\theta + f(\theta)\cos\theta)^2 =$ $(f'(\theta))^2\sin^2\theta + 2f'(\theta)f(\theta)\sin\theta\cos\theta + (f(\theta))^2\cos^2\theta$. $\therefore \left(\frac{dx}{d\theta}\right)^2 + \left(\frac{dy}{d\theta}\right)^2 = (f'(\theta))^2\left(\cos^2\theta + \sin^2\theta\right) + (f(\theta))^2\left(\cos^2\theta + \sin^2\theta\right)$

$= (f'(\theta))^2 + (f(\theta))^2 = r^2 + \left(\frac{dr}{d\theta}\right)^2$. Thus, $L = \int_\alpha^\beta \sqrt{\left(\frac{dx}{d\theta}\right)^2 + \left(\frac{dy}{d\theta}\right)^2}\, d\theta = \int_\alpha^\beta \sqrt{r^2 + \left(\frac{dr}{d\theta}\right)^2}\, d\theta$.

37. a) $r = 2f(\theta),\ \alpha \le \theta \le \beta \Rightarrow \frac{dr}{d\theta} = 2f'(\theta) \Rightarrow r^2 + \left(\frac{dr}{d\theta}\right)^2 = (2f(\theta))^2 + (2f'(\theta))^2 = 4\left((f(\theta))^2 + (f'(\theta))^2\right) \Rightarrow$

Length $= \int_\alpha^\beta \sqrt{4\left((f(\theta))^2 + (f'(\theta))^2\right)}\, d\theta = 2\int_\alpha^\beta \sqrt{(f(\theta))^2 + (f'(\theta))^2}\, d\theta$ or twice the length of $r = f(\theta)$,

$\alpha \le \theta \le \beta$.

b) Again $r = 2f(\theta) \Rightarrow r^2 + \left(\frac{dr}{d\theta}\right)^2 = (2f(\theta))^2 + (2f'(\theta))^2 = 4\left((f(\theta))^2 + (f'(\theta))^2\right) \Rightarrow$ Area $=$

$\int_\alpha^\beta 2\pi(2f(\theta))\sin\theta\sqrt{4\left((f(\theta))^2 + (f'(\theta))^2\right)}\, d\theta = 4\int_\alpha^\beta 2\pi f(\theta)\sqrt{(f(\theta))^2 + (f'(\theta))^2}\, d\theta$ or four times the area of the surface generated by revolving $r = f(\theta),\ \alpha \le \theta \le \beta$, about the x–axis.

10.P PRACTICE EXERCISES

1. $x^2 = -4y \Rightarrow y = -\frac{x^2}{4} \Rightarrow 4p = 4 \Rightarrow p = 1$

$\therefore$ Focus: (0,–1); Directrix: y = 1

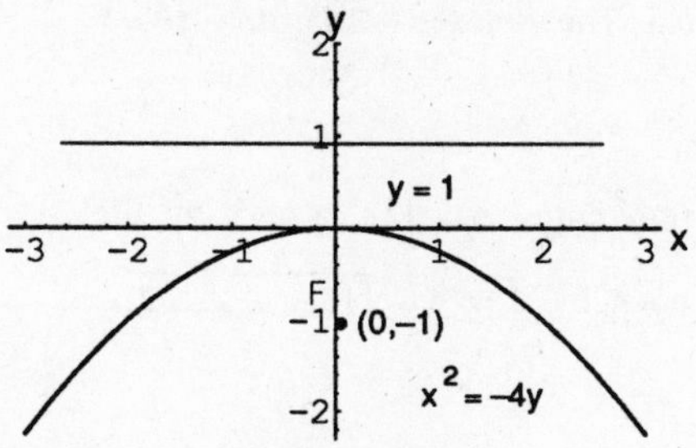

Graph 10.P.1

3. $y^2 = 3x \Rightarrow x = \frac{y^2}{3} \Rightarrow 4p = 3 \Rightarrow p = \frac{3}{4}$

$\therefore$ focus is $\left(\frac{3}{4}, 0\right)$, directrix is $x = -\frac{3}{4}$.

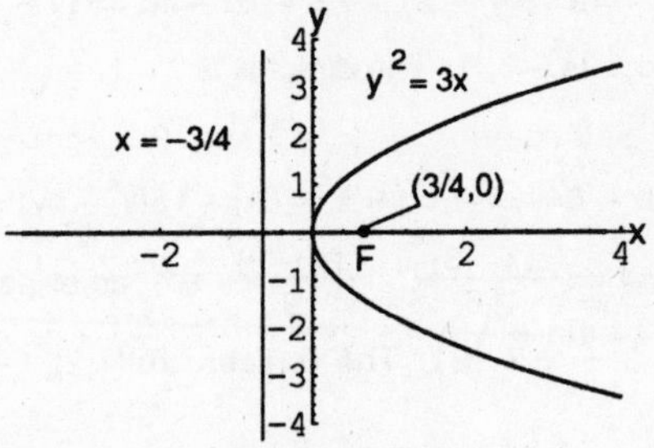

Graph 10.P.3

5. $16x^2 + 7y^2 = 112 \Rightarrow \frac{x^2}{7} + \frac{y^2}{16} = 1 \Rightarrow$

$c^2 = 16 - 7 = 9 \Rightarrow c = 3.\ e = \frac{c}{a} = \frac{3}{4}$

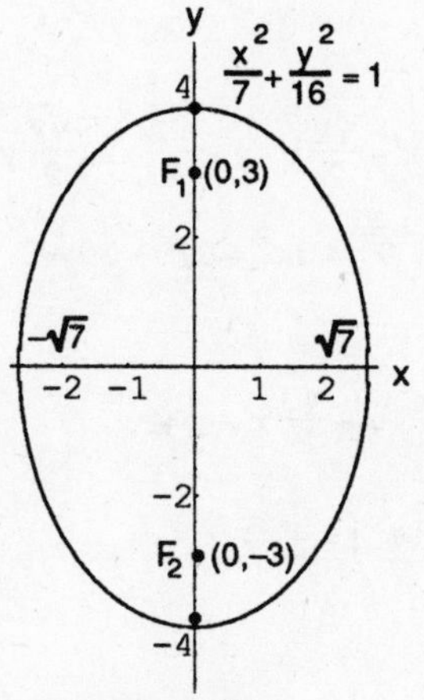

Graph 10.P.5

7. $3x^2 - y^2 = 3 \Rightarrow x^2 - \frac{y^2}{3} = 1 \Rightarrow c^2 = 1 + 3 = 4 \Rightarrow c = 2.$

$e = \frac{c}{a} = \frac{2}{1} = 2$. The asymptotes are $y = \pm\sqrt{3}\,x$.

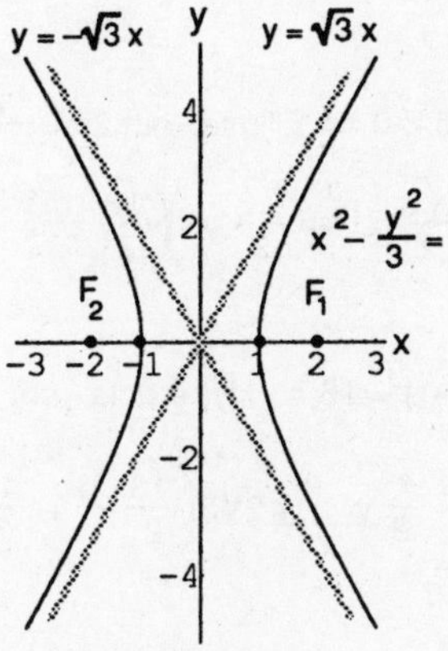

Graph 10.P.7

9. $x^2 = -12y \Rightarrow -\frac{x^2}{12} = y \Rightarrow 4p = 12 \Rightarrow p = 3 \Rightarrow$ focus is (0,–3), directrix is y = 3. Vertex is (0,0). $\therefore$ new vertex is (2,3), new focus is (2,0), new directrix is y = 6. The new equation is $(x - 2)^2 = -12(y - 3)$.

11. $\frac{x^2}{9} + \frac{y^2}{25} = 1 \Rightarrow a = 5, b = 3 \Rightarrow c = \sqrt{25 - 9} = 4 \Rightarrow$ foci are(0,±4), vertices are (0,±5), center is (0,0). $\therefore$ the new center is (–3,–5), new foci are (–3,–1), (–3,–9), new vertices are (–3,–10), (–3,0). The new equation is $\frac{(x+3)^2}{9} + \frac{(y+5)^2}{25} = 1$.

13. $\frac{y^2}{8} - \frac{x^2}{2} = 1 \Rightarrow a = 2\sqrt{2}, b = \sqrt{2} \Rightarrow c = \sqrt{8 + 2} = \sqrt{10} \Rightarrow$ foci are $\left(0, \pm\sqrt{10}\right)$, vertices are $\left(0, \pm 2\sqrt{2}\right)$, center is (0,0). The asymptotes are $y = \pm 2x$. $\therefore$ the new center is $\left(2, 2\sqrt{2}\right)$, the new foci are $\left(2, 2\sqrt{2} \pm \sqrt{10}\right)$, the new vertices are $\left(2, 4\sqrt{2}\right)$, (2,0). The new asymptotes are $y - 2\sqrt{2} = \pm 2(x - 2)$. The new equation is $\frac{\left(y - 2\sqrt{2}\right)^2}{8} - \frac{(x-2)^2}{2} = 1$.

15. $x^2 - 4x - 4y^2 = 0 \Rightarrow x^2 - 4x + 4 - 4y^2 = 4 \Rightarrow (x-2)^2 - 4y^2 = 4 \Rightarrow \frac{(x-2)^2}{4} - y^2 = 1$, a hyperbola. $a = 2, b = 1 \Rightarrow$ $c = \sqrt{1+4} = \sqrt{5}$. The center is (2,0). The vertices are (0,0) and (4,0). The foci are $\left(2 \pm \sqrt{5}, 0\right)$. The asymptotes are $y = \pm\frac{x-2}{2}$.

17. $y^2 - 2y + 16x = -49 \Rightarrow y^2 - 2y + 1 = -16x - 48 \Rightarrow (y-1)^2 = -16(x+3)$, a parabola. The vertex is (–3,1). $4p = 16 \Rightarrow$ $p = 4 \Rightarrow$ the focus is (–7,1). The directrix is $x = 1$.

19. $9x^2 + 16y^2 + 54x - 64y = -1 \Rightarrow 9(x^2 + 6x) + 16(y^2 - 4y) = -1 \Rightarrow 9(x^2 + 6x + 9) + 16(y^2 - 4y + 4) = 144 \Rightarrow 9(x+3)^2 +$ $16(y-2)^2 = 144 \Rightarrow \frac{(x+3)^2}{16} + \frac{(y-2)^2}{9} = 1$, an ellipse. The center is (–3,2). $a = 4, b = 3 \Rightarrow c = \sqrt{16-9} = \sqrt{7}$. The foci are $\left(-3 \pm \sqrt{7}, 2\right)$. The vertices are (1,2), (–7,2).

21. $x^2 + y^2 - 2x - 2y = 0 \Rightarrow x^2 - 2x + 1 + y^2 - 2y + 1 = 2 \Rightarrow (x-1)^2 + (y-1)^2 = 2$, a circle with center (1,1), radius $= \sqrt{2}$.

23. $B^2 - 4AC = 1 - 4(1)(1) = -3 < 0 \Rightarrow$ Ellipse

25. $B^2 - 4AC = 3^2 - 4(1)(2) = 1 > 0 \Rightarrow$ Hyperbola

27. $x^2 - 2xy + y^2 = 0 \Rightarrow (x-y)^2 = 0 \Rightarrow x - y = 0$ or $y = x$, a straignt line (a degenerate parabola).

29. $B^2 - 4AC = 1^2 - 4(2)(2) = -15 < 0 \Rightarrow$ Ellipse. $\cot 2\alpha = \frac{A-C}{B} = 0 \Rightarrow 2\alpha = \frac{\pi}{2} \Rightarrow \alpha = \frac{\pi}{4}$. $x = \frac{\sqrt{2}}{2}x' - \frac{\sqrt{2}}{2}y', y = \frac{\sqrt{2}}{2}x' + \frac{\sqrt{2}}{2}y'$ $\Rightarrow 2\left(\frac{\sqrt{2}}{2}x' - \frac{\sqrt{2}}{2}y'\right)^2 + \left(\frac{\sqrt{2}}{2}x' - \frac{\sqrt{2}}{2}y'\right)\left(\frac{\sqrt{2}}{2}x' + \frac{\sqrt{2}}{2}y'\right) + 2\left(\frac{\sqrt{2}}{2}x' + \frac{\sqrt{2}}{2}y'\right)^2 - 15 = 0 \Rightarrow 5x'^2 + 3y'^2 - 30 = 0$.

31. $B^2 - 4AC = \left(2\sqrt{3}\right)^2 - 4(1)(-1) = 16 \Rightarrow$ Hyperbola. $\cot 2\alpha = \frac{A-C}{B} = \frac{1}{\sqrt{3}} \Rightarrow 2\alpha = \frac{\pi}{3} \Rightarrow \alpha = \frac{\pi}{6}$. $x = \frac{\sqrt{3}}{2}x' - \frac{1}{2}y'$, $y = \frac{1}{2}x' + \frac{\sqrt{3}}{2}y' \Rightarrow \left(\frac{\sqrt{3}}{2}x' - \frac{1}{2}y'\right)^2 + 2\sqrt{3}\left(\frac{\sqrt{3}}{2}x' - \frac{1}{2}y'\right)\left(\frac{1}{2}x' + \frac{\sqrt{3}}{2}y'\right) - \left(\frac{1}{2}x' + \frac{\sqrt{3}}{2}y'\right)^2 = 4 \Rightarrow$ $2x'^2 - 2y'^2 = 4 \Rightarrow x'^2 - y'^2 = 2$

33. a) Around the x–axis: $9x^2 + 4y^2 = 36 \Rightarrow y^2 = 9 - \frac{9}{4}x^2 \Rightarrow y = \pm\sqrt{9 - \frac{9}{4}x^2}$, use the positive root.

$$V = 2\int_0^2 \pi\left(\sqrt{9 - \frac{9}{4}x^2}\right)^2 dx = 2\int_0^2 \pi\left(9 - \frac{9}{4}x^2\right) dx = 2\pi\left[9x - \frac{3}{4}x^3\right]_0^2 = 24\pi$$

b) Around the y–axis: $9x^2 + 4y^2 = 36 \Rightarrow x^2 = 4 - \frac{4}{9}y^2 \Rightarrow x = \pm\sqrt{4 - \frac{4}{9}y^2}$, use the positive root.

$$V = 2\int_0^3 \pi\left(\sqrt{4 - \frac{4}{9}y^2}\right)^2 dy = 2\int_0^3 \pi\left(4 - \frac{4}{9}y^2\right) dy = 2\pi\left[4y - \frac{4}{27}y^3\right]_0^3 = 16\pi$$

35. $x = \frac{t}{2}$, $y = t + 1 \Rightarrow 2x = t \Rightarrow y = 2x + 1$

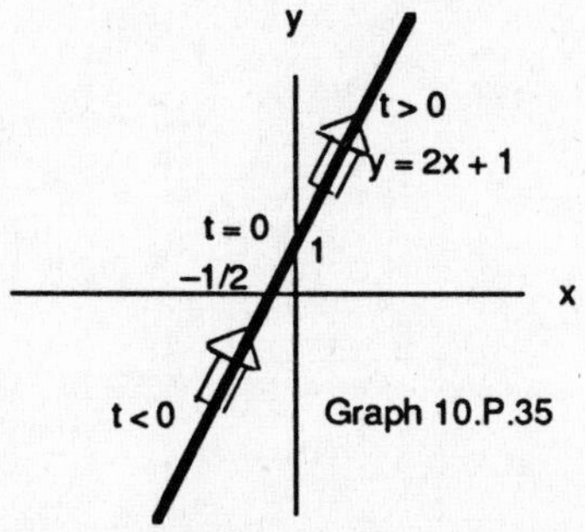

Graph 10.P.35

37. $x = \frac{1}{2}\tan t$, $y = \frac{1}{2}\sec t \Rightarrow x^2 = \frac{1}{4}\tan^2 t$, $y = \frac{1}{4}\sec^2 t \Rightarrow 4x^2 = \tan^2 t$, $4y^2 = \sec^2 t \Rightarrow 4x^2 + 1 = 4y^2 \Rightarrow 1 = 4y^2 - 4x^2$

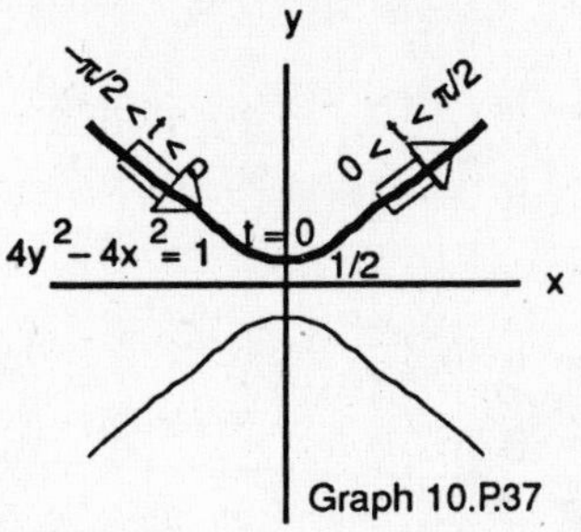

Graph 10.P.37

39. $x = -\cos t$, $y = \cos^2 t \Rightarrow y = (-x)^2 = x^2$

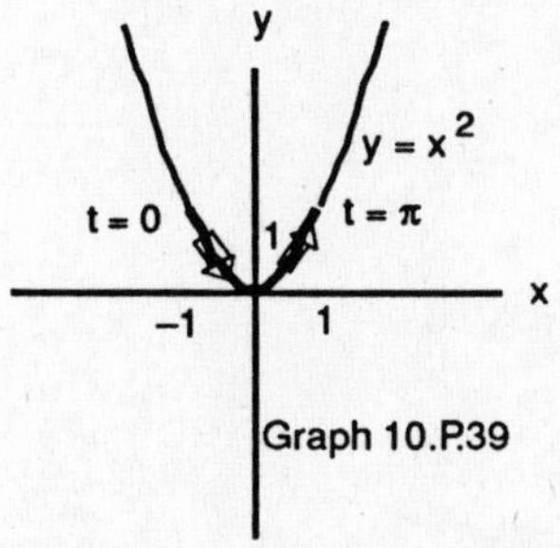

Graph 10.P.39

41. $16x^2 + 9y^2 = 144 \Rightarrow \frac{x^2}{9} + \frac{y^2}{16} = 1 \Rightarrow a = 3, b = 4 \Rightarrow x = 3\cos t, y = 4\sin t, 0 \le t \le 2\pi$

43. $x = \frac{1}{2}\tan t$, $y = \frac{1}{2}\sec t \Rightarrow \frac{dy}{dx} = \frac{dy/dt}{dx/dt} = \frac{\frac{1}{2}\sec t\tan t}{\frac{1}{2}\sec^2 t} = \frac{\tan t}{\sec t} = \sin t \Rightarrow \frac{dy}{dx}\left(\frac{\pi}{3}\right) = \sin\frac{\pi}{3} = \frac{\sqrt{3}}{2}$. $t = \frac{\pi}{3} \Rightarrow$ $x = \frac{1}{2}\tan\frac{\pi}{3} = \frac{\sqrt{3}}{2}$ and $y = \frac{1}{2}\sec t = 1 \Rightarrow y = \frac{\sqrt{3}}{2}x + \frac{1}{4}$. $\frac{d^2y}{dx^2} = \frac{dy'/dt}{dx/dt} = \frac{\cos t}{\frac{1}{2}\sec^2 t} = 2\cos^3 t \Rightarrow \frac{d^2y}{dx^2}\left(\frac{\pi}{3}\right) = 2\cos^3\frac{\pi}{3} = \frac{1}{4}$

45. $x = e^{2t} - \frac{t}{8}$, $y = e^t$, $0 \le t \le \ln 2 \Rightarrow \frac{dx}{dt} = 2e^{2t} - \frac{1}{8}$, $\frac{dy}{dt} = e^t \Rightarrow$ Length $= \int_0^{\ln 2}\sqrt{\left(2e^{2t} - \frac{1}{8}\right)^2 + \left(e^t\right)^2}\,dt =$

$$\int_0^{\ln 2}\sqrt{4e^{4t} + \frac{1}{2}e^{2t} + \frac{1}{64}}\,dt = \int_0^{\ln 2}\sqrt{\left(2e^{2t} + \frac{1}{8}\right)^2}\,dt = \int_0^{\ln 2}\left(2e^{2t} + \frac{1}{8}\right)dt = \left[e^{2t} + \frac{t}{8}\right]_0^{\ln 2} = 3 + \frac{\ln 2}{8}$$

47. $x = \frac{t^2}{2}, y = 2t, 0 \le t \le \sqrt{5} \Rightarrow \frac{dx}{dt} = t, \frac{dy}{dt} = 2 \Rightarrow \text{Area} = \int_0^{\sqrt{5}} 2\pi(2t)\sqrt{t^2 + 4}\, dt = \int_4^9 2\pi\, u^{1/2}\, du =$

$2\pi\left[\frac{2}{3} u^{3/2}\right]_4^9 = \frac{76\pi}{3}$ Let $u = t^2 + 4 \Rightarrow du = 2t\, dt$. $x = 0 \Rightarrow u = 4$, $x = \sqrt{5} \Rightarrow u = 9$

49.

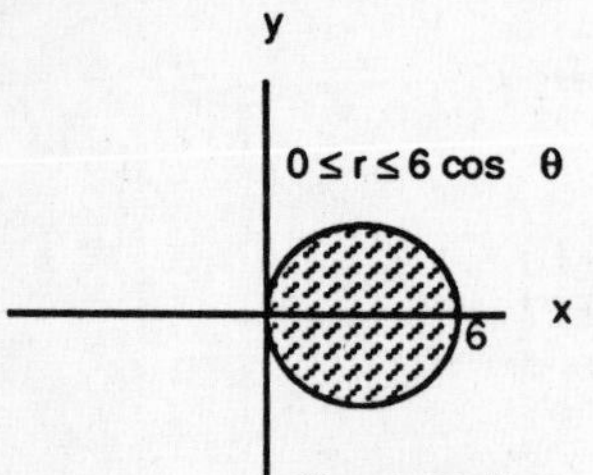

Graph 10.P.49

51. d 53. l 55. k 57. i

59. $r = \sin\theta, r = 1 + \sin\theta \Rightarrow \sin\theta = 1 + \sin\theta \Rightarrow \varnothing$. There are no points of intersection found by solving the system. The point of intersection (0,0) is found by graphing.

61. $r = 1 + \cos\theta, r = 1 - \cos\theta \Rightarrow 1 + \cos\theta = 1 - \cos\theta \Rightarrow 2\cos\theta = 0 \Rightarrow \cos\theta = 0 \Rightarrow \theta = \frac{\pi}{2}, \frac{3\pi}{2}$. $\theta = \frac{\pi}{2}, \frac{3\pi}{2} \Rightarrow r = 1$. The points of intersection are $\left(1, \frac{\pi}{2}\right), \left(1, \frac{3\pi}{2}\right)$. The point of intersection (0,0) is found by graphing. (See Graph 10.6.31 in this book.)

63. $r = 1 + \sin\theta$ and $r = -1 + \sin\theta$ intersect at all points of $r = 1 + \sin\theta$. This can be seen by graphing them.

65. $r = \sec\theta, r = 2\sin\theta \Rightarrow \sec\theta = 2\sin\theta \Rightarrow 1 = 2\sin\theta\cos\theta \Rightarrow 1 = \sin 2\theta \Rightarrow 2\theta = \frac{\pi}{2} \Rightarrow \theta = \frac{\pi}{4}$ $\Rightarrow r = 2\sin\frac{\pi}{4} = \sqrt{2} \Rightarrow$ the point of intersection is $\left(\sqrt{2}, \frac{\pi}{4}\right)$. No points are found by graphing.

67. $r^2 = \cos 2\theta \Rightarrow r = 0$ when $\cos 2\theta = 0 \Rightarrow 2\theta = \frac{\pi}{2}, \frac{3\pi}{2} \Rightarrow \theta = \frac{\pi}{4}, \frac{3\pi}{4}$. $\theta_1 = \frac{\pi}{4} \Rightarrow m_1 = \tan\frac{\pi}{4} = 1 \Rightarrow$ $y = x$ is one tangent line. $\theta_2 = \frac{3\pi}{4} \Rightarrow m_2 = \tan\frac{3\pi}{4} = -1 \Rightarrow y = -x$ is other tangent line.

69. Tips of the petals are at $\theta = \frac{\pi}{4}, \frac{3\pi}{4}, \frac{5\pi}{4}, \frac{7\pi}{4}$, $r = 1$ at those values of θ. Then for $\theta = \frac{\pi}{4}$, the line is $r\cos\left(\theta - \frac{\pi}{4}\right) = 1$; for $\theta = \frac{3\pi}{4}$, $r\cos\left(\theta - \frac{3\pi}{4}\right) = 1$; for $\theta = \frac{5\pi}{4}$, $r\cos\left(\theta - \frac{5\pi}{4}\right) = 1$; and for $\theta = \frac{7\pi}{4}$, $r\cos\left(\theta - \frac{7\pi}{4}\right) = 1$.

71. $r\cos\left(\theta + \frac{\pi}{3}\right) = 2\sqrt{3} \Rightarrow r\left(\cos\theta\cos\frac{\pi}{3} - \sin\theta\sin\frac{\pi}{3}\right) = 2\sqrt{3}$

$\Rightarrow \frac{1}{2}r\cos\theta - \frac{\sqrt{3}}{2}r\sin\theta = 2\sqrt{3} \Rightarrow r\cos\theta - \sqrt{3}\,r\sin\theta = 4\sqrt{3} \Rightarrow$

$x - \sqrt{3}\,y = 4\sqrt{3}$

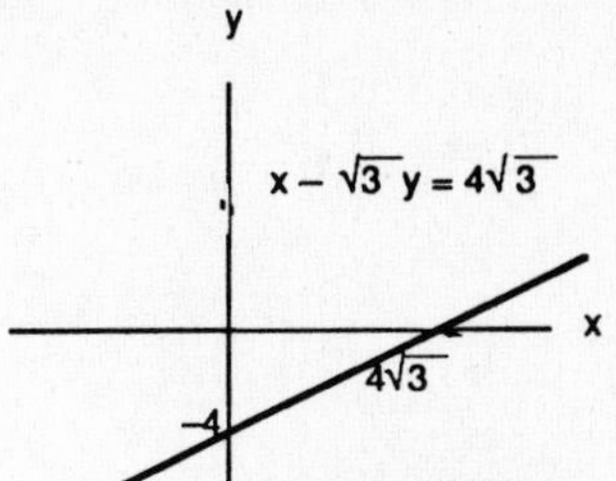

Graph 10.P.71

73. $r = 2\sec\theta \Rightarrow r = \frac{2}{\cos\theta} \Rightarrow r\cos\theta = 2 \Rightarrow x = 2$

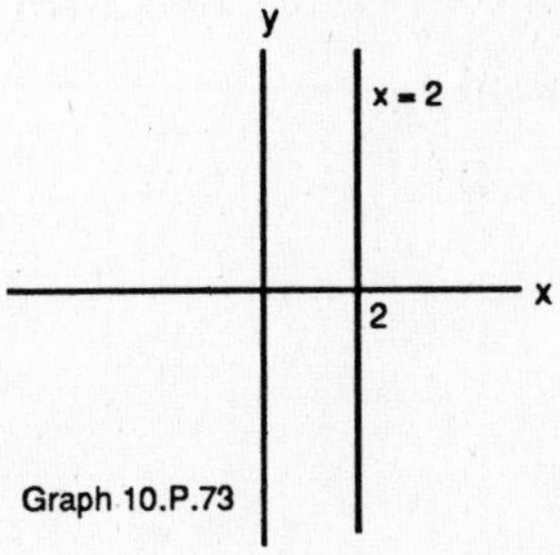

Graph 10.P.73

75. $r = -\frac{3}{2}\csc\theta \Rightarrow r\sin\theta = -\frac{3}{2} \Rightarrow y = -\frac{3}{2}$

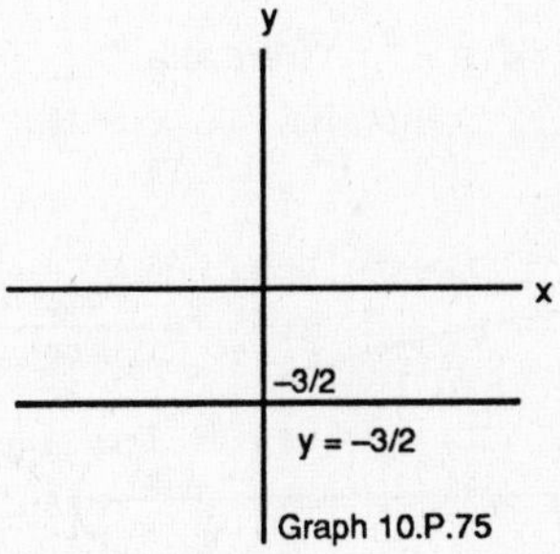

Graph 10.P.75

77. $x^2 + y^2 + 5y = 0 \Rightarrow x^2 + \left(y + \frac{5}{2}\right)^2 = \frac{25}{4} \Rightarrow C = \left(0, -\frac{5}{2}\right), a = \frac{5}{2} \Rightarrow$

$r = -5\sin\theta$ is the polar equation.

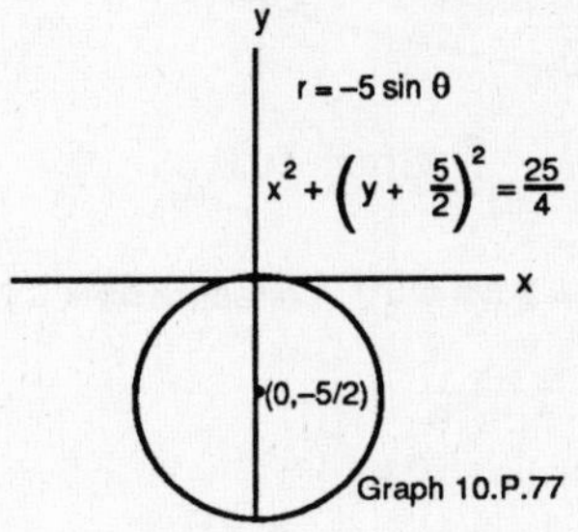

Graph 10.P.77

79. $x^2 + y^2 - 3x = 0 \Rightarrow \left(x - \frac{3}{2}\right)^2 + y^2 = \frac{9}{4} \Rightarrow C = \left(\frac{3}{2}, 0\right), a = \frac{3}{2} \Rightarrow$

$r = 3\cos\theta$ is the polar equation.

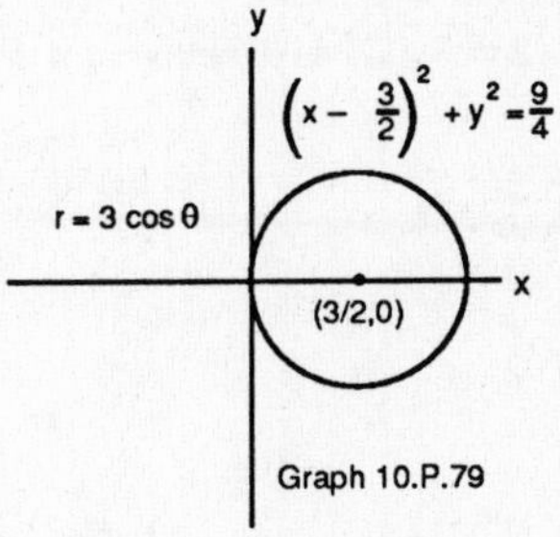

Graph 10.P.79

81. $r = -4\sin\theta \Rightarrow C = (0,-2),\ a = 2 \Rightarrow x^2 + (y+2)^2 = 4$ is the Cartesian equation.

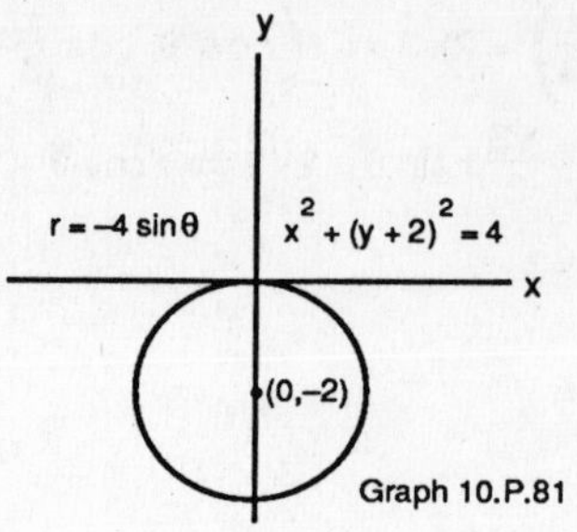

Graph 10.P.81

83. $r = 2\sqrt{2}\cos\theta \Rightarrow C = \left(\sqrt{2}, 0\right),\ a = \sqrt{2} \Rightarrow \left(x - \sqrt{2}\right)^2 + y^2 = 2$ is the Cartesian equation.

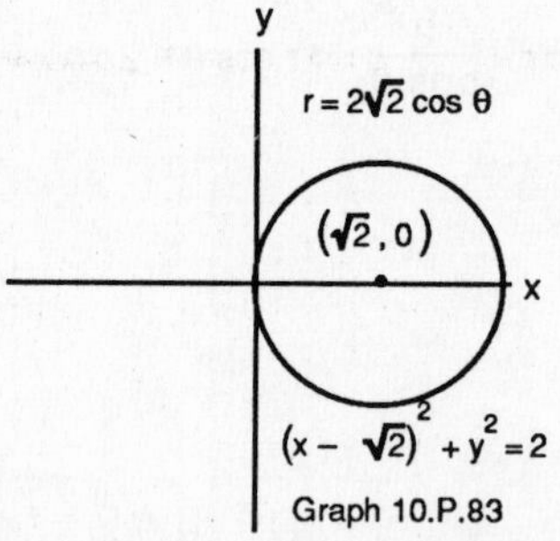

Graph 10.P.83

85. $r = \dfrac{2}{1 + \cos\theta} \Rightarrow e = 1 \Rightarrow$ Parabola

Vertex = (1,0)

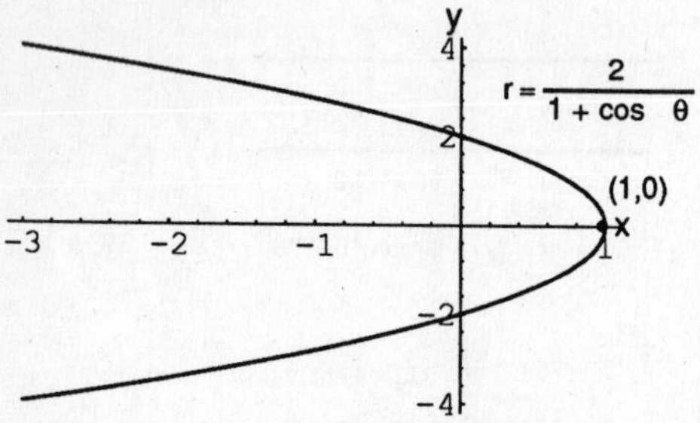

Graph 10.P.85

87. $r = \dfrac{6}{1 - 2\cos\theta} \Rightarrow e = 2 \Rightarrow$ Hyperbola

$ke = 6 \Rightarrow 2k = 6 \Rightarrow k = 3 \Rightarrow$ Vertices are $(2,\pi)$ and $(6,\pi)$.

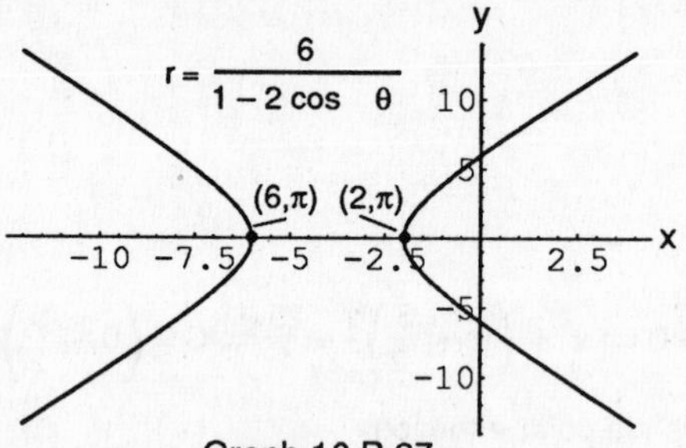

Graph 10.P.87

89. $e = 2,\ r\cos\theta = 2 \Rightarrow x = 2$ is directrix $\Rightarrow k = 2$ The conic is a hyperbola. $r = \dfrac{ke}{1 + e\cos\theta} \Rightarrow r = \dfrac{2(2)}{1 + 2\cos\theta}$

$\Rightarrow r = \dfrac{4}{1 + 2\cos\theta}$

91. $e = \dfrac{1}{2},\ r\sin\theta = 2 \Rightarrow y = 2$ is directrix $\Rightarrow k = 2$. The conic is an ellipse. $r = \dfrac{ke}{1 + e\sin\theta} \Rightarrow r = \dfrac{2\left(\frac{1}{2}\right)}{1 + \frac{1}{2}\sin\theta}$

$\Rightarrow r = \dfrac{2}{2 + \sin\theta}$

93. $A = 2\int_0^{\pi} \frac{1}{2} r^2\, d\theta = \int_0^{\pi} (2 - \cos\theta)^2\, d\theta = \int_0^{\pi} \left(4 - 2\cos\theta + \cos^2\theta\right) d\theta =$

$\int_0^{\pi} \left(4 - 2\cos\theta + \frac{1 + \cos 2\theta}{2}\right) d\theta = \int_0^{\pi} \left(\frac{9}{2} - 2\cos\theta + \frac{\cos 2\theta}{2}\right) d\theta = \left[\frac{9}{2}\theta - 2\sin\theta + \frac{\sin 2\theta}{4}\right]_0^{\pi} = \frac{9}{2}\pi$

95. $r = 1 + \cos 2\theta,\ r = 1 \Rightarrow 1 = 1 + \cos 2\theta \Rightarrow 0 = \cos 2\theta \Rightarrow 2\theta = \frac{\pi}{2} \Rightarrow \theta = \frac{\pi}{4}.$

$\therefore A = 4\int_0^{\pi/4} \frac{1}{2}\left((1 + \cos 2\theta)^2 - 1^2\right) d\theta = 2\int_0^{\pi/4} (1 + 2\cos 2\theta + \cos^2 2\theta - 1)\, d\theta =$

$2\int_0^{\pi/4} \left(2\cos 2\theta + \frac{1}{2} + \frac{\cos 4\theta}{2}\right) d\theta = 2\left[\sin 2\theta + \frac{1}{2}\theta + \frac{\sin 4\theta}{8}\right]_0^{\pi/4} = 2 + \frac{\pi}{4}$

97. $r = -1 + \cos\theta \Rightarrow \frac{dr}{d\theta} = -\sin\theta$

$\text{Length} = \int_0^{2\pi} \sqrt{(-1 + \cos\theta)^2 + (-\sin\theta)^2}\, d\theta = \int_0^{2\pi} \sqrt{2 - 2\cos\theta}\, d\theta =$

$\int_0^{2\pi} \sqrt{\frac{4(1 - \cos\theta)}{2}}\, d\theta = \int_0^{2\pi} 2\sin\frac{1}{2}\theta\, d\theta = \left[-4\cos\frac{1}{2}\theta\right]_0^{2\pi} = 8$

99. $r = 8\sin^3\left(\frac{\theta}{3}\right),\ 0 \le \theta \le \frac{\pi}{4} \Rightarrow \frac{dr}{d\theta} = 8\sin^2\left(\frac{\theta}{3}\right)\cos\left(\frac{\theta}{3}\right).\ r^2 + \left(\frac{dr}{d\theta}\right)^2 = \left(8\sin^3\left(\frac{\theta}{3}\right)\right)^2 + \left(8\sin^2\left(\frac{\theta}{3}\right)\cos\left(\frac{\theta}{3}\right)\right)^2 =$

$64\sin^4\left(\frac{\theta}{3}\right).\ \therefore L = \int_0^{\pi/4} \sqrt{64\sin^4(\theta/3)}\, d\theta = \int_0^{\pi/4} 8\sin^2\left(\frac{\theta}{3}\right) d\theta = \int_0^{\pi/4} 8\left(\frac{1 - \cos\left(\frac{2\theta}{3}\right)}{2}\right) d\theta =$

$\int_0^{\pi/4} \left(4 - 4\cos\left(\frac{2\theta}{3}\right)\right) d\theta = \left[4\theta - 6\sin\left(\frac{2\theta}{3}\right)\right]_0^{\pi/4} = 4\left(\frac{\pi}{4}\right) - 6\sin\left(\frac{2(\pi/4)}{3}\right) - 0 = \pi - 6\sin\frac{\pi}{6} = \pi - 3.$

101. $r = \sqrt{\cos 2\theta} \Rightarrow \frac{dr}{d\theta} = \frac{-\sin 2\theta}{\sqrt{\cos 2\theta}}.$ Surface Area $= \int_0^{\pi/4} 2\pi r\sin\theta\sqrt{r^2 + \left(\frac{dr}{d\theta}\right)^2}\, d\theta =$

$\int_0^{\pi/4} 2\pi\sqrt{\cos 2\theta}\,\sin\theta\sqrt{\left(\sqrt{\cos 2\theta}\right)^2 + \left(\frac{-\sin 2\theta}{\sqrt{\cos 2\theta}}\right)^2}\, d\theta =$

$\int_0^{\pi/4} 2\pi\sqrt{\cos 2\theta}\,\sin\theta\sqrt{\cos 2\theta + \frac{\sin^2 2\theta}{\cos 2\theta}}\, d\theta =$

$\int_0^{\pi/4} 2\pi\sqrt{\cos 2\theta}\,\sin\theta\sqrt{\frac{1}{\cos 2\theta}}\, d\theta = \int_0^{\pi/4} 2\pi\sin\theta\, d\theta = [2\pi(-\cos\theta)]_0^{\pi/4} = 2\pi\left(1 - \frac{\sqrt{2}}{2}\right)$

103. Each portion of the wave front will reflect to the other focus, and, since the wave front travels at a constant speed as it expands, the different portions of the wave will arrive at the second focus simultaneously.

105. The time for the bullet to hit the target remains constant, say $t = t_0$. Let the time it takes for sound to travel from the target to the listener be t_2. Since the listener hears the sounds simultaneously, $t_1 = t_0 + t_2$ where t_1 is the time for the sound to travel from the rifle to the listener.
If v is the velocity of sound, $vt_1 = vt_0 + vt_2$ or $vt_1 - vt_2 = vt_0$. vt_1 is the distance from the rifle to the listener, vt_2 is the distance from the target to the listener. $\therefore$ the difference of the distances is constant since vt_0 is constant $\Rightarrow$ the listener is on a branch of a hyperbola with foci at the rifle and the target. The branch is the one with the target as focus.

107. a) $r = \frac{k}{1 + e\cos\theta} \Rightarrow r + er\cos\theta = k \Rightarrow \sqrt{x^2 + y^2} + ex = k \Rightarrow \sqrt{x^2 + y^2} = k - ex \Rightarrow x^2 + y^2 = k^2 - 2kex + e^2x^2 \Rightarrow x^2 - e^2x^2 + y^2 + 2kex - k^2 = 0 \Rightarrow (1 - e^2)x^2 + y^2 + 2kex - k^2 = 0.$

b) $e = 0 \Rightarrow x^2 + y^2 - k^2 = 0 \Rightarrow$ Circle. $0 < e < 1 \Rightarrow 0 < e^2 < 1 \Rightarrow 0 < 1 - e^2 \Rightarrow x^2$ and y^2 have positive, unequal coefficients $\Rightarrow$ Ellipse.

$e = 1 \Rightarrow y^2 + 2kex - k^2 = 0 \Rightarrow$ Parabola

$e > 1 \Rightarrow e^2 > 1 \Rightarrow 1 - e^2 < 0 \Rightarrow$ the coefficient of x^2 is negative, the coefficient of y^2 is positive $\Rightarrow$ Hyperbola.

CHAPTER 11

VECTORS AND ANALYTIC GEOMETRY IN SPACE

11.1 VECTORS IN THE PLANE

1.

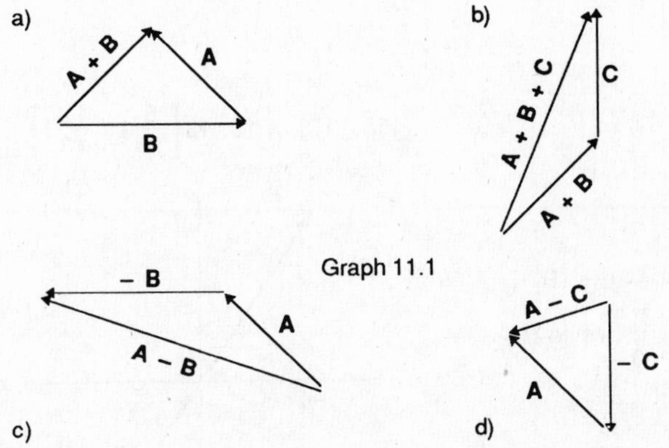

Graph 11.1

3. $\overrightarrow{DA} = -\overrightarrow{AD} = -(\mathbf{u} + \mathbf{w})$, $\overrightarrow{DP} = \frac{1}{2}\left[-\overrightarrow{AD}\right] = \frac{1}{2}\left[-(\mathbf{u} + \mathbf{w})\right] = -\left[\frac{\mathbf{u} + \mathbf{w}}{2}\right]$

5. a) $\mathbf{w} = \mathbf{u} + \mathbf{v}$ b) $\mathbf{v} = \mathbf{w} + (-\mathbf{u}) = \mathbf{w} - \mathbf{u}$

7. $\mathbf{a} = \mathbf{u} + \overrightarrow{BP} = \mathbf{u} + \frac{(\mathbf{w} - \mathbf{u})}{2} = \frac{2\,\mathbf{u} + \mathbf{w} - \mathbf{u}}{2} = \frac{\mathbf{u} + \mathbf{w}}{2}$

9. $(2\mathbf{i} - 7\mathbf{j}) + (\mathbf{i} + 6\mathbf{j}) = 3\mathbf{i} - \mathbf{j}$

11. $(-2\mathbf{i} + 6\mathbf{j}) - 2(\mathbf{i} + \mathbf{j}) + 3\mathbf{i} - 4\mathbf{j} = -\mathbf{i}$

13. $2\big((\ln 2)\mathbf{i} + \mathbf{j}\big) - \big((\ln 8)\mathbf{i} + \pi\mathbf{j}\big) = -\ln 2\,\mathbf{i} + (2 - \pi)\mathbf{j}$

15. $\overrightarrow{P_1P_2} = (2 - 1)\mathbf{i} + (-1 - 3)\mathbf{j} = \mathbf{i} - 4\mathbf{j}$

17. $\overrightarrow{AO} = -2\mathbf{i} - 3\mathbf{j}$

19. $\mathbf{u} = \frac{\sqrt{3}}{2}\mathbf{i} + \frac{1}{2}\mathbf{j}$, when $\theta = \frac{\pi}{6}$

$\mathbf{u} = -\frac{1}{2}\mathbf{i} + \frac{\sqrt{3}}{2}\mathbf{j}$, when $\theta = \frac{2\pi}{3}$

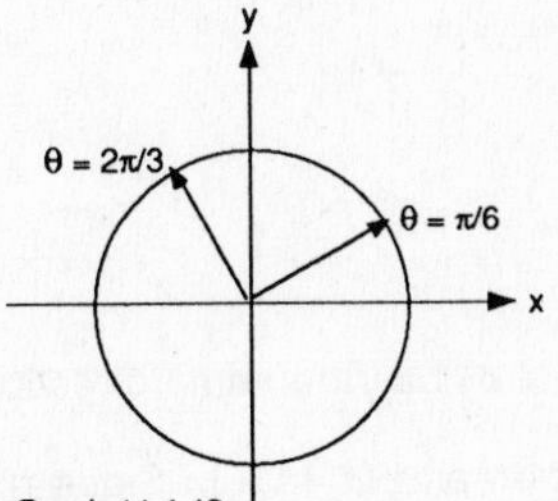

Graph 11.1.19

21. $\mathbf{u} = -\frac{\sqrt{2}}{2}\mathbf{i} - \frac{\sqrt{2}}{2}\mathbf{j}$

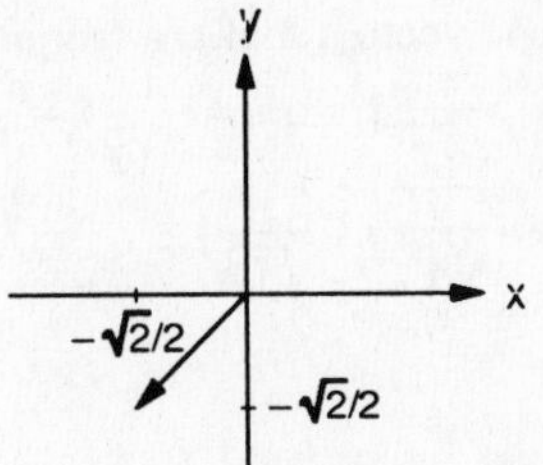

Graph 11.1.21

23. If $|x|$ is the magnitude of the x–component, then $\cos 30° = \frac{|x|}{|F|} \Rightarrow |x| = |F| \cos 30° = (10)\left(\frac{\sqrt{3}}{2}\right) = 5\sqrt{3}$ lb $\Rightarrow x = 5\sqrt{3}\ \mathbf{i}$.

If $|y|$ is the magnitude of the y–component, then $\sin 30° = \frac{|y|}{|F|} \Rightarrow |y| = |F| \sin 30° = (10)\left(\frac{1}{2}\right) = 5$ lb $\Rightarrow y = 5\,\mathbf{j}$.

25. If $|x|$ is the magnitude of the x–component, then $\cos 20° = \frac{|x|}{|F|} \Rightarrow |x| = |F| \cos 20° = (100) \cos 20° \approx 94$ lb $\Rightarrow$ $x \approx 94\,\mathbf{i}$.

If $|y|$ is the magnitude of the y–component, then $\sin 20° = \frac{|y|}{|F|} \Rightarrow |y| = |F| \sin 20° = (100) \sin 20° \approx 34.2$ lb $\Rightarrow y \approx -34.2\,\mathbf{j}$ the negative sign is indicated by the diagram.

27. $|\mathbf{i} + \mathbf{j}| = \sqrt{1^2 + 1^2} = \sqrt{2}$, $\mathbf{i} + \mathbf{j} = \sqrt{2}\left[\frac{1}{\sqrt{2}}\mathbf{i} + \frac{1}{\sqrt{2}}\mathbf{j}\right]$

29. $\left|\sqrt{3}\,\mathbf{i} + \mathbf{j}\right| = \sqrt{(\sqrt{3})^2 + 1^2} = 2$, $2\left[\frac{\sqrt{3}}{2}\mathbf{i} + \frac{1}{2}\mathbf{j}\right]$

31. $|5\mathbf{i} + 12\mathbf{j}| = 13$, $13\left[\frac{5}{13}\mathbf{i} + \frac{12}{13}\mathbf{j}\right]$

33. $\mathbf{A} = -4\mathbf{i} + 6\mathbf{j} \Rightarrow |\mathbf{A}| = \sqrt{(-4)^2 + 6^2} = \sqrt{52} = 2\sqrt{13}$ and $\mathbf{B} = 2\mathbf{i} - 3\mathbf{j} \Rightarrow |\mathbf{B}| = \sqrt{2^2 + (-3)^2} = \sqrt{13}$. Hence $\mathbf{A}$'s direction is $\frac{\mathbf{A}}{|\mathbf{A}|} = -\frac{2}{\sqrt{13}}\mathbf{i} + \frac{3}{\sqrt{3}}\mathbf{j}$ while $\mathbf{B}$'s is $\frac{\mathbf{B}}{|\mathbf{B}|} = \frac{2}{\sqrt{13}}\mathbf{i} - \frac{3}{\sqrt{13}}\mathbf{j} = -\frac{\mathbf{A}}{|\mathbf{A}|} \Rightarrow$ the opposite direction.

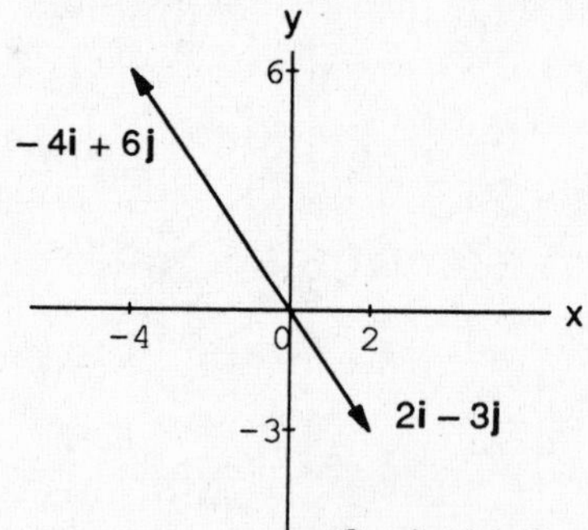

Graph 11.1.33

35. $\frac{2}{\sqrt{2}}(-\mathbf{i} - \mathbf{j})$, only one

37. The tangent of $y = f(x)$ at $x = x_o$ has the same direction as $\mathbf{i} + f'(x_o)\,\mathbf{j}$ and the normal has the same direction as $f'(x_o)\,\mathbf{i} - \mathbf{j}$. Since $f(x) = x^2 \Rightarrow f'(x) = 2x$ at $x = 2$ the tangent has the same direction as $\mathbf{i} + 4\,\mathbf{j}$ while the normal is in the direction of $4\,\mathbf{i} - \mathbf{j}$. Therefore, the unit vectors that are tangent and normal to the curve at $x = 2$ follow. $\mathbf{u} = \frac{1}{\sqrt{17}}\mathbf{i} + \frac{4}{\sqrt{17}}\mathbf{j}$, $-\mathbf{u} = -\frac{1}{\sqrt{17}}\mathbf{i} - \frac{4}{\sqrt{17}}\mathbf{j}$ and $\mathbf{n} = \frac{4}{\sqrt{17}}\mathbf{i} - \frac{1}{\sqrt{17}}\mathbf{j}$, $-\mathbf{n} = -\frac{4}{\sqrt{17}}\mathbf{i} + \frac{1}{\sqrt{17}}\mathbf{j}$

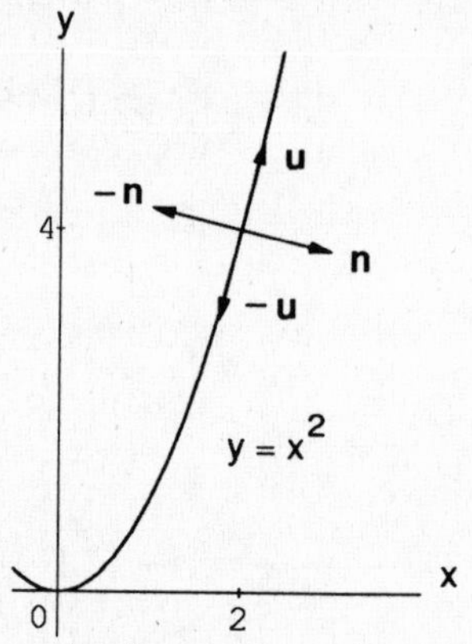

Graph 11.1.37

39. The tangent of $y = f(x)$ at (x_o, y_o) has the same direction as $\mathbf{i} + [y']_{(x_o, y_o)}\,\mathbf{j}$ and the normal has the same direction as $[y']_{(x_o, y_o)}\,\mathbf{i} - \mathbf{j}$. Since $x^2 + 2y^2 = 6 \Rightarrow 2x + 4yy' = 0 \Rightarrow y' = -\frac{x}{2y}$ at $x = (2,1)$ the tangent has the same direction as $\mathbf{i} - \mathbf{j}$ while the normal is in the direction of $\mathbf{i} + \mathbf{j}$. Therefore, the unit vectors that are tangent and normal to the curve at (2,1) follow. $\mathbf{u} = \frac{1}{\sqrt{2}}(\mathbf{i} - \mathbf{j})$, $-\mathbf{u} = \frac{1}{\sqrt{2}}(-\mathbf{i} + \mathbf{j})$ and $\mathbf{n} = \frac{1}{\sqrt{2}}(\mathbf{i} + \mathbf{j})$, $-\mathbf{n} = \frac{1}{\sqrt{2}}(-\mathbf{i} - \mathbf{j})$

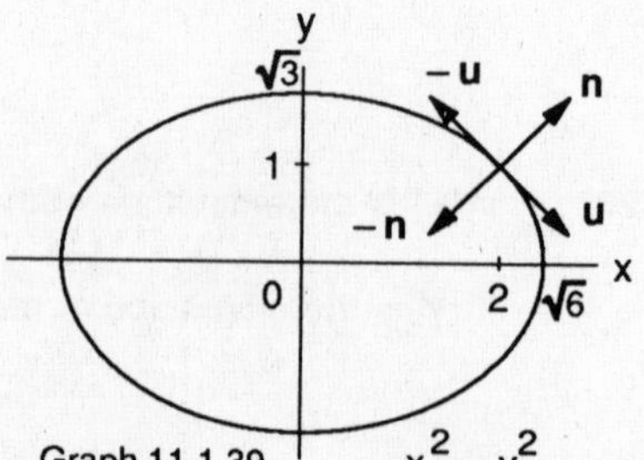

Graph 11.1.39

$\frac{x^2}{6} + \frac{y^2}{3} = 1$

41. The tangent of $y = f(x)$ at $x = x_o$ has the same direction as $\mathbf{i} + f'(x_o)\,\mathbf{j}$ and the normal has the same direction as $f'(x_o)\,\mathbf{i} - \mathbf{j}$. Since $f(x) = \tan^{-1} x \Rightarrow f'(x) = \frac{1}{1 + x^2}$ at $x = 1$ the tangent has the same direction as $\mathbf{i} + \frac{1}{2}\mathbf{j}$ while the normal is in the direction of $\frac{1}{2}\mathbf{i} - \mathbf{j}$. Therefore, the unit vectors that are tangent and normal to the curve at $x = 1$ follow. $\mathbf{u} = \frac{2}{\sqrt{5}}\mathbf{i} + \frac{1}{\sqrt{5}}\mathbf{j}$, $-\mathbf{u} = -\frac{2}{\sqrt{5}}\mathbf{i} - \frac{1}{\sqrt{5}}\mathbf{j}$ and $\mathbf{n} = -\frac{1}{\sqrt{5}}\mathbf{i} + \frac{2}{\sqrt{5}}\mathbf{j}$, $-\mathbf{n} = \frac{1}{\sqrt{5}}\mathbf{i} - \frac{2}{\sqrt{5}}\mathbf{j}$

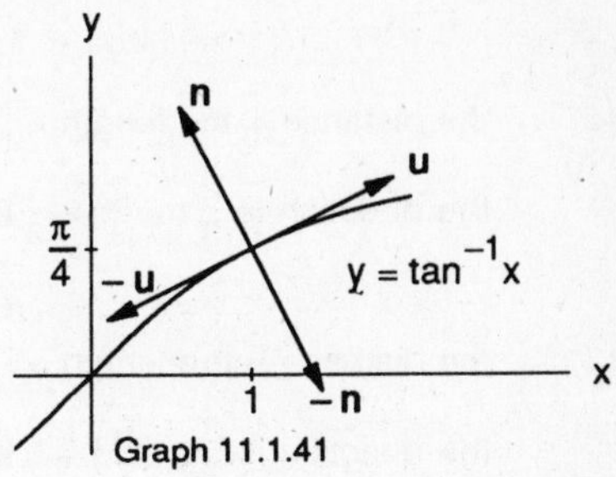

Graph 11.1.41

43. The slope of $-\mathbf{v} = -a\,\mathbf{i} - b\,\mathbf{j}$ is $(-b)/(-a) = b/a$, the same as the slope of $\mathbf{v}$.

11.2 CARTESIAN (RECTANGULAR) COORDINATES AND VECTORS IN SPACE

1. a line through the point (2,3,0) parallel to the z–axis

3. the x–axis

5. the circle, $x^2 + y^2 = 4$ in the xy–plane

7. the circle, $x^2 + z^2 = 4$ in the xz–plane

9. the circle, $y^2 + z^2 = 1$ in the yz–plane

11. the circle, $x^2 + y^2 = 16$ in the xy–plane

13. a) the first quadrant of the xy–plane b) the fourth quadrant of the xy–plane

15. a) a solid sphere of radius 1 centered at the origin
 b) all points which are greater than 1 unit from the origin

17. a) the upper hemisphere of radius 1 centered at the origin
 b) the solid upper hemisphere of radius 1 centered at the origin

19. a) $x = 3$ b) $y = -1$ c) $z = -2$

21. a) $z = 1$ b) $x = 3$ c) $y = -1$

23. a) $x^2 + (y-2)^2 = 4,\ z = 0$ b) $(y-2)^2 + z^2 = 4,\ x = 0$ c) $x^2 + z^2 = 4,\ y = 2$

25. a) $y = 3,\ z = -1$ b) $x = 1,\ z = -1$ c) $x = 1,\ y = 3$

27. $x^2 + y^2 + z^2 = 25,\ z = 3$

29. $0 \le z \le 1$

31. $z \le 0$

33. a) $(x-1)^2 + (y-1)^2 + (z-1)^2 < 1$ b) $(x-1)^2 + (y-1)^2 + (z-1)^2 > 1$

35. length $= |2\mathbf{i} + \mathbf{j} - 2\mathbf{k}| = \sqrt{2^2 + 1^2 + (-2)^2} = 3$, the direction is $\frac{2}{3}\mathbf{i} + \frac{1}{3}\mathbf{j} - \frac{2}{3}\mathbf{k} \Rightarrow 2\mathbf{i} + \mathbf{j} - 2\mathbf{k} = 3\left[\frac{2}{3}\mathbf{i} + \frac{1}{3}\mathbf{j} - \frac{2}{3}\mathbf{k}\right]$

37. length $= |\mathbf{i} + 4\mathbf{j} - 8\mathbf{k}| = \sqrt{1 + 16 + 64} = 9$, the direction is $\frac{1}{9}\mathbf{i} + \frac{4}{9}\mathbf{j} - \frac{8}{9}\mathbf{k} \Rightarrow \mathbf{i} + 4\mathbf{j} - 8\mathbf{k} = 9\left[\frac{1}{9}\mathbf{i} + \frac{4}{9}\mathbf{j} - \frac{8}{9}\mathbf{k}\right]$

39. length $= |5\mathbf{k}| = \sqrt{25} = 5$, the direction is $\mathbf{k} \Rightarrow 5\mathbf{k} = 5\,[\mathbf{k}]$

41. length $= \left|\frac{3}{5}\mathbf{i} + \frac{4}{5}\mathbf{k}\right| = \sqrt{\frac{9}{25} + \frac{16}{25}} = 1$, the direction is $\frac{3}{5}\mathbf{i} + \frac{4}{5}\mathbf{k} \Rightarrow \frac{3}{5}\mathbf{i} + \frac{4}{5}\mathbf{k} = 1\left[\frac{3}{5}\mathbf{i} + \frac{4}{5}\mathbf{k}\right]$

43. length $= \left|\frac{1}{\sqrt{6}}\mathbf{i} - \frac{1}{\sqrt{6}}\mathbf{j} - \frac{1}{\sqrt{6}}\mathbf{k}\right| = \sqrt{3\left(\frac{1}{\sqrt{6}}\right)^2} = \sqrt{\frac{1}{2}}$, the direction is $\frac{1}{\sqrt{3}}\mathbf{i} - \frac{1}{\sqrt{3}}\mathbf{j} - \frac{1}{\sqrt{3}}\mathbf{k} \Rightarrow \frac{1}{\sqrt{6}}\mathbf{i} - \frac{1}{\sqrt{6}}\mathbf{j} - \frac{1}{\sqrt{6}}\mathbf{k} = \sqrt{\frac{1}{2}}\left[\frac{1}{\sqrt{3}}\mathbf{i} - \frac{1}{\sqrt{3}}\mathbf{j} - \frac{1}{\sqrt{3}}\mathbf{k}\right]$

45. the distance = the length = $\left|\overrightarrow{P_1P_2}\right| = |2\mathbf{i} + 2\mathbf{j} - \mathbf{k}| = \sqrt{2^2 + 2^2 + (-1)^2} = 3$, $2\mathbf{i} + 2\mathbf{j} - \mathbf{k} = 3\left[\frac{2}{3}\mathbf{i} + \frac{2}{3}\mathbf{j} - \frac{1}{3}\mathbf{k}\right] \Rightarrow$ the direction is $\frac{2}{3}\mathbf{i} + \frac{2}{3}\mathbf{j} - \frac{1}{3}\mathbf{k}$, the midpoint is (2,2,1/2)

47. the distance = the length = $\left|\overrightarrow{P_1P_2}\right| = |3\mathbf{i} - 6\mathbf{j} + 2\mathbf{k}| = \sqrt{9 + 36 + 4} = 7$, $3\mathbf{i} - 6\mathbf{j} + 2\mathbf{k} = 7\left[\frac{3}{7}\mathbf{i} - \frac{6}{7}\mathbf{j} + \frac{2}{7}\mathbf{k}\right] \Rightarrow$ the direction is $\frac{3}{7}\mathbf{i} - \frac{6}{7}\mathbf{j} + \frac{2}{7}\mathbf{k}$, the midpoint is (5/2,1,6)

49. the distance = the length = $\left|\overrightarrow{P_1P_2}\right| = |2\mathbf{i} - 2\mathbf{j} - 2\mathbf{k}| = \sqrt{3 \cdot 2^2} = 2\sqrt{3}$, $2\mathbf{i} - 2\mathbf{j} - 2\mathbf{k} = 2\sqrt{3}\left[\frac{1}{\sqrt{3}}\mathbf{i} - \frac{1}{\sqrt{3}}\mathbf{j} - \frac{1}{\sqrt{3}}\mathbf{k}\right] \Rightarrow$ the direction is $\frac{1}{\sqrt{3}}\mathbf{i} - \frac{1}{\sqrt{3}}\mathbf{j} - \frac{1}{\sqrt{3}}\mathbf{k}$, the midpoint is (1,– 1,– 1)

51. a) $2\mathbf{i}$ b) $-\sqrt{3}\mathbf{k}$ c) $\frac{3}{10}\mathbf{j} + \frac{2}{5}\mathbf{k}$ d) $6\mathbf{i} - 2\mathbf{j} + 3\mathbf{k}$

53. $\frac{\mathbf{A}}{|\mathbf{A}|} = \frac{1}{13}\mathbf{A} = \frac{1}{13}[12\mathbf{i} - 5\mathbf{k}] = \frac{12}{13}\mathbf{i} - \frac{5}{13}\mathbf{k}$, where $|\mathbf{A}| = \sqrt{12^2 + 5^2} = \sqrt{169} = 13$; the desired vector is $\frac{7}{13}(12\mathbf{i} - 5\mathbf{k})$

55. $|\mathbf{A}| = |2\,\mathbf{i} - 3\,\mathbf{j} + 6\,\mathbf{k}| = \sqrt{2^2 + (-3)^2 + 6^2} = \sqrt{49} = 7$, $\mathbf{u} = \left[\frac{2}{7}\mathbf{i} - \frac{3}{7}\mathbf{j} + \frac{6}{7}\mathbf{k}\right] \Rightarrow$ the desired vector is $-5\,\mathbf{u} = -\frac{10}{7}\mathbf{i} + \frac{15}{7}\mathbf{j} - \frac{30}{7}\mathbf{k}$

57. center (– 2,0,2), radius $2\sqrt{2}$

59. center $(\sqrt{2},\sqrt{2},-\sqrt{2})$, radius $\sqrt{2}$

61. $(x-1)^2 + (y-2)^2 + (z-3)^2 = 14$

63. $(x+2)^2 + y^2 + z^2 = 3$

65. $x^2 + y^2 + z^2 + 4x - 4z = 0 \Rightarrow \left(x^2 + 4x + 4\right) + \left(y^2\right) + \left(z^2 - 4z + 4\right) = 4 + 4 \Rightarrow (x+2)^2 + y^2 + (z-2)^2 = \left(\sqrt{8}\right)^2 \Rightarrow$ the center is at (– 2,0,2) and the radius is $\sqrt{8}$

67. $2x^2 + 2y^2 + 2z^2 + x + y + z = 9 \Rightarrow x^2 + \frac{1}{2}x + y^2 + \frac{1}{2}y + z^2 + \frac{1}{2}z = \frac{9}{2} \Rightarrow \left(x^2 + \frac{1}{2}x + \frac{1}{16}\right) + \left(y^2 + \frac{1}{2}y + \frac{1}{16}\right) + \left(z^2 + \frac{1}{2}z + \frac{1}{16}\right) = \frac{9}{2} + \frac{3}{16} = \frac{75}{16} \Rightarrow \left(x + \frac{1}{4}\right)^2 + \left(y + \frac{1}{4}\right)^2 + \left(z + \frac{1}{4}\right)^2 = \left(\frac{5\sqrt{3}}{4}\right)^2 \Rightarrow$ the center is at (– 1/4,– 1/4,– 1/4) and the radius is $\frac{5\sqrt{3}}{4}$

69. a) the distance between (x,y,z) and (x,0,0) is $\sqrt{y^2 + z^2}$
b) the distance between (x,y,z) and (0,y,0) is $\sqrt{x^2 + z^2}$
c) the distance between (x,y,z) and (0,0,z) is $\sqrt{x^2 + y^2}$

11.3 DOT PRODUCTS

	A·B	|A|	|B|	cos θ	|B| cos θ	$Proj_A$ B
1.	−25	5	5	−1	−5	$-2\mathbf{i}+4\mathbf{j}-\sqrt{5}\mathbf{k}$
3.	25	15	5	$\frac{1}{3}$	$\frac{5}{3}$	$\frac{1}{9}[10\mathbf{i}+11\mathbf{j}-2\mathbf{k}]$
5.	0	$\sqrt{53}$	1	0	0	**0**
7.	2	$\sqrt{34}$	$\sqrt{3}$	$\frac{2}{\sqrt{3}\sqrt{34}}$	$\frac{2}{\sqrt{34}}$	$\frac{1}{17}[5\mathbf{j}-3\mathbf{k}]$
9.	10	$\sqrt{13}$	$\sqrt{26}$	$\frac{10}{13\sqrt{2}}$	$\frac{10}{\sqrt{13}}$	$\frac{10}{13}[3\mathbf{i}+2\mathbf{j}]$
11.	$\sqrt{3}-\sqrt{2}$	$\sqrt{2}$	3	$\frac{\sqrt{3}-\sqrt{2}}{3\sqrt{2}}$	$\frac{\sqrt{3}-\sqrt{2}}{\sqrt{2}}$	$\frac{\sqrt{3}-\sqrt{2}}{2}[-\mathbf{i}+\mathbf{j}]$

13. $\mathbf{B}=\left(\frac{\mathbf{A}\cdot\mathbf{B}}{\mathbf{A}\cdot\mathbf{A}}\mathbf{A}\right)+\left(\mathbf{B}-\frac{\mathbf{A}\cdot\mathbf{B}}{\mathbf{A}\cdot\mathbf{A}}\mathbf{A}\right)=\frac{3}{2}[\mathbf{i}+\mathbf{j}]+\left[(3\mathbf{j}+4\mathbf{k})-\frac{3}{2}(\mathbf{i}+\mathbf{j})\right]=\left[\frac{3}{2}\mathbf{i}+\frac{3}{2}\mathbf{j}\right]+\left[-\frac{3}{2}\mathbf{i}+\frac{3}{2}\mathbf{j}+4\mathbf{k}\right]$, where $\mathbf{A}\cdot\mathbf{B}=3$ and $\mathbf{A}\cdot\mathbf{A}=2$

15. $\mathbf{B}=\left(\frac{\mathbf{A}\cdot\mathbf{B}}{\mathbf{A}\cdot\mathbf{A}}\mathbf{A}\right)+\left(\mathbf{B}-\frac{\mathbf{A}\cdot\mathbf{B}}{\mathbf{A}\cdot\mathbf{A}}\mathbf{A}\right)=\frac{14}{3}[\mathbf{i}+2\mathbf{j}-\mathbf{k}]+\left[(8\mathbf{i}+4\mathbf{j}-12\mathbf{k})-\left(\frac{14}{3}\mathbf{i}+\frac{28}{3}\mathbf{j}-\frac{14}{3}\mathbf{k}\right)\right]=\left[\frac{14}{3}\mathbf{i}+\frac{28}{3}\mathbf{j}-\frac{14}{3}\mathbf{k}\right]+\left[\frac{10}{3}\mathbf{i}-\frac{16}{3}\mathbf{j}-\frac{22}{3}\mathbf{k}\right]$, where $\mathbf{A}\cdot\mathbf{B}=28$ and $\mathbf{A}\cdot\mathbf{A}=6$

17. $(\mathbf{i}+2\mathbf{j})\cdot\left((x-2)\mathbf{i}+(y-1)\mathbf{j}\right)=0\Rightarrow x+2y=4$

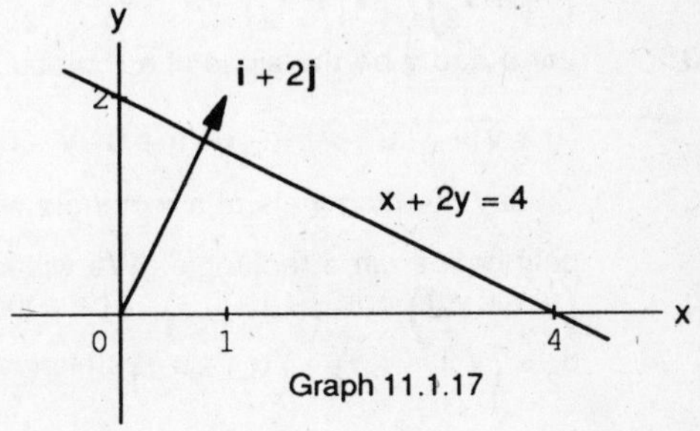

Graph 11.1.17

19. $(-2\mathbf{i}+\mathbf{j})\cdot\left((x+2)\mathbf{i}+(y+7)\mathbf{j}\right)=0\Rightarrow -2x-4+y+7=0\Rightarrow$
$-2x+y=-3$

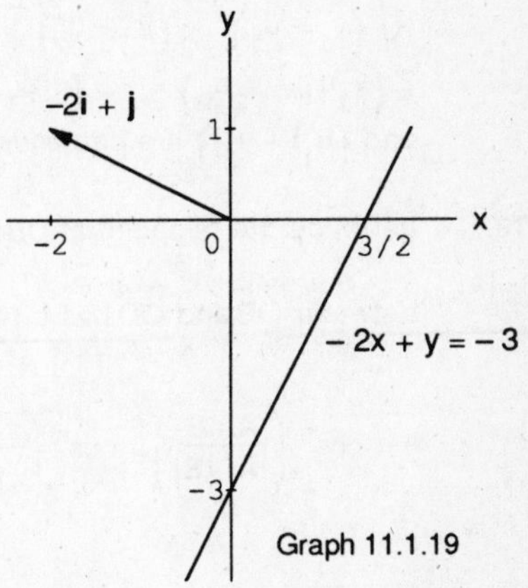

Graph 11.1.19

21. distance $= \left|\text{proj}_{\mathbf{N}} \overrightarrow{PS}\right| = \left|\dfrac{\mathbf{N}\cdot\overrightarrow{PS}}{|\mathbf{N}|}\right| = \left|\dfrac{(\mathbf{i}+3\mathbf{j})\cdot(2\mathbf{i}+6\mathbf{j})}{\sqrt{1^2+3^2}}\right| = \left|\dfrac{20}{\sqrt{10}}\right| = 2\sqrt{10}$, where S(2,8), P(0,2) and $\mathbf{N} = \mathbf{i} + 3\mathbf{j}$

23. distance $= \left|\text{proj}_{\mathbf{N}} \overrightarrow{PS}\right| = \left|\dfrac{\mathbf{N}\cdot\overrightarrow{PS}}{|\mathbf{N}|}\right| = \left|\dfrac{(\mathbf{i}+\mathbf{j})\cdot(\mathbf{i}+\mathbf{j})}{\sqrt{1+1}}\right| = \sqrt{2}$, where S(2,1), P(1,0) and $\mathbf{N} = \mathbf{i} + \mathbf{j}$

25. The distance between the parallel lines is $\text{proj}_{\mathbf{N}} \overrightarrow{P_1P_2}$ where $\mathbf{N}$ is normal to the given lines and each line contains only one of the points P_1 or P_2. Let P_1 be located at (1,2), P_2 at (1,1) and $\mathbf{N} = 3\mathbf{i} + 2\mathbf{j}$. The distance is

$$\left|\text{proj}_{\mathbf{N}} \overrightarrow{P_1P_2}\right| = \left|\frac{\overrightarrow{P_1P_2}\cdot\mathbf{N}}{\mathbf{N}\cdot\mathbf{N}}\mathbf{N}\right| = \left|\frac{(-\mathbf{j})\cdot(3\mathbf{i}+2\mathbf{j})}{(3\mathbf{i}+2\mathbf{j})\cdot(3\mathbf{i}+2\mathbf{j})}(3\mathbf{i}+2\mathbf{j})\right| = \left|\frac{-2}{9+4}(3\mathbf{i}+2\mathbf{j})\right| = \frac{2}{13}\sqrt{9+4} = \frac{2}{\sqrt{13}}.$$

27. $\mathbf{A}\cdot\mathbf{B} = \left(\frac{1}{\sqrt{3}}\right)(0) + \left(-\frac{1}{\sqrt{3}}\right)\left(\frac{1}{\sqrt{2}}\right) + \left(\frac{1}{\sqrt{3}}\right)\left(\frac{1}{\sqrt{2}}\right) = 0$, $\mathbf{A}\cdot\mathbf{C} = \left(\frac{1}{\sqrt{3}}\right)\left(-\frac{2}{\sqrt{6}}\right) + \left(-\frac{1}{\sqrt{3}}\right)\left(-\frac{1}{\sqrt{6}}\right) + \left(\frac{1}{\sqrt{3}}\right)\left(\frac{1}{\sqrt{6}}\right) = 0$ and $\mathbf{B}\cdot\mathbf{C} =$ $(0)\left(-\frac{2}{\sqrt{6}}\right) + \left(\frac{1}{\sqrt{2}}\right)\left(-\frac{1}{\sqrt{6}}\right) + \left(\frac{1}{\sqrt{2}}\right)\left(\frac{1}{\sqrt{6}}\right) = 0 \Rightarrow$ that $\mathbf{A}$, $\mathbf{B}$ and $\mathbf{C}$ are mutually orthogonal

29. a) Since $|\cos\theta| \le 1$, we have $|\mathbf{u}\cdot\mathbf{v}| = |\mathbf{u}|\,|\mathbf{v}|\,|\cos\theta| \le |\mathbf{u}|\,|\mathbf{v}|\,(1) = |\mathbf{u}|\,|\mathbf{v}|$.

b) We have equality precisely when $|\cos\theta| = 1$ or when one or both of $\mathbf{u}$ and $\mathbf{v}$ is $\mathbf{0}$. In the case of nonzero vectors, we have equality when $\theta = 0$ or π, i.e., when the vectors are parallel.

31. $\mathbf{v}\cdot\mathbf{u}_1 = (a\,\mathbf{u}_1 + b\,\mathbf{u}_2)\cdot\mathbf{u}_1 = a\,\mathbf{u}_1\cdot\mathbf{u}_1 + b\,\mathbf{u}_2\cdot\mathbf{u}_1 = a\,|\mathbf{u}_1|^2 + b\,(\mathbf{u}_2\cdot\mathbf{u}_1) = a(1)^2 + b\,(0) = a$

33. The sum of two vectors of equal length is *always* orthogonal to their difference, as we can see from the equation $(\mathbf{v}_1+\mathbf{v}_2)\cdot(\mathbf{v}_1-\mathbf{v}_2) = \mathbf{v}_1\cdot\mathbf{v}_1 + \mathbf{v}_2\cdot\mathbf{v}_1 - \mathbf{v}_1\cdot\mathbf{v}_2 - \mathbf{v}_2\cdot\mathbf{v}_2 = |\mathbf{v}_1|^2 - |\mathbf{v}_2|^2$

35. Let $\mathbf{u}$ and $\mathbf{v}$ be the sides of a rhombus. Where the diagonals $\mathbf{d}_1 = \mathbf{u} + \mathbf{v}$ and $\mathbf{d}_2 = -\mathbf{u} + \mathbf{v}$. $\mathbf{d}_1\cdot\mathbf{d}_2 =$ $(\mathbf{u}+\mathbf{v})\cdot(-\mathbf{u}+\mathbf{v}) = -\mathbf{u}\cdot\mathbf{u} + \mathbf{u}\cdot\mathbf{v} - \mathbf{u}\cdot\mathbf{v} + \mathbf{v}\cdot\mathbf{v} = |\mathbf{v}|^2 - |\mathbf{u}|^2 = 0$ because $|\mathbf{u}| = |\mathbf{v}|$, a rhombus has equal sides.

37. Clearly the diagonals of a rectangle are equal in length. What is not as obvious is the statement that equal diagonals only happen in a rectangle. We will show this by letting the opposite sides of a parallelogram be the vectors $(v_1\mathbf{i} + v_2\mathbf{j})$ and $(u_1\mathbf{i} + u_2\mathbf{j})$. The equal diagonals of the parallellogram are $\mathbf{d}_1 = (v_1\mathbf{i} + v_2\mathbf{j}) + (u_1\mathbf{i} + u_2\mathbf{j})$ and $\mathbf{d}_2 = (v_1\mathbf{i} + v_2\mathbf{j}) - (u_1\mathbf{i} + u_2\mathbf{j})$. Hence $|\mathbf{d}_1| = |\mathbf{d}_2| \Rightarrow |(v_1\mathbf{i} + v_2\mathbf{j}) + (u_1\mathbf{i} + u_2\mathbf{j})| = |(v_1\mathbf{i} + v_2\mathbf{j}) - (u_1\mathbf{i} + u_2\mathbf{j})| \Rightarrow$ $|(v_1+u_1)\mathbf{i} + (v_2+u_2)\mathbf{j}| = |(v_1-u_1)\mathbf{i} + (v_2-u_2)\mathbf{j}| \Rightarrow \sqrt{(v_1+u_1)^2 + (v_2+u_2)^2} =$ $\sqrt{(v_1-u_1)^2 + (v_2-u_2)^2} \Rightarrow v_1^2 + 2v_1u_1 + u_1^2 + v_2^2 + 2v_2u_2 + u_2^2 = v_1^2 - 2v_1u_1 + u_1^2 + v_2^2 - 2v_2u_2 + u_2^2 \Rightarrow$ $2(v_1u_1 + v_2u_2) = -2(v_1u_1 + v_2u_2) \Rightarrow v_1u_1 + v_2u_2 = 0 \Rightarrow (v_1\mathbf{i} + v_2\mathbf{j})\cdot(u_1\mathbf{i} + u_2\mathbf{j}) = 0 \Rightarrow$ the vectors $(v_1\mathbf{i} + v_2\mathbf{j})$ and $(u_1\mathbf{i} + u_2\mathbf{j})$ are perpendicular and the parallelogram must be a rectangle.

39. Let M be the midpoint of OB. By the Pythagorean Theorem $OB = \sqrt{1^2+1^2} = \sqrt{2}$ and $OM = \frac{\sqrt{2}}{2}$. Hence the angle θ between $\overrightarrow{OB}$ and $\overrightarrow{OD}$ has a tangent of $\dfrac{DM}{OB} = \dfrac{1}{\frac{\sqrt{2}}{2}} = \dfrac{2}{\sqrt{2}} = \sqrt{2}$. Therefore, $\tan\theta = \sqrt{2} \Rightarrow \theta = \tan^{-1}\sqrt{2} \approx 54.7°$.

41. $\theta = \cos^{-1}\left(\dfrac{\mathbf{A}\cdot\mathbf{B}}{|\mathbf{A}|\,|\mathbf{B}|}\right) = \cos^{-1}\left(\dfrac{(2)(1)+(1)(2)+(0)(-1)}{\sqrt{2^2+1^2+0^2}\sqrt{1^2+2^2+(-1)^2}}\right) = \cos^{-1}\left(\dfrac{4}{\sqrt{5}\sqrt{6}}\right) = \cos^{-1}\left(\dfrac{4}{\sqrt{30}}\right) \approx 0.75$ rad

43. $\theta = \cos^{-1}\left(\frac{\mathbf{A}\cdot\mathbf{B}}{|\mathbf{A}|\,|\mathbf{B}|}\right) = \cos^{-1}\left(\frac{(\sqrt{3})(\sqrt{3}) + (-7)(1) + (0)(-2)}{\sqrt{(\sqrt{3})^2 + (-7)^2 + 0^2}\sqrt{(\sqrt{3})^2 + (1)^2 + (-2)^2}}\right) = \cos^{-1}\left(\frac{3-7}{\sqrt{52}\,\sqrt{8}}\right) =$

$\cos^{-1}\left(\frac{-1}{\sqrt{26}}\right) \approx 1.77$ rad

45. $\overrightarrow{AB} = 3\mathbf{i} + \mathbf{j} - 3\mathbf{k}$, $\overrightarrow{AC} = 2\mathbf{i} - 2\mathbf{j}$, $\overrightarrow{BA} = -3\mathbf{i} - \mathbf{j} + 3\mathbf{k}$, $\overrightarrow{CA} = -2\mathbf{i} + 2\mathbf{j}$, $\overrightarrow{CB} = \mathbf{i} + 3\mathbf{j} - 3\mathbf{k}$, $\overrightarrow{BC} = -\mathbf{i} - 3\mathbf{j} + 3\mathbf{k} \Rightarrow \angle A =$ $\cos^{-1}\left(\frac{\overrightarrow{AB}\cdot\overrightarrow{AC}}{\left|\overrightarrow{AB}\right|\left|\overrightarrow{AC}\right|}\right) = \cos^{-1}\left(\frac{4}{\sqrt{152}}\right) \approx 1.24 \text{ rad} \approx 71.07°$, $\angle B = \cos^{-1}\left(\frac{\overrightarrow{BA}\cdot\overrightarrow{BC}}{\left|\overrightarrow{BA}\right|\left|\overrightarrow{BC}\right|}\right) =$ $\cos^{-1}\left(\frac{15}{19}\right) \approx 0.66 \text{ rad} \approx 37.86°$, $\angle C = \cos^{-1}\left(\frac{\overrightarrow{CA}\cdot\overrightarrow{CB}}{\left|\overrightarrow{CA}\right|\left|\overrightarrow{CB}\right|}\right) = \cos^{-1}\left(\frac{4}{\sqrt{152}}\right) \approx 1.24 \text{ rad} \approx 71.07°$

47. $\theta = \cos^{-1}\left(\frac{\mathbf{A}\cdot\mathbf{B}}{|\mathbf{A}|\,|\mathbf{B}|}\right) = \cos^{-1}\left(\frac{2}{\sqrt{2}\sqrt{3}}\right) \approx 0.62 \text{ rad} \approx 35.26°$, where $\mathbf{A} = \mathbf{i} + \mathbf{k}$ and $\mathbf{B} = \mathbf{i} + \mathbf{j} + \mathbf{k}$

49. $P(0,0,0)$, $Q(1,1,1)$ and $\mathbf{F} = 5\mathbf{k} \Rightarrow \overrightarrow{PQ} = \mathbf{i} + \mathbf{j} + \mathbf{k}$, $W = \mathbf{F}\cdot\overrightarrow{PQ} = (5\mathbf{k})\cdot(\mathbf{i} + \mathbf{j} + \mathbf{k}) = 5 \text{ N}\cdot\text{m} = 5 \text{ J}$

51. $W = |\mathbf{F}|\left|\overrightarrow{PQ}\right|\cos\theta = (200)(20)\left(\cos\frac{\pi}{6}\right) = 2000\sqrt{3} \approx 3464.10 \text{ N}\cdot\text{m} = 3464.10 \text{ J}$

53. The angle between the corresponding normals is equal to the angle between the corresponding tangents. $\theta = \cos^{-1}\left(\frac{\mathbf{N}_1\cdot\mathbf{N}_2}{|\mathbf{N}_1|\,|\mathbf{N}_2|}\right) = \cos^{-1}\left(\frac{1}{\sqrt{2}}\right) = 45° = \pi/4$, where $\mathbf{N}_1 = 3\mathbf{i} + \mathbf{j}$, $\mathbf{N}_2 = 2\mathbf{i} - \mathbf{j}$.

55. The angle between the corresponding normals is equal to the angle between the corresponding tangents. $\theta = \cos^{-1}\left(\frac{\mathbf{N}_1\cdot\mathbf{N}_2}{|\mathbf{N}_1|\,|\mathbf{N}_2|}\right) = \cos^{-1}\left(\frac{\sqrt{3}+\sqrt{3}}{\sqrt{4}\,\sqrt{4}}\right) = \cos^{-1}\left(\frac{2\sqrt{3}}{4}\right) = 30° = \pi/6$, where $\mathbf{N}_1 = \sqrt{3}\,\mathbf{i} - \mathbf{j}$, $\mathbf{N}_2 = \mathbf{i} - \sqrt{3}\,\mathbf{j}$.

57. The angle between the corresponding normals is equal to the angle between the corresponding tangents. $\theta = \cos^{-1}\left(\frac{\mathbf{N}_1\cdot\mathbf{N}_2}{|\mathbf{N}_1|\,|\mathbf{N}_2|}\right) = \cos^{-1}\left(\frac{3+4}{\sqrt{9+16}\,\sqrt{1+1}}\right) = \cos^{-1}\left(\frac{7}{5\sqrt{2}}\right) \approx 0.14$ rad, where $\mathbf{N}_1 = 3\,\mathbf{i} - 4\,\mathbf{j}$, $\mathbf{N}_2 = \mathbf{i} - \mathbf{j}$.

59. The angle between the corresponding normals is equal to the angle between the corresponding tangents. The points of intersection are $\left(-\frac{\sqrt{3}}{2}, \frac{3}{4}\right)$ and $\left(\frac{\sqrt{3}}{2}, \frac{3}{4}\right)$. At $\left(-\frac{\sqrt{3}}{2}, \frac{3}{4}\right)$ the tangent line for $f(x) = x^2$ is $y - \frac{3}{4} =$ $f'\left(-\frac{3}{2}\right)\left(x - \left(-\frac{\sqrt{3}}{2}\right)\right) \Rightarrow y = -\sqrt{3}\left(x + \frac{\sqrt{3}}{2}\right) + \frac{3}{4} \Rightarrow y = -\sqrt{3}\,x - \frac{3}{4}$ and the tangent line for $f(x) = \left(\frac{3}{2}\right) - x^2$ is $y - \frac{3}{4} = f'\left(-\frac{\sqrt{3}}{2}\right)\left(x - \left(-\frac{\sqrt{3}}{2}\right)\right) \Rightarrow y = \sqrt{3}\left(x + \frac{\sqrt{3}}{2}\right) + \frac{3}{4} = \sqrt{3}\,x + \frac{9}{4}$. The corresponding normals are $\mathbf{N}_1 = \sqrt{3}\,\mathbf{i} +$ $\mathbf{j}$ and $\mathbf{N}_2 = -\sqrt{3}\,\mathbf{i} + \mathbf{j}$. The angle at $\left(-\frac{\sqrt{3}}{2}, \frac{3}{4}\right)$ is $\theta = \cos^{-1}\left(\frac{\mathbf{N}_1\cdot\mathbf{N}_2}{|\mathbf{N}_1|\,|\mathbf{N}_2|}\right) = \cos^{-1}\left(\frac{-3+1}{\sqrt{4}\,\sqrt{4}}\right) = \cos^{-1}\left(-\frac{1}{2}\right) = \frac{2\pi}{3}$, the angle is either $\frac{\pi}{3}$ or $\frac{2\pi}{3}$. At $\left(\frac{\sqrt{3}}{2}, \frac{3}{4}\right)$ the tangent line for $f(x) = x^2$ is $y = \sqrt{3}\left(x + \frac{\sqrt{3}}{2}\right) + \frac{3}{4} = \sqrt{3}\,x + \frac{9}{4}$ and the tangent line for $f(x) = \left(\frac{3}{2}\right) - x^2$ is $y = -\sqrt{3}\left(x + \frac{\sqrt{3}}{2}\right) + \frac{3}{4} = -\sqrt{3}\,x - \frac{3}{4}$. The corresponding normals are $\mathbf{N}_1 = \sqrt{3}\,\mathbf{i} - \mathbf{j}$ and $\mathbf{N}_2 = -\sqrt{3}\,\mathbf{i} - \mathbf{j}$. The angle at $\left(\frac{\sqrt{3}}{2}, \frac{3}{4}\right)$ is $\theta = \cos^{-1}\left(\frac{\mathbf{N}_1\cdot\mathbf{N}_2}{|\mathbf{N}_1|\,|\mathbf{N}_2|}\right) = \cos^{-1}\left(\frac{-3+1}{\sqrt{4}\,\sqrt{4}}\right) =$ $\cos^{-1}\left(-\frac{1}{2}\right) = \frac{2\pi}{3}$, the angle is either $\frac{\pi}{3}$ or $\frac{2\pi}{3}$.

61. The intersection of $y = x^3$ and $x = y^2$ don't occur where $y < 0$. Therefore, we will consider $y = x^3$ and $y = \sqrt{x}$. The points of intersection for the curves $y = x^3$ and $y = \sqrt{x}$ are (0,0) and (1,1). At (0,0) the tangent line for $y = x^3$ is $y = 0$ and the tangent line for $y = \sqrt{x}$ is $x = 0$. Therefore, the angle of intersection at (0,0) is $\frac{\pi}{2}$. At (1,1) the tangent line for $y = x^3$ is $y = 3x - 2$ and the tangent line for $y = \sqrt{x}$ is $y = \frac{1}{2}x + \frac{1}{2}$. The corresponding normal vectors are: $\mathbf{N}_1 = -3\mathbf{i} + \mathbf{j}$, $\mathbf{N}_2 = -\frac{1}{2}\mathbf{i} + \mathbf{j}$. $\theta = \cos^{-1}\left(\frac{\mathbf{N}_1 \cdot \mathbf{N}_2}{|\mathbf{N}_1|\,|\mathbf{N}_2|}\right) = \cos^{-1}\frac{1}{\sqrt{2}} = \frac{\pi}{4}$; the angle is either $45° = \frac{\pi}{4}$ or $135° = \frac{3\pi}{4}$.

11.4 CROSS PRODUCTS

1. $\mathbf{A} \times \mathbf{B} = \begin{vmatrix} \mathbf{i} & \mathbf{j} & \mathbf{k} \\ 2 & -2 & -1 \\ 1 & 0 & -1 \end{vmatrix} = 3\left[\frac{2}{3}\mathbf{i} + \frac{1}{3}\mathbf{j} + \frac{2}{3}\mathbf{k}\right] \Rightarrow$ length = 3 and the direction is $\frac{2}{3}\mathbf{i} + \frac{1}{3}\mathbf{j} + \frac{2}{3}\mathbf{k}$

$\mathbf{B} \times \mathbf{A} = \begin{vmatrix} \mathbf{i} & \mathbf{j} & \mathbf{k} \\ 1 & 0 & -1 \\ 2 & -2 & -1 \end{vmatrix} = -3\left[\frac{2}{3}\mathbf{i} + \frac{1}{3}\mathbf{j} + \frac{2}{3}\mathbf{k}\right] \Rightarrow$ length = 3 and the direction is $-\frac{2}{3}\mathbf{i} - \frac{1}{3}\mathbf{j} - \frac{2}{3}\mathbf{k}$

3. $\mathbf{A} \times \mathbf{B} = \begin{vmatrix} \mathbf{i} & \mathbf{j} & \mathbf{k} \\ 2 & -2 & 4 \\ -1 & 1 & -2 \end{vmatrix} = \mathbf{0} \Rightarrow$ length = 0 and has no direction

$\mathbf{B} \times \mathbf{A} = \begin{vmatrix} \mathbf{i} & \mathbf{j} & \mathbf{k} \\ -1 & 1 & -2 \\ 2 & -2 & 4 \end{vmatrix} = \mathbf{0} \Rightarrow$ length = 0 and has no direction

5. $\mathbf{A} \times \mathbf{B} = \begin{vmatrix} \mathbf{i} & \mathbf{j} & \mathbf{k} \\ 2 & 0 & 0 \\ 0 & -3 & 0 \end{vmatrix} = -6\,[\mathbf{k}] \Rightarrow$ length = 6 and the direction is $-\mathbf{k}$

$\mathbf{B} \times \mathbf{A} = \begin{vmatrix} \mathbf{i} & \mathbf{j} & \mathbf{k} \\ 0 & -3 & 0 \\ 2 & 0 & 0 \end{vmatrix} = 6\,[\mathbf{k}] \Rightarrow$ length = 6 and the direction is $\mathbf{k}$

7. $\mathbf{A} \times \mathbf{B} = \begin{vmatrix} \mathbf{i} & \mathbf{j} & \mathbf{k} \\ -8 & -2 & -4 \\ 2 & 2 & 1 \end{vmatrix} = [6\mathbf{i} - 12\mathbf{k}] \Rightarrow$ length = $6\sqrt{5}$ and the direction is $\frac{1}{\sqrt{5}}\mathbf{i} - \frac{2}{\sqrt{5}}\mathbf{k}$

$\mathbf{B} \times \mathbf{A} = \begin{vmatrix} \mathbf{i} & \mathbf{j} & \mathbf{k} \\ 2 & 2 & 1 \\ -8 & -2 & -4 \end{vmatrix} = -[6\mathbf{i} - 12\mathbf{k}] \Rightarrow$ length = $6\sqrt{5}$ and the direction is $-\frac{1}{\sqrt{5}}\mathbf{i} + \frac{2}{\sqrt{5}}\mathbf{k}$

9. $\mathbf{A} \times \mathbf{B} = \begin{vmatrix} \mathbf{i} & \mathbf{j} & \mathbf{k} \\ 1 & 0 & 0 \\ 0 & 1 & 0 \end{vmatrix} = \mathbf{k}$

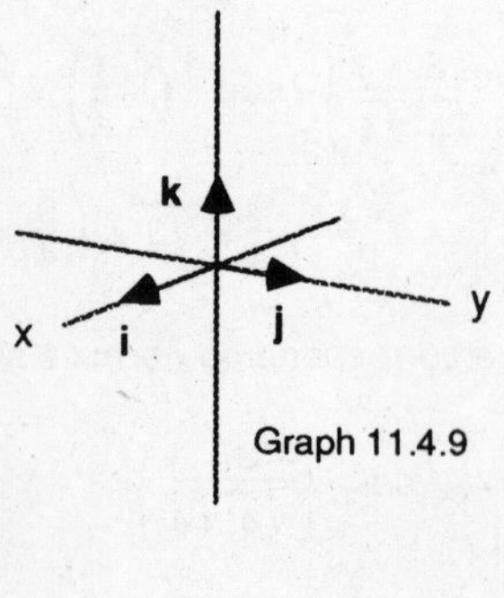

Graph 11.4.9

11. $\mathbf{A} \times \mathbf{B} = \begin{vmatrix} \mathbf{i} & \mathbf{j} & \mathbf{k} \\ 1 & 0 & -1 \\ 0 & 1 & 1 \end{vmatrix} = \mathbf{i} - \mathbf{j} + \mathbf{k}$

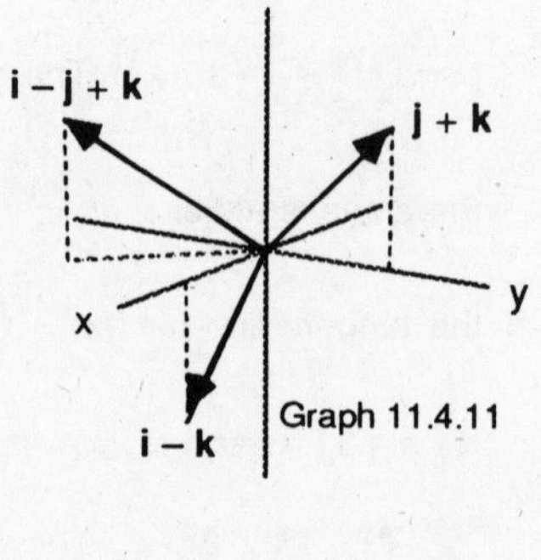

Graph 11.4.11

13. $\mathbf{A} \times \mathbf{B} = \begin{vmatrix} \mathbf{i} & \mathbf{j} & \mathbf{k} \\ 1 & 1 & 0 \\ 1 & -1 & 0 \end{vmatrix} = -2\mathbf{k}$

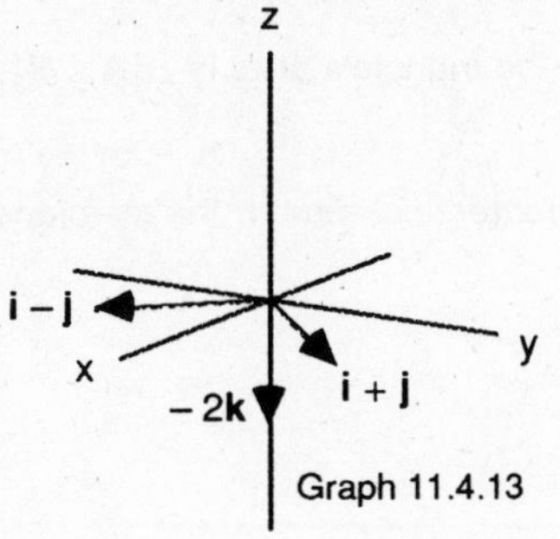

Graph 11.4.13

15. a) $\dfrac{\left|\overrightarrow{PQ} \times \overrightarrow{PR}\right|}{2} = \dfrac{\sqrt{64+16+16}}{2} = 2\sqrt{6}$

b) $\pm\dfrac{\overrightarrow{PQ} \times \overrightarrow{PR}}{\left|\overrightarrow{PQ} \times \overrightarrow{PR}\right|} = \pm\left[\dfrac{2}{\sqrt{6}}\mathbf{i} + \dfrac{1}{\sqrt{6}}\mathbf{j} + \dfrac{1}{\sqrt{6}}\mathbf{k}\right]$

17. a) $\dfrac{\left|\overrightarrow{PQ} \times \overrightarrow{PR}\right|}{2} = \dfrac{\sqrt{2}}{2}$

b) $\pm\dfrac{\overrightarrow{PQ} \times \overrightarrow{PR}}{\left|\overrightarrow{PQ} \times \overrightarrow{PR}\right|} = \pm\left[\dfrac{-1}{\sqrt{2}}\mathbf{i} + \dfrac{1}{\sqrt{2}}\mathbf{j}\right]$

19. a) $\mathbf{A} \cdot \mathbf{B} = -6,\ \mathbf{A} \cdot \mathbf{C} = -81,\ \mathbf{B} \cdot \mathbf{C} = 18 \Rightarrow$ none

b) $\mathbf{A} \times \mathbf{B} = \begin{vmatrix} \mathbf{i} & \mathbf{j} & \mathbf{k} \\ 5 & -1 & 1 \\ 0 & 1 & -5 \end{vmatrix} \neq \mathbf{0},\ \mathbf{A} \times \mathbf{C} = \begin{vmatrix} \mathbf{i} & \mathbf{j} & \mathbf{k} \\ 5 & -1 & 1 \\ -15 & 3 & -3 \end{vmatrix} = \mathbf{0},\ \mathbf{B} \times \mathbf{C} = \begin{vmatrix} \mathbf{i} & \mathbf{j} & \mathbf{k} \\ 0 & 1 & -5 \\ -15 & 3 & -3 \end{vmatrix} \neq \mathbf{0} \Rightarrow$ **A** and **C** are parallel

21. $\left|\overrightarrow{PQ} \times \mathbf{F}\right| = \left|\overrightarrow{PQ}\right| |\mathbf{F}| \sin(60^\circ) = 10\sqrt{3}$ ft · lb

23. If $\mathbf{A} = a_1\mathbf{i} + a_2\mathbf{j} + a_3\mathbf{k}$, $\mathbf{B} = b_1\mathbf{i} + b_2\mathbf{j} + b_3\mathbf{k}$, and $\mathbf{C} = c_1\mathbf{i} + c_2\mathbf{j} + c_3\mathbf{k}$, then $\mathbf{A} \cdot (\mathbf{B} \times \mathbf{C}) = \begin{vmatrix} a_1 & a_2 & a_3 \\ b_1 & b_2 & b_3 \\ c_1 & c_2 & c_3 \end{vmatrix}$, $\mathbf{B} \cdot (\mathbf{C} \times \mathbf{A}) = \begin{vmatrix} b_1 & b_2 & b_3 \\ c_1 & c_2 & c_3 \\ a_1 & a_2 & a_3 \end{vmatrix}$ and $\mathbf{C} \cdot (\mathbf{A} \times \mathbf{B}) = \begin{vmatrix} c_1 & c_2 & c_3 \\ a_1 & a_2 & a_3 \\ b_1 & b_2 & b_3 \end{vmatrix}$ which all have the same value, since the interchanging of two pair of rows in a determinant does not change its value. The volume is $\left|(\mathbf{A} \times \mathbf{B}) \cdot \mathbf{C}\right| = \begin{vmatrix} 2 & 0 & 0 \\ 0 & 2 & 0 \\ 0 & 0 & 2 \end{vmatrix} = 8$.

25. $\left|(\mathbf{A} \times \mathbf{B}) \cdot \mathbf{C}\right| = \begin{vmatrix} 2 & 1 & 0 \\ 2 & -1 & 1 \\ 1 & 0 & 2 \end{vmatrix} = -7$. For details about verification, see exercise 23.

27. $\dfrac{1}{2}\left|\overrightarrow{AB} \times \overrightarrow{AC}\right| = \dfrac{\sqrt{4+16+16}}{2} = 3$ see example 4 section 11.4

29. a) $\text{proj}_{\mathbf{B}}\mathbf{A} = \dfrac{\mathbf{A} \cdot \mathbf{B}}{\mathbf{B} \cdot \mathbf{B}}\mathbf{B}$ b) $(\pm)(\mathbf{A} \times \mathbf{B})$ c) $(\pm)(\mathbf{A} \times \mathbf{B}) \times \mathbf{C}$ d) $\left|(\mathbf{A} \times \mathbf{B}) \cdot \mathbf{C}\right|$

31. No, **B** need not equal **C**. For example, $\mathbf{i} + \mathbf{j} \neq -\mathbf{i} + \mathbf{j}$, but
$\mathbf{i} \times (\mathbf{i} + \mathbf{j}) = \mathbf{i} \times \mathbf{i} + \mathbf{i} \times \mathbf{j} = \mathbf{0} + \mathbf{k} = \mathbf{k}$ and $\mathbf{i} \times (-\mathbf{i} + \mathbf{j}) = -\mathbf{i} \times \mathbf{i} + \mathbf{i} \times \mathbf{j} = \mathbf{0} + \mathbf{k} = \mathbf{k}$.

33. Let $\overrightarrow{AB} = (0-1)\mathbf{i} + (1-0)\mathbf{j} = -\mathbf{i} + \mathbf{j}$; $\overrightarrow{AD} = (0-1)\mathbf{i} + (-1-0)\mathbf{j} = -\mathbf{i} - \mathbf{j}$, $\overrightarrow{AB} \times \overrightarrow{AD} = (-\mathbf{i} + \mathbf{j}) \times (-\mathbf{i} - \mathbf{j}) = 2\mathbf{k}$.
Area $= |2\mathbf{k}| = 2$.

35. Let $\overrightarrow{AB} = (2+1)\mathbf{i} + (0-2)\mathbf{j} = 3\mathbf{i} - 2\mathbf{j}$; $\overrightarrow{AD} = (4+1)\mathbf{i} + (3-2)\mathbf{j} = 5\mathbf{i} + \mathbf{j}$, $\overrightarrow{AB} \times \overrightarrow{AD} = (3\mathbf{i} - 2\mathbf{j}) \times (5\mathbf{i} + \mathbf{j}) = 13\mathbf{k}$.
Area $= |13\mathbf{k}| = 13$.

37. Let $\overrightarrow{AB} = (-2-0)\mathbf{i} + (3-0)\mathbf{j} = -2\mathbf{i} + 3\mathbf{j}$; $\overrightarrow{AC} = (3-0)\mathbf{i} + (1-0)\mathbf{j} = 3\mathbf{i} + \mathbf{j}$, $\overrightarrow{AB} \times \overrightarrow{AC} = (-2\mathbf{i} + 3\mathbf{j}) \times (3\mathbf{i} + \mathbf{j}) = -11\mathbf{k}$.
Area $= \frac{1}{2}|-11\mathbf{k}| = \frac{11}{2}$.

39. Let $\overrightarrow{AB} = (1+5)\mathbf{i} + (-2-3)\mathbf{j} = 6\mathbf{i} - 5\mathbf{j}$; $\overrightarrow{AC} = (6+5)\mathbf{i} + (-2-3)\mathbf{j} = 11\mathbf{i} - 5\mathbf{j}$, $\overrightarrow{AB} \times \overrightarrow{AC} = (6\mathbf{i} - 5\mathbf{j}) \times (11\mathbf{i} - 5\mathbf{j}) = 25\mathbf{k}$. Area $= \frac{1}{2}|25\mathbf{k}| = \frac{25}{2}$.

41. If $\mathbf{A} = a_1\mathbf{i} + a_2\mathbf{j}$ and $\mathbf{B} = b_1\mathbf{i} + b_2\mathbf{j}$, then $\mathbf{A} \times \mathbf{B} = \begin{vmatrix} \mathbf{i} & \mathbf{j} & \mathbf{k} \\ a_1 & a_2 & 0 \\ b_1 & b_2 & 0 \end{vmatrix} = \begin{vmatrix} a_1 & a_2 \\ b_1 & b_2 \end{vmatrix} \mathbf{k}$ and the triangle's area is $\frac{1}{2}|\mathbf{A} \times \mathbf{B}| = \pm\frac{1}{2}\begin{vmatrix} a_1 & a_2 \\ b_1 & b_2 \end{vmatrix}$. The applicable sign is (+) if the acute angle from **A** to **B** runs counterclockwise in the xy–plane, and (–) if it runs clockwise.

11.5 LINES AND PLANES IN SPACE

1. the direction $\mathbf{i} + \mathbf{j} + \mathbf{k}$ and $P(3,-4,-1) \Rightarrow x = 3 + t,\ y = -4 + t,\ z = -1 + t$

3. the direction $\overrightarrow{PQ} = 5\mathbf{i} + 5\mathbf{j} - 5\mathbf{k}$ and $P(-2,0,3) \Rightarrow x = -2 + 5t,\ y = 5t,\ z = 3 - 5t$

5. the direction $2\mathbf{j} + \mathbf{k}$ and $P(0,0,0) \Rightarrow x = 0,\ y = 2t,\ z = t$

7. the direction $\mathbf{k}$ and $P(1,1,1) \Rightarrow x = 1,\ y = 1,\ z = 1 + t$

9. the direction $\mathbf{i} + 2\mathbf{j} + 2\mathbf{k}$ and $(0,-7,0) \Rightarrow x = t,\ y = -7 + 2t,\ z = 2t$

11. the direction $\mathbf{i}$ and $P(0,0,0) \Rightarrow x = t,\ y = 0,\ z = 0$

13. the direction $\overrightarrow{PQ} = \mathbf{i} + \mathbf{j} + 3/2\mathbf{k}$ and $P(0,0,0) \Rightarrow x = t,\ y = t,\ z = 3/2\ t$ where $0 \le t \le 1$

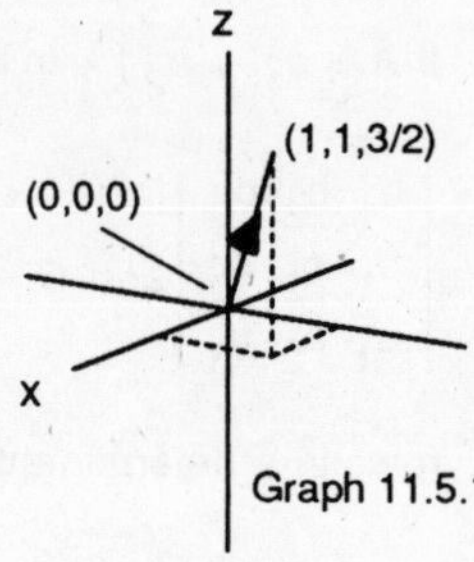

Graph 11.5.13

15. the direction $\overrightarrow{PQ} = \mathbf{j}$ and $P(1,0,0) \Rightarrow x = 1,\ y = 1 + t,\ z = 0$, where $-1 \le t \le 0$

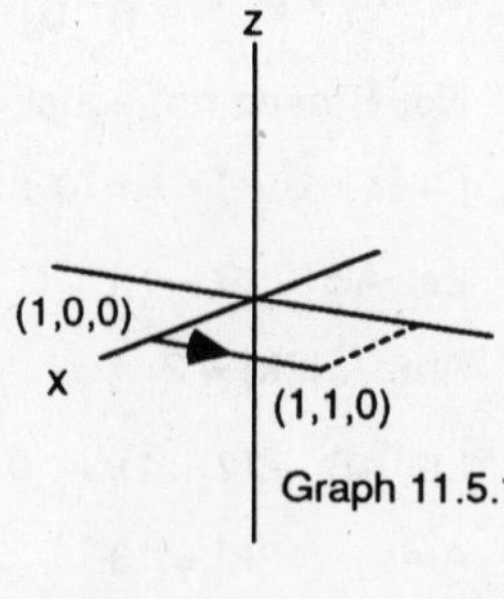

Graph 11.5.15

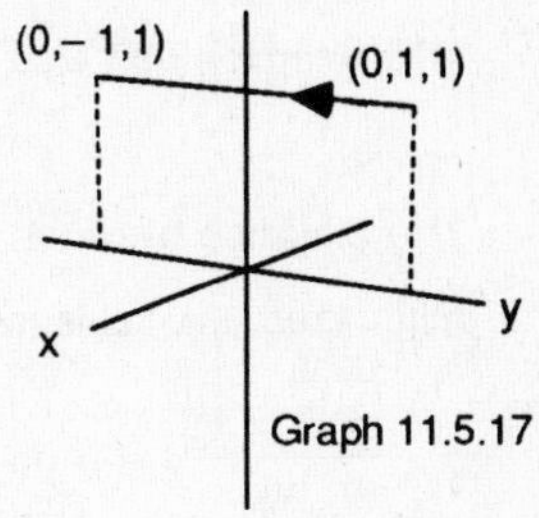

Graph 11.5.17

17. the direction $\overrightarrow{PQ} = -2\mathbf{j}$ and $P(0,1,1) \Rightarrow x = 0,\ y = 1 - 2t,\ z = 1$, where $0 \le t \le 1$

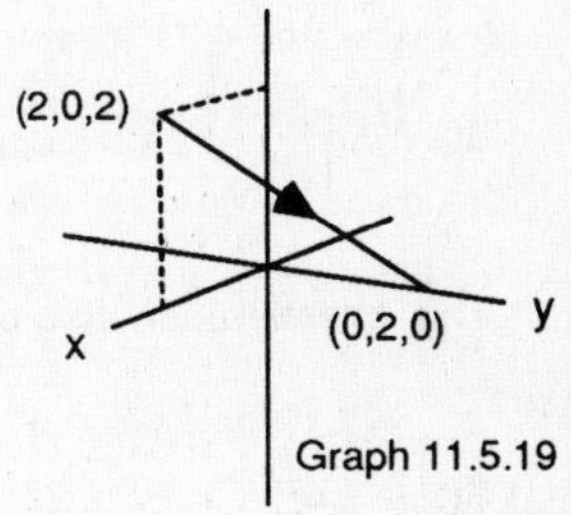

Graph 11.5.19

19. the direction $\overrightarrow{PQ} = (0-2)\mathbf{i} + (2-0)\mathbf{j} + (0-2)\mathbf{k} = -2\mathbf{i} + 2\mathbf{j} - 2\mathbf{k}$ and $P(2,0,2) \Rightarrow x = 2 - 2t,\ y = 2t,\ z = 2 - 2t$, where $0 \le t \le 1$

21. $3(x) + (-2)(y-2) + (-1)(z+1) = 0 \Rightarrow 3x - 2y - z = -3$

23. $\overrightarrow{PQ} = \mathbf{i} - \mathbf{j} + 3\mathbf{k},\ \overrightarrow{PS} = -\mathbf{i} - 3\mathbf{j} + 2\mathbf{k} \Rightarrow \overrightarrow{PQ} \times \overrightarrow{PS} = \begin{vmatrix} \mathbf{i} & \mathbf{j} & \mathbf{k} \\ 1 & -1 & 3 \\ -1 & -3 & 2 \end{vmatrix} = 7\mathbf{i} - 5\mathbf{j} - 4\mathbf{k}$, the normal;

$(x-2)(7) + (y-0)(-5) + (z-2)(-4) = 0 \Rightarrow 7x - 5y - 4z = 6$

25. $\mathbf{N} = \mathbf{i} + 3\mathbf{j} + 4\mathbf{k},\ P(2,4,5) \Rightarrow (x-2)(1) + (y-4)(3) + (z-5)(4) = 0 \Rightarrow x + 3y + 4z = 34$

27. The vector $2\mathbf{i} - 4\mathbf{j} + \mathbf{k}$ determined by $(1,2,3)$ and a point on the line $(-1,6,2)$ and a vector in the direction of the line $3\mathbf{i} + \mathbf{k}$ are in the desired plane. The cross product of these vectors has the same direction as the normal of the plane. $\mathbf{N} = \begin{vmatrix} \mathbf{i} & \mathbf{j} & \mathbf{k} \\ 3 & 0 & 1 \\ 2 & -4 & 1 \end{vmatrix} = 4\mathbf{i} - \mathbf{j} - 12\mathbf{k}$. Therefore, the desired plane is $\mathbf{N} \cdot \left[(x-1)\mathbf{i} + (y-2)\mathbf{j} + (z-3)\mathbf{k}\right] = 0 \Rightarrow$

$4(x-1) - 1(y-2) - 12(z-3) = 0 \Rightarrow 4x - 4 - y + 2 - 12z + 36 = 0 \Rightarrow 4x - y - 12z = -34$.

29. $x = 2 + 2t,\ y = -4 - t,\ z = 7 + 3t;\ x = -2 - t,\ y = -2 + (1/2)t,\ z = 1 - (3/2)t$

31. The distance between $(4t,-2t,2t)$ and $(0,0,12)$ is $d = \sqrt{(4t)^2 + (-2t)^2 + (2t-12)^2}$. If $f(t) =$

$(4t)^2 + (-2t)^2 + (2t-12)^2$ is minimized, then d is minimized. $f'(t) = 0 \Rightarrow t = 1 \Rightarrow d = \sqrt{16+4+100} = 2\sqrt{30}$.

This exercise can also be done with formula (7), $d = \dfrac{|\overrightarrow{PS} \times \mathbf{v}|}{|\mathbf{v}|}$, on page 726 of the text.

33. The distance between (2 + 2t,1 + 6t,3) and (2,1,3) is $d = \sqrt{(2t)^2 + (6t)^2}$. If $f(t) = (2t)^2 + (6t)^2$ is minimized, then d is minimized. $f'(t) = 0 \Rightarrow t = 0 \Rightarrow d = 0$. This exercise can also be done with formula (7), $d = \dfrac{|\overrightarrow{PS} \times \mathbf{v}|}{|\mathbf{v}|}$, on page 726 of the text.

35. The distance between (4 – t,3 + 2t,– 5 + 3t) and (3,– 1,4) is $d = \sqrt{(1-t)^2 + (4+2t)^2 + (3t-9)^2}$. If $f(t) = (1-t)^2 + (4+2t)^2 + (3t-9)^2$ is minimized, then d is minimized. $f'(t) = 0 \Rightarrow t = \frac{10}{7} \Rightarrow$

$d = \sqrt{\left(\frac{3}{7}\right)^2 + \left(\frac{48}{7}\right)^2 + \left(\frac{33}{7}\right)^2} = \frac{\sqrt{3402}}{7} = \frac{9\sqrt{42}}{7}$. This exercise can also be done with formula (7), $d = \dfrac{|\overrightarrow{PS} \times \mathbf{v}|}{|\mathbf{v}|}$, on page 726 of the text.

This exercise can also be done with formula (7), $d = \dfrac{|\overrightarrow{PS} \times \mathbf{v}|}{|\mathbf{v}|}$, on page 726 of the text.

37. S(2,– 3,4), x + 2y + 2z = 13 and P(13,0,0) is on the plane $\Rightarrow \overrightarrow{PS} = -11\mathbf{i} - 3\mathbf{j} + 4\mathbf{k}$, $\mathbf{N} = \mathbf{i} + 2\mathbf{j} + 2\mathbf{k}$;

$d = \left|\overrightarrow{PS} \cdot \frac{\mathbf{N}}{|\mathbf{N}|}\right| = \left|\frac{-11-6+8}{\sqrt{1+4+4}}\right| = 3$

39. S(0,1,1), 4y + 3z = – 12 and P(0,– 3,0) is on the plane $\Rightarrow \overrightarrow{PS} = 4\mathbf{j} + \mathbf{k}$, $\mathbf{N} = 4\mathbf{j} + 3\mathbf{k}$;

$d = \left|\overrightarrow{PS} \cdot \frac{\mathbf{N}}{|\mathbf{N}|}\right| = \left|\frac{16+3}{\sqrt{16+9}}\right| = \frac{19}{5}$

41. S(0,– 1,0), 2x + y + 2z = 4 and P(2,0,0) is on the plane $\Rightarrow \overrightarrow{PS} = -2\mathbf{i} - \mathbf{j}$, $\mathbf{N} = 2\mathbf{i} + \mathbf{j} + 2\mathbf{k}$;

$d = \left|\overrightarrow{PS} \cdot \frac{\mathbf{N}}{|\mathbf{N}|}\right| = \left|\frac{-4-1+0}{\sqrt{4+1+4}}\right| = \frac{5}{3}$

43. $2(1-t) - (3t) + 3(1+t) = 6 \Rightarrow t = -1/2 \Rightarrow (3/2, -3/2, 1/2)$

45. $1(1+2t) + 1(1+5t) + 1(3t) = 2 \Rightarrow t = 0 \Rightarrow (1,1,0)$

47. $\mathbf{N}_1 = \mathbf{i} + \mathbf{j}$, $\mathbf{N}_2 = 2\mathbf{i} + \mathbf{j} - 2\mathbf{k} \Rightarrow \theta = \cos^{-1}\left(\frac{\mathbf{N}_1 \cdot \mathbf{N}_2}{|\mathbf{N}_1|\,|\mathbf{N}_2|}\right) = \cos^{-1}\left(\frac{2+1}{\sqrt{2}\sqrt{9}}\right) = \cos^{-1}\frac{1}{\sqrt{2}} = \frac{\pi}{4}$

49. $\mathbf{N}_1 = 2\mathbf{i} + 2\mathbf{j} + 2\mathbf{k}$, $\mathbf{N}_2 = 2\mathbf{i} - 2\mathbf{j} - \mathbf{k} \Rightarrow \theta = \cos^{1}\left(\frac{\mathbf{N}_1 \cdot \mathbf{N}_2}{|\mathbf{N}_1|\,|\mathbf{N}_2|}\right) = \cos^{-1}\left(\frac{4-4-2}{\sqrt{12}\sqrt{9}}\right) = \cos^{-1}\left(\frac{-1}{3\sqrt{3}}\right) \approx 1.76 \text{ rad} \approx 101.1^\circ$

51. $\mathbf{N}_1 = 2\mathbf{i} + 2\mathbf{j} - \mathbf{k}$, $\mathbf{N}_2 = \mathbf{i} + 2\mathbf{j} + \mathbf{k} \Rightarrow \theta = \cos^{-1}\left(\frac{\mathbf{N}_1 \cdot \mathbf{N}_2}{|\mathbf{N}_1|\,|\mathbf{N}_2|}\right) = \cos^{-1}\left(\frac{2+4-1}{\sqrt{9}\sqrt{6}}\right) = \cos^{-1}\left(\frac{5}{3\sqrt{6}}\right) \approx 0.82 \text{ rad} \approx 47.12°$

53. $\mathbf{N}_1 = \mathbf{i} + \mathbf{j} + \mathbf{k}$, $\mathbf{N}_2 = \mathbf{i} + \mathbf{j} \Rightarrow \mathbf{N}_1 \times \mathbf{N}_2 = \begin{vmatrix} \mathbf{i} & \mathbf{j} & \mathbf{k} \\ 1 & 1 & 1 \\ 1 & 1 & 0 \end{vmatrix} = -\mathbf{i} + \mathbf{j}$, the direction of the desired line; (1,1,– 1) is on both planes; the desired line is $x = 1 - t$, $y = 1 + t$, $z = -1$

55. $\mathbf{N}_1 = \mathbf{i} - 2\mathbf{j} + 4\mathbf{k}$, $\mathbf{N}_2 = \mathbf{i} + \mathbf{j} - 2\mathbf{k} \Rightarrow \mathbf{N}_1 \times \mathbf{N}_2 = \begin{vmatrix} \mathbf{i} & \mathbf{j} & \mathbf{k} \\ 1 & -2 & 4 \\ 1 & 1 & -2 \end{vmatrix} = 6\mathbf{j} + 3\mathbf{k}$, the direction of the desired line; (4,3,1) is on both planes; the desired line is $x = 4$, $y = 3 + 6t$, $z = 1 + 3t$

57. $x = 0 \Rightarrow t = -\frac{1}{2}, y = -\frac{1}{2}, z = -\frac{3}{2} \Rightarrow \left(0, -\frac{1}{2}, -\frac{3}{2}\right)$; $y = 0 \Rightarrow t = -1, x = -1, z = -3 \Rightarrow (-1, 0, -3)$; $z = 0 \Rightarrow t = 0 \Rightarrow x = 1, y = -1 \Rightarrow (1, -1, 0)$

59. With substitution of the line into the plane we have $2(1 - 2t) + (2 + 5t) - (-3t) = 8 \Rightarrow 2 - 4t + 2 + 5t + 3t = 8 \Rightarrow 4t + 4 = 8 \Rightarrow t = 1 \Rightarrow$ the point (– 1,7,– 3) is contained in both the line and plane and hence they are not parallel.

61. The cross product of $\mathbf{i} + \mathbf{j} - \mathbf{k}$ and $-4\mathbf{i} + 2\mathbf{j} - 2\mathbf{k}$ has the same direction as the normal of the plane. $\mathbf{N} = \begin{vmatrix} \mathbf{i} & \mathbf{j} & \mathbf{k} \\ 1 & 1 & -1 \\ -4 & 2 & -2 \end{vmatrix} = 6\mathbf{j} + 6\mathbf{k}$. Select a point in either line, such as (– 1,2,1). If the point (x,y,z) is on the plane, then the vector $(x + 1)\mathbf{i} + (y - 2)\mathbf{j} + (z - 1)\mathbf{k}$ is also on the plane. Therefore, the desired plane is $\left((x + 1)\mathbf{i} + (y - 2)\mathbf{j} + (z - 1)\mathbf{k}\right) \cdot \mathbf{N} = 0 \Rightarrow 0(x + 1) + 6(y - 2) + 6(z - 1) = 0 \Rightarrow 6y - 12 + 6z - 6 = 0 \Rightarrow 6y + 6z + 18 \Rightarrow y + z = 3$.

63. L1 & L2: $x = x$ and $y = y \Rightarrow 3 + 2t = 1 + 4s$ and $-1 + 4t = 1 + 2s \Rightarrow \begin{cases} 2t - 4s = -2 \\ 4t - 2s = 2 \end{cases} \Rightarrow \begin{cases} 2t - 4s = -2 \\ 2t - s = 1 \end{cases} \Rightarrow$ $-3s = -3 \Rightarrow s = 1$ and $t = 1 \Rightarrow$ on L1, $z = 1$ and on L2, $z = 1$. Therefore, L1 and L2 intersect at (5,3,1).

L2 & L3: The direction of L2, $\frac{1}{6}(4\mathbf{i} + 2\mathbf{j} + 4\mathbf{k}) = \frac{1}{3}(2\mathbf{i} + \mathbf{j} + 2\mathbf{k})$, is the same as the direction of L3, $\frac{1}{3}(2\mathbf{i} + \mathbf{j} + 2\mathbf{k})$, and hence L2 and L3 are parallel.

L1 & L3: $x = x$ and $y = y \Rightarrow 3 + 2t = 3 + 2r$ and $-1 + 4t = 2 + r \Rightarrow \begin{cases} 2t - 2r = 0 \\ 4t - r = 3 \end{cases} \Rightarrow \begin{cases} t - r = 0 \\ 4t - r = 3 \end{cases} \Rightarrow 3t = 3 \Rightarrow$ $t = 1$ and $r = 1 \Rightarrow$ on L1, $z = 2$ while on L3, $z = 0 \Rightarrow$ L1 and L2 do not intersect. The direction of L1 is $\frac{1}{\sqrt{21}}(2\mathbf{i} + 4\mathbf{j} - \mathbf{k})$ while the direction of L3 is $\frac{1}{3}(2\mathbf{i} + \mathbf{j} + 2\mathbf{k})$ and neither is a multiple of the other.

Therefore, L1 and L3 are skew.

11.6 SURFACES IN SPACE

1. d, ellipsoid of revolution
3. a, elliptic cylinder
5. l, hyperbolic paraboloid
7. b, elliptic cylinder
9. k, hyperbolic paraboloid
11. h, elliptic cone
13. $x^2 + y^2 = 4$

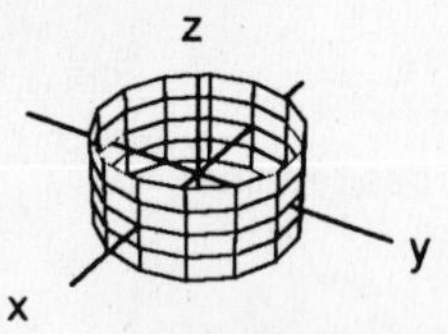

Graph 11.6.13

15. $z = y^2 - 1$

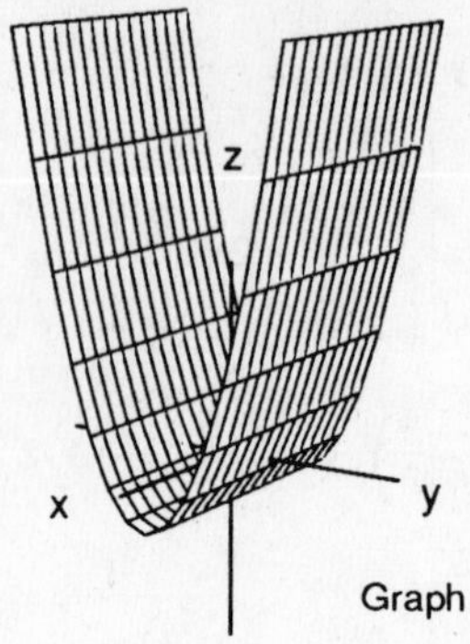

Graph 11.6.15

17. $x^2 + 4z^2 = 16$

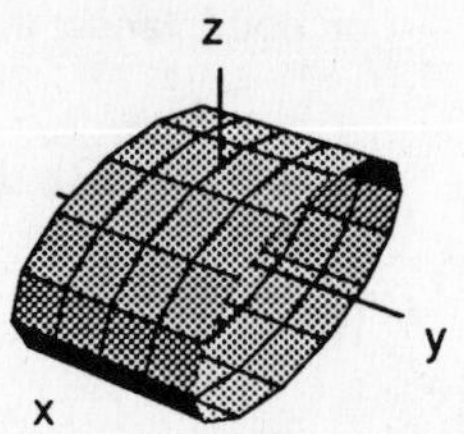

Graph 11.6.17

19. $z^2 - y^2 = 1$

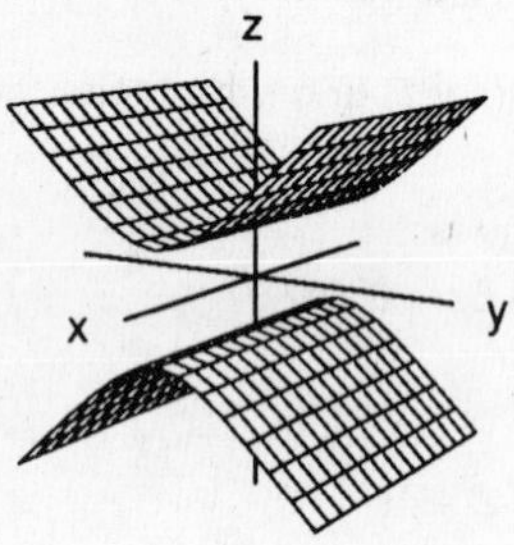

Graph 11.6.19

21. $9x^2 + y^2 + z^2 = 9$

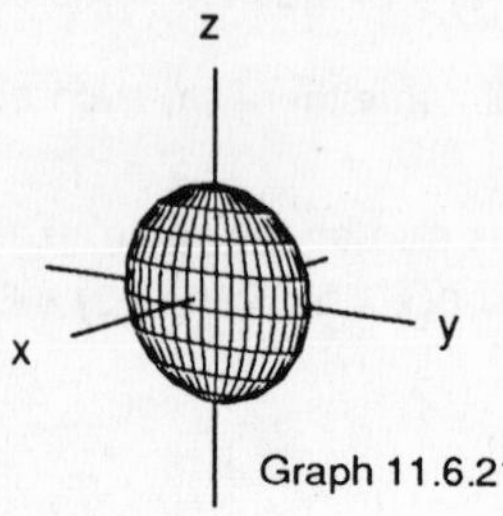

Graph 11.6.21

23. $4x^2 + 9y^2 + 4z^2 = 36$

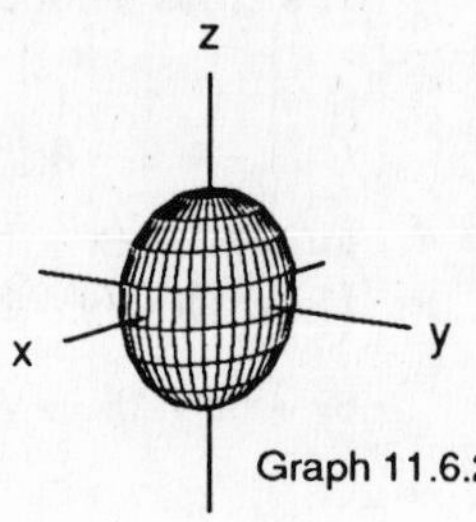

Graph 11.6.23

25. $x^2 + 4y^2 = z$

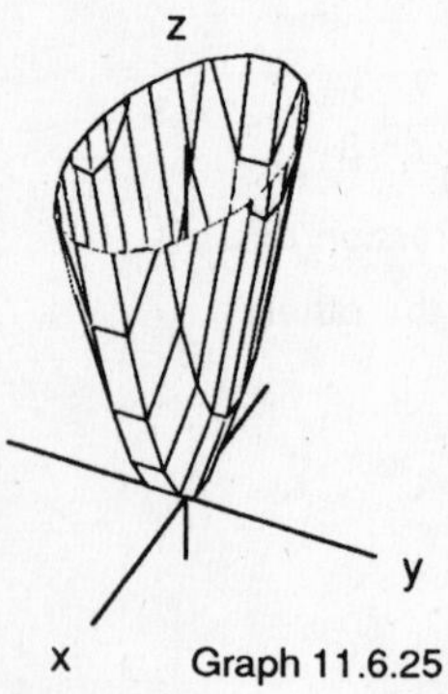

Graph 11.6.25

27. $z = 8 - x^2 - y^2$

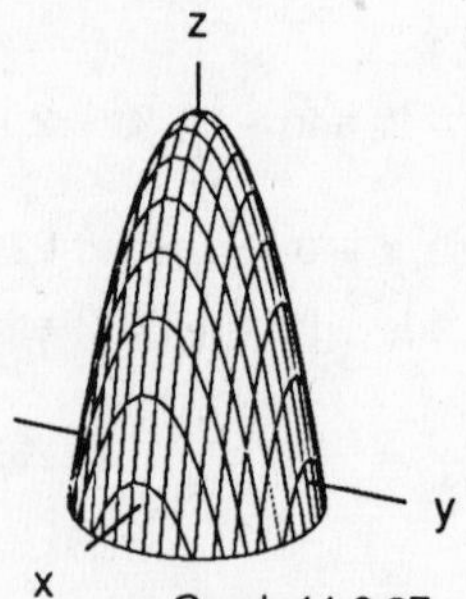

Graph 11.6.27

29. $x = 4 - 4y^2 - z^2$

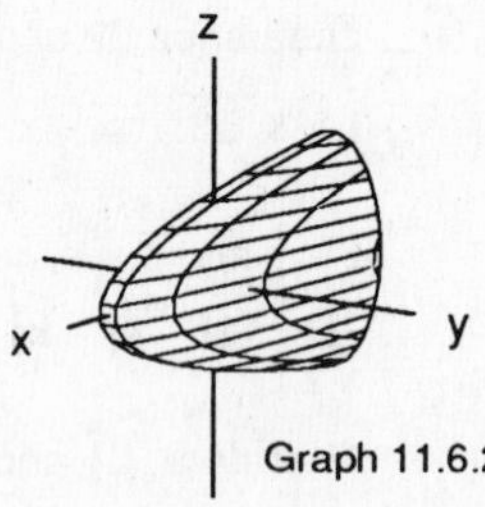

Graph 11.6.29

31. $x^2 + y^2 = z^2$

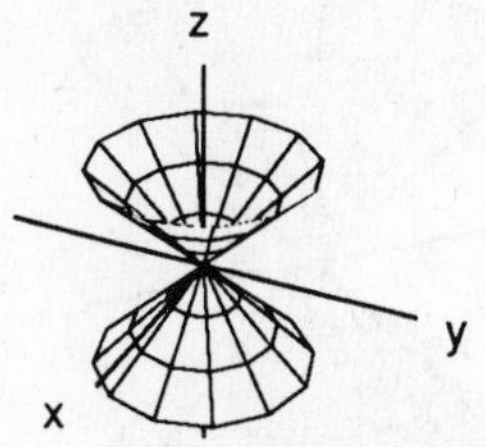

Graph 11.6.31

33. $4x^2 + 9z^2 = 9y^2$

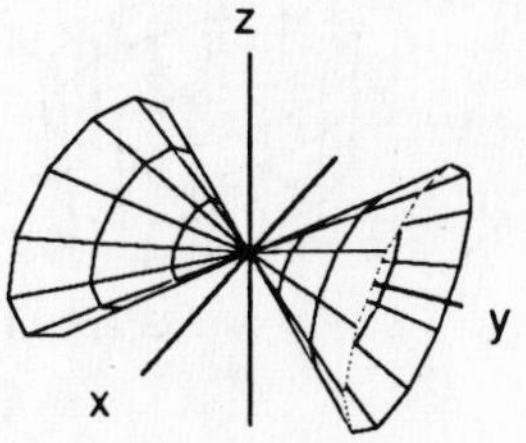

Graph 11.6.33

35. $x^2 + y^2 - z^2 = 1$

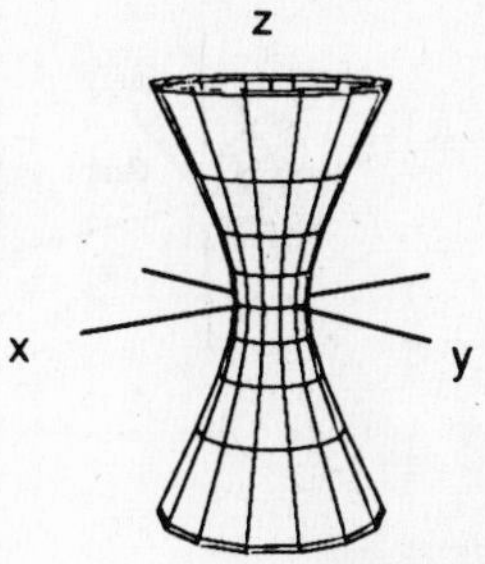

Graph 11.6.35

37. $\left(y^2/4\right) + \left(z^2/9\right) - \left(x^2/4\right) = 1$

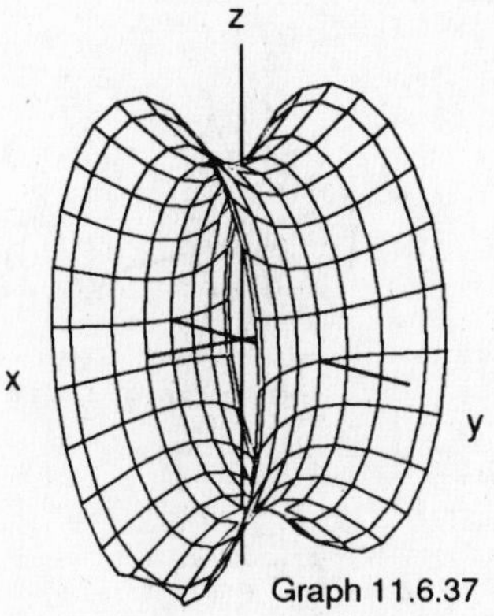

Graph 11.6.37

39. $z^2 - x^2 - y^2 = 1$

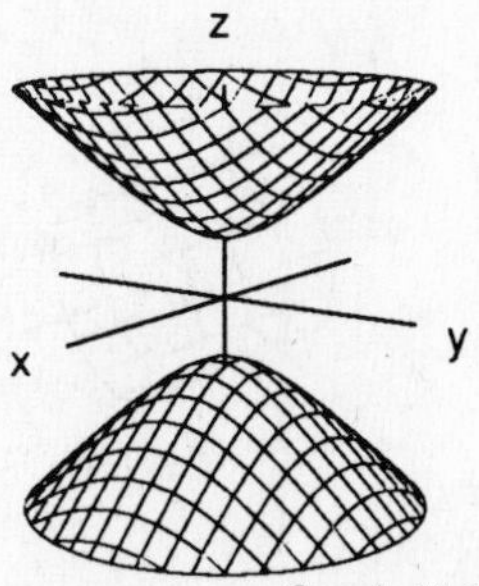

Graph 11.6.39

41. $x^2 - y^2 - \left(z^2/4\right) = 1$

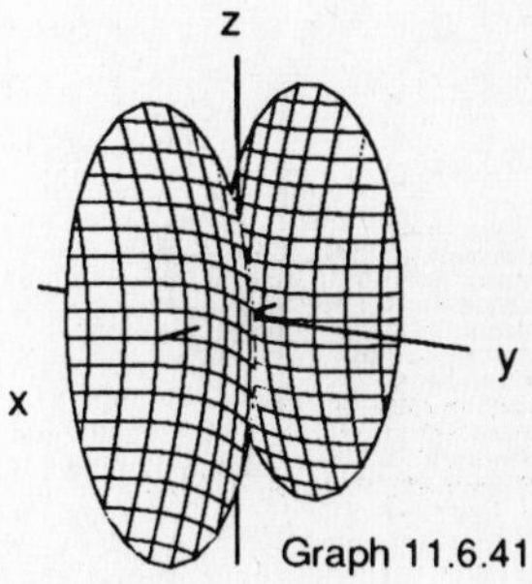

Graph 11.6.41

43. $y^2 - x^2 = z$

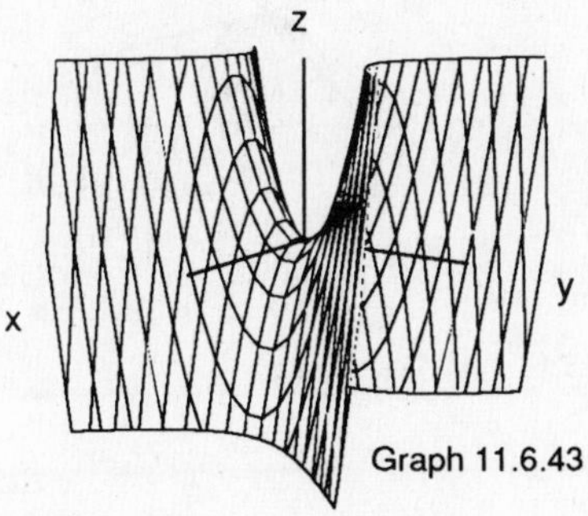

Graph 11.6.43

45. $x^2 + y^2 + z^2 = 4$

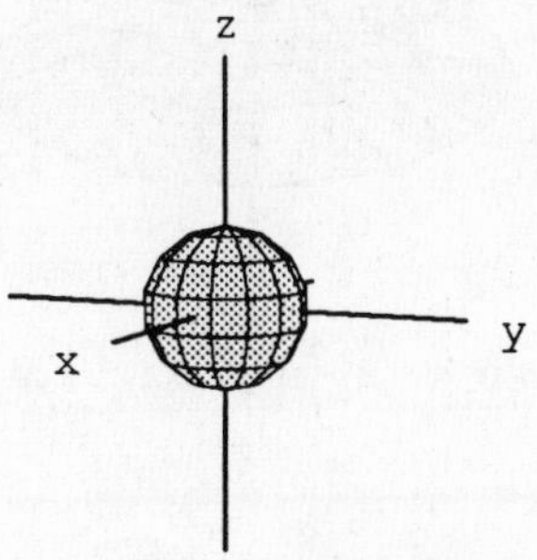

Graph 11.6.45

47. $z = 1 + y^2 - x^2$

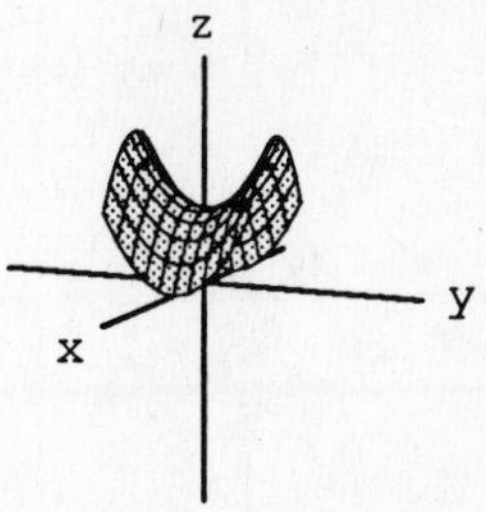

Graph 11.6.47

49. $y = -\left(x^2 + z^2\right)$

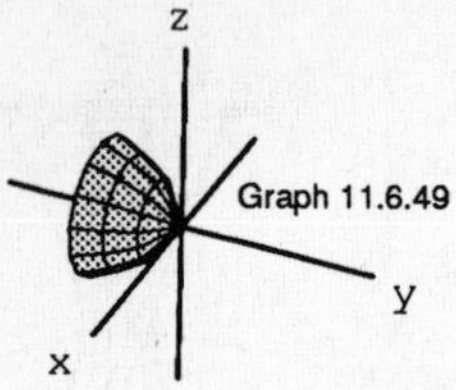

Graph 11.6.49

51. $16x^2 + 4y^2 = 1$

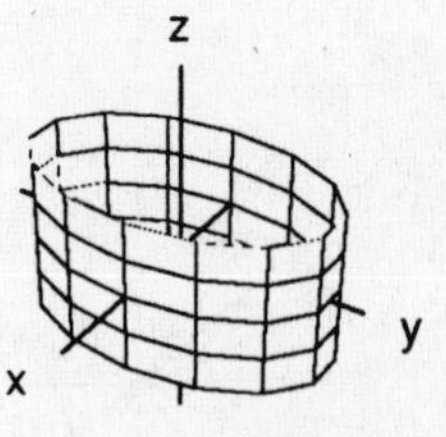

Graph 11.6.51

53. $x^2 + y^2 - z^2 = 4$

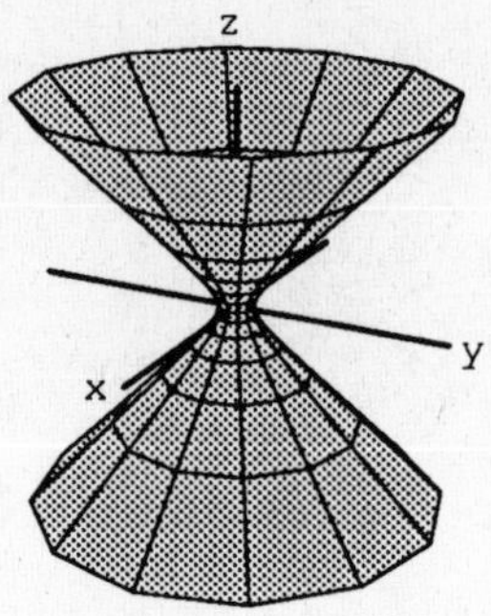

Graph 11.6.53

55. $x^2 + z^2 = y$

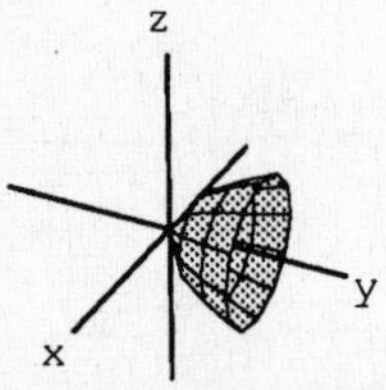

Graph 11.6.55

57. $x^2 + z^2 = 1$

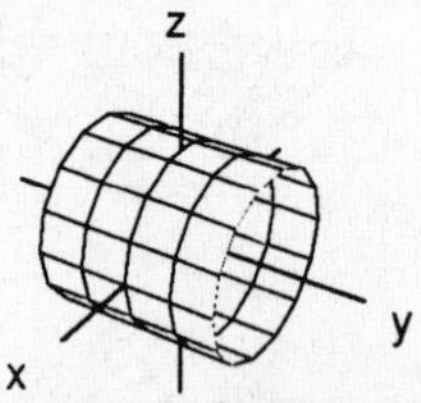

Graph 11.6.57

59. $16y^2 + 9z^2 = 4x^2$

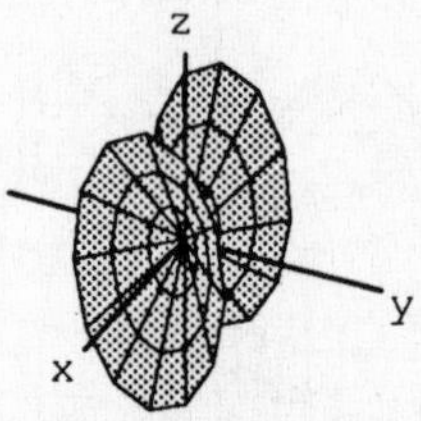

Graph 11.6.59

61. $9x^2 + 4y^2 + z^2 = 36$

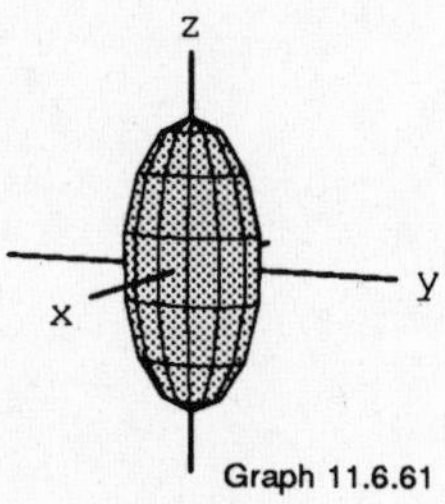

Graph 11.6.61

63. $x^2 + y^2 - 16z^2 = 16$

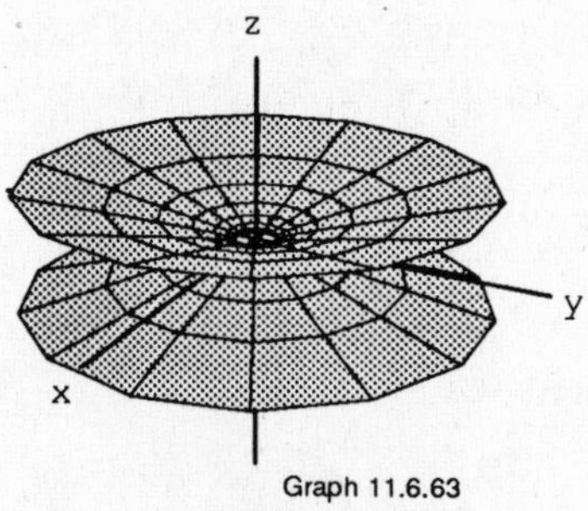

Graph 11.6.63

65. $z = -\left(x^2 + y^2\right)$

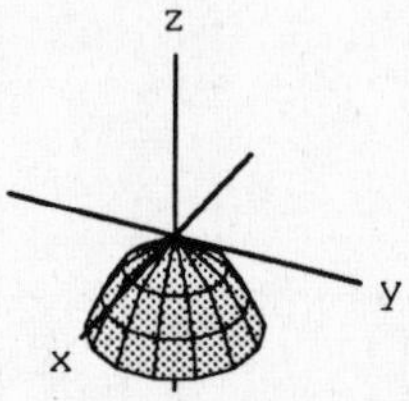

Graph 11.6.65

67. $x^2 - 4y^2 = 1$

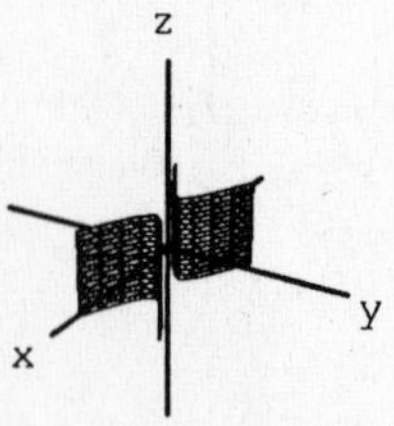

Graph 11.6.67

69. $4y^2 + z^2 - 4x^2 = 4$

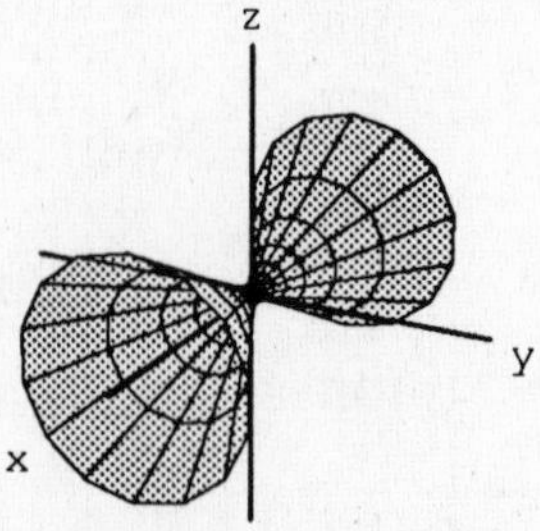

Graph 11.6.69

71. $x^2 + y^2 = z$

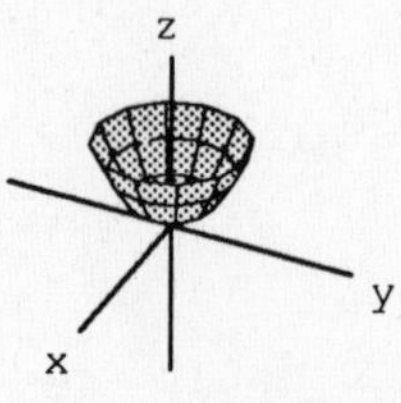

Graph 11.6.71

73. $yz = 1$

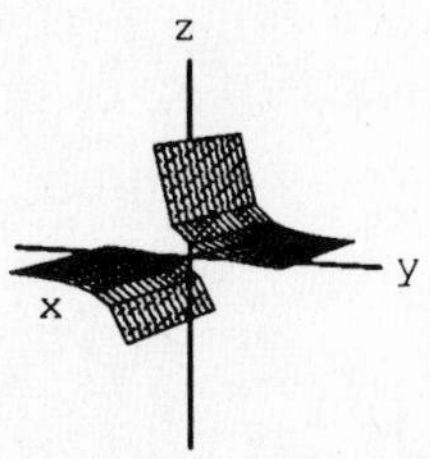

Graph 11.6.73

75. $9x^2 + 16y^2 = 4z^2$

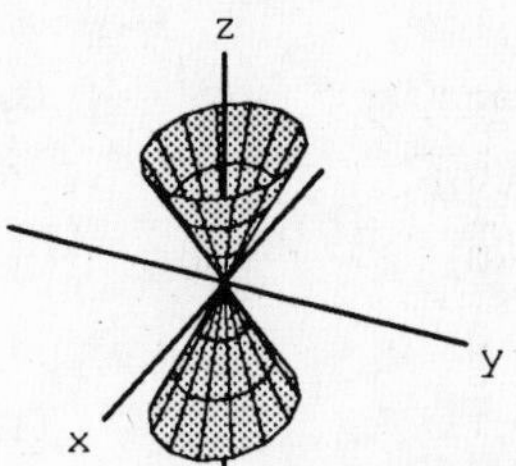

Graph 11.6.75

77. a) If $x^2 + \frac{y^2}{4} + \frac{z^2}{9} = 1$ and $z = c$, then $x^2 + \frac{y^2}{4} = \frac{9-c^2}{9} \Rightarrow \frac{x^2}{\frac{9-c^2}{9}} + \frac{y^2}{\frac{4(9-c^2)}{9}} = 1 \Rightarrow A = ab\pi =$

$\pi\left(\frac{\sqrt{9-c^2}}{3}\right)\left(\frac{2\sqrt{9-c^2}}{3}\right) = \frac{2\pi\left(9-c^2\right)}{9}$

b) From part a) each slice has the area $\frac{2\pi\left(9-z^2\right)}{9}$ where $-3 \le z \le 3$. $\therefore$ $V = 2\int_0^3 \frac{2\pi}{9}\left(9-z^2\right)\,dz =$

$\frac{4\pi}{9}\int_0^3 \left(9-z^2\right)\,dz = \frac{4\pi}{9}\left[9z - \frac{z^3}{3}\right]_0^3 = 8\pi$

c) $\frac{x^2}{a^2} + \frac{y^2}{b^2} + \frac{z^2}{c^2} = 1 \Rightarrow \frac{x^2}{\frac{a^2\left(c^2-z^2\right)}{c^2}} + \frac{y^2}{\frac{a^2\left(c^2-z^2\right)}{c^2}} = 1$ and $V = 2\int_0^c \frac{\pi ab}{c^2}\left(c^2-z^2\right)\,dz = \frac{4\pi abc}{3}$. If $r = a = b = c$,

then $V = \frac{4\pi r^3}{3}$ the volume of a sphere.

79. $z = y^2$

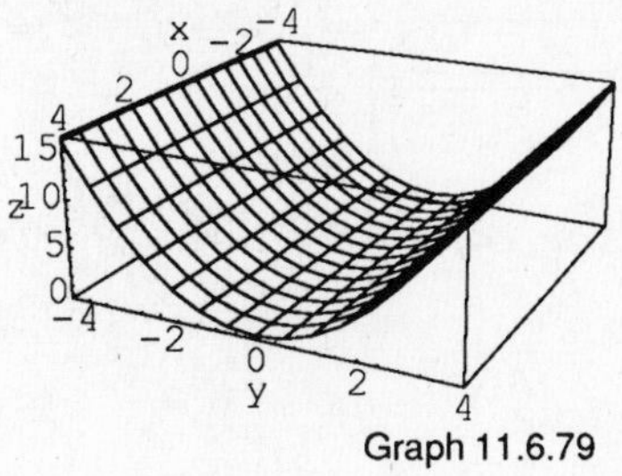

Graph 11.6.79

81. $z = x^2 + y^2$

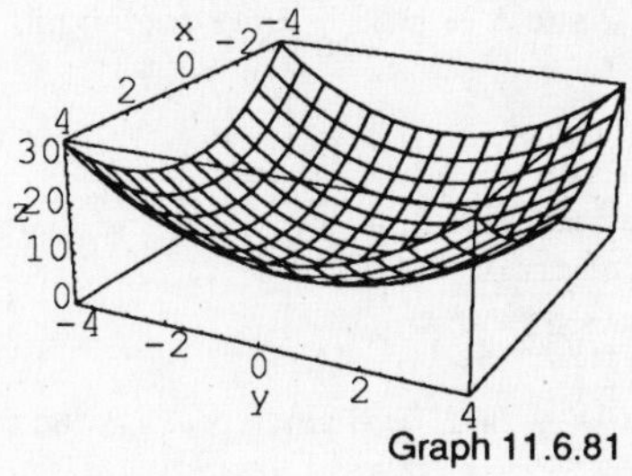

Graph 11.6.81

83. $z = \sqrt{1 - x^2}$

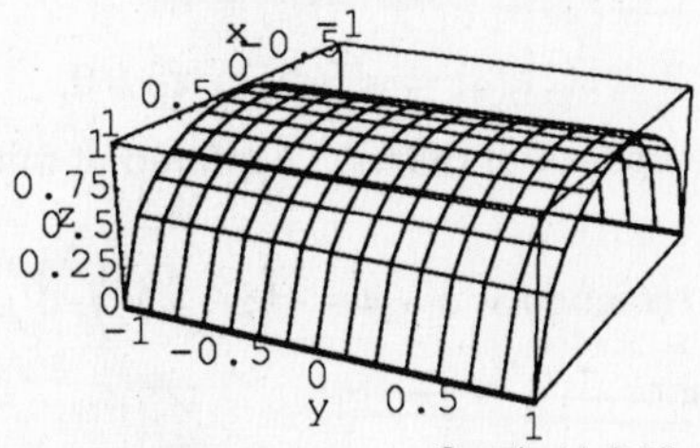

Graph 11.6.83

85. $z = \sqrt{x^2 + 2y^2 + 4}$

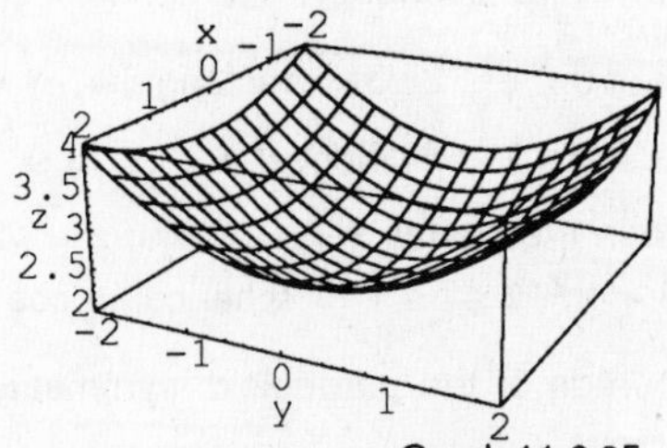

Graph 11.6.85

11.7 CYLINDRICAL AND SPHERICAL COORDINATES

	Rectangular	Cylindrical	Spherical
1.	(0,0,0)	(0,0,0)	(0,0,0)
3.	(0,1,0)	$(1,\pi/2,0)$	$(1,\pi/2,\pi/2)$
5.	(1,0,0)	(1,0,0)	$(1,\pi/2,0)$
7.	(0,1,1)	$(1,\pi/2,1)$	$(\sqrt{2},\pi/4,\pi/2)$
9.	$(0,-2\sqrt{2},0)$	$(2\sqrt{2},3\pi/2,0)$	$(2\sqrt{2},\pi/2,3\pi/2)$

11. $r = 0 \Rightarrow$ rectangular, $x^2 + y^2 = 0$; spherical, $\phi = 0$ and $\phi = \pi$; the z–axis

13. $z = 0 \Rightarrow$ cylindrical, $z = 0$; spherical, $\phi = \frac{\pi}{2}$; the xy–plane

15. $z = \sqrt{x^2+y^2},\ z \le 1 \Rightarrow$ cylindrical, $z = r,\ 0 \le r \le 1$; spherical, $\phi = \tan^{-1}\frac{\sqrt{x^2+y^2}}{z} = \tan^{-1}1 = \frac{\pi}{4}$, $\rho = \sqrt{x^2+y^2+z^2} = \sqrt{x^2+y^2+x^2+y^2} = \sqrt{2\left(x^2+y^2\right)} = \sqrt{2z^2} = \sqrt{2}\,|z| \le \sqrt{2} \Rightarrow 0 \le \rho \le \sqrt{2}$; a (finite) cone

17. $\rho \sin\phi \cos\theta = 0 \Rightarrow$ rectangular, $x = 0$; cylindrical $\theta = \pi/2$; the yz–plane

19. $x^2 + y^2 + z^2 = 4 \Rightarrow$ cylindrical, $r^2 + z^2 = 4$; spherical, $\rho = 2$; a sphere centered at the origin with a radius of 2

21. $\rho = 5\cos\phi \Rightarrow$ rectangular, $\sqrt{x^2+y^2+z^2} = 5\cos\left(\cos^{-1}\left(\frac{z}{\sqrt{x^2+y^2+z^2}}\right)\right) \Rightarrow \sqrt{x^2+y^2+z^2} = 5\frac{z}{\sqrt{x^2+y^2+z^2}} \Rightarrow$ $x^2+y^2+z^2 = 5z \Rightarrow x^2+y^2+z^2-5z+\frac{25}{4} = \frac{25}{4} \Rightarrow x^2+y^2+\left(z-\frac{5}{2}\right)^2 = \left(\frac{5}{2}\right)^2$; cylindrical, $r^2+\left(z-\frac{5}{2}\right)^2 = \left(\frac{5}{2}\right)^2$, a sphere of radius $\frac{5}{2}$ centered at (0,0,5/2) (rectangular)

23. $r = \csc\theta \Rightarrow$ rectangular, $r = \frac{r}{y} \Rightarrow y = 1$; spherical, $\rho\sin\phi = \csc\theta \Rightarrow \rho = \frac{1}{\sin\phi\sin\theta}$; the plane $y = 1$

25. $\rho = \sqrt{2}\sec\phi \Rightarrow \rho = \frac{\sqrt{2}}{\cos\phi} \Rightarrow$ rectangular, $\sqrt{x^2+y^2+z^2} = \frac{\sqrt{2}}{\cos\left(\cos^{-1}\left(\frac{z}{\sqrt{x^2+y^2+z^2}}\right)\right)} \Rightarrow \sqrt{x^2+y^2+z^2} = \frac{\sqrt{2}}{\frac{z}{\sqrt{x^2+y^2+z^2}}} \Rightarrow z\sqrt{x^2+y^2+z^2} = \sqrt{2}\sqrt{x^2+y^2+z^2} \Rightarrow \sqrt{x^2+y^2+z^2}\left(z-\sqrt{2}\right) = 0 \Rightarrow x^2+y^2+z^2 = 0$ or $z = \sqrt{2}$, $x^2+y^2+z^2 \ne 0$ since $\rho = \sqrt{2}\sec\phi \ne 0$; cylindrical $z = \sqrt{2}$; the plane $z = \sqrt{2}$

27. $x^2+y^2+(z-1)^2 = 1,\ z \le 1 \Rightarrow$ cylindrical, $r^2+(z-1)^2 = 1 \Rightarrow r^2+z^2-2z+1 = 1 \Rightarrow r^2+z^2 = 2z,\ z \le 1$; spherical, $x^2+y^2+z^2-2z = 0 \Rightarrow \rho^2 - 2\rho\cos\phi = 0$ $\rho(\rho-2\cos\phi) = 0 \Rightarrow \rho = 0$ or $\rho = 2\cos\phi \Rightarrow \rho = 2\cos\phi,\ \pi/4 \le \phi \le \pi/2$; the lower hemisphere of the sphere centered at (0,0,1) (rectangular) with a radius of 1

29. $\rho = 3,\ \pi/3 \le \phi \le 2\pi/3 \Rightarrow$ rectangular, $\sqrt{x^2+y^2+z^2} = 3$, $3\cos(\pi/3) \ge z \ge 3\cos(2\pi/3) \Rightarrow x^2+y^2+z^2 = 9$, $-3/2 \le z \le 3/2$; cylindrical, $r^2+z^2 = 9,\ -3/2 \le z \le 3/2$; the portion of the sphere of radius 3 centered at the origin between the planes $z = -3/2$ and $z = 3/2$

31. $z = 4-4r^2,\ 0 \le r \le 1 \Rightarrow$ spherical, $\rho\cos\phi = 4-4\rho^2(\sin\phi)^2,\ 0 \le \phi \le \pi/2$; rectangular $z = 4-4\left(x^2+y^2\right),\ 0 \le z \le 4$; the portion of the paroboloid, symmetric to the z–axiz, above the xy–plane

33. $\phi = 3\pi/4$, $0 \le \rho \le \sqrt{2} \Rightarrow$ rectangular, $\cos \phi = \cos \frac{3\pi}{4} = \frac{z}{\sqrt{x^2 + y^2 + z^2}} \Rightarrow -\frac{1}{\sqrt{2}} = \frac{z}{\sqrt{x^2 + y^2 + z^2}} \Rightarrow \sqrt{x^2 + y^2 + z^2} =$ $-\sqrt{2}\, z \Rightarrow x^2 + y^2 + z^2 = 2z^2 \Rightarrow x^2 + y^2 - z^2 = 0$ but $z \le 0 \Rightarrow z = -\sqrt{x^2 + y^2}$, $0 \ge z \ge \sqrt{2} \cos \frac{3\pi}{4} \Rightarrow z = -\sqrt{x^2 + y^2}$, $-1 \le z \le 0$; cylindrical $x^2 + y^2 - z^2 = 0 \Rightarrow r^2 - z^2 = 0 \Rightarrow (r + z)(r - z) = 0 \Rightarrow r = -z$ or $r = z$ but $r \ge 0$ and $z \le 0 \Rightarrow$ $r = -z$; a cone whose vertex is at the origin, the base is the circle $x^2 + y^2 = 1$ in the plane $z = -1$

35. $z + r^2\cos 2\theta = 0 \Rightarrow z + r^2\left(\cos^2\theta - \sin^2\theta\right) = 0 \Rightarrow z + (r \cos \theta)^2 - (r \sin \theta)^2 = 0 \Rightarrow$ rectangular, $z + x^2 - y^2 = 0 \Rightarrow$ $z = y^2 - x^2$; spherical, $z + r^2\cos 2\theta = 0 \Rightarrow \rho \cos \phi + (\rho \sin \phi)^2\cos 2\theta = 0 \Rightarrow \rho(\cos \phi + \rho \sin^2\phi \cos 2\theta) = 0 \Rightarrow \cos \phi +$ $\rho \sin^2\phi \cos 2\theta = 0$; a hyperbolic paraboloid

37. Right circular cylinder parallel to the z–axis generated by the circle $r = -2 \sin \theta$ in the $r\theta$–plane

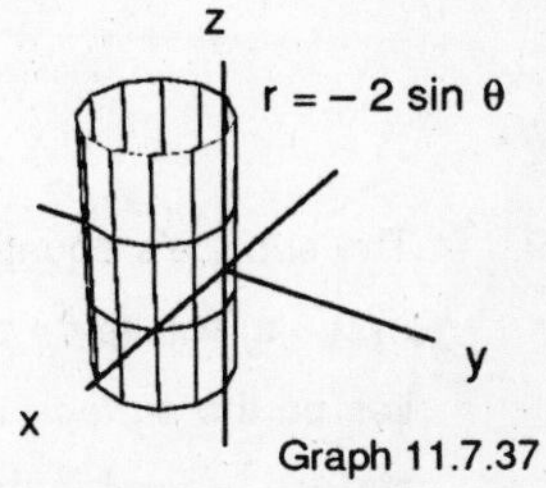

Graph 11.7.37

39. Cylinder of lines parallel to the z–axis generated by the cardioid $r = 1 - \cos \theta$ in the $r\theta$–plane

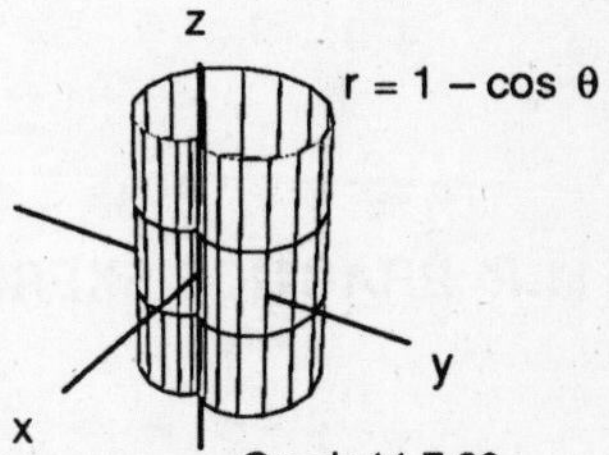

Graph 11.7.39

41. Cardioid of revolution symmetric abot the z–axis, cusp at the origin pointing down

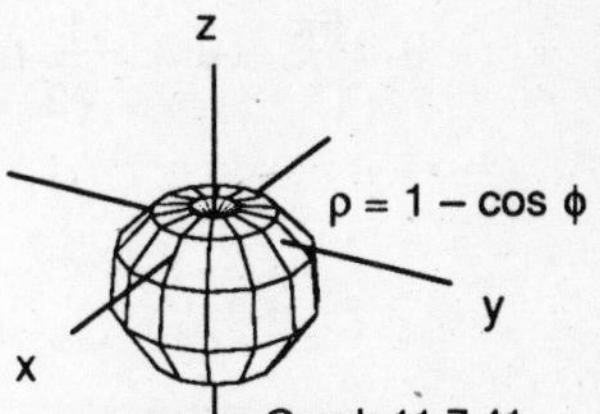

Graph 11.7.41

43. $r^2 + z^2 = 4r\cos\theta + 6r\sin\theta + 2z \Rightarrow x^2 + y^2 + z^2 = 4x + 6y + 2z \Rightarrow \left(x^2 - 4x + 4\right) + \left(y^2 - 6y + 9\right) + \left(z^2 - 2z + 1\right) = 14 \Rightarrow (x-2)^2 + (y-3)^2 + (z-1)^2 = 14 \Rightarrow$ the center is located at (2,3,1)

45. a) $z = c \Rightarrow \rho\cos\phi = c \Rightarrow \rho = \frac{c}{\csc\phi} \Rightarrow \rho = c\sec\phi$

b) The xy–plane is perpendicular to the z–axis, hence $\phi = \pi/2$.

47. a) A plane perpendicular to the x–axis has the form x = a in rectangulae coordinates. Now $x = a \Rightarrow r\cos\theta = a \Rightarrow r = \frac{a}{\cos\theta} \Rightarrow r = c\sec\theta$ in cylindrical coordinates.

b) A plane perpendicular to the y–axis has the form y = b in rectangular coordinates. Now $y = b \Rightarrow r\sin\theta = b \Rightarrow r = \frac{b}{\sin\theta} \Rightarrow r = b\csc\theta$ in cylindrical coordinates.

49. The surface's equation $r = f(z)$ tells us that the point $(r,\theta,z) = (f(z),\theta,z)$ will lie on the surface for all θ. In particular $(f(z),\theta + \pi,z)$ lies on the surface whenever $(f(z),\theta,z)$ lies on the surface, so the surface is symmetric with respect to the z–axis.

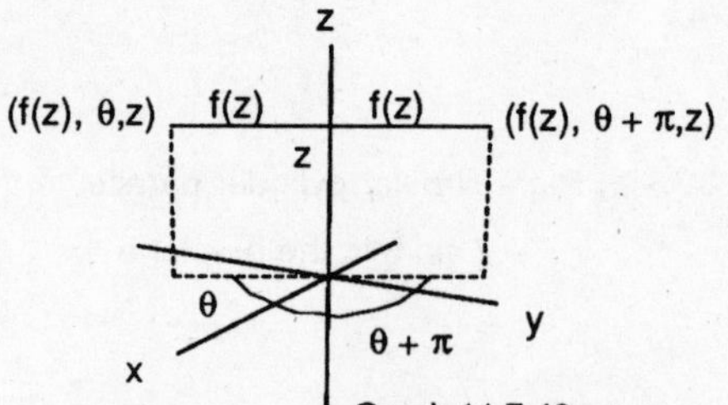

Graph 11.7.49

11.P PRACTICE EXERCISES

1. $\theta = 0 \Rightarrow \mathbf{u} = \mathbf{i};\ \theta = \frac{\pi}{2} \Rightarrow \mathbf{u} = \mathbf{j};\ \theta = \frac{2\pi}{3} \Rightarrow \mathbf{u} = -\frac{1}{2}\mathbf{i} + \frac{\sqrt{3}}{2}\mathbf{j};$
$\theta = \frac{5\pi}{4} \Rightarrow \mathbf{u} = -\frac{1}{\sqrt{2}}\mathbf{i} - \frac{1}{\sqrt{2}}\mathbf{j};\ \theta = \frac{5\pi}{3} \Rightarrow \mathbf{u} = \frac{1}{2}\mathbf{i} - \frac{\sqrt{3}}{2}\mathbf{j}$

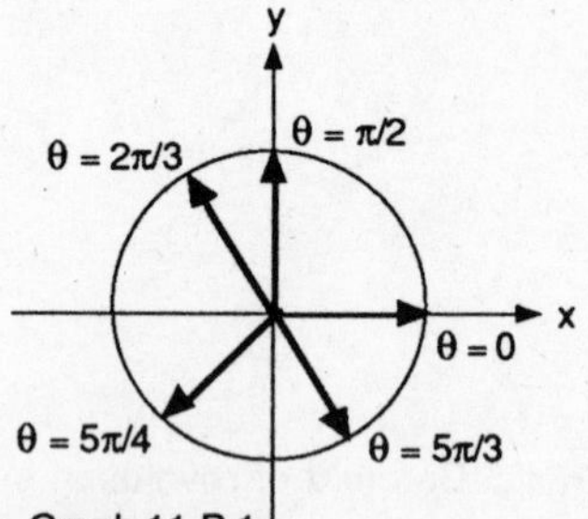

Graph 11.P.1

3. $y = \tan x \Rightarrow [y']_{\pi/4} = \left[\sec^2 x\right]_{\pi/4} = 2 = \frac{2}{1} \Rightarrow \mathbf{T} = \mathbf{i} + 2\mathbf{j} \Rightarrow$ the unit tangents are $\pm\left(\frac{1}{\sqrt{5}}\mathbf{i} + \frac{2}{\sqrt{5}}\mathbf{j}\right)$ and the unit normals are $\pm\left(-\frac{2}{\sqrt{5}}\mathbf{i} + \frac{1}{\sqrt{5}}\mathbf{j}\right)$

5. length $= \left|\sqrt{2}\mathbf{i} + \sqrt{2}\mathbf{j}\right| = \sqrt{2+2} = 2,\ \sqrt{2}\mathbf{i} + \sqrt{2}\mathbf{j} = 2\left[\frac{1}{\sqrt{2}}\mathbf{i} + \frac{1}{\sqrt{2}}\mathbf{j}\right] \Rightarrow$ the direction is $\frac{1}{\sqrt{2}}\mathbf{i} + \frac{1}{\sqrt{2}}\mathbf{j}$

7. length $= |2\mathbf{i} - 3\mathbf{j} + 6\mathbf{k}| = \sqrt{4+9+36} = 7,\ 2\mathbf{i} - 3\mathbf{j} + 6\mathbf{k} = 7\left[\frac{2}{7}\mathbf{i} - \frac{3}{7}\mathbf{j} + \frac{6}{7}\mathbf{k}\right] \Rightarrow$ the direction is $= \frac{2}{7}\mathbf{i} - \frac{3}{7}\mathbf{j} + \frac{6}{7}\mathbf{k}$

9. $2\frac{\mathbf{A}}{|\mathbf{A}|} = 2\frac{(4\mathbf{i} - \mathbf{j} + 4\mathbf{k})}{\sqrt{4^2 + (-1)^2 + 4^2}} = 2\frac{(4\mathbf{i} - \mathbf{j} + 4\mathbf{k})}{\sqrt{33}} = \frac{8}{\sqrt{33}}\mathbf{i} - \frac{2}{\sqrt{33}}\mathbf{j} + \frac{8}{\sqrt{33}}\mathbf{k}$

11. a) $\overrightarrow{BD} = \overrightarrow{AD} - \overrightarrow{AB}$

b) $\overrightarrow{AP} = \overrightarrow{AB} + \frac{1}{2}\overrightarrow{BD} = \overrightarrow{AB} + \frac{1}{2}\left(\overrightarrow{AD} - \overrightarrow{AB}\right) = \overrightarrow{AB} + \frac{1}{2}\overrightarrow{AD} - \frac{1}{2}\overrightarrow{AB} = \frac{1}{2}\overrightarrow{AB} + \frac{1}{2}\overrightarrow{AD} = \frac{1}{2}\left(\overrightarrow{AB} + \overrightarrow{AD}\right)$

c) $\overrightarrow{PC} = \overrightarrow{AC} - \overrightarrow{AP} = \left(\overrightarrow{AB} + \overrightarrow{AD}\right) - \frac{1}{2}\left(\overrightarrow{AB} + \overrightarrow{AD}\right) = \frac{1}{2}\left(\overrightarrow{AB} + \overrightarrow{AD}\right) = \overrightarrow{AP} \Rightarrow$ P is the midpoint of AC.

13.

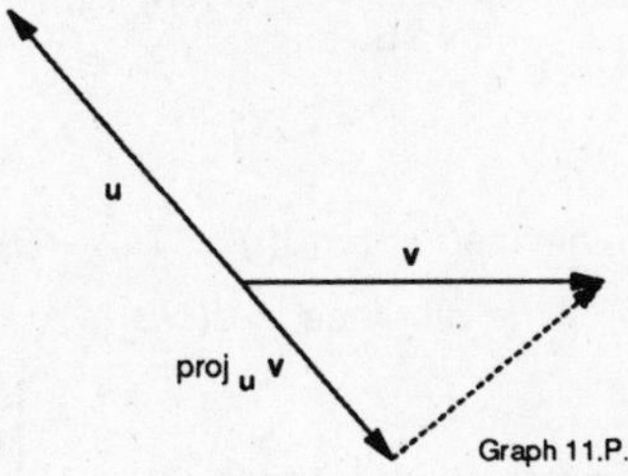

Graph 11.P.13

15. $|\mathbf{A}| = \sqrt{2},\ |\mathbf{B}| = 3,\ \mathbf{A}\cdot\mathbf{B} = 3,\ \mathbf{B}\cdot\mathbf{A} = 3,\ \mathbf{A}\times\mathbf{B} = \begin{vmatrix} \mathbf{i} & \mathbf{j} & \mathbf{k} \\ 1 & 1 & 0 \\ 2 & 1 & -2 \end{vmatrix} = -2\mathbf{i} + 2\mathbf{j} - \mathbf{k},\ \mathbf{B}\times\mathbf{A} = \begin{vmatrix} \mathbf{i} & \mathbf{j} & \mathbf{k} \\ 2 & 1 & -2 \\ 1 & 1 & 0 \end{vmatrix} = 2\mathbf{i} - 2\mathbf{j} + \mathbf{k},$

$|\mathbf{A}\times\mathbf{B}| = \sqrt{4+4+1} = 3,\ \theta = \cos^{-1}\left(\frac{\mathbf{A}\cdot\mathbf{B}}{|\mathbf{A}|\,|\mathbf{B}|}\right) = \cos^{-1}\left(\frac{1}{\sqrt{2}}\right) = \frac{\pi}{4},\ \text{comp}_{\mathbf{A}}\,\mathbf{B} = \frac{\mathbf{A}\cdot\mathbf{B}}{|\mathbf{A}|} = \frac{3}{\sqrt{2}},\ \text{proj}_{\mathbf{A}}\,\mathbf{B} = \frac{\mathbf{A}\cdot\mathbf{B}}{\mathbf{A}\cdot\mathbf{A}}\mathbf{A} = \frac{3}{2}[\mathbf{i} + \mathbf{j}]$

17. $\mathbf{B} = \left(\frac{\mathbf{A}\cdot\mathbf{B}}{\mathbf{A}\cdot\mathbf{A}}\mathbf{A}\right) + \left(\mathbf{B} - \frac{\mathbf{A}\cdot\mathbf{B}}{\mathbf{A}\cdot\mathbf{A}}\mathbf{A}\right) = \frac{4}{3}[2\mathbf{i} + \mathbf{j} - \mathbf{k}] + \left[(\mathbf{i} + \mathbf{j} - 5\mathbf{k}) - \frac{4}{3}(2\mathbf{i} + \mathbf{j} - \mathbf{k})\right] = \frac{4}{3}[2\mathbf{i} + \mathbf{j} - \mathbf{k}] - \frac{1}{3}[5\mathbf{i} + \mathbf{j} + 11\mathbf{k}],$

where $\mathbf{A}\cdot\mathbf{B} = 8$ and $\mathbf{A}\cdot\mathbf{A} = 6$

19. $\mathbf{A}\times\mathbf{B} = \begin{vmatrix} \mathbf{i} & \mathbf{j} & \mathbf{k} \\ 1 & 0 & 0 \\ 1 & 1 & 0 \end{vmatrix} = \mathbf{k}$

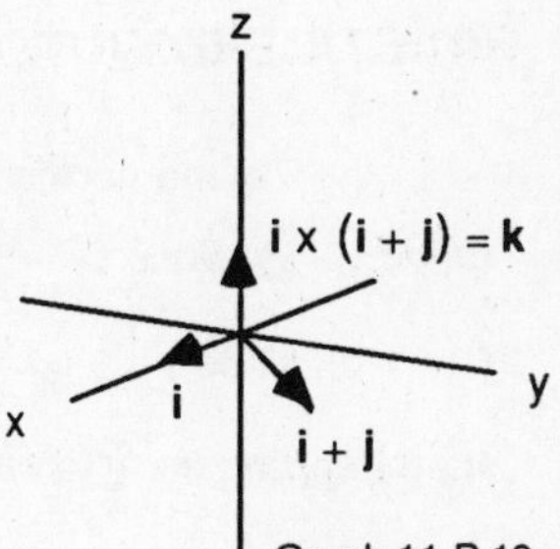

Graph 11.P.19

21. Let $\mathbf{A} = a_1\mathbf{i} + a_2\mathbf{j} + a_3\mathbf{k}$ and $\mathbf{B} = b_1\mathbf{j} + b_2\mathbf{j} + b_3\mathbf{k}$. $|\mathbf{A} + \mathbf{B}|^2 + |\mathbf{A} - \mathbf{B}|^2 =$

$\left(\sqrt{(a_1+b_1)^2 + (a_2+b_2)^2 + (a_3+b_3)^2}\right)^2 + \left(\sqrt{(a_1-b_1)^2 - (a_2-b_2)^2 - (a_3-b_3)^2}\right)^2 = (a_1+b_1)^2 +$
$(a_2+b_2)^2 + (a_3+b_3)^2 + (a_1-b_1)^2 - (a_2-b_2)^2 - (a_3-b_3)^2 = a_1^2 + 2a_1b_1 + b_1^2 + a_2^2 + 2a_2b_2 + b_2^2 +$
$a_3^2 + 2a_3b_3 + b_3^2 + a_1^2 - 2a_1b_1 + b_1^2 + a_2^2 - 2a_2b_2 + b_2^2 + a_3^2 - 2a_3b_3 + b_3^2 = 2\left(a_1^2 + a_2^2 + a_3^2\right) +$
$2\left(b_1^2 + b_2^2 + b_3^2\right) = 2|\mathbf{A}|^2 + 2|\mathbf{B}|^2.$

23. Let $\mathbf{v} = v_1\mathbf{i} + v_2\mathbf{j} + v_3\mathbf{k}$ and $\mathbf{w} = w_1\mathbf{i} + w_2\mathbf{j} + w_3\mathbf{k}$. $|\mathbf{v} - 2\mathbf{w}|^2 = \left|(v_1\mathbf{i} + v_2\mathbf{j} + v_3\mathbf{k}) - 2(w_1\mathbf{i} + w_2\mathbf{j} + w_3\mathbf{k})\right|^2 =$ $\left|(v_1 - 2w_1)\mathbf{i} + (v_2 - 2w_2)\mathbf{j} + (v_3 - 2w_3)\mathbf{k}\right|^2 = \left(\sqrt{(v_1 - 2w_1)^2 + (v_2 - 2w_2)^2 + (v_3 - 2w_3)^2}\right)^2 =$ $\left(v_2^{\,2} + v_2^{\,2} + v_3^{\,2}\right) - 4\left(v_1w_1 + v_2w_2 + v_3w_3\right) + 4\left(w_1^{\,2} + w_2^{\,2} + w_3^{\,2}\right) = |\mathbf{v}|^2 - 4\,\mathbf{v}\cdot\mathbf{w} + 4|\mathbf{w}|^2 = 2^2 -$ $4\,|\mathbf{v}|\,|\mathbf{w}|\cos\theta + 4(3)^2 = 4 - 4(2)(3)\left(\cos\frac{\pi}{3}\right) + 36 = 40 - 24\left(\frac{1}{2}\right) = 40 - 12 = 28 \Rightarrow |\mathbf{v} - 2\mathbf{w}| = \sqrt{28} = 2\sqrt{7}$.

25. The desired vector is $\mathbf{N} \times \mathbf{v}$ or $\mathbf{v} \times \mathbf{N}$. $\mathbf{N} \times \mathbf{v}$ is perpendicular to both $\mathbf{N}$ and $\mathbf{v}$ and, therefore, also parallel with the plane.

27. $d = \left|\overrightarrow{PS}\,\frac{\mathbf{N}}{|\mathbf{N}|}\right| = \left|\frac{\overrightarrow{PS}\cdot\mathbf{N}}{|\mathbf{N}|}\right| = \left|\frac{(\mathbf{i} + 3\mathbf{j})\cdot(3\mathbf{i} + 4\mathbf{j})}{\sqrt{3^2 + 4^2}}\right| = \frac{|(1)(3) + (3)(4)|}{\sqrt{3^2 + 4^2}} = \frac{15}{\sqrt{25}} = 3$, where $P(2,-1)$, a point on the line,and $S(3,2)$

29. The desired distance $d(t) = \sqrt{(-t-2)^2 + (t-2)^2 + (t-1)^2}$ is minimized when $f(t) = (-t-2)^2 + (t-2)^2 + (t-1)^2$ is minimized. $f'(t) = 2(t + 2) + 2(t - 2) + 2(t - 1) = 0 \Rightarrow t = \frac{1}{3}$. $\therefore$ the distance is $d(1/3) =$ $\sqrt{\left(\frac{7}{3}\right)^2 + \left(\frac{-5}{3}\right)^2 + \left(\frac{-2}{3}\right)^2} = \frac{\sqrt{78}}{3}$. This exercise can also be done with formula (7), $d = \frac{\left|\overrightarrow{PS} \times \mathbf{v}\right|}{|\mathbf{v}|}$, on page 726 of the text.

31. $x = 1 - 3t,\ y = 2,\ z = 3 + 7t$

33. $S(6,0,-6)$, $P(4,0,0)$ is on $x - y = 4 \Rightarrow \overrightarrow{PS} = 2\mathbf{i} - 6\mathbf{k}$ and $\mathbf{N} = \mathbf{i} - \mathbf{j} \Rightarrow d = \left|\frac{\mathbf{N}\cdot\overrightarrow{PS}}{|\mathbf{N}|}\right| = \left|\frac{2}{\sqrt{2}}\right| = \sqrt{2}$

35. $(3,-2,1)$ and $\mathbf{N} = 2\mathbf{i} + \mathbf{j} - \mathbf{k} \Rightarrow 2(x - 3) + 1(y + 2) + (-1)(z - 1) = 0 \Rightarrow 2x + y - z = 3$

37. $P(1,-1,2)$, $Q(2,1,3)$ and $R(-1,2,-1) \Rightarrow \overrightarrow{PQ} = \mathbf{i} + 2\mathbf{j} + \mathbf{k}$, $\overrightarrow{PR} = -2\mathbf{i} + 3\mathbf{j} - 3\mathbf{k}$ and $\overrightarrow{PQ} \times \overrightarrow{PR} = \begin{vmatrix} \mathbf{i} & \mathbf{j} & \mathbf{k} \\ 1 & 2 & 1 \\ -2 & 3 & -3 \end{vmatrix} =$ $-9\mathbf{i} + \mathbf{j} + 7\mathbf{k}$, the normal of the plane; $-9(x - 1) + 1(y + 1) + 7(z - 2) = 0 \Rightarrow -9x + y + 7z = 4$

39. $(0,-\frac{1}{2},-\frac{3}{2})$, since $t = -\frac{1}{2}$, $y = -\frac{1}{2}$ and $z = -\frac{3}{2}$ when $x = 0$; $(-1,0,-3)$, since $t = -1$, $x = -1$ and $z = -3$ when $y = 0$; $(1,-1,0)$, since $t = 0$, $x = 1$ and $y = -1$ when $z = 0$

41. $\mathbf{N}_1 = \mathbf{i}$ and $\mathbf{N}_2 = \mathbf{i} + \mathbf{j} + \sqrt{2}\mathbf{k} \Rightarrow$ the desired angle is $\cos^{-1}\left(\frac{\mathbf{N}_1\cdot\mathbf{N}_2}{|\mathbf{N}_1|\,|\mathbf{N}_2|}\right) = \cos^{-1}\left(\frac{1}{2}\right) = \frac{\pi}{3}$

43. The direction of the line is $\mathbf{N}_1 \times \mathbf{N}_2 = \begin{vmatrix} \mathbf{i} & \mathbf{j} & \mathbf{k} \\ 1 & 2 & 1 \\ 1 & -1 & 2 \end{vmatrix} = 5\mathbf{i} - \mathbf{j} - 3\mathbf{k}$. Since, the point $(-5,3,0)$ is on both planes the desired line is $x = -5 + 5t$, $y = 3 - t$, $z = -3t$.

45. a) The corresponding normals are $\mathbf{N}_1 = 3\mathbf{i} + 6\mathbf{k}$ and $\mathbf{N}_2 = 2\mathbf{i} + 2\mathbf{j} - \mathbf{k}$ and since $\mathbf{N}_1 \cdot \mathbf{N}_2 = (3)(2) + (0)(2) + (6)(-1) = 6 + 0 - 6 = 0 \Rightarrow$ the planes are orthogonal

b) The line of intersection is parallel with $\mathbf{N}_1 \times \mathbf{N}_2 = \begin{vmatrix} \mathbf{i} & \mathbf{j} & \mathbf{k} \\ 3 & 0 & 6 \\ 2 & 2 & -1 \end{vmatrix} = -12\mathbf{i} + 15\mathbf{j} + 6\mathbf{k}$. Now to find a point in the intersection, solve $\begin{cases} 3x + 6z = 1 \\ 2x + 2y - z = 3 \end{cases} \Rightarrow \begin{cases} 3x + 6z = 1 \\ 12x + 12y - 6z = 18 \end{cases} \Rightarrow 15x + 12y = 19 \Rightarrow x = 0$ and $y = \frac{19}{12} \Rightarrow (0,19/12,1/6)$ is a point on the line we seek. Therefore, the line is $x = -12t$, $y = \frac{19}{12} + 15t$ and $z = \frac{1}{6} + 6t$.

47. Yes, $\mathbf{v}$ is parallel to the plane. The reason for this is the plane's normal $2\mathbf{i} + \mathbf{j}$ and the line's direction, $\frac{1}{\sqrt{2^2 + (-4)^2 + 1^2}}(2\mathbf{i} - 4\mathbf{j} + \mathbf{k}) = \frac{1}{\sqrt{21}}(2\mathbf{i} - 4\mathbf{j} + \mathbf{k})$, are perpendicular since $(2\mathbf{i} + \mathbf{j}) \cdot \frac{1}{\sqrt{21}}(2\mathbf{i} - 4\mathbf{j} + \mathbf{k}) = 0 \Rightarrow$ the line and plane must be parallel.

49. $W = \mathbf{F} \cdot \overrightarrow{PQ} = |\mathbf{F}| \left|\overrightarrow{PQ}\right| \cos\theta = (160\ \text{N})(250\ \text{m}) \cos\frac{\pi}{6} \approx 34641\ \text{N} \cdot \text{m} = 34641\ \text{J}$

51. a) $\text{area} = |\mathbf{A} \times \mathbf{B}| = \left\| \begin{matrix} \mathbf{i} & \mathbf{j} & \mathbf{k} \\ 1 & 1 & -1 \\ 2 & 1 & 1 \end{matrix} \right\| = |2\mathbf{i} - 3\mathbf{j} - \mathbf{k}| = \sqrt{4 + 9 + 1} = \sqrt{14}$

b) $\text{volume} = \mathbf{A} \cdot (\mathbf{B} \times \mathbf{C}) = \begin{vmatrix} 1 & 1 & -1 \\ 2 & 1 & 1 \\ -1 & -2 & 3 \end{vmatrix} = 1$

53. a) true, $\sqrt{\mathbf{A} \cdot \mathbf{A}} = \sqrt{|\mathbf{A}|\,|\mathbf{A}| \cos 0} = \sqrt{|\mathbf{A}|^2} = |\mathbf{A}|$ b) not always true, $(2\mathbf{i}) \cdot (2\mathbf{i}) = 4$ while $|2\mathbf{i}| = \sqrt{2^2} = 2$

c) true, $\mathbf{A} \times \mathbf{0} = \mathbf{n}\,|\mathbf{A}|\,|\mathbf{0}| \sin\theta = \mathbf{0} = \mathbf{n}\,|\mathbf{0}|\,|\mathbf{A}| \sin\theta = \mathbf{0} \times \mathbf{A}$ d) true, $\mathbf{A} \times (-\mathbf{A}) = \mathbf{n}\,|\mathbf{A}|\,|-\mathbf{A}| \sin\pi = \mathbf{0}$

e) not always true, they may have opposite directions f) true, by the Vector Distributive Law

g) true, $\mathbf{B}$ is perpendicular to $(\mathbf{A} \times \mathbf{B})$, therefore $(\mathbf{A} \times \mathbf{B}) \cdot \mathbf{B} = 0$ h) true, a property of the Triple Scalar Product

55. $x^2 + y^2 + z^2 = 4$

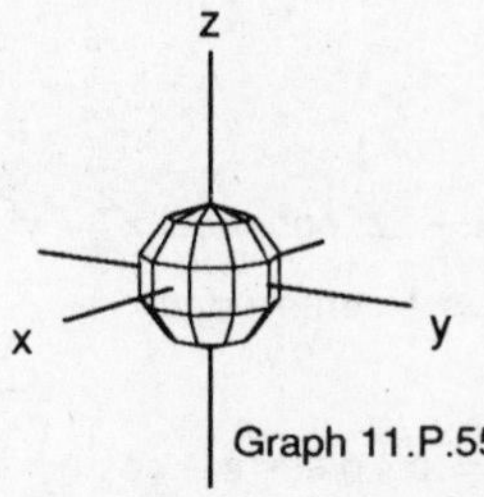

Graph 11.P.55

57. $4x^2 + 4y^2 + z^2 = 4$

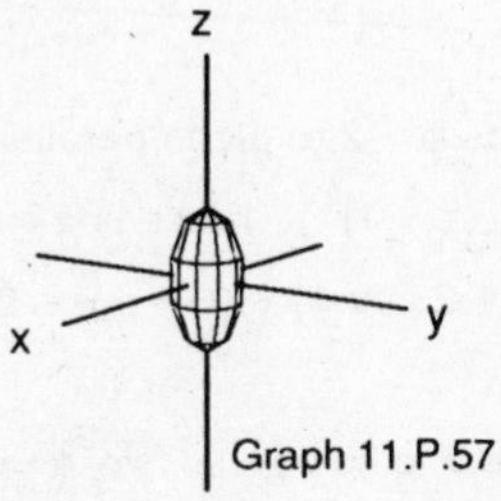

Graph 11.P.57

59. $z = -\left(x^2 + y^2\right)$

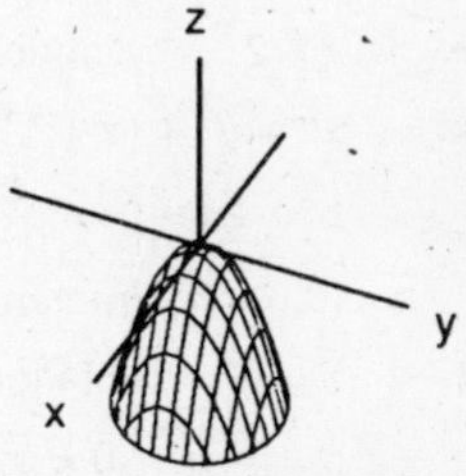

Graph 11.P.59

61. $x^2 + y^2 = z^2$

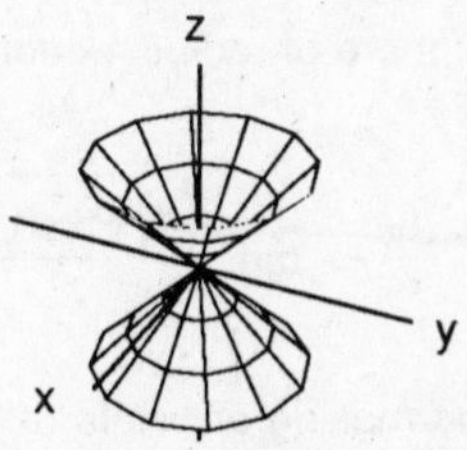

Graph 11.P.61

63. $x^2 + y^2 - z^2 = 4$

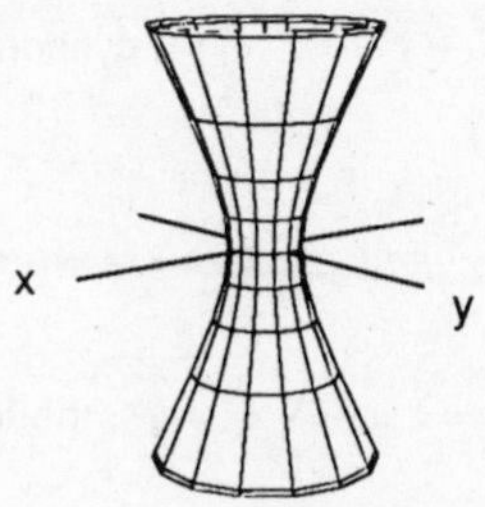

Graph 11.P.63

65. $y^2 - x^2 - z^2 = 1$

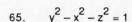

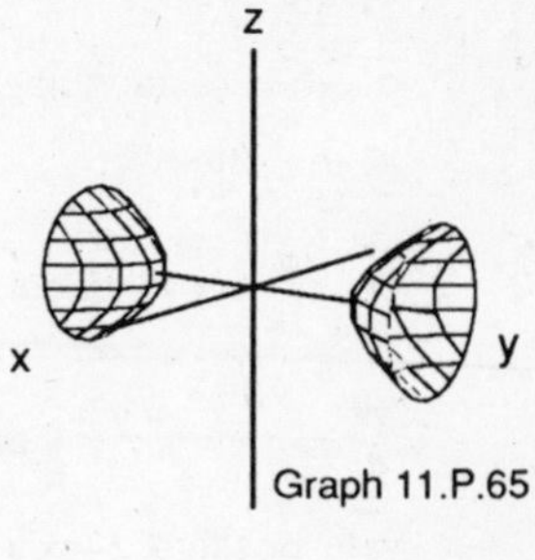

Graph 11.P.65

67. The y–axis in the xy–plane; the yz–plane in three dimensional space

69. The circle centered at (0,0) with a radius of 2 in the xy–plane; the cylinder parallel with the z–axis in three dimensional space with the circle as a generating curve

71. The parabola $x = y^2$ in the xy–plane; the cylinder parallel with the z–axis in three dimensional space with the parabola as a generating curve

73. A cardioid in the $r\theta$–plane; a cylinder parallel with the z–axis in three dimensional space with the cardioid as a generating curve

75. A horizontal lemniscate of length $2\sqrt{2}$ in the $r\theta$–plane; the cylinder parallel with the z–axis in three dimensional space with the lemniscate as a generating curve

77. The sphere with a radius of 2 centered at the origin

79. The upper nappe of a cone having its vertex at the origin making a $\frac{\pi}{6}$ angle with the z–axis

81. The upper hemisphere of a sphere of radius 1 centered at the origin

	Rectangular	Cylindrical	Spherical
83.	(1,0,0)	(1,0,0)	$(1,\pi/2,0)$
85.	(0,1,1)	$(1,\pi/2,1)$	$(\sqrt{2},\pi/4,\pi/2)$
87.	(– 1,0,– 1)	$(1,\pi,-1)$	$(\sqrt{2},3\pi/4,\pi)$

89. $z = 2 \Rightarrow$ cylindrical, $z = 2$; spherical, $\rho \cos \phi = 2$; a plane parallel with the xy–plane

91. $x^2 + y^2 + (z + 1)^2 = 1 \Rightarrow$ cylindrical, $r^2 + (z + 1)^2 = 1 \Rightarrow r^2 + z^2 + 2z + 1 = 1 \Rightarrow r^2 + z^2 = -2z$; spherical $x^2 + y^2 + z^2 + 2z = 0 \Rightarrow \rho^2 + 2\rho \cos \phi = 0 \Rightarrow \rho(\rho + 2 \cos \phi) = 0 \Rightarrow \rho = -2 \cos \phi$ a sphere of radius 1 centered at (0,0,– 1) (rectangular)

93. $z = r^2 \Rightarrow$ rectangular, $z = x^2 + y^2$; spherical, $\rho \cos \phi = \rho^2 \sin^2\phi \Rightarrow \rho^2\sin^2\phi - \rho \cos \phi = 0 \Rightarrow \rho\left(\rho \sin^2\phi - \cos \phi\right) = 0 \Rightarrow \rho = \frac{\cos \phi}{\sin^2\phi}$, $0 < \phi \le \frac{\pi}{2}$, $\rho = 0$ is not needed because at $\phi = \frac{\pi}{2}$ the origin is included, a circular paraboloid symmetric to the z–axis opening upward with its vertex at the origin

95. $r = 7 \sin \theta \Rightarrow$ rectangular, $r = 7 \sin \theta \Rightarrow r = 7\left(\frac{y}{r}\right) \Rightarrow r^2 = 7y \Rightarrow x^2 + y^2 - 7y = 0 \Rightarrow x^2 + y^2 - 7y + \frac{49}{4} = \frac{49}{4} \Rightarrow x^2 + \left(y - \frac{7}{2}\right)^2 = \left(\frac{7}{2}\right)^2$; spherical, $r = 7 \sin \theta \Rightarrow \rho \sin \phi = 7 \sin \theta$, a circular cylinder parallel to the z–axis generated by the circle

97. $\rho = 4 \Rightarrow$ rectangular, $\sqrt{x^2 + y^2 + z^2} = 4 \Rightarrow x^2 + y^2 + z^2 = 16$; cylindrical, $r^2 + z^2 = 16$, a sphere of radius 4 centered at the origin

99. $\phi = \frac{3\pi}{4} \Rightarrow$ cylindrical, $\tan^{-1}\left(\frac{r}{z}\right) = \frac{3\pi}{4} \Rightarrow \frac{r}{z} = -1 \Rightarrow r = -z \Rightarrow z = -r$, $r \ge 0$; rectangular, $\phi = \frac{3\pi}{4} \Rightarrow \tan^{-1}\left(\frac{\sqrt{x^2 + y^2}}{z}\right) = \frac{3\pi}{4} \Rightarrow \frac{\sqrt{x^2 + y^2}}{z} = \tan \frac{3\pi}{4} \Rightarrow \frac{\sqrt{x^2 + y^2}}{z} = -1 \Rightarrow z = -\sqrt{x^2 + y^2}$, the lower nappe of a cone making an angle of $3\pi/4$ with positive z–axis with vertex at the origin

CHAPTER 12

VECTOR–VALUED FUNCTIONS

12.1 VECTOR–VALUED FUNCTIONS AND SPACE CURVES

1. $\mathbf{r} = (2\cos t)\,\mathbf{i} + (3\sin t)\,\mathbf{j} + 4t\,\mathbf{k} \Rightarrow \mathbf{v} = \frac{d\mathbf{r}}{dt} = (-2\sin t)\,\mathbf{i} + (3\cos t)\,\mathbf{j} + 4\,\mathbf{k}$

   $\mathbf{a} = \frac{d^2\mathbf{r}}{dt^2} = (-2\cos t)\,\mathbf{i} - (3\sin t)\,\mathbf{j}$. Speed: $\left|\mathbf{v}\left(\frac{\pi}{2}\right)\right| = \sqrt{\left(-2\sin\frac{\pi}{2}\right)^2 + \left(3\cos\frac{\pi}{2}\right)^2 + 4^2} = 2\sqrt{5}$

   Direction: $\dfrac{\mathbf{v}\left(\frac{\pi}{2}\right)}{\left|\mathbf{v}\left(\frac{\pi}{2}\right)\right|} = \left(-\frac{2}{2\sqrt{5}}\sin\frac{\pi}{2}\right)\mathbf{i} + \left(\frac{3}{2\sqrt{5}}\cos\frac{\pi}{2}\right)\mathbf{j} + \frac{4}{2\sqrt{5}}\mathbf{k} = -\frac{1}{\sqrt{5}}\mathbf{i} + \frac{2}{\sqrt{5}}\mathbf{k}$

   $\mathbf{v}\left(\frac{\pi}{2}\right) = 2\sqrt{5}\left[-\frac{1}{\sqrt{5}}\mathbf{i} + \frac{2}{\sqrt{5}}\mathbf{k}\right]$

3. $\mathbf{r} = (t+1)\,\mathbf{i} + (t^2-1)\,\mathbf{j} + 2t\,\mathbf{k} \Rightarrow \mathbf{v} = \frac{d\mathbf{r}}{dt} = \mathbf{i} + 2t\,\mathbf{j} + 2\,\mathbf{k}$, $\mathbf{a} = \frac{d^2\mathbf{r}}{dt^2} = 2\,\mathbf{j}$. Speed: $|\mathbf{v}(1)| = \sqrt{1^2 + (2(1))^2 + 2^2} = 3$.

   Direction: $\dfrac{\mathbf{v}(1)}{|\mathbf{v}(1)|} = \dfrac{\mathbf{i} + 2(1)\,\mathbf{j} + 2\,\mathbf{k}}{3} = \frac{1}{3}\mathbf{i} + \frac{2}{3}\mathbf{j} + \frac{2}{3}\mathbf{k}$. $\mathbf{v}(1) = 3\left[\frac{1}{3}\mathbf{i} + \frac{2}{3}\mathbf{j} + \frac{2}{3}\mathbf{k}\right]$

5. $\mathbf{r} = (2\ln(t+1))\,\mathbf{i} + t^2\,\mathbf{j} + \frac{t^2}{2}\,\mathbf{k} \Rightarrow \mathbf{v} = \frac{d\mathbf{r}}{dt} = \left(\frac{2}{t+1}\right)\mathbf{i} + 2t\,\mathbf{j} + t\,\mathbf{k}$, $\mathbf{a} = \frac{d^2\mathbf{r}}{dt^2} = \left(\frac{-2}{(t+1)^2}\right)\mathbf{i} + 2\,\mathbf{j} + \mathbf{k}$

   Speed: $|\mathbf{v}(1)| = \sqrt{\left(\frac{2}{1+1}\right)^2 + (2(1))^2 + 1^2} = \sqrt{6}$. Direction: $\dfrac{\mathbf{v}(1)}{|\mathbf{v}(1)|} = \dfrac{\left(\frac{2}{1+1}\right)\mathbf{i} + 2(1)\,\mathbf{j} + (1)\,\mathbf{k}}{\sqrt{6}}$

   $= \frac{1}{\sqrt{6}}\mathbf{i} + \frac{2}{\sqrt{6}}\mathbf{j} + \frac{1}{\sqrt{6}}\mathbf{k}$. $\mathbf{v}(1) = \sqrt{6}\left[\frac{1}{\sqrt{6}}\mathbf{i} + \frac{2}{\sqrt{6}}\mathbf{j} + \frac{1}{\sqrt{6}}\mathbf{k}\right]$

7. $\mathbf{r} = e^t\,\mathbf{i} + \frac{2}{9}e^{2t}\,\mathbf{j} \Rightarrow \mathbf{v} = \frac{d\mathbf{r}}{dt} = e^t\,\mathbf{i} + \frac{4}{9}e^{2t}\,\mathbf{j}$, $\mathbf{a} = \frac{d^2\mathbf{r}}{dt^2} = e^t\,\mathbf{i} + \frac{8}{9}e^{2t}\,\mathbf{j}$. Speed: $|\mathbf{v}(\ln 3)| =$

   $\sqrt{\left(e^{\ln 3}\right)^2 + \left(\frac{4}{9}e^{2\ln 3}\right)^2} = 5$ Direction: $\dfrac{\mathbf{v}(\ln 3)}{|\mathbf{v}(\ln 3)|} = \dfrac{e^{\ln 3}\,\mathbf{i} + \frac{4}{9}e^{2\ln 3}\,\mathbf{j}}{5} = \frac{3}{5}\mathbf{i} + \frac{4}{5}\mathbf{j}$.

   $\mathbf{v}(\ln 3) = 5\left[\frac{3}{5}\mathbf{i} + \frac{4}{5}\mathbf{j}\right]$

9. $\mathbf{v} = 3\,\mathbf{i} + \sqrt{3}\,\mathbf{j} + 2t\,\mathbf{k}$, $\mathbf{a} = 2\,\mathbf{k}$. $\mathbf{v}(0) = 3\,\mathbf{i} + \sqrt{3}\,\mathbf{j}$, $\mathbf{a}(0) = 2\,\mathbf{k} \Rightarrow |\mathbf{v}(0)| = \sqrt{3^2 + \left(\sqrt{3}\right)^2 + 0^2} = \sqrt{12}$,

   $|\mathbf{a}(0)| = \sqrt{2^2} = 2$. $\mathbf{v}(0)\cdot\mathbf{a}(0) = 0 \Rightarrow \cos\theta = \frac{0}{2\sqrt{12}} = 0 \Rightarrow \theta = \frac{\pi}{2}$

11. $\mathbf{v} = \frac{2t}{t^2+1}\,\mathbf{i} + \frac{1}{t^2+1}\,\mathbf{j} + t\left(t^2+1\right)^{-1/2}\,\mathbf{k}$, $\mathbf{a} = \frac{-2t^2+2}{\left(t^2+1\right)^2}\,\mathbf{i} - \frac{2t}{\left(t^2+1\right)^2}\,\mathbf{j} + \frac{1}{\left(t^2+1\right)^{3/2}}\,\mathbf{k}$. $\mathbf{v}(0) = \mathbf{j}$,

    $\mathbf{a}(0) = 2\,\mathbf{i} + \mathbf{k} \Rightarrow |\mathbf{v}(0)| = 1$, $|\mathbf{a}(0)| = \sqrt{2^2 + 1^2} = \sqrt{5}$. $\mathbf{v}(0)\cdot\mathbf{a}(0) = 0 \Rightarrow \cos\theta = \frac{0}{1\sqrt{5}} = 0 \Rightarrow \theta = \frac{\pi}{2}$

13. $\mathbf{v} = (1-\cos t)\,\mathbf{i} + (\sin t)\,\mathbf{j}$, $\mathbf{a} = (\sin t)\,\mathbf{i} + (\cos t)\,\mathbf{j} \Rightarrow \mathbf{v}\cdot\mathbf{a} = \sin t(1-\cos t) + \sin t(\cos t) = \sin t$.

    $\mathbf{v}\cdot\mathbf{a} = 0 \Rightarrow \sin t = 0 \Rightarrow t = 0, \pi, 2\pi$

15. $\int_0^1 \left(t^3\,\mathbf{i} + 7\,\mathbf{j} + (t+1)\,\mathbf{k}\right) dt = \left[\frac{t^4}{4}\right]_0^1 \mathbf{i} + [7t]_0^1 \mathbf{j} + \left[\frac{t^2}{2} + t\right]_0^1 \mathbf{k} = \frac{1}{4}\mathbf{i} + 7\mathbf{j} + \frac{3}{2}\mathbf{k}$

17. $\int_{-\pi/4}^{\pi/4} \left((\sin t)\,\mathbf{i} + (1+\cos t)\,\mathbf{j} + (\sec^2 t)\,\mathbf{k}\right) dt = [-\cos t]_{-\pi/4}^{\pi/4} \mathbf{i} + [t + \sin t]_{-\pi/4}^{\pi/4} \mathbf{j} + [\tan t]_{-\pi/4}^{\pi/4} \mathbf{k} =$
$\left(\frac{\pi + 2\sqrt{2}}{2}\right)\mathbf{j} + 2\,\mathbf{k}$

19. $\int_1^4 \left(\frac{1}{t}\,\mathbf{i} + \frac{1}{5-t}\,\mathbf{j} + \frac{1}{2t}\,\mathbf{k}\right) dt = [\ln t]_1^4 \mathbf{i} + [-\ln(5-t)]_1^4 \mathbf{j} + \left[\frac{1}{2}\ln t\right]_1^4 \mathbf{k} = (\ln 4)\,\mathbf{i} + (\ln 4)\,\mathbf{j} + (\ln 2)\,\mathbf{k}$

21. $\mathbf{v} = (\cos t)\,\mathbf{i} - (\sin t)\,\mathbf{j}$, $\mathbf{a} = -(\sin t)\,\mathbf{i} - (\cos t)\,\mathbf{j} \Rightarrow$ For $t = \frac{\pi}{4}$, $\mathbf{v}\left(\frac{\pi}{4}\right) = \frac{\sqrt{2}}{2}\mathbf{i} - \frac{\sqrt{2}}{2}\mathbf{j}$, $\mathbf{a}\left(\frac{\pi}{4}\right) = -\frac{\sqrt{2}}{2}\mathbf{i} - \frac{\sqrt{2}}{2}\mathbf{j}$;
For $t = \frac{\pi}{2}$, $\mathbf{v}\left(\frac{\pi}{2}\right) = -\mathbf{j}$, $\mathbf{a}\left(\frac{\pi}{2}\right) = -\mathbf{i}$

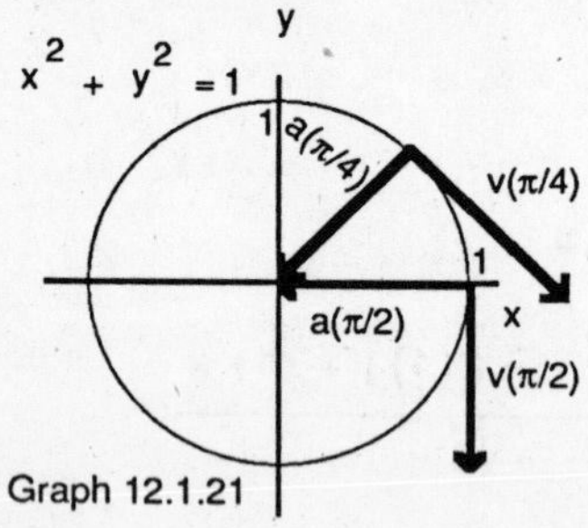

Graph 12.1.21

23. $\mathbf{v} = (1 - \cos t)\,\mathbf{i} + (\sin t)\,\mathbf{j}$, $\mathbf{a} = (\sin t)\,\mathbf{i} + (\cos t)\,\mathbf{j} \Rightarrow$ For $t = \pi$, $\mathbf{v}(\pi) = 2\,\mathbf{i}$, $\mathbf{a}(\pi) = -\mathbf{j}$; For $t = \frac{3\pi}{2}$,
$\mathbf{v}\left(\frac{3\pi}{2}\right) = \mathbf{i} - \mathbf{j}$, $\mathbf{a}\left(\frac{3\pi}{2}\right) = -\mathbf{i}$

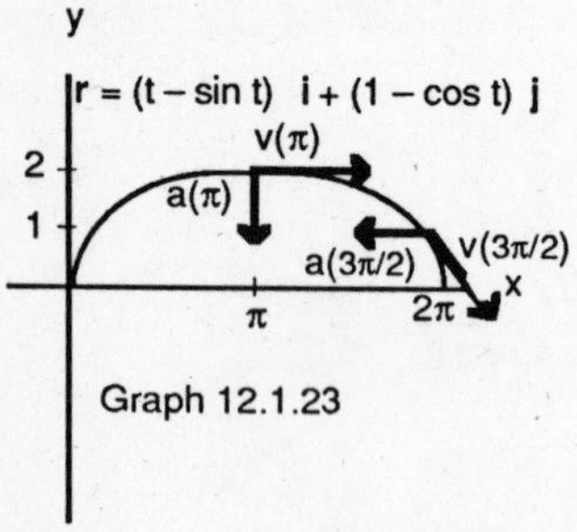

Graph 12.1.23

25. $\mathbf{r} = \int (-t\,\mathbf{i} - t\,\mathbf{j} - t\,\mathbf{k})\,dt = -\frac{t^2}{2}\mathbf{i} - \frac{t^2}{2}\mathbf{j} - \frac{t^2}{2}\mathbf{k} + \mathbf{C}$. $\mathbf{r}(0) = 0\,\mathbf{i} - 0\,\mathbf{j} - 0\,\mathbf{k} + \mathbf{C} = \mathbf{i} + 2\,\mathbf{j} + 3\,\mathbf{k} \Rightarrow$
$\mathbf{C} = \mathbf{i} + 2\,\mathbf{j} + 3\,\mathbf{k}$. $\therefore\ \mathbf{r} = \left(-\frac{t^2}{2} + 1\right)\mathbf{i} + \left(-\frac{t^2}{2} + 2\right)\mathbf{j} + \left(-\frac{t^2}{2} + 3\right)\mathbf{k}$

27. $\mathbf{r} = \int \left(\frac{3}{2}(t+1)^{1/2}\,\mathbf{i} + e^{-t}\,\mathbf{j} + \frac{1}{t+1}\,\mathbf{k}\right) dt = (t+1)^{3/2}\,\mathbf{i} - e^{-t}\,\mathbf{j} + \ln(t+1)\,\mathbf{k} + \mathbf{C}$. $\mathbf{r}(0) = \mathbf{k} \Rightarrow$
$(0+1)^{3/2}\,\mathbf{i} - e^{-0}\,\mathbf{j} + \ln(0+1)\,\mathbf{k} + \mathbf{C} = \mathbf{k} \Rightarrow \mathbf{C} = -\mathbf{i} + \mathbf{j} + \mathbf{k}$.
$\therefore\ \mathbf{r} = \left((t+1)^{3/2} - 1\right)\mathbf{i} + \left(1 - e^{-t}\right)\mathbf{j} + (1 + \ln(t+1))\,\mathbf{k}$

29. $\frac{d\mathbf{r}}{dt} = \int (-32\,\mathbf{k})\,dt = -32t\,\mathbf{k} + \mathbf{C}_1$. $\frac{d\mathbf{r}}{dt}(0) = 8\,\mathbf{i} + 8\,\mathbf{j} \Rightarrow -32(0)\,\mathbf{k} + \mathbf{C}_1 = 8\,\mathbf{i} + 8\,\mathbf{j} \Rightarrow \mathbf{C}_1 = 8\,\mathbf{i} + 8\,\mathbf{j}$

$\therefore \frac{d\mathbf{r}}{dt} = 8\,\mathbf{i} + 8\,\mathbf{j} - 32t\,\mathbf{k}$. $\mathbf{r} = \int (8\,\mathbf{i} + 8\,\mathbf{j} - 32t\,\mathbf{k})\,dt = 8t\,\mathbf{i} + 8t\,\mathbf{j} - 16t^2\,\mathbf{k} + \mathbf{C}_2$. $\mathbf{r}(0) = 100\,\mathbf{k} \Rightarrow$

$8(0)\,\mathbf{i} + 8(0)\,\mathbf{j} - 16(0)^2\,\mathbf{k} + \mathbf{C}_2 = 100\,\mathbf{k} \Rightarrow \mathbf{C}_2 = 100\,\mathbf{k}$. $\therefore\ \mathbf{r} = 8t\,\mathbf{i} + 8t\,\mathbf{j} + (100 - 16t^2)\,\mathbf{k}$

31. $\mathbf{v} = (1 - \cos t)\,\mathbf{i} + (\sin t)\,\mathbf{j}$, $\mathbf{a} = (\sin t)\,\mathbf{i} + (\cos t)\,\mathbf{j}$. $|\mathbf{v}|^2 = (1 - \cos t)^2 + \sin^2 t = 2 - 2\cos t$. $|\mathbf{v}|^2$ is at a max when $\cos t = -1 \Rightarrow t = \pi, 3\pi, 5\pi$, etc. At these values of t, $|\mathbf{v}|^2 = 4 \Rightarrow \max|\mathbf{v}| = \sqrt{4} = 2$. $|\mathbf{v}|^2$ is at a min when $\cos t = 1 \Rightarrow t = 0, 2\pi, 4\pi$, etc. At these values of t, $|\mathbf{v}|^2 = 0 \Rightarrow \min|\mathbf{v}| = 0$.
$|\mathbf{a}|^2 = \sin^2 t + \cos^2 t = 1$ for every $t \Rightarrow \max|\mathbf{a}| = \min|\mathbf{a}| = \sqrt{1} = 1$.

33. The length of $\mathbf{r}$ is constant (it equals the radius of the sphere). Hence $\mathbf{r} \cdot (d\mathbf{r}/dt) = 0$.

35. $\mathbf{r}(t) = (\sin t)\mathbf{i} + (t^2 - \cos t)\mathbf{j} + e^t\,\mathbf{k} \Rightarrow \mathbf{v}(t) = (\cos t)\mathbf{i} + (2t + \sin t)\mathbf{j} + e^t\,\mathbf{k}$. $t_0 = 0 \Rightarrow \mathbf{v}_0 = \mathbf{i} + \mathbf{k}$. $P_0 = (0,-1,1)$.
$\therefore$ $x = 0 + t = t$, $y = -1$, $z = 1 + t$ are the parametric equations of the tangent line.

37. $\mathbf{r}(t) = (a \sin t)\mathbf{i} + (a \cos t)\mathbf{j} + bt\,\mathbf{k} \Rightarrow \mathbf{v}(t) = (a \cos t)\mathbf{i} - (a \sin t)\mathbf{j} + b\,\mathbf{k}$. $t_0 = 2\pi \Rightarrow \mathbf{v}_0 = a\,\mathbf{i} + b\,\mathbf{k}$. $P_0 = (0,a,2b\pi)$.
$\therefore$ $x = 0 + at = at$, $y = a$, $z = 2b\pi + bt$ are the parametric equations of the tangent line.

39. Let $\mathbf{r} = \mathbf{C}$, a constant vector. Then $\mathbf{r} = c_1\,\mathbf{i} + c_2\,\mathbf{j} + c_3\,\mathbf{k}$ where c_1, c_2, c_3 are Real Numbers.
$\frac{d\mathbf{r}}{dt} = 0\,\mathbf{i} + 0\,\mathbf{j} + 0\,\mathbf{k} = 0$

41. Let $\mathbf{u} = f_1(t)\,\mathbf{i} + f_2(t)\,\mathbf{j} + f_3(t)\,\mathbf{k}$, $\mathbf{v} = g_1(t)\,\mathbf{i} + g_2(t)\,\mathbf{j} + g_3(t)\,\mathbf{k}$. Then $\mathbf{u} + \mathbf{v} = (f_1(t) + g_1(t))\,\mathbf{i} + (f_2(t) + g_2(t))\,\mathbf{j} +$
$(f_3(t) + g_3(t))\,\mathbf{k} \Rightarrow \frac{d}{dt}(\mathbf{u} + \mathbf{v}) = (f'_1(t) + g'_1(t))\,\mathbf{i} + (f'_2(t) + g'_2(t))\,\mathbf{j} + (f'_3(t) + g'_3(t))\,\mathbf{k} =$
$(f'_1(t)\,\mathbf{i} + f'_2(t)\,\mathbf{j} + f'_3(t)\,\mathbf{k}) + (g'_1(t)\,\mathbf{i} + g'_2(t)\,\mathbf{j} + g'_3(t)\,\mathbf{k} = \frac{d\mathbf{u}}{dt} + \frac{d\mathbf{v}}{dt}$.
$\mathbf{u} - \mathbf{v} = (f_1(t) - g_1(t))\,\mathbf{i} + (f_2(t) - g_2(t))\,\mathbf{j} + (f_3(t) - g_3(t))\,\mathbf{k} \Rightarrow \frac{d}{dt}(\mathbf{u} - \mathbf{v}) = (f'_1(t) - g'_1(t))\,\mathbf{i} + (f'_2(t) - g'_2(t))\,\mathbf{j} +$
$(f'_3(t) - g'_3(t))\,\mathbf{k} = (f'_1(t)\,\mathbf{i} + f'_2(t)\,\mathbf{j} + f'_3(t)\,\mathbf{k}) - (g'_1(t)\,\mathbf{i} + g'_2(t)\,\mathbf{j} + g'_3(t)\,\mathbf{k}) = \frac{d\mathbf{u}}{dt} - \frac{d\mathbf{v}}{dt}$.

43. a) Let u, $\mathbf{r}$ be continuous on [a,b]. Then $\lim_{t \to t_0} u(t)\,\mathbf{r}(t) = \lim_{t \to t_0} [u(t)\,f(t)\,\mathbf{i} + u(t)\,g(t)\,\mathbf{j} + u(t)\,h(t)\,\mathbf{k}] = u(t_0)\,f(t_0)\,\mathbf{i} +$
$u(t_0)\,g(t_0)\,\mathbf{j} + u(t_0)\,h(t_0)\,\mathbf{k} = u(t_0)\,\mathbf{r}(t_0)$

b) Let u, $\mathbf{r}$ be differentiable. Then $\frac{d}{dt}(u\mathbf{r}) = \frac{d}{dt}[u(t)\,f(t)\,\mathbf{i} + u(t)\,g(t)\,\mathbf{j} + u(t)\,h(t)\,\mathbf{k}] = \left(\frac{du}{dt} f(t) + u(t)\frac{df}{dt}\right)\mathbf{i} +$
$\left(\frac{du}{dt} g(t) + u(t)\frac{dg}{dt}\right)\mathbf{j} + \left(\frac{du}{dt} h(t) + u(t)\frac{dh}{dt}\right)\mathbf{k} = (f(t)\,\mathbf{i} + g(t)\,\mathbf{j} + h(t)\,\mathbf{k})\frac{du}{dt} + u(t)\left(\frac{df}{dt} + \frac{dg}{dt} + \frac{dh}{dt}\right) =$
$\mathbf{r}\frac{du}{dt} + u\frac{d\mathbf{r}}{dt}$

12.2 MODELING PROJECTILE MOTION

1. $x = (v_0 \cos \alpha)t \Rightarrow (21\text{ km})\left(\frac{1000\text{ m}}{1\text{ km}}\right) = 840\text{ m/s}(\cos 60°)t \Rightarrow t = \frac{21\,000\text{ m}}{(840\text{ km/s})(\cos 60°)} = 50$ seconds

3. a) $t = \frac{2v_0 \sin \alpha}{g} = \frac{2(500\text{ m/s})\sin 45°}{9.8\text{ m/s}^2} \approx 72.2$ seconds. $R = \frac{v_0^2}{g}\sin 2\alpha = \frac{(500\text{ m/s})^2}{9.8\text{ m/s}^2}(\sin 2(45°)) =$
25 510.2 m

b) $x = (v_0 \cos \alpha)t \Rightarrow 5000\text{ m} = (500\text{ m/s})(\cos 45°)t \Rightarrow t = \frac{5000\text{ m}}{(500\text{ m/s})\cos 45°} \approx 14.14$ s

3. (Continued)

$y = (v_0 \sin \alpha)t - \frac{1}{2}gt^2 \Rightarrow y \approx (500 \text{ m/s})(\sin 45°)(14.14 \text{ s}) - \frac{1}{2}(9.8 \text{ m/s}^2)(14.14 \text{ s})^2 \approx 4020.3 \text{ m}$

c) $y_{max} = \frac{(v_0 \sin \alpha)^2}{2g} = \frac{((500 \text{ m/s})\sin 45°)^2}{2(9.8 \text{ m/s}^2)} \approx 6377.6 \text{ m}$

5. $R = \frac{v_0^2}{g} \sin 2\alpha = \frac{v_0^2}{g}(2 \sin \alpha \cos \alpha) = \frac{v_0^2}{g}(2 \cos(90° - \alpha) \sin(90° - \alpha)) = \frac{v_0^2}{g}(\sin 2(90° - \alpha))$

7. $R = \frac{v_0^2}{g} \sin 2\alpha \Rightarrow 10 \text{ m} = \frac{v_0^2}{9.8 \text{ m/s}^2} \sin 2(45°) \Rightarrow v_0^2 = 98 \text{ m}^2/\text{s}^2 \Rightarrow v_0 \approx 9.9 \text{ m/s}.$

$6 \text{ m} \approx \frac{(9.9 \text{ m/s})^2}{9.8 \text{ m/s}^2} \sin 2\alpha \sin 2\alpha \approx 0.59999 \Rightarrow 2\alpha \approx 36.87° \text{ or } 143.12° \; \alpha \approx 18.44° \text{ or } 71.56°$

9. $R = \frac{v_0^2}{g} \sin 2\alpha \Rightarrow 746.4 \text{ ft}) = \frac{v_0^2}{32 \text{ ft/sec}^2} \sin 2(9°) \Rightarrow v_0^2 \approx 77\,292.84 \text{ ft}^2/\text{sec}^2 \Rightarrow v_0 \approx 278.01 \text{ ft/sec} \approx$ 189.6 mph

11. $y_{max} = \frac{(v_0 \sin \alpha)^2}{2g} \Rightarrow \frac{3}{4} y_{max} = \frac{3(v_0 \sin \alpha)^2}{8g}$. $y = (v_0 \sin \alpha)t - \frac{1}{2}gt^2 \Rightarrow \frac{3(v_0 \sin \alpha)^2}{8g} =$
$(v_0 \sin \alpha)t - \frac{1}{2}gt^2 \Rightarrow 3(v_0 \sin \alpha)^2 = (8gv_0 \sin \alpha)t - 4g^2t^2 \Rightarrow 4g^2t^2 - (8gv_0 \sin \alpha)t + 3(v_0 \sin \alpha)^2 = 0$
$\Rightarrow 2gt - 3v_0 \sin \alpha = 0 \text{ or } 2gt - v_0 \sin \alpha = 0 \Rightarrow t = \frac{3v_0 \sin \alpha}{2g} \text{ or } t = \frac{v_0 \sin \alpha}{2g}$. Since the time it takes to reach y_{max} is $t_{max} = \frac{v_0 \sin \alpha}{g}$, then the time it takes the projectile to reach $\frac{3}{4}$ of y_{max} is $t = \frac{v_0 \sin \alpha}{2g}$ or $\frac{1}{2} t_{max}$.

13. $x = (v_0 \cos \alpha)t \Rightarrow 135 \text{ ft} = (90 \text{ ft/sec})(\cos 30°)t \Rightarrow t \approx 1.732 \text{ sec}.$ $y = (v_0 \sin \alpha)t - \frac{1}{2}gt^2 \Rightarrow$
$y \approx (90 \text{ ft/sec})(\sin 30°)(1.732 \text{ sec}) - \frac{1}{2}(32 \text{ ft/sec}^2)(1.732 \text{ sec})^2 \Rightarrow y \approx 29.94 \text{ ft}$. The golf ball will clip the leaves at the top.

15. $x = 0 + (44 \cos 40°)\,t \approx 33.706\,t,\; y = 6.5 + (44 \sin 40°)\,t - 16t^2 \approx 6.5 + 28.283\,t - 16t^2.\; y = 0 \Rightarrow$
$t \approx \frac{28.283 + \sqrt{(28.283)^2 + 416}}{32} \approx 1.9735$ sec, the positive answer. Then $x \approx 33.706(1.9735)$
≈ 66.51 ft $\Rightarrow$ the difference in distances is about $66.51 - 66.43 = 0.08$ ft or 1 inch.

17. $x = (v_0 \cos \alpha)\,t \Rightarrow 315 \text{ ft} = (v_0 \cos 20°)t \Rightarrow v_0 = \frac{315}{t \cos 20°}$. $y = (v_0 \sin \alpha)t - \frac{1}{2}gt^2 \Rightarrow$
$34 \text{ ft} = \frac{315}{t \cos 20°}(t \sin 20°) - \frac{1}{2}(32)t^2 \Rightarrow 34 = 315 \tan 20° - 16t^2 \Rightarrow t^2 \approx 5.04 \text{ sec}^2 \Rightarrow t \approx 2.25 \text{ sec}$
$t \approx 2.25 \text{ sec } \alpha v_0 = \frac{315}{(2.25)\cos 20°} \approx 148.98 \text{ ft/sec}$

19. Height of the Marble A, R units downrange: $x = (v_0 \cos \alpha)\,t$ and $x = R \Rightarrow R = (v_0 \cos \alpha)\,t \Rightarrow$
$t = \frac{R}{v_0 \cos \alpha}$. $y = (v_0 \sin \alpha)t - \frac{1}{2}gt^2 \Rightarrow y = (v_0 \sin \alpha)\left(\frac{R}{v_0 \cos \alpha}\right) - \frac{1}{2}g\left(\frac{R}{v_0 \cos \alpha}\right)^2 \Rightarrow$
$y = R \tan \alpha - \frac{1}{2}g\left(\frac{R^2}{v_0^2 \cos^2 \alpha}\right)$ is the height of Marble A after $t = \frac{R}{v_0 \cos \alpha}$ seconds.

Height of the Marble B, at $t = \frac{R}{v_0 \cos \alpha}$ seconds: $y = R \tan \alpha - \frac{1}{2}gt^2 = R \tan \alpha - \frac{1}{2}g\left(\frac{R}{v_0 \cos \alpha}\right)^2 =$

19. (Continued)

$R\tan\alpha - \frac{1}{2}g\left(\frac{R^2}{v_0^2\cos^2\alpha}\right)$ which is the height of Marble A. ∴ They collide regardless of the initial velocity.

21. $\frac{d\mathbf{r}}{dt} = \int(-g\,\mathbf{j})\,dt = -gt\,\mathbf{j} + \mathbf{C}_1$. $\frac{d\mathbf{r}}{dt}(0) = (v_0\cos\alpha)\,\mathbf{i} + (v_0\sin\alpha)\,\mathbf{j} \Rightarrow -g(0)\,\mathbf{j} + \mathbf{C}_1 = (v_0\cos\alpha)\,\mathbf{i} +$ $(v_0\sin\alpha)\,\mathbf{j} \Rightarrow \mathbf{C}_1 = (v_0\cos\alpha)\,\mathbf{i} + (v_0\sin\alpha)\,\mathbf{j} \Rightarrow \frac{d\mathbf{r}}{dt} = (v_0\cos\alpha)\,\mathbf{i} + (v_0\sin\alpha - gt)\,\mathbf{j}$.

$\mathbf{r} = \int\left((v_0\cos\alpha)\,\mathbf{i} + (v_0\sin\alpha - gt)\,\mathbf{j}.\right)dt = (v_0t\cos\alpha)\,\mathbf{i} + (v_0t\sin\alpha - \frac{1}{2}gt^2)\,\mathbf{j} + \mathbf{C}_2$.

$\mathbf{r}(0) = x_0\,\mathbf{i} + y_0\,\mathbf{j} \Rightarrow (v_0(0)\cos\alpha)\,\mathbf{i} + (v_0(0)\sin\alpha - \frac{1}{2}g(0)^2)\,\mathbf{j} + \mathbf{C}_2 = x_0\,\mathbf{i} + y_0\,\mathbf{j} \Rightarrow$

$\mathbf{C}_2 = x_0\,\mathbf{i} + y_0\,\mathbf{j}$. ∴ $\mathbf{r} = (x_0 + v_0t\cos\alpha)\,\mathbf{i} + (y_0 + v_0t\sin\alpha - \frac{1}{2}gt^2)\,\mathbf{j} \Rightarrow x = x_0 + v_0t\cos\alpha$,

$y = y_0 + v_0t\sin\alpha - \frac{1}{2}gt^2$

23. The horizontal distance from the archer to the center of the cauldron is 90 ft ⇒ the horizontal distance from the archer to the rim of the cauldron is 84 ft. ∴ $x = x_0 + (v_0\cos\theta)t \Rightarrow 84 \approx 0 + 43.66t \Rightarrow t \approx 1.92$ sec.

The vertical distance at this time is $y = y_0 + (v_0\sin\theta)t - \frac{1}{2}gt^2 = 6 + 65.97(1.92) - \frac{1}{2}(32)(1.92)^2 \approx 73.68$ ft.

∴ the arrow clears the rim.

12.3 ARC LENGTH PARAMETRIZATION

1. $\mathbf{r} = (2\cos t)\,\mathbf{i} + (2\sin t)\,\mathbf{j} + \sqrt{5}\,t\,\mathbf{k} \Rightarrow \mathbf{v} = (-2\sin t)\,\mathbf{i} + (2\cos t)\,\mathbf{j} + \sqrt{5}\,\mathbf{k} \Rightarrow$

$|\mathbf{v}| = \sqrt{(-2\sin t)^2 + (2\cos t)^2 + (\sqrt{5})^2} = \sqrt{4\sin^2 t + 4\cos^2 t + 5} = 3$.

$\mathbf{T} = \frac{\mathbf{v}}{|\mathbf{v}|} = \left(-\frac{2}{3}\sin t\right)\mathbf{i} + \left(\frac{2}{3}\cos t\right)\mathbf{j} + \frac{\sqrt{5}}{3}\mathbf{k}$. Length $= \int_0^\pi |\mathbf{v}|\,dt = \int_0^\pi 3\,dt = [3t]_0^\pi = 3\pi$

3. $\mathbf{r} = t\,\mathbf{i} + \frac{2}{3}t^{3/2}\,\mathbf{k} \Rightarrow \mathbf{v} = \mathbf{i} + t^{1/2}\,\mathbf{k} \Rightarrow |\mathbf{v}| = \sqrt{1^2 + (t^{1/2})^2} = \sqrt{1+t}$. $\mathbf{T} = \frac{\mathbf{v}}{|\mathbf{v}|} = \frac{1}{\sqrt{1+t}}\mathbf{i} + \frac{\sqrt{t}}{\sqrt{1+t}}\mathbf{k}$

Length $= \int_0^8 \sqrt{1+t}\,dt = \left[\frac{2}{3}(1+t)^{3/2}\right]_0^8 = \frac{52}{3}$

5. $\mathbf{r} = (2+t)\,\mathbf{i} - (t+1)\,\mathbf{j} + t\,\mathbf{k} \Rightarrow \mathbf{v} = \mathbf{i} - \mathbf{j} + \mathbf{k} \Rightarrow |\mathbf{v}| = \sqrt{1^2 + (-1)^2 + 1^2} = \sqrt{3}$. $\mathbf{T} = \frac{\mathbf{v}}{|\mathbf{v}|} = \frac{1}{\sqrt{3}}\mathbf{i} - \frac{1}{\sqrt{3}}\mathbf{j} + \frac{1}{\sqrt{3}}\mathbf{k}$

Length $= \int_0^3 \sqrt{3}\,dt = [\sqrt{3}\,t]_0^3 = 3\sqrt{3}$

7. $\mathbf{r} = (t\cos t)\,\mathbf{i} + (t\sin t)\,\mathbf{j} + \frac{2\sqrt{2}}{3}t^{3/2}\,\mathbf{k} \Rightarrow \mathbf{v} = (\cos t - t\sin t)\,\mathbf{i} + (\sin t + t\cos t)\,\mathbf{j} + (\sqrt{2}\,t^{1/2})\,\mathbf{k} \Rightarrow$

$|\mathbf{v}| = \sqrt{(\cos t - t\sin t)^2 + (\sin t + t\cos t)^2 + (\sqrt{2}\,t^{1/2})^2} = \sqrt{1 + t^2 + 2t} = \sqrt{(t+1)^2} =$

$|t+1| = t+1$ if $t \ge 0$. $\mathbf{T} = \frac{\mathbf{v}}{|\mathbf{v}|} = \left(\frac{\cos t - t\sin t}{t+1}\right)\mathbf{i} + \left(\frac{\sin t + t\cos t}{t+1}\right)\mathbf{j} + \left(\frac{\sqrt{2}\,t^{1/2}}{t+1}\right)\mathbf{k}$.

7. (Continued)

Length = $\displaystyle\int_0^{\pi} (t+1)\,dt = \left[\frac{t^2}{2}+t\right]_0^{\pi} = \frac{\pi^2}{2}+\pi$

9. $\mathbf{r} = (4\cos t)\,\mathbf{i} + (4\sin t)\,\mathbf{j} + 3t\,\mathbf{k} \Rightarrow \mathbf{v} = (-4\sin t)\,\mathbf{i} + (4\cos t)\,\mathbf{j} + 3\,\mathbf{k} \Rightarrow |\mathbf{v}| = \sqrt{(-4\sin t)^2 + (4\cos t)^2 + 3^2}$

$= \sqrt{25} = 5.$ $s(t) = \displaystyle\int_0^t 5\,d\lambda = 5t$ Length $= s\left(\frac{\pi}{2}\right) = \frac{5\pi}{2}$

11. $\mathbf{r} = (e^t\cos t)\,\mathbf{i} + (e^t\sin t)\,\mathbf{j} + e^t\,\mathbf{k} \Rightarrow \mathbf{v} = (e^t\cos t - e^t\sin t)\,\mathbf{i} + (e^t\sin t + e^t\cos t)\,\mathbf{j} + e^t\,\mathbf{k} \Rightarrow$

$|\mathbf{v}| = \sqrt{(e^t\cos t - e^t\sin t)^2 + (e^t\sin t + e^t\cos t)^2 + (e^t)^2} = \sqrt{3e^{2t}} = \sqrt{3}\,e^t$

$s(t) = \displaystyle\int_0^t \sqrt{3}\,e^{\lambda}\,d\lambda = \sqrt{3}\,e^t - \sqrt{3}.$ Length $= s(0) - s(-\ln 4) = 0 - \left(\sqrt{3}\,e^{-\ln 4} - \sqrt{3}\right) = \frac{3\sqrt{3}}{4}$

13. $\mathbf{r} = (\sqrt{2}\,t)\,\mathbf{i} + (\sqrt{2}\,t)\,\mathbf{j} + (1-t^2)\,\mathbf{k} \Rightarrow \mathbf{v} = \sqrt{2}\,\mathbf{i} + \sqrt{2}\,\mathbf{j} - 2t\,\mathbf{k} \Rightarrow |\mathbf{v}| = \sqrt{(\sqrt{2})^2 + (\sqrt{2})^2 + (-2t)^2} = \sqrt{4+4t^2}$

$= 2\sqrt{1+t^2}$ Length $= \displaystyle\int_0^1 2\sqrt{1+t^2}\,dt = \left[2\left(\frac{t}{2}\sqrt{1+t^2} + \frac{1}{2}\ln\left(t+\sqrt{1+t^2}\right)\right)\right]_0^1 = \sqrt{2} + \ln\left(1+\sqrt{2}\right)$

15. a) $\mathbf{r} = (\cos t)\,\mathbf{i} + (\sin t)\,\mathbf{j} + (1-\cos t)\,\mathbf{k},\ 0 \le t \le 2\pi \Rightarrow x = \cos t,\ y = \sin t,\ z = 1 - \cos t.$ Then $x^2 + y^2 = \cos^2 t + \sin^2 t = 1$, a right circular cylinder z–axis as the axis, radius = 1. ∴ $P(\cos t, \sin t, 1-\cos t)$ lie on the cylinder $x^2 + y^2 = 1$. $t = 0 \Rightarrow P(1,0,0)$ is on the curve. $t = \frac{\pi}{2} \Rightarrow Q(0,1,1)$ is on the curve. $t = \pi \Rightarrow R(-1,0,2)$ is on the curve. Then $\vec{PQ} = -\mathbf{i} + \mathbf{j} + \mathbf{k}$, $\vec{PR} = -2\mathbf{i} + 2\mathbf{k}$. Thus, $\vec{PQ} \times \vec{PR} = \begin{vmatrix} \mathbf{i} & \mathbf{j} & \mathbf{k} \\ -1 & 1 & 1 \\ -2 & 0 & 2 \end{vmatrix} =$ $2\mathbf{i} + 2\mathbf{k}$ is a vector normal to the plane of P, Q, and R. Then the plane containing P, Q, and R has an equation $2x + 2z = 2(1) + 2(0)$ or $x + z = 1$. Any point on the curve will satisfy this equation since $x + z = \cos t + (1 - \cos t) = 1$. ∴ any point on the curve lies on the intersection of the cylinder $x^2 + y^2 = 1$ and the plane $x + z = 1 \Rightarrow$ the curve is an ellipse.

b)

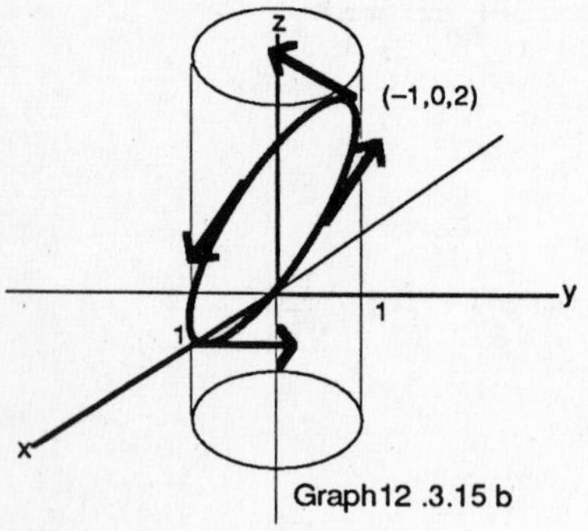

Graph12 .3.15 b

$\mathbf{v} = (-\sin t)\,\mathbf{i} + (\cos t)\,\mathbf{j} + (\sin t)\,\mathbf{k} \Rightarrow$

$|\mathbf{v}| = \sqrt{\sin^2 t + \cos^2 t + \sin^2 t} = \sqrt{1 + \sin^2 t} \Rightarrow$

$\mathbf{T} = \dfrac{\mathbf{v}}{|\mathbf{v}|} = \dfrac{(-\sin t)\,\mathbf{i} + (\cos t)\,\mathbf{j} + (\sin t)\,\mathbf{k}}{\sqrt{1+\sin^2 t}} \Rightarrow \mathbf{T}_{t=0} = \mathbf{j},\ \mathbf{T}_{t=\pi/2} =$

$\dfrac{-\mathbf{i} + \mathbf{k}}{\sqrt{2}},\ \mathbf{T}_{t=\pi} = -\mathbf{j},\ \mathbf{T}_{t=3\pi/2} = \dfrac{\mathbf{i} - \mathbf{k}}{\sqrt{2}}$

15. c)

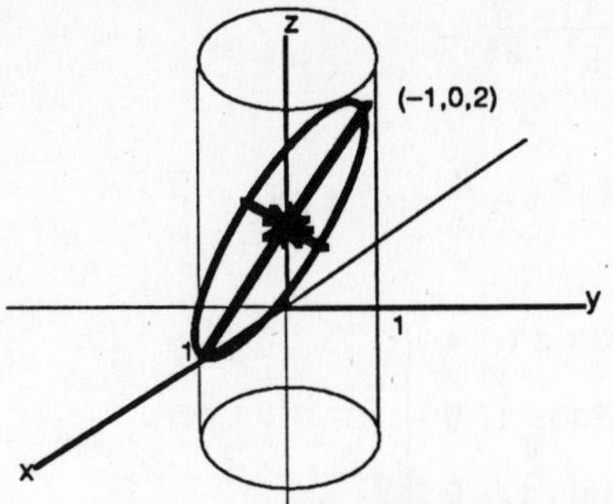

Graph 12.3.15 c

$\mathbf{a} = (-\cos t)\,\mathbf{i} - (\sin t)\,\mathbf{j} + (\cos t)\,\mathbf{k}$. $\mathbf{N} = \mathbf{i} + \mathbf{k}$ is normal to the plane $x + z = 1$. $\mathbf{N} \cdot \mathbf{a} = -\cos t + \cos t = 0 \Rightarrow \mathbf{a}$ is orthogonal to $\mathbf{N} \Rightarrow \mathbf{a}$ is parallel to the plane. $\mathbf{a}_{t=0} = -\mathbf{i} + \mathbf{k}$, $\mathbf{a}_{t=\pi/2} = -\mathbf{j}$, $\mathbf{a}_{t=\pi} = \mathbf{i} - \mathbf{k}$, $\mathbf{a}_{t=3\pi/2} = \mathbf{j}$

d) $|\mathbf{v}| = \sqrt{1 + \sin^2 t}$ (See part b). $\therefore\ L = \int_0^{2\pi} \sqrt{1 + \sin^2 t}\ dt$

e) $L \approx 7.64$ (by *Mathematica*)

12.4 CURVATURE, TORSION, AND THE TNB FRAME

1. $\mathbf{r} = t\,\mathbf{i} + \ln(\cos t)\,\mathbf{j} \Rightarrow \mathbf{v} = \mathbf{i} + \frac{-\sin t}{\cos t}\mathbf{j} = \mathbf{i} - \tan t\,\mathbf{j} \Rightarrow |\mathbf{v}| = \sqrt{1^2 + (-\tan t)^2} = \sqrt{\sec^2 t} = |\sec t| = \sec t$ since $-\frac{\pi}{2} < t < \frac{\pi}{2}$. $\mathbf{T} = \frac{\mathbf{v}}{|\mathbf{v}|} = \frac{1}{\sec t}\mathbf{i} - \frac{\tan t}{\sec t}\mathbf{j} = \cos t\,\mathbf{i} - \sin t\,\mathbf{j}$. $\frac{d\mathbf{T}}{dt} = -\sin t\,\mathbf{i} - \cos t\,\mathbf{j} \Rightarrow$
$\left|\frac{d\mathbf{T}}{dt}\right| = \sqrt{(-\sin t)^2 + (-\cos t)^2} = 1$. $\mathbf{N} = \frac{d\mathbf{T}/dt}{|d\mathbf{T}/dt|} = (-\sin t)\,\mathbf{i} - (\cos t)\,\mathbf{j}$. $\mathbf{a} = (-\sec^2 t)\,\mathbf{j} \Rightarrow$

$$\mathbf{v} \times \mathbf{a} = \begin{vmatrix} \mathbf{i} & \mathbf{j} & \mathbf{k} \\ 1 & -\tan t & 0 \\ 0 & -\sec^2 t & 0 \end{vmatrix} = (-\sec^2 t)\,\mathbf{k}.\ |\mathbf{v} \times \mathbf{a}| = \sqrt{(-\sec^2 t)^2} = \sec^2 t \Rightarrow \kappa = \frac{|\mathbf{v} \times \mathbf{a}|}{|\mathbf{v}|^3} = \frac{\sec^2 t}{\sec^3 t} = \cos t$$

3. $\mathbf{r} = (2t + 3)\,\mathbf{i} + \left(5 - t^2\right)\mathbf{j} \Rightarrow \mathbf{v} = 2\,\mathbf{i} - 2t\,\mathbf{j} \Rightarrow |\mathbf{v}| = \sqrt{2^2 + (-2t)^2} = 2\sqrt{1 + t^2}$. $\mathbf{T} = \frac{\mathbf{v}}{|\mathbf{v}|} = \frac{2}{2\sqrt{1 + t^2}}\mathbf{i} + \frac{-2t}{2\sqrt{1 + t^2}}\mathbf{j} = \frac{1}{\sqrt{1 + t^2}}\mathbf{i} - \frac{t}{\sqrt{1 + t^2}}\mathbf{j}$. $\frac{d\mathbf{T}}{dt} = = \frac{-t}{\left(\sqrt{1 + t^2}\right)^3}\mathbf{i} - \frac{1}{\left(\sqrt{1 + t^2}\right)^3}\mathbf{j} \Rightarrow$
$\left|\frac{d\mathbf{T}}{dt}\right| = \sqrt{\left(\frac{-t}{\left(\sqrt{1 + t^2}\right)^3}\right)^2 + \left(-\frac{1}{\left(\sqrt{1 + t^2}\right)^3}\right)^2} = \sqrt{\frac{1}{(1 + t^2)^2}} = \frac{1}{1 + t^2}$. $\mathbf{N} = \frac{d\mathbf{T}/dt}{|d\mathbf{T}/dt|} = \frac{-t}{\sqrt{1 + t^2}}\mathbf{i}$
$-\frac{1}{\sqrt{1 + t^2}}\mathbf{j}$. $\mathbf{a} = -2\,\mathbf{j} \Rightarrow$

$$\mathbf{v} \times \mathbf{a} = \begin{vmatrix} \mathbf{i} & \mathbf{j} & \mathbf{k} \\ 2 & -2t & 0 \\ 0 & -2 & 0 \end{vmatrix} = -4\,\mathbf{k} \Rightarrow |\mathbf{v} \times \mathbf{a}| = \sqrt{(-4)^2} = 4.\ \kappa = \frac{|\mathbf{v} \times \mathbf{a}|}{|\mathbf{v}|^3} =$$

$$\frac{4}{\left(2\sqrt{1 + t^2}\right)^3} = \frac{1}{2\left(\sqrt{1 + t^2}\right)^3}$$

5. $\mathbf{r} = (3\sin t)\,\mathbf{i} + (3\cos t)\,\mathbf{j} + 4t\,\mathbf{k} \Rightarrow \mathbf{v} = (3\cos t)\,\mathbf{i} + (-3\sin t)\,\mathbf{j} + 4\,\mathbf{k} \Rightarrow |\mathbf{v}| = \sqrt{(3\cos t)^2 + (-3\sin t)^2 + 4^2}$
$= \sqrt{25} = 5$. $\mathbf{T} = \frac{\mathbf{v}}{|\mathbf{v}|} = \frac{3\cos t}{5}\mathbf{i} - \frac{3\sin t}{5}\mathbf{j} + \frac{4}{5}\mathbf{k} \Rightarrow \frac{d\mathbf{T}}{dt} = \left(-\frac{3}{5}\sin t\right)\mathbf{i} - \left(\frac{3}{5}\cos t\right)\mathbf{j}$
$\left|\frac{d\mathbf{T}}{dt}\right| = \sqrt{\left(-\frac{3}{5}\sin t\right)^2 + \left(-\frac{3}{5}\cos t\right)^2} = \frac{3}{5}$. $\mathbf{N} = \frac{d\mathbf{T}/dt}{|d\mathbf{T}/dt|} = (-\sin t)\,\mathbf{i} - (\cos t)\,\mathbf{j}$

$$\mathbf{a} = (-3\sin t)\,\mathbf{i} + (-3\cos t)\,\mathbf{j} \Rightarrow \mathbf{v} \times \mathbf{a} = \begin{vmatrix} \mathbf{i} & \mathbf{j} & \mathbf{k} \\ 3\cos t & -3\sin t & 4 \\ -3\sin t & -3\cos t & 0 \end{vmatrix} = (12\cos t)\,\mathbf{i} - (12\sin t)\,\mathbf{j} - 9\,\mathbf{k} \Rightarrow$$

5. (Continued)

$|\mathbf{v} \times \mathbf{a}| = \sqrt{(12\cos t)^2 + (-12\sin t)^2 + (-9)^2} = \sqrt{225} = 15.$ $\kappa = \dfrac{|\mathbf{v} \times \mathbf{a}|}{|\mathbf{v}|^3} = \dfrac{15}{5^3} = \dfrac{3}{25}$

$$\mathbf{B} = \mathbf{T} \times \mathbf{N} = \begin{vmatrix} \mathbf{i} & \mathbf{j} & \mathbf{k} \\ \frac{3}{5}\cos t & -\frac{3}{5}\sin t & \frac{4}{5} \\ -\sin t & -\cos t & 0 \end{vmatrix} = \left(\frac{4}{5}\cos t\right)\mathbf{i} - \left(\frac{4}{5}\sin t\right)\mathbf{j} + \left(-\frac{3}{5}\cos^2 t - \frac{3}{5}\sin^2 t\right)\mathbf{k} =$$

$$\left(\frac{4}{5}\cos t\right)\mathbf{i} - \left(\frac{4}{5}\sin t\right)\mathbf{j} - \frac{3}{5}\mathbf{k}.\ \dot{\mathbf{a}} = (-3\cos t)\,\mathbf{i} + (3\sin t)\,\mathbf{j} \Rightarrow \tau = \frac{\begin{vmatrix} 3\cos t & -3\sin t & 4 \\ -3\sin t & -3\cos t & 0 \\ -3\cos t & 3\sin t & 0 \end{vmatrix}}{|\mathbf{v} \times \mathbf{a}|^2} =$$

$$\frac{-36\sin^2 t - 36\cos^2 t}{15^2} = -\frac{4}{25}$$

7. $\mathbf{r} = (e^t\cos t)\,\mathbf{i} + (e^t\sin t)\,\mathbf{j} + 2\,\mathbf{k} \Rightarrow \mathbf{v} = (e^t\cos t - e^t\sin t)\,\mathbf{i} + (e^t\sin t + e^t\cos t)\,\mathbf{j} \Rightarrow$

$|\mathbf{v}| = \sqrt{(e^t\cos t - e^t\sin t)^2 + (e^t\sin t + e^t\cos t)^2} = \sqrt{2e^{2t}} = e^t\sqrt{2}.$

$\mathbf{a} = \left(e^t(\cos t - \sin t) + e^t(-\sin t - \cos t)\right)\mathbf{i} + \left(e^t(\sin t + \cos t) + e^t(\cos t - \sin t)\right)\mathbf{j} =$

$$(-2e^t\sin t)\,\mathbf{i} + (2e^t\cos t)\,\mathbf{j} \Rightarrow \mathbf{v} \times \mathbf{a} = \begin{vmatrix} \mathbf{i} & \mathbf{j} & \mathbf{k} \\ e^t\cos t - e^t\sin t & e^t\sin t + e^t\cos t & 0 \\ -2e^t\sin t & 2e^t\cos t & 0 \end{vmatrix} = (2e^{2t})\,\mathbf{k} \Rightarrow$$

$|\mathbf{v} \times \mathbf{a}| = \sqrt{(2e^{2t})^2} = 2e^{2t}.$ $\kappa = \dfrac{|\mathbf{v} \times \mathbf{a}|}{|\mathbf{v}|^3} = \dfrac{2e^{2t}}{\left(e^t\sqrt{2}\right)^3} = \dfrac{1}{e^t\sqrt{2}}.$

$$\mathbf{T} = \frac{\mathbf{v}}{|\mathbf{v}|} = \left(\frac{e^t\cos t - e^t\sin t}{e^t\sqrt{2}}\right)\mathbf{i} + \left(\frac{e^t\cos t + e^t\sin t}{e^t\sqrt{2}}\right)\mathbf{j} \Rightarrow \frac{d\mathbf{T}}{dt} = \left(\frac{-\sin t - \cos t}{\sqrt{2}}\right)\mathbf{i} + \left(\frac{\cos t - \sin t}{\sqrt{2}}\right)\mathbf{j} \Rightarrow$$

$$\left|\frac{d\mathbf{T}}{dt}\right| = \sqrt{\left(\frac{-\sin t - \cos t}{\sqrt{2}}\right)^2 + \left(\frac{\cos t - \sin t}{\sqrt{2}}\right)^2} = 1.\ \mathbf{N} = \frac{d\mathbf{T}/dt}{|d\mathbf{T}/dt|} = \left(\frac{-\cos t - \sin t}{\sqrt{2}}\right)\mathbf{i} + \left(\frac{-\sin t + \cos t}{\sqrt{2}}\right)\mathbf{j}$$

$\dot{\mathbf{a}} = (-2e^t\sin t - 2e^t\cos t)\,\mathbf{i} + (2e^t\cos t - 2e^t\sin t)\,\mathbf{j} \Rightarrow$

$$\tau = \frac{\begin{vmatrix} e^t\cos t - e^t\sin t & e^t\sin t + e^t\cos t & 0 \\ -2e^t\sin t & 2e^t\cos t & 0 \\ -2e^t\sin t - 2e^t\cos t & 2e^t\cos t - 2e^t\sin t & 0 \end{vmatrix}}{|\mathbf{v} \times \mathbf{a}|^2} = 0.$$

$$\mathbf{B} = \mathbf{T} \times \mathbf{N} = \begin{vmatrix} \mathbf{i} & \mathbf{j} & \mathbf{k} \\ \frac{\cos t - \sin t}{\sqrt{2}} & \frac{\sin t + \cos t}{\sqrt{2}} & 0 \\ \frac{-\cos t - \sin t}{\sqrt{2}} & \frac{-\sin t + \cos t}{\sqrt{2}} & 0 \end{vmatrix} = \mathbf{k}$$

9. $\mathbf{r}(t) = (t^3/3)\,\mathbf{i} + (t^2/2)\,\mathbf{j},\ t > 0 \Rightarrow \mathbf{v} = t^2\mathbf{i} + t\mathbf{j} \Rightarrow |\mathbf{v}| = \sqrt{t^4 + t^2} = t\sqrt{t^2+1}$ since $t > 0$. $\mathbf{T} = \frac{\mathbf{v}}{|\mathbf{v}|} = \frac{t^2\mathbf{i} + t\mathbf{j}}{t\sqrt{t^2+1}} =$

$$\frac{t}{\sqrt{t^2+1}}\mathbf{i} + \frac{\mathbf{j}}{\sqrt{t^2+1}} \Rightarrow \frac{d\mathbf{T}}{dt} = \frac{\mathbf{i}}{(t^2+1)^{3/2}} - \frac{t\,\mathbf{j}}{(t^2+1)^{3/2}} \Rightarrow \left|\frac{d\mathbf{T}}{dt}\right| = \sqrt{\left(\frac{1}{(t^2+1)^{3/2}}\right)^2 + \left(\frac{-t}{(t^2+1)^{3/2}}\right)^2} =$$

$$\sqrt{\frac{1+t^2}{(t^2+1)^3}} = \frac{1}{t^2+1}.\ \mathbf{N} = \frac{d\mathbf{T}/dt}{|d\mathbf{T}/dt|} = \frac{\mathbf{i}}{\sqrt{t^2+1}} - \frac{t\,\mathbf{j}}{\sqrt{t^2+1}}$$

$$\mathbf{a} = 2t\mathbf{i} + \mathbf{j} \Rightarrow \mathbf{v}\times\mathbf{a} = \begin{vmatrix} \mathbf{i} & \mathbf{j} & \mathbf{k} \\ t^2 & t & 0 \\ 2t & 1 & 0 \end{vmatrix} = -t^2\mathbf{k} \Rightarrow |\mathbf{v}\times\mathbf{a}| = \sqrt{(-t^2)^2} = t^2.\ \therefore\ \kappa = \frac{|\mathbf{v}\times\mathbf{a}|}{|\mathbf{v}|^3} = \frac{t^2}{\left(t\sqrt{t^2+1}\right)^3} =$$

$$\frac{1}{t(t^2+1)^{3/2}}.\ \mathbf{B} = \mathbf{T}\times\mathbf{N} = \begin{vmatrix} \mathbf{i} & \mathbf{j} & \mathbf{k} \\ \frac{t}{\sqrt{t^2+1}} & \frac{1}{\sqrt{t^2+1}} & 0 \\ \frac{1}{\sqrt{t^2+1}} & \frac{-t}{\sqrt{t^2+1}} & 0 \end{vmatrix} = -\mathbf{k}.\ \ \dot{\mathbf{a}} = 2\mathbf{i} \Rightarrow \tau = \frac{\begin{vmatrix} t^2 & t & 0 \\ 2t & 1 & 0 \\ 2 & 0 & 0 \end{vmatrix}}{|\mathbf{v}\times\mathbf{a}|^2} = 0$$

11. $\mathbf{r} = t\mathbf{i} + \left(a\cosh\frac{t}{a}\right)\mathbf{j},\ a > 0 \Rightarrow \mathbf{v} = \mathbf{i} + \left(\sinh\frac{t}{a}\right)\mathbf{j} \Rightarrow |\mathbf{v}| = \sqrt{1 + \sinh^2\left(\frac{t}{a}\right)} = \sqrt{\cosh^2\left(\frac{t}{a}\right)} = \cosh\frac{t}{a}$

$$\mathbf{T} = \frac{\mathbf{v}}{|\mathbf{v}|} = \frac{\mathbf{i} + \left(\sinh\frac{t}{a}\right)\mathbf{j}}{\cosh\left(\frac{t}{a}\right)} = \left(\operatorname{sech}\frac{t}{a}\right)\mathbf{i} + \left(\tanh\frac{t}{a}\right)\mathbf{j} \Rightarrow \frac{d\mathbf{T}}{dt} = \left(-\frac{1}{a}\operatorname{sech}\frac{t}{a}\tanh\frac{t}{a}\right)\mathbf{i} + \left(\frac{1}{a}\operatorname{sech}^2\frac{t}{a}\right)\mathbf{j} \Rightarrow$$

$$\left|\frac{d\mathbf{T}}{dt}\right| = \sqrt{\frac{1}{a^2}\operatorname{sech}^2\frac{t}{a}\tanh^2\frac{t}{a} + \frac{1}{a^2}\operatorname{sech}^4\frac{t}{a}} = \frac{1}{a}\operatorname{sech}\frac{t}{a} \Rightarrow \mathbf{N} = \frac{d\mathbf{T}/dt}{|d\mathbf{T}/dt|} =$$

$$\frac{\left(-\frac{1}{a}\operatorname{sech}\frac{t}{a}\tanh\frac{t}{a}\right)\mathbf{i} + \left(\frac{1}{a}\operatorname{sech}^2\frac{t}{a}\right)\mathbf{j}}{\frac{1}{a}\operatorname{sech}\frac{t}{a}} = \left(-\tanh\frac{t}{a}\right)\mathbf{i} + \left(\operatorname{sech}\frac{t}{a}\right)\mathbf{j}.\ \ \mathbf{a} = \left(\frac{1}{a}\cosh\frac{t}{a}\right)\mathbf{j} \Rightarrow$$

$$\mathbf{v}\times\mathbf{a} = \begin{vmatrix} \mathbf{i} & \mathbf{j} & \mathbf{k} \\ 1 & \sinh\frac{t}{a} & 0 \\ 0 & \frac{1}{a}\cosh\frac{t}{a} & 0 \end{vmatrix} = \left(\frac{1}{a}\cosh\frac{t}{a}\right)\mathbf{k} \Rightarrow |\mathbf{v}\times\mathbf{a}| = \sqrt{\frac{1}{a^2}\cosh^2\frac{t}{a}} = \frac{1}{a}\cosh\frac{t}{a} \Rightarrow$$

$$\kappa = \frac{|\mathbf{v}\times\mathbf{a}|}{|\mathbf{v}|^3} = \frac{\frac{1}{a}\cosh\frac{t}{a}}{\cosh^3\frac{t}{a}} = \frac{1}{a\cosh^2\frac{t}{a}}.\ \mathbf{B} = \mathbf{T}\times\mathbf{N} = \begin{vmatrix} \mathbf{i} & \mathbf{j} & \mathbf{k} \\ \operatorname{sech}\frac{t}{a} & \tanh\frac{t}{a} & 0 \\ -\tanh\frac{t}{a} & \operatorname{sech}\frac{t}{a} & 0 \end{vmatrix} = \mathbf{k}$$

$$\dot{\mathbf{a}} = \frac{1}{a^2}\sinh\frac{t}{a}\mathbf{j} \Rightarrow \tau = \frac{\begin{vmatrix} 1 & \sinh\frac{t}{a} & 0 \\ 0 & \frac{1}{a}\cosh\frac{t}{a} & 0 \\ 0 & \frac{1}{a^2}\sinh\frac{t}{a} & 0 \end{vmatrix}}{|\mathbf{v}\times\mathbf{a}|^2} = 0$$

13. $\mathbf{r} = (2t+3)\,\mathbf{i} + (t^2-1)\,\mathbf{j} \Rightarrow \mathbf{v} = 2\,\mathbf{i} + 2t\,\mathbf{j} \Rightarrow |\mathbf{v}| = \sqrt{2^2 + (2t)^2} = 2\sqrt{1+t^2}$

$a_T = 2\left(\frac{1}{2}\right)(1+t^2)^{-1/2}(2t) = \frac{2t}{\sqrt{1+t^2}}$; $\mathbf{a} = 2\,\mathbf{j} \Rightarrow |\mathbf{a}| = 2 \Rightarrow a_N = \sqrt{|\mathbf{a}|^2 - a_T^{\,2}} = \sqrt{2^2 - \left(\frac{2t}{\sqrt{1+t^2}}\right)^2}$

$= \frac{2}{\sqrt{1+t^2}}$. $\therefore \mathbf{a} = \frac{2t}{\sqrt{1+t^2}}\mathbf{T} + \frac{2}{\sqrt{1+t^2}}\mathbf{N}$.

15. $\mathbf{r} = (a\cos t)\,\mathbf{i} + (a\sin t)\,\mathbf{j} + bt\,\mathbf{k} \Rightarrow \mathbf{v} = (-a\sin t)\,\mathbf{i} + (a\cos t)\,\mathbf{j} + b\,\mathbf{k} \Rightarrow |\mathbf{v}| = \sqrt{(-a\sin t)^2 + (a\cos t)^2 + b^2}$
$= \sqrt{a^2+b^2}$. $a_T = 0$. $\mathbf{a} = (-a\cos t)\,\mathbf{i} + (-a\sin t)\,\mathbf{j} \Rightarrow |\mathbf{a}| = \sqrt{(-a\cos t)^2 + (-a\sin t)^2} = \sqrt{a^2} = |a|$.

$a_N = \sqrt{|\mathbf{a}|^2 - a_T^{\,2}} = \sqrt{|a|^2 - 0^2} = |a| = |a|$. $\therefore\ \mathbf{a} = (0)\,\mathbf{T} + |a|\,\mathbf{N} = |a|\,\mathbf{N}$

17. $\mathbf{r} = (t+1)\,\mathbf{i} + 2t\,\mathbf{j} + t^2\mathbf{k} \Rightarrow \mathbf{v} = \mathbf{i} + 2\,\mathbf{j} + 2t\,\mathbf{k} \Rightarrow |\mathbf{v}| = \sqrt{1^2 + 2^2 + (2t)^2} =$
$\sqrt{5+4t^2}$. $a_T = \frac{1}{2}\left(5+4t^2\right)^{-1/2}(8t) = 4t\left(5+4t^2\right)^{-1/2} \Rightarrow a_T(1) = \frac{4}{\sqrt{9}} = \frac{4}{3}$. $\mathbf{a} = 2\,\mathbf{k} \Rightarrow \mathbf{a}(1) = 2\,\mathbf{k} \Rightarrow$

$|\mathbf{a}(1)| = 2$. $a_N = \sqrt{|\mathbf{a}|^2 - a_T^{\,2}} = \sqrt{2^2 - \left(\frac{4}{3}\right)^2} = \frac{2\sqrt{5}}{3}$. $\therefore\ \mathbf{a}(1) = \frac{4}{3}\mathbf{T} + \frac{2\sqrt{5}}{3}\mathbf{N}$

19. $\mathbf{r} = t^2\,\mathbf{i} + (t + \frac{1}{3}t^3)\,\mathbf{j} + (t - \frac{1}{3}t^3)\,\mathbf{k} \Rightarrow \mathbf{v} = 2t\,\mathbf{i} + (1+t^2)\,\mathbf{j} + (1-t^2)\,\mathbf{k} \Rightarrow |\mathbf{v}| = \sqrt{(2t)^2 + (1+t^2)^2 + (1-t^2)^2}$
$= \sqrt{2}\left(1+t^2\right)$. $a_T = 2t\sqrt{2} \Rightarrow a_T(0) = 0$. $\mathbf{a} = 2\,\mathbf{i} + 2t\,\mathbf{j} - 2t\,\mathbf{k} \Rightarrow$

$\mathbf{a}(0) = 2\,\mathbf{i} \Rightarrow |\mathbf{a}(0)| = 2$. $a_N = \sqrt{|\mathbf{a}|^2 - a_T^{\,2}} = \sqrt{2^2 - 0^2} = 2$. $\therefore\ \mathbf{a}(0) = (0)\,\mathbf{T} + 2\,\mathbf{N} = 2\,\mathbf{N}$

21. $\mathbf{r} = (\cos t)\,\mathbf{i} + (\sin t)\,\mathbf{j} - \mathbf{k} \Rightarrow \mathbf{v} = (-\sin t)\,\mathbf{i} + (\cos t)\,\mathbf{j} \Rightarrow |\mathbf{v}| = \sqrt{(-\sin t)^2 + (\cos t)^2} = 1$. $\mathbf{T} = \frac{\mathbf{v}}{|\mathbf{v}|} =$

$(-\sin t)\,\mathbf{i} + (\cos t)\,\mathbf{j} \Rightarrow \mathbf{T}\left(\frac{\pi}{4}\right) = -\frac{\sqrt{2}}{2}\mathbf{i} + \frac{\sqrt{2}}{2}\mathbf{j}$. $\frac{d\mathbf{T}}{dt} = (-\cos t)\,\mathbf{i} - (\sin t)\,\mathbf{j} \Rightarrow \left|\frac{d\mathbf{T}}{dt}\right| = \sqrt{(-\cos t)^2 + (-\sin t)^2}$

$= 1$. $\mathbf{N} = \frac{d\mathbf{T}/dt}{|d\mathbf{T}/dt|} = (-\cos t)\,\mathbf{i} - (\sin t)\,\mathbf{j} \Rightarrow \mathbf{N}\left(\frac{\pi}{4}\right) = -\frac{\sqrt{2}}{2}\mathbf{i} - \frac{\sqrt{2}}{2}\mathbf{j}$. $\mathbf{r}\left(\frac{\pi}{4}\right) = \frac{\sqrt{2}}{2}\mathbf{i} + \frac{\sqrt{2}}{2}\mathbf{j} - \mathbf{k}$

$$\mathbf{B} = \mathbf{T} \times \mathbf{N} = \begin{vmatrix} \mathbf{i} & \mathbf{j} & \mathbf{k} \\ -\sin t & \cos t & 0 \\ -\cos t & -\sin t & 0 \end{vmatrix} = \mathbf{k} \Rightarrow \mathbf{B}\left(\frac{\pi}{4}\right) = \mathbf{k}.\ P = \left(\frac{\sqrt{2}}{2}, \frac{\sqrt{2}}{2}, -1\right)\left(\text{see } \mathbf{r}\left(\frac{\pi}{4}\right)\right), \text{ the}$$

osculating plane is $z = -1$ since $\mathbf{B}$ is the normal vector and $(0)x + (0)y + (1)z = (0)\left(\frac{\sqrt{2}}{2}\right) + (0)\left(\frac{\sqrt{2}}{2}\right) + (1)(-1)$.

The normal plane is $-x + y = 0$ since $\mathbf{T}$ is the normal vector and $-\frac{\sqrt{2}}{2}x + \frac{\sqrt{2}}{2}y + (0)z = \left(-\frac{\sqrt{2}}{2}\right)\left(\frac{\sqrt{2}}{2}\right) +$
$\left(\frac{\sqrt{2}}{2}\right)\left(\frac{\sqrt{2}}{2}\right) + (-1)(0) \Rightarrow -\frac{\sqrt{2}}{2}x + \frac{\sqrt{2}}{2}y = 0$. The rectifying plane is $x + y = \sqrt{2}$ since $\mathbf{N}$ is the normal vector and $-\frac{\sqrt{2}}{2}x - \frac{\sqrt{2}}{2}y + (0)z = \left(-\frac{\sqrt{2}}{2}\right)\left(\frac{\sqrt{2}}{2}\right) - \left(\frac{\sqrt{2}}{2}\right)\left(\frac{\sqrt{2}}{2}\right) + (-1)(0) \Rightarrow -\frac{\sqrt{2}}{2}x - \frac{\sqrt{2}}{2}y = -1$.

23. Yes, if the car is moving around a circle at a constant speed, the acceleration points along $\mathbf{N}$ toward the center of the circle and is of constant magnitude, but not $\mathbf{0}$.

25. If acceleration is perpendicular to the velocity, then $a_T = 0 \Rightarrow |\mathbf{v}|$ is constant.

27. $a_N = t$, $|\mathbf{v}| = t$ (from Example 6) $\Rightarrow t = \kappa t^2 \Rightarrow \kappa = \frac{1}{t} \Rightarrow \rho = \frac{1}{\kappa} = t$

29. $\kappa = \frac{a}{a^2 + b^2} \Rightarrow \frac{d\kappa}{da} = \frac{-a^2 + b^2}{(a^2 + b^2)^2}$. If $\frac{d\kappa}{da} = 0$, then $-a^2 + b^2 = 0 \Rightarrow a = \pm b$. When $a = b$ $(b > 0)$, $\frac{d\kappa}{da} > 0$ if $a < b$ and $\frac{d\kappa}{da} < 0$ if $a > b$. $\therefore$ κ is at a maximum when $a = b \Rightarrow \kappa(b) = \frac{b}{b^2 + b^2} = \frac{1}{2b}$, the maximum value of κ.

31. From Example 4, the curvature of the helix $\mathbf{r}(t) = (a \cos t)\mathbf{i} + (a \sin t)\mathbf{j} + bt\,\mathbf{k}$, $a,b \ge 0$ is $\frac{a}{a^2 + b^2}$. Also $|\mathbf{v}| = \sqrt{a^2 + b^2}$. In $\mathbf{r}(t) = (3 \cos t)\mathbf{i} + (3 \sin t)\mathbf{j} + t\,\mathbf{k}$, $0 \le t \le 4\pi$, $a = 3$, $b = 1 \Rightarrow \kappa = \frac{3}{3^2 + 1^2} = \frac{3}{10}$, $|\mathbf{v}| = \sqrt{10} \Rightarrow$

$$K = \int_0^{4\pi} \frac{3}{10}\sqrt{10}\, dt = \left[\frac{3}{\sqrt{10}}\, t\right]_0^{4\pi} = \frac{12\pi}{\sqrt{10}}.$$

33. $\mathbf{r} = t\,\mathbf{i} + (\sin t)\,\mathbf{j} \Rightarrow \mathbf{v} = \mathbf{i} + (\cos t)\,\mathbf{j} \Rightarrow |\mathbf{v}| = \sqrt{1^2 + (\cos t)^2} = \sqrt{1 + \cos^2 t} \Rightarrow \left|\mathbf{v}\left(\frac{\pi}{2}\right)\right| = \sqrt{1 + \cos^2\left(\frac{\pi}{2}\right)}$ $= 1$. $\mathbf{a} = (-\sin t)\,\mathbf{j} \Rightarrow \mathbf{v} \times \mathbf{a} = \begin{vmatrix} \mathbf{i} & \mathbf{j} & \mathbf{k} \\ 1 & \cos t & 0 \\ 0 & -\sin t & 0 \end{vmatrix} = (-\sin t)\,\mathbf{k} \Rightarrow |\mathbf{v} \times \mathbf{a}| = \sqrt{(-\sin t)^2} = |\sin t| \Rightarrow$

$|\mathbf{v} \times \mathbf{a}|\left(\frac{\pi}{2}\right) = \left|\sin\left(\frac{\pi}{2}\right)\right| = 1 \Rightarrow \kappa = \frac{|\mathbf{v} \times \mathbf{a}|}{|\mathbf{v}|^3} = \frac{1}{1^3} = 1$. $\therefore$ $\rho = \frac{1}{1} = 1 \Rightarrow$ center is $\left(\frac{\pi}{2}, 0\right)$, $r = 1 \Rightarrow$
$\left(x - \frac{\pi}{2}\right)^2 + y^2 = 1$

35. a) $\mathbf{r} = x\,\mathbf{i} + f(x)\,\mathbf{j} \Rightarrow \mathbf{v} = \mathbf{i} + f'(x)\,\mathbf{j} \Rightarrow \mathbf{a} = f''(x)\,\mathbf{j}$. $\mathbf{v} \times \mathbf{a} = \begin{vmatrix} \mathbf{i} & \mathbf{j} & \mathbf{k} \\ 1 & f'(x) & 0 \\ 0 & f''(x) & 0 \end{vmatrix} = f''(x)\,\mathbf{k} \Rightarrow$

$|\mathbf{v} \times \mathbf{a}| = \sqrt{(f''(x))^2} = |f''(x)|$. $|\mathbf{v}| = \sqrt{1^2 + (f'(x))^2} = \sqrt{1 + (f'(x))^2}$. Then $\kappa = \frac{|\mathbf{v} \times \mathbf{a}|}{|\mathbf{v}|^3} =$
$\frac{|f''(x)|}{\left(\sqrt{1 + (f'(x))^2}\right)^3} = \frac{|f''|}{\left(1 + (f')^2\right)^3}$

b) $y = \ln(\cos x) \Rightarrow \frac{dy}{dx} = \frac{1}{\cos x}(-\sin x) = -\tan x \Rightarrow \frac{d^2y}{dx^2} = -\sec^2 x$. Then $\kappa = \frac{|-\sec^2 x|}{\left(\sqrt{1 + (-\tan x)^2}\right)^3} =$
$\frac{\sec^2 x}{|\sec^3 x|} = \frac{\sec^2 x}{\text{sex}^3 x} = \frac{1}{\sec x} = \cos x$ since $-\frac{\pi}{2} < x < \frac{\pi}{2}$.

c) If $x = x_0$ gives a point of inflection, then $f''(x_0) = 0$ (since f is twice differentiable) $\Rightarrow \kappa = 0$.

37. $y = x^2 \Rightarrow f'(x) = 2x$, $f''(x) = 2 \Rightarrow \kappa = \frac{|2|}{\left(1 + (2x)^2\right)^{3/2}} = \frac{2}{\left(1 + 4x^2\right)^{3/2}}$ (Compare with κ in Exercise 32 b, with $x = t$.)

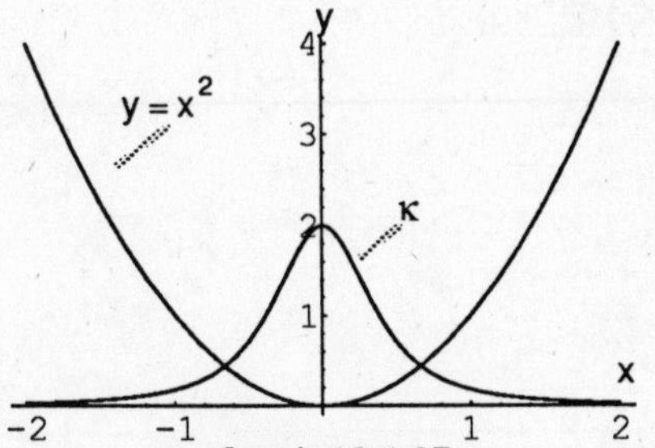

Graph 12.4.37

39. $y = \sin x \Rightarrow f'(x) = \cos x,\ f''(x) = -\sin x \Rightarrow \kappa = \dfrac{|-\sin x|}{(1 + \cos^2 x)^{3/2}} = \dfrac{|\sin x|}{(1 + \cos^2 x)^{3/2}}$

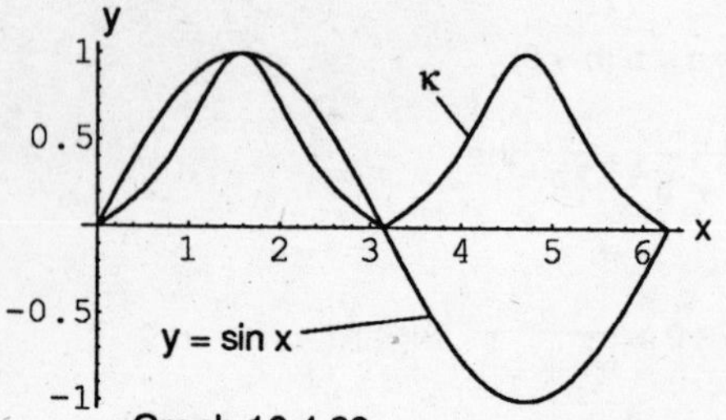

Graph 12.4.39

12.5 PLANETARY MOTION AND SATELLITES

1. $\dfrac{T^2}{a^3} = \dfrac{4\pi^2}{GM} \Rightarrow T^2 = \dfrac{4\pi^2}{GM}a^3 \Rightarrow T^2 = \dfrac{4\pi^2}{(6.6720 \text{ X } 10^{-11}\text{Nm}^2\text{kg}^{-2})(5.975 \text{ X } 10^{24} \text{ kg})}(6\ 808\ 000 \text{ m})^3 \approx$
$3.125 \text{ X } 10^7 \text{ sec}^2 \Rightarrow T \approx \sqrt{3125 \text{ X } 10^4 \text{ sec}^2} \approx 55.90 \text{ X } 10^2 \text{ sec} \approx 93.17$ minutes.

3. 92.25 minutes = 5535 seconds. $\dfrac{T^2}{a^3} = \dfrac{4\pi^2}{GM} \Rightarrow a^3 = \dfrac{GM}{4\pi^2}T^2 \Rightarrow a^3 =$
$\dfrac{(6.6720 \text{ X } 10^{-11}\text{Nm}^2\text{kg}^{-2})(5.975 \text{ X } 10^{24}\text{kg})}{4\pi^2}(5535 \text{ s})^2 = 3.094 \text{ X } 10^{20} \text{ m}^3 \Rightarrow a \approx \sqrt[3]{3.094 \text{ X } 10^{20} \text{ m}^3}$
$= 6.764 \text{ X } 10^6 \text{ m} \approx 6764$ km. Mean distance from center of the Earth $= \dfrac{12758 \text{ km} + 183 \text{ km} + 589 \text{ km}}{2}$
$= 6765$ km

5. $a = 22030 \text{ km} = 2.203 \text{ X } 10^7$ m. $T^2 = \dfrac{4\pi^2}{GM}a^3 \Rightarrow T^2 =$
$\dfrac{4\pi^2}{(6.670 \text{ X } 10^{-11}\text{Nm}^2\text{kg}^{-2})(6.418 \text{ X } 10^{23} \text{ kg})}(2.203 \text{ X } 10^7\text{s})^3 \approx 9.857\text{X } 10^9 \text{ sec}^2 \Rightarrow T \approx$
$\sqrt{9.857 \text{ X } 10^8 \text{ sec}^2} \approx 9.928 \text{ X } 10^4 \text{ sec} \approx 1655$ minutes.

7. $T = 1477.4$ minutes = 88644 seconds. $a^3 = \dfrac{GMT^2}{4\pi^2} =$
$\dfrac{(6.6720 \text{ X } 10^{-11}\text{Nm}^2\text{kg}^{-2})(6.418 \text{ X } 10^{23}\text{kg})(88644 \text{ s})^2}{4\pi^2} = 8.523 \text{ X } 10^{21} \text{ m}^3 \Rightarrow$
$a \approx \sqrt[3]{8.523 \text{ X } 10^{21} \text{ m}^3} \approx 2.043 \text{ X } 10^7 \text{ m} = 20430$ km

9. $r = \dfrac{GM}{v^2} \Rightarrow v^2 = \dfrac{GM}{r} \Rightarrow |v| = \sqrt{\dfrac{GM}{r}} = \sqrt{\dfrac{(6.6720 \text{ X } 10^{-11} \text{ Nm}^2\text{kg}^{-2})(5.975 \text{ X } 10^{24} \text{ kg})}{r}}$
$\approx 1.9966 \text{ X } 10^7\, r^{-1/2}$ m/s

11. $e = \frac{r_0 v_0^2}{GM} - 1 \Rightarrow v_0^2 = \frac{GM(e+1)}{r_0} \Rightarrow v_0 = \sqrt{\frac{GM(e+1)}{r_0}}$

Circle: $e = 0 \Rightarrow v_0 = \sqrt{\frac{GM}{r_0}}$

Ellipse: $0 < e < 1 \Rightarrow \sqrt{\frac{GM}{r_0}} < v_0 < \sqrt{\frac{2GM}{r_0}}$

Parabola: $e = 1 \Rightarrow v_0 = \sqrt{\frac{2GM}{r_0}}$

Hyperbola: $e > 1 \Rightarrow v_0 > \sqrt{\frac{2GM}{r_0}}$

12.P PRACTICE EXERCISES

1. $\mathbf{r} = (4\cos t)\,\mathbf{i} + (\sqrt{2}\sin t)\,\mathbf{j} \Rightarrow x = 4\cos t \Rightarrow x^2 = 16\cos^2 t$

$y = \sqrt{2}\sin t \Rightarrow y^2 = 2\sin^2 t \Rightarrow 8y^2 = 16\sin^2 t \Rightarrow$
$x^2 + 8y^2 = 16 \Rightarrow \frac{x^2}{16} + \frac{y^2}{2} = 1.$ $t = 0 \Rightarrow x = 4, y = 0;$
$t = \frac{\pi}{4} \Rightarrow x = 2\sqrt{2}, y = 1.$ $\mathbf{v} = (-4\sin t)\,\mathbf{i} + (\sqrt{2}\cos t)\,\mathbf{j}$
$\Rightarrow \mathbf{v}(0) = \sqrt{2}\,\mathbf{j}, \mathbf{v}\left(\frac{\pi}{4}\right) = -2\sqrt{2}\,\mathbf{i} + \mathbf{j}.$
$\mathbf{a} = (-4\cos t)\,\mathbf{i} + (-\sqrt{2}\sin t)\,\mathbf{j} \Rightarrow \mathbf{a}(0) = -4\,\mathbf{i}, \mathbf{a}\left(\frac{\pi}{4}\right) =$
$-2\sqrt{2}\,\mathbf{i} - \mathbf{j}.$

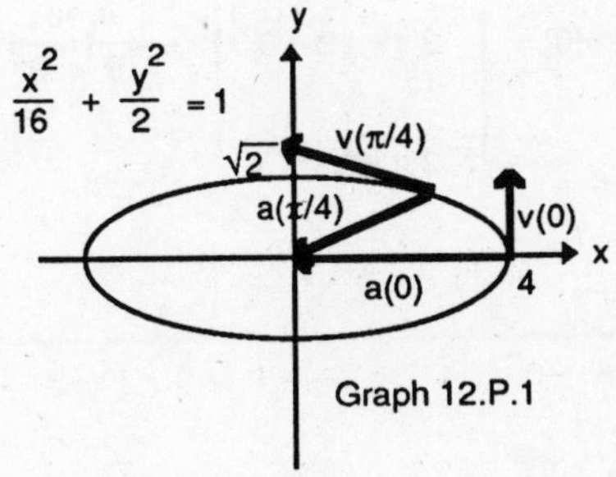

Graph 12.P.1

3. $\int_0^1 [(3+6t)\,\mathbf{i} + (4+8t)\,\mathbf{j} + (6\pi\cos\pi t)\,\mathbf{k}]\,dt = \left[3t + 3t^2\right]_0^1 \mathbf{i} + \left[4t + 4t^2\right]_0^1 \mathbf{j} + \left[6\sin\pi t\right]_0^1 \mathbf{k} = 6\,\mathbf{i} + 8\,\mathbf{j}$

5. $\mathbf{r} = \int ((-\sin t)\,\mathbf{i} + (\cos t)\,\mathbf{j} + \mathbf{k})\,dt = (\cos t)\,\mathbf{i} + (\sin t)\,\mathbf{j} + t\,\mathbf{k} + \mathbf{C}.$ $\mathbf{r}(0) = \mathbf{j} \Rightarrow (\cos 0)\,\mathbf{i} + (\sin 0)\,\mathbf{j} + (0)\,\mathbf{k}$
$+ \mathbf{C} = \mathbf{j} \Rightarrow \mathbf{C} = \mathbf{j} - \mathbf{i} \Rightarrow \mathbf{r} = ((\cos t) - 1)\,\mathbf{i} + ((\sin t) + 1)\,\mathbf{j} + t\,\mathbf{k}$

7. $\frac{d\mathbf{r}}{dt} = \int 2\,\mathbf{j}\,dt = 2t\,\mathbf{j} + \mathbf{C}_1.$ $\frac{d\mathbf{r}}{dt}(0) = \mathbf{k} \Rightarrow 2(0)\,\mathbf{j} + \mathbf{C}_1 = \mathbf{k} \Rightarrow \mathbf{C}_1 = \mathbf{k}.$ $\therefore \frac{d\mathbf{r}}{dt} = 2t\,\mathbf{j} + \mathbf{k}.$
$\mathbf{r} = \int (2t\,\mathbf{j} + \mathbf{k})\,dt = t^2\,\mathbf{j} + t\,\mathbf{k} + \mathbf{C}_2.$ $\mathbf{r}(0) = \mathbf{i} \Rightarrow (0^2)\,\mathbf{j} + (0)\,\mathbf{k} + \mathbf{C}_2 = \mathbf{i} \Rightarrow \mathbf{C}_2 = \mathbf{i}$
$\therefore \mathbf{r} = \mathbf{i} + t^2\,\mathbf{j} + t\,\mathbf{k}$

9. $\mathbf{r} = (2\cos t)\,\mathbf{i} + (2\sin t)\,\mathbf{j} + t^2\,\mathbf{k} \Rightarrow \mathbf{v} = (-2\sin t)\,\mathbf{i} + (2\cos t)\,\mathbf{j} + 2t\,\mathbf{k} \Rightarrow$

$|\mathbf{v}| = \sqrt{(-2\sin t)^2 + (2\cos t)^2 + (2t)^2} = 2\sqrt{1+t^2}.$ Length $= \int_0^{\pi/4} 2\sqrt{1+t^2}\,dt =$

$\left[t\sqrt{1+t^2} + \ln\left|t + \sqrt{1+t^2}\right|\right]_0^{\pi/4} = \frac{\pi}{4}\sqrt{1 + \frac{\pi^2}{16}} + \ln\left(\frac{\pi}{4} + \sqrt{1 + \frac{\pi^2}{16}}\right)$

11. $\mathbf{r} = \frac{4}{9}(1+t)^{3/2}\,\mathbf{i} + \frac{4}{9}(1-t)^{3/2}\,\mathbf{j} + \frac{1}{3}t\,\mathbf{k} \Rightarrow \mathbf{v} = \frac{2}{3}(1+t)^{1/2}\,\mathbf{i} - \frac{2}{3}(1-t)^{1/2}\,\mathbf{j} + \frac{1}{3}\mathbf{k} \Rightarrow$

$|\mathbf{v}| = \sqrt{\left(\frac{2}{3}(1+t)^{1/2}\right)^2 + \left(-\frac{2}{3}(1-t)^{1/2}\right)^2 + \left(\frac{1}{3}\right)^2} = 1.\ \mathbf{T} = \frac{2}{3}(1+t)^{1/2}\,\mathbf{i} - \frac{2}{3}(1-t)^{1/2} + \frac{1}{3}\mathbf{k} \Rightarrow$

$\mathbf{T}(0) = \frac{2}{3}\mathbf{i} - \frac{2}{3}\mathbf{j} + \frac{1}{3}\mathbf{k}.\ \frac{d\mathbf{T}}{dt} = \frac{1}{3}(1+t)^{-1/2}\,\mathbf{i} + \frac{1}{3}(1-t)^{-1/2}\,\mathbf{j} \Rightarrow$

$\left|\frac{d\mathbf{T}}{dt}\right| = \sqrt{\left(\frac{1}{3}(1+t)^{-1/2}\right)^2 + \left(\frac{1}{3}(1-t)^{-1/2}\right)^2} = \frac{1}{3}\sqrt{\frac{2}{1-t^2}}.\quad \frac{d\mathbf{T}}{dt}(0) = \frac{1}{3}\mathbf{i} + \frac{1}{3}\mathbf{j} \Rightarrow \left|\frac{d\mathbf{T}}{dt}(0)\right| = \frac{\sqrt{2}}{3}.$

$\therefore\ \mathbf{N}(0) = \frac{1}{\sqrt{2}}\mathbf{i} + \frac{1}{\sqrt{2}}\mathbf{j}.\ \mathbf{B}(0) = \mathbf{T}(0) \times \mathbf{N}(0) = \begin{vmatrix} \mathbf{i} & \mathbf{j} & \mathbf{k} \\ \frac{2}{3} & -\frac{2}{3} & \frac{1}{3} \\ \frac{1}{\sqrt{2}} & \frac{1}{\sqrt{2}} & 0 \end{vmatrix} = -\frac{1}{3\sqrt{2}}\mathbf{i} + \frac{1}{3\sqrt{2}}\mathbf{j} + \frac{4}{3\sqrt{2}}\mathbf{k}$

$\mathbf{a} = \frac{1}{3}(1+t)^{-1/2}\,\mathbf{i} + \frac{1}{3}(1-t)^{-1/2}\,\mathbf{j} \Rightarrow \mathbf{a}(0) = \frac{1}{3}\mathbf{i} + \frac{1}{3}\mathbf{j}.\ \mathbf{v}(0) = \frac{2}{3}\mathbf{i} - \frac{2}{3}\mathbf{j} + \frac{1}{3}\mathbf{k}.$

$\therefore\ \mathbf{v}(0) \times \mathbf{a}(0) = \begin{vmatrix} \mathbf{i} & \mathbf{j} & \mathbf{k} \\ \frac{2}{3} & -\frac{2}{3} & \frac{1}{3} \\ \frac{1}{3} & \frac{1}{3} & 0 \end{vmatrix} = -\frac{1}{9}\mathbf{i} + \frac{1}{9}\mathbf{j} + \frac{4}{9}\mathbf{k} \Rightarrow |\mathbf{v} \times \mathbf{a}| = \frac{\sqrt{2}}{3} \Rightarrow \kappa = \frac{|\mathbf{v} \times \mathbf{a}|}{|\mathbf{v}|^3} = \frac{\sqrt{2}/3}{1^3} = \frac{\sqrt{2}}{3}.$

$\dot{\mathbf{a}} = -\frac{1}{6}(1+t)^{-3/2}\,\mathbf{i} + \frac{1}{6}(1-t)^{-3/2}\,\mathbf{j} \Rightarrow \dot{\mathbf{a}}(0) = -\frac{1}{6}\mathbf{i} + \frac{1}{6}\mathbf{j} \Rightarrow \tau = \frac{\begin{vmatrix} \frac{2}{3} & -\frac{2}{3} & \frac{1}{3} \\ \frac{1}{3} & \frac{1}{3} & 0 \\ -\frac{1}{6} & \frac{1}{6} & 0 \end{vmatrix}}{|\mathbf{v} \times \mathbf{a}|^2} = \frac{1/27}{(\sqrt{2}/3)^2} = \frac{1}{6}$

$t = 0 \Rightarrow \left(\frac{4}{9}, \frac{4}{9}, 0\right)$ is the point on the curve.

13. $\mathbf{r} = t\,\mathbf{i} + \frac{1}{2}e^{2t}\,\mathbf{j},\ t = \ln 2 \Rightarrow \mathbf{v} = \mathbf{i} + e^{2t}\,\mathbf{j} \Rightarrow |\mathbf{v}| = \sqrt{1 + e^{4t}} \Rightarrow \mathbf{T} = \frac{\mathbf{v}}{|\mathbf{v}|} = \frac{\mathbf{i} + e^{2t}\,\mathbf{j}}{\sqrt{1 + e^{4t}}} = \frac{1}{\sqrt{1 + e^{4t}}}\mathbf{i} + \frac{e^{2t}}{\sqrt{1 + e^{4t}}}\mathbf{j}$

$\Rightarrow \frac{d\mathbf{T}}{dt} = \frac{-2e^{4t}}{(1 + e^{4t})^{3/2}}\mathbf{i} + \frac{2e^{2t}}{(1 + e^{4t})^{3/2}}\mathbf{j} \Rightarrow \left|\frac{d\mathbf{T}}{dt}\right| = \sqrt{\frac{4e^{8t} + 4e^{4t}}{(1 + e^{4t})^3}} = \frac{2e^{2t}}{1 + e^{4t}} \Rightarrow \mathbf{N} = \frac{d\mathbf{T}/dt}{|d\mathbf{T}/dt|} = \frac{-e^{2t}}{(1 + e^{4t})^{1/2}}\mathbf{i} +$

$\frac{1}{(1 + e^{4t})^{1/2}}\mathbf{j}.\ \therefore\ \mathbf{T}(\ln 2) = \frac{1}{\sqrt{17}}\mathbf{i} + \frac{4}{\sqrt{17}}\mathbf{j},\ \mathbf{N}(\ln 2) = -\frac{4}{\sqrt{17}}\mathbf{i} + \frac{1}{\sqrt{17}}\mathbf{j}.\ \mathbf{B}(\ln 2) = \mathbf{T}(\ln 2) \times \mathbf{N}(\ln 2) =$

$\begin{vmatrix} \mathbf{i} & \mathbf{j} & \mathbf{k} \\ \frac{1}{\sqrt{17}} & \frac{4}{\sqrt{17}} & 0 \\ -\frac{4}{\sqrt{17}} & \frac{1}{\sqrt{17}} & 0 \end{vmatrix} = \mathbf{k}.\ \mathbf{a} = 2e^{2t}\,\mathbf{j} \Rightarrow \mathbf{a}(\ln 2) = 8\,\mathbf{j}.\ \mathbf{v}(\ln 2) = \mathbf{i} + 4\,\mathbf{j} \Rightarrow \mathbf{v}(\ln 2) \times \mathbf{a}(\ln 2) = \begin{vmatrix} \mathbf{i} & \mathbf{j} & \mathbf{k} \\ 1 & 4 & 0 \\ 0 & 8 & 0 \end{vmatrix}$

$= 8\,\mathbf{k} \Rightarrow |\mathbf{v}(\ln 2) \times \mathbf{a}(\ln 2)| = 8.\ |\mathbf{v}(\ln 2)| = \sqrt{17}. \Rightarrow \kappa = \frac{8}{(\sqrt{17})^3} = \frac{8}{17\sqrt{17}}.\ \dot{\mathbf{a}} = 4e^{2t}\,\mathbf{j} \Rightarrow \dot{\mathbf{a}}(\ln 2) = 16\,\mathbf{j} \Rightarrow$

$\tau = \frac{\begin{vmatrix} 1 & 4 & 0 \\ 0 & 8 & 0 \\ 0 & 16 & 0 \end{vmatrix}}{|\mathbf{v} \times \mathbf{a}|^2} = 0.\ t = \ln 2 \Rightarrow (\ln 2, 2, 0)$ is on the curve.

15. $\mathbf{r} = (2 + 3t + 3t^2)\,\mathbf{i} + (4t + 4t^2)\,\mathbf{j} - (6\cos t)\,\mathbf{k} \Rightarrow \mathbf{v} = (3 + 6t)\,\mathbf{i} + (4 + 8t)\,\mathbf{j} + (6\sin t)\,\mathbf{k} \Rightarrow$

$|\mathbf{v}| = \sqrt{(3 + 6t)^2 + (4 + 8t)^2 + (6\sin t)^2} = \sqrt{25 + 100t + 100t^2 + 36\sin^2 t}$

$\frac{d|\mathbf{v}|}{dt} = \frac{1}{2}(25 + 100t + 100t^2 + 36\sin^2 t)^{-1/2}(100 + 200t + 72\sin t\cos t) \Rightarrow a_T(0) = \frac{d|\mathbf{v}|}{dt}(0) = 10.$

$\mathbf{a} = 6\,\mathbf{i} + 8\,\mathbf{j} + (6\cos t)\,\mathbf{k} \Rightarrow |\mathbf{a}| = \sqrt{6^2 + 8^2 + (6\cos t)^2} = \sqrt{100 + 36\cos^2 t} \Rightarrow |\mathbf{a}(0)| = \sqrt{136} = 2\sqrt{34}$

$a_N = \sqrt{|\mathbf{a}|^2 - a_T^{\,2}} = \sqrt{(2\sqrt{34})^2 - 10^2} = \sqrt{36} = 6.\ \therefore\ \mathbf{a}(0) = 10\,\mathbf{T} + 6\,\mathbf{N}$

17. $\mathbf{r} = (\sin t)\,\mathbf{i} + (\sqrt{2}\cos t)\,\mathbf{j} + (\sin t)\,\mathbf{k} \Rightarrow \mathbf{v} = (\cos t)\,\mathbf{i} - (\sqrt{2}\sin t)\,\mathbf{j} + (\cos t)\,\mathbf{k} \Rightarrow$

$|\mathbf{v}| = \sqrt{(\cos t)^2 + (-\sqrt{2}\sin t)^2 + (\cos t)^2} = \sqrt{2} \Rightarrow \mathbf{T} = \frac{\mathbf{v}}{|\mathbf{v}|} = \frac{(\cos t)\,\mathbf{i} - (\sqrt{2}\sin t)\,\mathbf{j} + (\cos t)\,\mathbf{k}}{\sqrt{2}} =$

$\left(\frac{1}{\sqrt{2}}\cos t\right)\mathbf{i} - (\sin t)\,\mathbf{j} + \left(\frac{1}{\sqrt{2}}\cos t\right)\mathbf{k}.\ \frac{d\mathbf{T}}{dt} = \left(-\frac{1}{\sqrt{2}}\sin t\right)\mathbf{i} - (\cos t)\,\mathbf{j} - \left(\frac{1}{\sqrt{2}}\sin t\right)\mathbf{k} \Rightarrow$

$\left|\frac{d\mathbf{T}}{dt}\right| = \sqrt{\left(-\frac{1}{\sqrt{2}}\sin t\right)^2 + (-\cos t)^2 + \left(-\frac{1}{\sqrt{2}}\sin t\right)^2} = 1.\ \mathbf{N} = \frac{d\mathbf{T}/dt}{|d\mathbf{T}/dt|} = \left(-\frac{1}{\sqrt{2}}\sin t\right)\mathbf{i} - (\cos t)\,\mathbf{j} -$

$$\left(\frac{1}{\sqrt{2}}\sin t\right)\mathbf{k}.\ \mathbf{B} = \mathbf{T} \times \mathbf{N} = \begin{vmatrix} \mathbf{i} & \mathbf{j} & \mathbf{k} \\ \frac{1}{\sqrt{2}}\cos t & -\sin t & \frac{1}{\sqrt{2}}\cos t \\ -\frac{1}{\sqrt{2}}\sin t & -\cos t & -\frac{1}{\sqrt{2}}\sin t \end{vmatrix} = \frac{1}{\sqrt{2}}\mathbf{i} - \frac{1}{\sqrt{2}}\mathbf{k}$$

$$\mathbf{a} = (-\sin t)\,\mathbf{i} - (\sqrt{2}\cos t)\,\mathbf{j} - (\sin t)\,\mathbf{k} \Rightarrow \mathbf{v} \times \mathbf{a} = \begin{vmatrix} \mathbf{i} & \mathbf{j} & \mathbf{k} \\ \cos t & -\sqrt{2}\sin t & \cos t \\ -\sin t & -\sqrt{2}\cos t & -\sin t \end{vmatrix} = \sqrt{2}\,\mathbf{i} - \sqrt{2}\,\mathbf{k}$$

$|\mathbf{v} \times \mathbf{a}| = \sqrt{(\sqrt{2})^2 + (-\sqrt{2})^2} = \sqrt{4} = 2 \Rightarrow \kappa = \frac{|\mathbf{v} \times \mathbf{a}|}{|\mathbf{v}|^3} = \frac{2}{(\sqrt{2})^3} = \frac{1}{\sqrt{2}}.$

$$\dot{\mathbf{a}} = (-\cos t)\,\mathbf{i} + (\sqrt{2}\sin t)\,\mathbf{j} - (\cos t)\,\mathbf{k} \Rightarrow \tau = \frac{\begin{vmatrix} \cos t & -\sqrt{2}\sin t & \cos t \\ -\sin t & -\sqrt{2}\cos t & -\sin t \\ -\cos t & \sqrt{2}\sin t & -\cos t \end{vmatrix}}{|\mathbf{v} \times \mathbf{a}|^2} == \frac{0}{|\mathbf{v} \times \mathbf{a}|^2} = 0$$

19. $\mathbf{r} = (e^t\cos t)\,\mathbf{i} + (e^t\sin t)\,\mathbf{j} \Rightarrow \mathbf{v} = (e^t\cos t - e^t\sin t)\,\mathbf{i} + (e^t\sin t + e^t\cos t)\,\mathbf{j} \Rightarrow$

$\mathbf{a} = (e^t\cos t - e^t\sin t - e^t\sin t - e^t\cos t)\,\mathbf{i} + (e^t\sin t + e^t\cos t + e^t\cos t - e^t\sin t)\,\mathbf{j}$

$= (-2e^t\sin t)\,\mathbf{i} + (2e^t\cos t)\,\mathbf{j}$. Let θ be the angle between $\mathbf{r}$ and $\mathbf{a}$.

Then $\theta = \cos^{-1}\left(\frac{\mathbf{r} \cdot \mathbf{a}}{|\mathbf{r}||\mathbf{a}|}\right) = \cos^{-1}\left(\frac{-2e^{2t}\sin t\cos t + 2e^{2t}\sin t\cos t}{\sqrt{(e^t\cos t)^2 + (e^t\sin t)^2}\sqrt{(-2e^t\sin t)^2 + (2e^t\cos t)^2}}\right) =$

$\cos^{-1}\left(\frac{0}{2e^{2t}}\right) = \cos^{-1}0 = \frac{\pi}{2}$ for all t.

21. $\mathbf{r} = 2\,\mathbf{i} + \left(4\sin\frac{t}{2}\right)\mathbf{j} + \left(3 - \frac{t}{\pi}\right)\mathbf{k} \Rightarrow \mathbf{r} \cdot (\mathbf{i} - \mathbf{j}) = 2(1) + \left(4\sin\frac{t}{2}\right)(-1).\ \mathbf{r} \cdot (\mathbf{i} - \mathbf{j}) = 0 \Rightarrow 2 - 4\sin\frac{t}{2} = 0 \Rightarrow$

$\sin\frac{t}{2} = \frac{1}{2} \Rightarrow \frac{t}{2} = \frac{\pi}{6} \Rightarrow t = \frac{\pi}{3}$ (for the first time).

23. $y_{max} = y_0 + \frac{(v_0 \sin \alpha)^2}{2g} = 7\text{ ft} + \frac{((80\text{ ft/sec})(\sin 45°))^2}{2(32\text{ ft/sec}^2)} \approx 57\text{ ft}$

25. $R = \frac{v_0^2}{g}\sin 2\alpha \Rightarrow 109.5\text{ ft} = \frac{v_0^2}{32\text{ ft/sec}^2}(\sin 2(45°)) \Rightarrow v_0^2 = 3504\text{ ft}^2/\text{sec}^2 \Rightarrow v_0 = \sqrt{3504\text{ ft}^2/\text{sec}^2}$

≈ 59.19 ft/sec

27. $\mathbf{r} = e^t\mathbf{i} + (\sin t)\mathbf{j} + (\ln(1-t))\mathbf{k} \Rightarrow \mathbf{v} = e^t\mathbf{i} + (\cos t)\mathbf{j} - \frac{1}{1-t}\mathbf{k} \Rightarrow \mathbf{v}(0) = \mathbf{i} + \mathbf{j} - \mathbf{k}$. $\mathbf{r}(0) = \mathbf{i} \Rightarrow (1,0,0)$ is on the line. $\therefore$ $x = 1 + t$, $y = t$, $z = -t$ are the parametric equations of the line.

29. $\mathbf{v} = 3\mathbf{i} + 4\mathbf{j}$, $\mathbf{a} = 5\mathbf{i} + 15\mathbf{j} \Rightarrow \mathbf{v} \times \mathbf{a} = \begin{vmatrix} \mathbf{i} & \mathbf{j} & \mathbf{k} \\ 3 & 4 & 0 \\ 5 & 15 & 0 \end{vmatrix} = 25\mathbf{k} \Rightarrow |\mathbf{v} \times \mathbf{a}| = \sqrt{25^2} = 25$. $|\mathbf{v}| = \sqrt{3^2 + 4^2}$ $= 5$. $\therefore$ $\kappa = \frac{|\mathbf{v} \times \mathbf{a}|}{|\mathbf{v}|^3} = \frac{25}{5^3} = \frac{1}{5}$.

31. $\Delta ACB \approx \Delta BCD \Rightarrow \frac{DC}{BC} = \frac{BC}{AC} \Rightarrow$

$\frac{y_0}{6380\text{ miles}} = \frac{6380\text{ miles}}{(6380 + 437)\text{ miles}} \Rightarrow$

$y_0 = \frac{6380^2\text{ miles}^2}{6817\text{ miles}} \approx 5971$ miles

$VA = \int_{5971}^{6817} 2\pi x\sqrt{1 + \left(\frac{dx}{dy}\right)^2}\,dy =$

$2\pi \int_{5971}^{6817} \sqrt{6380^2 - y^2}\,\frac{6380}{\sqrt{6380^2 - y^2}}\,dy =$

$2\pi \int_{5971}^{6817} 6380\,dy = 2\pi\left[6380y\right]_{5971}^{6817} \approx 16\,395\,469\text{ km}^2$. Percentage visible $\approx \frac{16\,395\,469\text{ km}^2}{4\pi(6380\text{ km})^2} \approx 3.21\%$

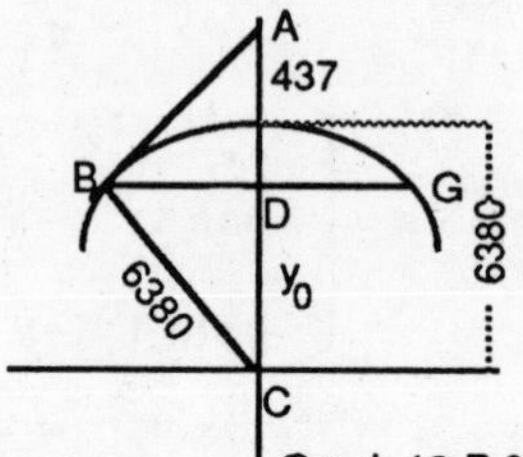

Graph 12.P.31

33. a) Given $f(x) = x - 1 - \frac{1}{2}\sin x = 0$, $f(0) = -1$ and $f(2) = 2 - 1 - \frac{1}{2}\sin 2 \geq \frac{1}{2}$ since $|\sin 2| \leq 1$. Since f is continuous on [0,2], the intermediate value theorem implies there is a root between 0 and 2.

b) Root ≈ 1.49870113

CHAPTER 13

PARTIAL DERIVATIVES

13.1 FUNCTIONS OF SEVERAL INDEPENDENT VARIABLES

1. Domain: All points in the xy–plane; Range: All Real Numbers
   Level curves are straight lines parallel to the line $y = x$.

3. Domain: Set of all (x,y) so that $(x,y) \neq (0,0)$; Range: All Real Numbers
   Level curves are circles with center $(0,0)$ and radii > 0.

5. Domain: All points in the xy–plane; Range: All Real Numbers
   Level curves are hyperbolas with the x- and y-axes as asymptotes when $f(x,y) \neq 0$, and the x- and y-axes when $f(x,y) = 0$.

7. Domain: All points in the xy–plane; Range: $z \geq 0$
   Level curves: for $f(x,y) = 0$, $(0,0)$; for $f(x,y) = c \neq 0$, ellipses with center $(0,0)$ and major and minor axes along the x- and y-axes, respectively.

9. a)

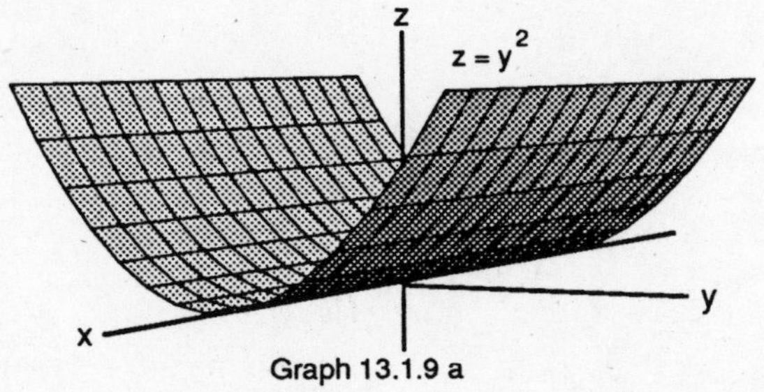

Graph 13.1.9 a

b)

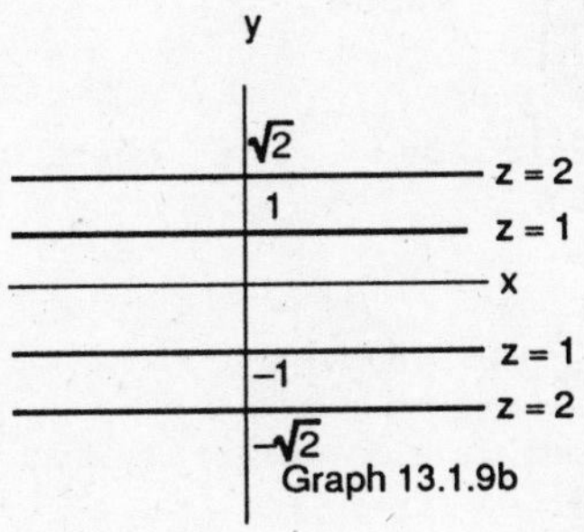

Graph 13.1.9b

11. a)

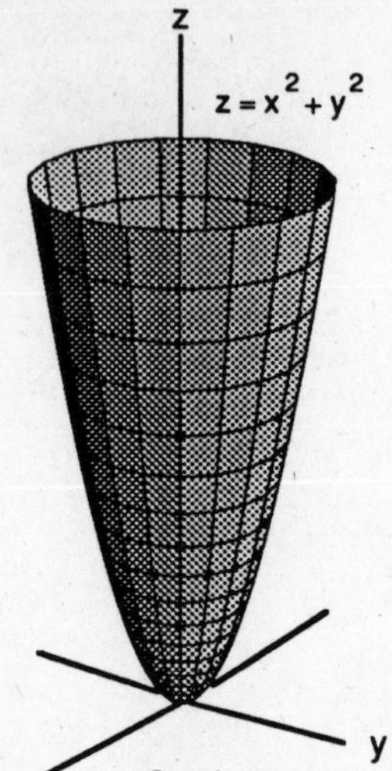

Graph 13.1.11 a

b)

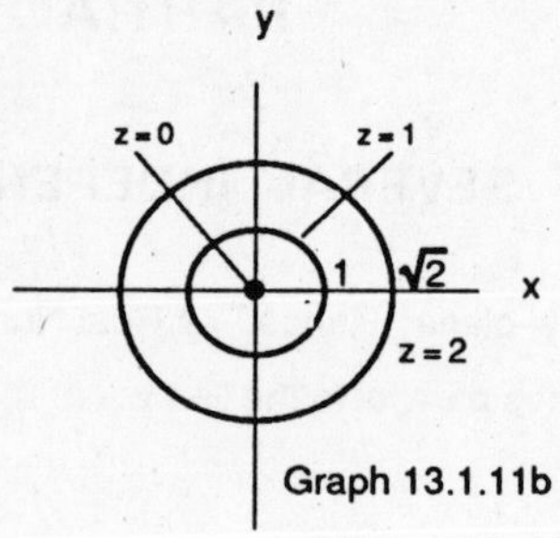

Graph 13.1.11b

13. a)

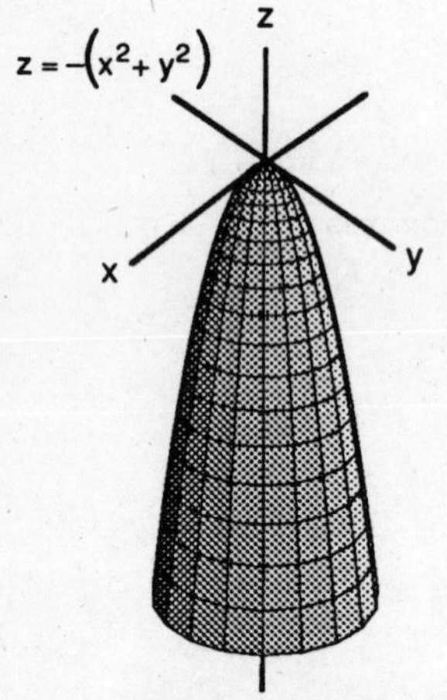

Graph 13.1.13 a

b)

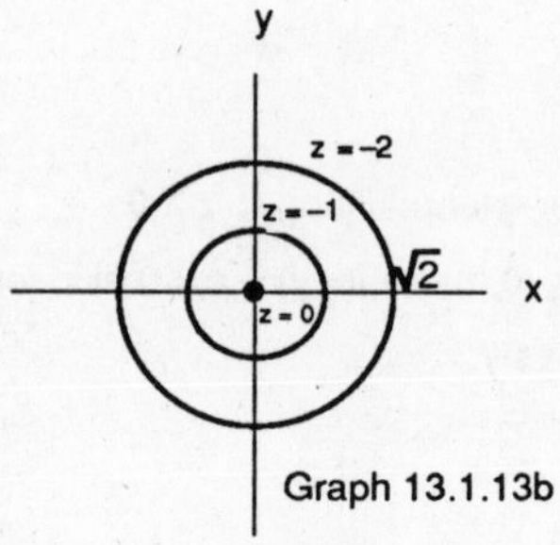

Graph 13.1.13b

15. a)

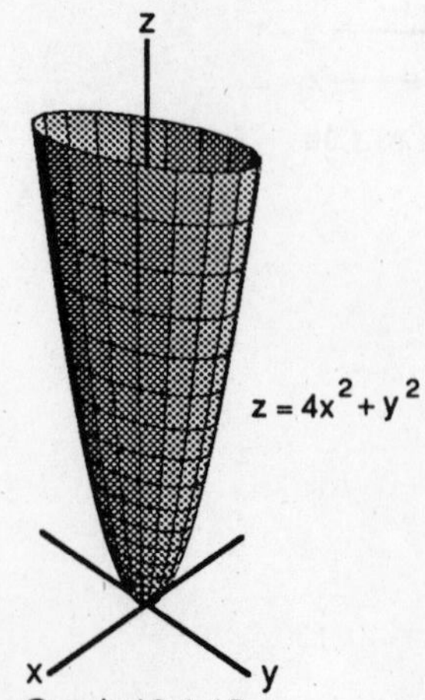

Graph 13.1.15 a

b)

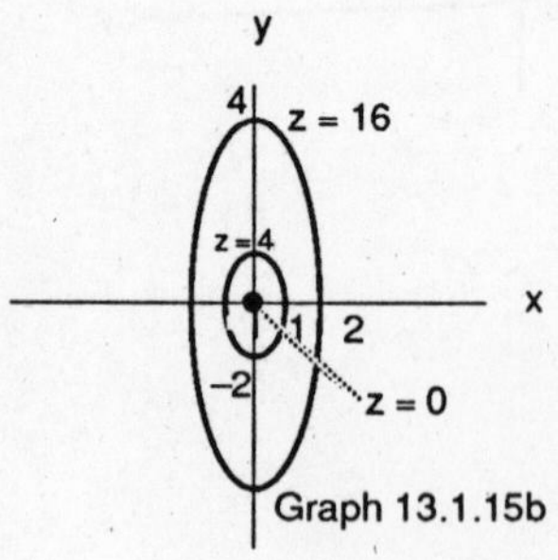

Graph 13.1.15b

17. f

19. a

21. d

23.

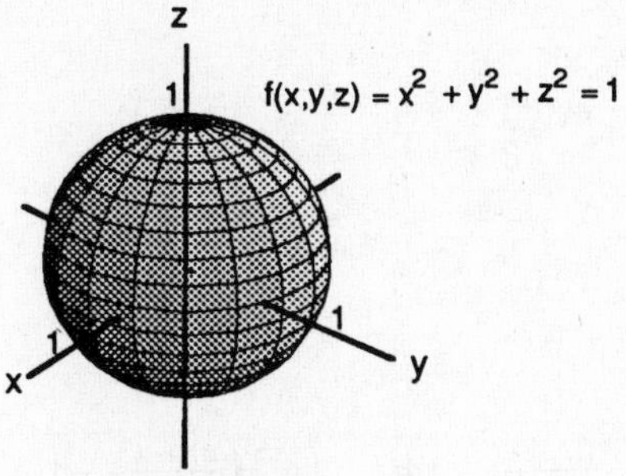

$f(x,y,z) = x^2 + y^2 + z^2 = 1$

Graph 13.1.23

25.

$f(x,y,z) = x + z = 1$

Graph 13.1.25

27.

$f(x,y,z) = x^2 + y^2 = 1$

Graph 13.1.27

29.

$f(x,y,z) = z - x^2 - y^2 = 1$

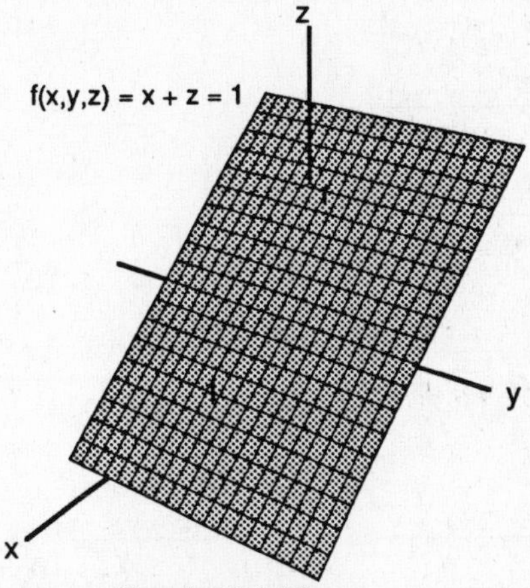

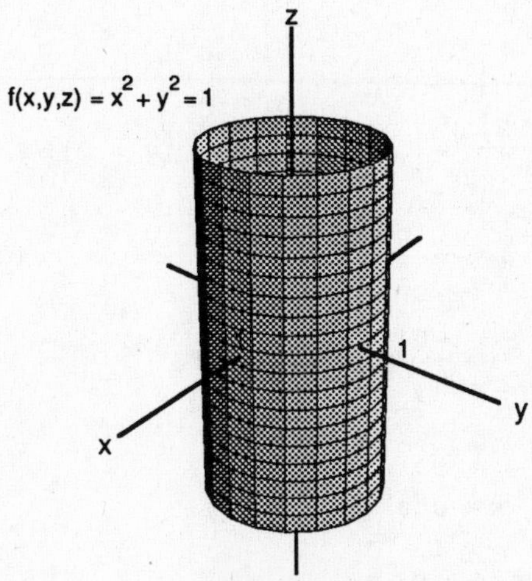

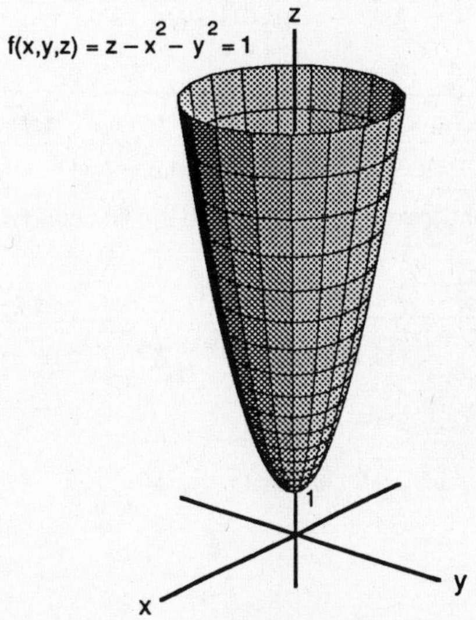

Graph 13.1.29

31. $f(x,y) = 16 - x^2 - y^2$ and $\left(2\sqrt{2}, 2\sqrt{2}\right) \Rightarrow z = 16 - \left(2\sqrt{2}\right)^2 - \left(2\sqrt{2}\right)^2 = 0 \Rightarrow 0 = 16 - x^2 - y^2 \Rightarrow x^2 + y^2 = 16$

33. $f(x,y,z) = \sqrt{x-y} - \ln z$ at $(3,-1,1) \Rightarrow w = \sqrt{x-y} - \ln z$. At $(3,-1,1)$, $w = \sqrt{3-(-1)} - \ln 1 = 2$.

$\therefore 2 = \sqrt{x-y} - \ln z \Rightarrow \ln z = \sqrt{x-y} - 2 \Rightarrow z = e^{\sqrt{x-y} - 2}$

35. $f(x,y,z) = xyz$ and $x = 20 - t$, $y = t$, $z = 20 \Rightarrow w = (20-t)(t)(20)$ along the line $\Rightarrow w = 400t - 20t^2 \Rightarrow$

$\dfrac{dw}{dt} = 400 - 40t$. Let $\dfrac{dw}{dt} = 0 \Rightarrow 400 - 40t = 0 \Rightarrow t = 10$. $\dfrac{d^2w}{dt^2} = -40$ for all $t \Rightarrow$ maximum at $t = 10 \Rightarrow$

$x = 20 - 10 = 10$, $y = 10$, $z = 20 \Rightarrow$ maximum of f along the line is $f(10,10,20) = (10)(10)(20) = 2000$.

37. $w = 4\left(\frac{Th}{d}\right)^{1/2} = 4\left(\frac{(290\text{ k})(16.8\text{ km})}{5\text{ k/km}}\right)^{1/2} = 124.86$ km. ∴ must be 62.43 km south of Nantucket.

13.2 LIMITS AND CONTINUITY

1. $\lim_{(x,y)\to(0,0)} \frac{3x^2 - y^2 + 5}{x^2 + y^2 + 2} = \frac{5}{2}$

3. $\lim_{(x,y)\to(0,\ln 2)} e^{x-y} = \frac{1}{2}$

5. $\lim_{(x,y)\to(3,4)} \sqrt{x^2 + y^2 - 1} = \sqrt{24} = 2\sqrt{6}$

7. $\lim_{(x,y)\to(0,\pi/4)} \sec x \tan y = 1$

9. $\lim_{(x,y)\to(1,1)} \cos\left(\sqrt[3]{|xy| - 1}\right) = 1$

11. $\lim_{(x,y)\to(0,0)} \frac{e^y \sin x}{x} = \lim_{(x,y)\to(0,0)} \left(e^y\right)\left(\frac{\sin x}{x}\right) = 1$

13. $\lim_{\substack{(x,y)\to(1,1)\\ x \neq y}} \frac{x^2 - 2xy + y^2}{x - y} = \lim_{(x,y)\to(1,1)} \frac{(x-y)^2}{x - y} = \lim_{(x,y)\to(1,1)} (x - y) = 0$

15. $\lim_{\substack{(x,y)\to(1,1)\\ x \neq 1}} \frac{xy - y - 2x + 2}{x - 1} = \lim_{\substack{(x,y)\to(1,1)\\ x \neq 1}} \frac{(x-1)(y-2)}{x - 1} = \lim_{(x,y)\to(1,1)} (y - 2) = -1$

17. $\lim_{(x,y)\to(0,0)} \frac{x - y + 2\sqrt{x} - 2\sqrt{y}}{\sqrt{x} - \sqrt{y}} = \lim_{(x,y)\to(0,0)} \frac{\left(\sqrt{x} - \sqrt{y}\right)\left(\sqrt{x} + \sqrt{y} + 2\right)}{\sqrt{x} - \sqrt{y}} = \lim_{(x,y)\to(0,0)} \left(\sqrt{x} + \sqrt{y} + 2\right) = 2.$

Note: (x,y) must approach (0,0) through the first quadrant onlyx, $x \neq y$.

19. $\lim_{(x,y)\to(2,0)} \frac{\sqrt{2x - y} - 2}{2x - y - 4} = \lim_{\substack{(x,y)\to(2,0)\\ 2x - y \neq 4}} \frac{\sqrt{2x - y} - 2}{\left(\sqrt{2x - y} + 2\right)\left(\sqrt{2x - y} - 2\right)} = \lim_{\substack{(x,y)\to(2,0)\\ 2x - y \neq 4}} \frac{1}{\sqrt{2x - y} + 2} = \frac{1}{4}$

21. $\lim_{P\to(2,3,-6)} \sqrt{x^2 + y^2 + z^2} = 7$

23. $\lim_{P\to(3,3,0)} \left(\sin^2 x + \cos^2 y + \sec^2 z\right) = \sin^2 3 + \cos^2 3 + \sec^2 0 = 2$

25. $\lim_{P\to(-1/4,\pi/2,2)} \tan^{-1}(xyz) = \tan^{-1}\left(-\frac{\pi}{4}\right)$

27 a) Continuous at all (x,y)

b) Continuous at all (x,y) except (0,0)

29. a) Continuous at all (x,y) except where $x = 0$ or $y = 0$

b) Continuous at all (x,y)

31. a) Continuous at all (x,y,z)

b) Continuous at all (x,y,z) except the interior of the cylinder $x^2 + y^2 = 1$

33. a) Continuous at all (x,y,z) so that $(x,y,z) \neq (x,y,0)$

b) Continuous at all (x,y,z) except those on the cylinder $x^2 + z^2 = 1$

35. $\lim\limits_{\substack{(x,y)\to(0,0)\\ \text{along } y = x\\ x > 0}} -\frac{x}{\sqrt{x^2+y^2}} = \lim\limits_{x\to 0} -\frac{x}{\sqrt{x^2+x^2}} = \lim\limits_{x\to 0} -\frac{x}{\sqrt{2}\,|x|} = \lim\limits_{x\to 0} -\frac{x}{\sqrt{2}\,x} = \lim\limits_{x\to 0} -\frac{1}{\sqrt{2}} = -\frac{1}{\sqrt{2}}$

$\lim\limits_{\substack{(x,y)\to(0,0)\\ \text{along } y = x\\ x < 0}} \frac{x}{\sqrt{x^2+y^2}} = \lim\limits_{x\to 0} -\frac{x}{\sqrt{2}\,|x|} = \lim\limits_{x\to 0} -\frac{x}{\sqrt{2}(-x)} = \lim\limits_{x\to 0} \frac{1}{\sqrt{2}} = -\frac{1}{\sqrt{2}}$

∴ consider paths along y = x where x > 0 or x < 0.

37. $\lim\limits_{\substack{(x,y)\to(0,0)\\ \text{along } y = kx^2}} \frac{x^4 - y^2}{x^4 + y^2} = \lim\limits_{x\to 0} \frac{x^4 - (kx^2)^2}{x^4 + (kx^2)^2} = \lim\limits_{x\to 0} \frac{x^4 - k^2x^4}{x^4 + k^2x^4} = \frac{1 - k^2}{1 + k^2} \Rightarrow$ different limits for different values of k. ∴ consider paths along $y = kx^2$, k a constant.

39. $\lim\limits_{\substack{(x,y)\to(0,0)\\ \text{along } y = kx\\ k \neq -1}} \frac{x - y}{x + y} = \lim\limits_{x\to 0} \frac{x - kx}{x + kx} = \frac{1 - k}{1 + k} \Rightarrow$ different limits for different values of k. ∴ consider paths along y = kx, k a constant, $k \neq -1$.

41. $\lim\limits_{\substack{(x,y)\to(0,0)\\ \text{along } y = kx^2\\ k \neq 0}} \frac{x^2 + y}{y} = \lim\limits_{x\to 0} \frac{x^2 + kx^2}{kx^2} = \frac{1 + k}{k} \Rightarrow$ different limits for different values of k. ∴ consider paths along $y = kx^2$, k a constant, $k \neq 0$.

43. No. The limit depends only on the values f(x,y) has when $(x,y) \neq (x_0,y_0)$.

45. $\lim\limits_{(x,y)\to(0,0)} 1 - \frac{x^2y^2}{3} = 1$, $\lim\limits_{(x,y)\to(0,0)} 1 = 1 \Rightarrow \lim\limits_{(x,y)\to(0,0)} \frac{\tan^{-1} xy}{xy} = 1$, by the Sandwich Theorem.

47. The limit is 0 since $\left|\sin\left(\frac{1}{x}\right)\right| \le 1 \Rightarrow -1 \le \sin\left(\frac{1}{x}\right) \le 1 \Rightarrow -y \le y \sin\left(\frac{1}{x}\right) \le y,\ y \ge 0; -y \ge y \sin\left(\frac{1}{x}\right) \ge y,\ y \le 0.$

As $(x,y) \to (0,0)$, $-y$ and y approach $0 \Rightarrow y \sin\left(\frac{1}{x}\right) \to 0$, by the Sandwich Theorem.

49. a) $f(x,y)\big|_{y = mx} = \frac{2m}{1 + m^2} = \frac{2 \tan\theta}{1 + \tan^2\theta} = \sin 2\theta$. The value of f(x,y) is $\sin 2\theta$ where $\tan\theta = m$ along $y = mx$.

b) Since $f(x,y)\big|_{y = mx} = \sin 2\theta$ and since $-1 \le \sin 2\theta \le 1$ for every θ, $\lim\limits_{(x,y)\to(0,0)} f(x,y)$ varies from -1 to 1 along y = mx.

13.3 PARTIAL DERIVATIVES

1. $\frac{\partial f}{\partial x} = 2, \frac{\partial f}{\partial y} = 0$

3. $\frac{\partial f}{\partial x} = 0, \frac{\partial f}{\partial y} = 0$

5. $\frac{\partial f}{\partial x} = y - 1, \frac{\partial f}{\partial y} = x$

7. $\frac{\partial f}{\partial x} = 2x - y, \frac{\partial f}{\partial y} = -x + 2y$

9. $\frac{\partial f}{\partial x} = 2x(5y-1)^3,$
$\frac{\partial f}{\partial y} = 15x^2(5y-1)^2$

11. $\frac{\partial f}{\partial x} = x(x^2+y^2)^{-1/2},$
$\frac{\partial f}{\partial y} = y(x^2+y^2)^{-1/2}$

13. $\frac{\partial f}{\partial x} = \frac{-y^2-1}{(xy-1)^2}, \frac{\partial f}{\partial y} = \frac{-x^2-1}{(xy-1)^2}$

15. $\frac{\partial f}{\partial x} = e^x \ln y, \frac{\partial f}{\partial y} = \frac{e^x}{y}$

17. $\frac{\partial f}{\partial x} = e^x \sin(y+1), \frac{\partial f}{\partial y} = e^x \cos(y+1)$

19. $f_x(x,y,z) = y + z, f_y(x,y,z) = x + z, f_z(x,y,z) = y + x$

21. $f_x(x,y,z) = -x(x^2+y^2+z^2)^{-3/2}, f_y(x,y,z) = -y(x^2+y^2+z^2)^{-3/2}, f_z(x,y,z) = -z(x^2+y^2+z^2)^{-3/2}$

23. $f_x(x,y,z) = \cos(x+yz), f_y(x,y,z) = z\cos(x+yz), f_z(x,y,z) = y\cos(x+yz)$

25. $\frac{\partial f}{\partial t} = -2\pi \sin(2\pi t - \alpha), \frac{\partial f}{\partial \alpha} = \sin(2\pi t - \alpha)$

27. $\frac{\partial h}{\partial \rho} = \sin\phi\cos\theta, \frac{\partial h}{\partial \phi} = \rho\cos\phi\cos\theta, \frac{\partial h}{\partial \theta} = -\rho\sin\phi\sin\theta$

29. $\frac{\partial W}{\partial P} = V, \frac{\partial W}{\partial V} = P + \frac{\delta v^2}{2g}, \frac{\partial W}{\partial \delta} = \frac{Vv^2}{2g}, \frac{\partial W}{\partial v} = \frac{V\delta v}{g}, \frac{\partial W}{\partial g} = -\frac{V\delta v^2}{2g^2}$

31. $\frac{\partial f}{\partial x} = y + 1, \frac{\partial f}{\partial y} = x + 1, \frac{\partial^2 f}{\partial x^2} = 0, \frac{\partial^2 f}{\partial y^2} = 0, \frac{\partial^2 f}{\partial y \partial x} = \frac{\partial^2 f}{\partial x \partial y} = 1$

33. $\frac{\partial g}{\partial x} = 2xy + y\cos x, \frac{\partial g}{\partial y} = x^2 - \sin y + \sin x, \frac{\partial^2 g}{\partial x^2} = 2y - y\sin x, \frac{\partial^2 g}{\partial y^2} = -\cos y, \frac{\partial^2 g}{\partial y \partial x} = \frac{\partial^2 g}{\partial x \partial y} = 2x + \cos x$

35. $\frac{\partial r}{\partial x} = \frac{1}{x+y}, \frac{\partial r}{\partial y} = \frac{1}{x+y}, \frac{\partial^2 r}{\partial x^2} = \frac{-1}{(x+y)^2}, \frac{\partial^2 r}{\partial y^2} = \frac{-1}{(x+y)^2}, \frac{\partial^2 r}{\partial y \partial x} = \frac{\partial^2 r}{\partial x \partial y} = \frac{-1}{(x+y)^2}$

37. $\frac{\partial w}{\partial x} = \frac{2}{2x+3y}, \frac{\partial w}{\partial y} = \frac{3}{2x+3y}, \frac{\partial^2 w}{\partial y \partial x} = \frac{-6}{(2x+3y)^2}$ and $\frac{\partial^2 w}{\partial x \partial y} = \frac{-6}{(2x+3y)^2}$

39. $\frac{\partial w}{\partial x} = y^2 + 2xy^3 + 3x^2y^4, \frac{\partial w}{\partial y} = 2xy + 3x^2y^2 + 4x^3y^3, \frac{\partial^2 w}{\partial y \partial x} = 2y + 6xy^2 + 12x^2y^3$ and
$\frac{\partial^2 w}{\partial x \partial y} = 2y + 6xy^2 + 12x^2y^3$

41. a) x first b) y first c) x first d) x first e) y first f) y first

43. $x^3z + z^3x - 2yz = 0 \Rightarrow 3x^2z + x^3\frac{\partial z}{\partial x} + 3xz^2\frac{\partial z}{\partial x} + z^3 - 2y\frac{\partial z}{\partial x} = 0 \Rightarrow$ at (1,1,1), $3 + \frac{\partial z}{\partial x} + 3\frac{\partial z}{\partial x} + 1 - 2\frac{\partial z}{\partial x} = 0 \Rightarrow$
$4 + 2\frac{\partial z}{\partial x} = 0 \Rightarrow \frac{\partial z}{\partial x} = -2$

45. $\frac{\partial f}{\partial x} = 2x, \frac{\partial f}{\partial y} = 2y, \frac{\partial f}{\partial z} = -4z \Rightarrow \frac{\partial^2 f}{\partial x^2} = 2, \frac{\partial^2 f}{\partial y^2} = 2, \frac{\partial^2 f}{\partial z^2} = -4 \Rightarrow \frac{\partial^2 f}{\partial x^2} + \frac{\partial^2 f}{\partial y^2} + \frac{\partial^2 f}{\partial z^2} = 2 + 2 + (-4) = 0$

47. $\frac{\partial f}{\partial x} = -2e^{-2y}\sin 2x, \frac{\partial f}{\partial y} = -2e^{-2y}\cos 2x, \frac{\partial^2 f}{\partial x^2} = -4e^{-2y}\cos 2x, \frac{\partial^2 f}{\partial y^2} = 4e^{-2y}\cos 2x$
$\therefore \frac{\partial^2 f}{\partial x^2} + \frac{\partial^2 f}{\partial y^2} = -4e^{-2y}\cos 2x + 4e^{-2y}\cos 2x = 0$

49. $\frac{\partial f}{\partial x} = -\frac{1}{2}(x^2 + y^2 + z^2)^{-3/2}(2x) = -x(x^2 + y^2 + x^2)^{-3/2}, \frac{\partial f}{\partial y} = -\frac{1}{2}(x^2 + y^2 + z^2)^{-3/2}(2y) = -y(x^2 + y^2 + z^2)^{-3/2}$
$\frac{\partial f}{\partial z} = -\frac{1}{2}(x^2 + y^2 + z^2)^{-3/2}(2z) = -z(x^2 + y^2 + z^2)^{-3/2}. \frac{\partial^2 f}{\partial x^2} = -(x^2 + y^2 + z^2)^{-3/2} + 3x^2(x^2 + y^2 + z^2)^{-5/2}$
$\frac{\partial^2 f}{\partial y^2} = -(x^2 + y^2 + z^2)^{-3/2} + 3y^2(x^2 + y^2 + z^2)^{-5/2}, \frac{\partial^2 f}{\partial z^2} = -(x^2 + y^2 + z^2)^{-3/2} + 3z^2(x^2 + y^2 + z^2)^{-5/2}$
$\therefore \frac{\partial^2 f}{\partial x^2} + \frac{\partial^2 f}{\partial y^2} + \frac{\partial^2 f}{\partial z^2} = (-(x^2 + y^2 + z^2)^{-3/2} + 3x^2(x^2 + y^2 + z^2)^{-5/2}) +$
$(-(x^2 + y^2 + z^2)^{-3/2} + 3y^2(x^2 + y^2 + z^2)^{-5/2}) + (-(x^2 + y^2 + z^2)^{-3/2} + 3z^2(x^2 + y^2 + z^2)^{-5/2}) =$
$-3(x^2 + y^2 + z^2)^{-3/2} + (3x^2 + 3y^2 + 3z^2)(x^2 + y^2 + z^2)^{-5/2} = 0$

51. $\frac{\partial w}{\partial x} = \cos(x + ct), \frac{\partial w}{\partial t} = c\cos(x + ct). \frac{\partial^2 w}{\partial x^2} = -\sin(x + ct), \frac{\partial^2 w}{\partial t^2} = -c^2\sin(x + ct)$
$\therefore \frac{\partial^2 w}{\partial t^2} = c^2(-\sin(x + ct)) = c^2\frac{\partial^2 w}{\partial x^2}$

53. $\frac{\partial w}{\partial x} = \cos(x + ct) - 2\sin(2x + 2ct), \frac{\partial w}{\partial t} = c\cos(x + ct) - 2c\sin(2x + 2ct). \frac{\partial^2 w}{\partial x^2} = -\sin(x + ct) -$
$4\cos(2x + 2ct), \frac{\partial^2 w}{\partial t^2} = -c^2\sin(x + ct) - 4c^2\cos(2x + 2ct) \quad \therefore \frac{\partial^2 w}{\partial t^2} = c^2(-\sin(x + ct) - 4\cos(2x + 2ct)) =$
$c^2\frac{\partial^2 w}{\partial x^2}$

55. $\frac{\partial w}{\partial x} = 2\sec^2(2x - 2ct), \frac{\partial w}{\partial t} = -2c\sec^2(2x - 2ct). \frac{\partial^2 w}{\partial x^2} = 8\sec^2(2x - 2ct)\tan(2x - 2ct)$,
$\frac{\partial^2 w}{\partial t^2} = 8c^2\sec^2(2x - 2ct)\tan(2x - 2ct) \quad \therefore \frac{\partial^2 w}{\partial t^2} = c^2(8\sec^2(2x - 2ct)\tan(2x - 2ct)) = c^2\frac{\partial^2 w}{\partial x^2}$

13.4 DIFFERENTIABILITY, LINEARIZATION, AND DIFFERENTIALS

1. a) $f(0,0) = 1$, $f_x(x,y) = 2x \Rightarrow f_x(0,0) = 0$, $f_y(x,y) = 2y \Rightarrow f_y(0,0) = 0 \Rightarrow L(x,y) = 1 + 0(x-0) + 0(y-0) = 1$

   b) $f(1,1) = 3$, $f_x(1,1) = 2$, $f_y(1,1) = 2 \Rightarrow L(x,y) = 3 + 2(x-1) + 2(y-1) = 2x + 2y - 1$

3. a) $f(0,0) = 1$, $f_x(x,y) = e^x \cos y \Rightarrow f_x(0,0) = 1$, $f_y(x,y) = -e^x \sin y \Rightarrow f_y(0,0) = 0 \Rightarrow L(x,y) = 1 + 1(x-0) + 0(y-0) = 1 + x$

   b) $f\left(0,\frac{\pi}{2}\right) = 0$, $f_x\left(0,\frac{\pi}{2}\right) = 0$, $f_y\left(0,\frac{\pi}{2}\right) = -1 \Rightarrow L(x,y) = 0 + 0(x-0) - 1\left(y - \frac{\pi}{2}\right) = -y + \frac{\pi}{2}$

5. a) $f(0,0) = 5$, $f_x(x,y) = 3$ for all (x,y), $f_y(x,y) = -4$ for all $(x,y) \Rightarrow L(x,y) = 5 + 3(x-0) - 4(y-0) = 5 + 3x - 4y$

   b) $f(1,1) = 4$, $f_x(1,1) = 3$, $f_y(1,1) = -4 \Rightarrow L(x,y) = 4 + 3(x-1) - 4(y-1) = 3x - 4y + 5$

7. $f(2,1) = 3$, $f_x(x,y) = 2x - 3y \Rightarrow f_x(2,1) = 1$, $f_y(x,y) = -3x \Rightarrow f_y(2,1) = -6 \Rightarrow L(x,y) = 3 + 1(x-2) - 6(y-1) = 7 + x - 6y$. $f_{xx}(x,y) = 2$, $f_{yy}(x,y) = 0$, $f_{xy}(x,y) = -3 \Rightarrow M = 3$. $\therefore |E(x,y)| \le \frac{1}{2}(3)(|x-2| + |y-1|)^2 \le \frac{3}{2}(0.1 + 0.1)^2 = 0.06$

9. $f(0,0) = 1$, $f_x(x,y) = \cos y \Rightarrow f_x(0,0) = 1$, $f_y(x,y) = 1 - x \sin y \Rightarrow f_y(0,0) = 1 \Rightarrow L(x,y) = 1 + 1(x-0) + 1(y-0) = x + y + 1$. $f_{xx}(x,y) = 0$, $f_{yy}(x,y) = -x \cos y$, $f_{xy}(x,y) = -\sin y \Rightarrow M = 1$.
   $\therefore |E(x,y)| \le \frac{1}{2}(1)(|x| + |y|)^2 \le \frac{1}{2}(0.2 + 0.2)^2 = 0.08$

11. $f(0,0) = 1$, $f_x(x,y) = e^x \cos y \Rightarrow f_x(0,0) = 1$, $f_y(x,y) = -e^x \sin y \Rightarrow f_y(0,0) = 0 \Rightarrow L(x,y) = 1 + 1(x-0) + 0(y-0) = 1 + x$. $f_{xx}(x,y) = e^x \cos y$, $f_{yy}(x,y) = -e^x \cos y$, $f_{xy}(x,y) = -e^x \sin y$. $|x| \le 0.1 \Rightarrow -0.1 \le x \le 0.1$, $|y| \le 0.1 \Rightarrow -0.1 \le y \le 0.1. \Rightarrow$ max of $|f_{xx}(x,y)|$ on R is $e^{0.1} \cos(0.1) \le 1.11$, max of $|f_{yy}(x,y)|$ on R is $e^{0.1} \cos(0.1) \le 1.11$, max of $|f_{xy}(x,y)|$ on R is $e^{0.1} \sin(0.1) \le 0.002 \Rightarrow M = 1.11$.
    $\therefore |E(x,y)| \le \frac{1}{2}(1.11)(|x| + |y|)^2 \le 0.555(0.1 + 0.1)^2 = 0.0222$

13. Let the width, w, be the long side. Then $A = lw \Rightarrow dA = A_l\, dl + A_w\, dw \Rightarrow dA = w\, dl + l\, dw$. Since $w > l$, dA is more sensitive to a change in w than l. $\therefore$ pay more attention to the width.

15. $T_x(x,y) = e^y + e^{-y}$, $T_y(x,y) = x(e^y - e^{-y}) \Rightarrow dT = T_x(x,y)\, dx + T_y(x,y)\, dy = (e^y + e^{-y})dx + x(e^y - e^{-y})dy \Rightarrow dT\big|_{(2,\ln 2)} = 2.5\, dx + 3.0\, dy$. If $|dx| \le 0.1$, $|dy| \le 0.02$, then the maximum possible error (estimate) $\le 2.5(0.1) + 3.0(0.02) = 0.31$ in magnitude.

17. $V_r = 2\pi rh$, $V_h = \pi r^2 \Rightarrow dV = V_r\, dr + V_h\, dh \Rightarrow dV = 2\pi rh\, dr + \pi r^2\, dh \Rightarrow dV\big|_{(5,12)} = 120\pi\, dr + 25\pi\, dh$. Since $|dr| \le 0.1$ cm, $|dh| \le 0.1$ cm, $dV \le 120\pi(0.1) + 25\pi(0.1) = 14.5\pi\ \text{cm}^3$. $V(5,12) = 300\pi\ \text{cm}^3 \Rightarrow$ Maximum percentage error $= \pm\frac{14.5\pi}{300\pi} \times 100 = \pm 4.83\%$

19. $df = f_x(x,y)\,dx + f_y(x,y)\,dy = 3x^2y^4dx + 4x^3y^3dy \Rightarrow df|_{(1,1)} = 3\,dx + 4\,dy$. Let $dx = dy \Rightarrow df = 7\,dx$.
$|df| \le 0.1 \Rightarrow 7|dx| \le 0.1 \Rightarrow |dx| \le \frac{0.1}{7} \approx 0.014$. $\therefore$ for the square, let $|x-1| \le 0.014$, $|y-1| \le 0.014$

21. $dR = \left(\frac{R}{R_1}\right)^2 dR_1 + \left(\frac{R}{R_2}\right)^2 dR_2$ (See Exercise 20 above). R_1 changes from 20 to 20.1 ohms $\Rightarrow dR_1 =$ 0.1 ohms, R_2 changes from 25 to 24.9 ohms $\Rightarrow dR_2 = -0.1$ ohms. $\frac{1}{R} = \frac{1}{R_1} + \frac{1}{R_2} \Rightarrow R = \frac{100}{9}$ ohms.
$dR|_{(20,25)} = \frac{(100/9)^2}{(20)^2}(0.1) + \frac{(100/9)^2}{(25)^2}(-0.1) = 0.011$ ohms $\Rightarrow$ Percentage change $= \frac{dR}{R}\Big|_{(20,25)} \times 100$
$= \frac{0.011}{100/9} \times 100 \approx 0.099\%$ or about 0.1%.

23. a) $f(1,1,1) = 3$, $f_x(1,1,1) = y+z|_{(1,1,1)} = 2$, $f_y(1,1,1) = x+z|_{(1,1,1)} = 2$, $f_z(1,1,1) = y+x|_{(1,1,1)} = 2 \Rightarrow$
$L(x,y,z) = 2x + 2y + 2z - 3$
b) $f(1,0,0) = 0$, $f_x(1,0,0) = 0$, $f_y(1,0,0) = 1$, $f_z(1,0,0) = 1 \Rightarrow L(x,y,z) = y + z$
c) $f(0,0,0) = 0$, $f_x(0,0,0) = 0$, $f_y(0,0,0) = 0$, $f_z(0,0,0) = 0 \Rightarrow L(x,y,z) = 0$

25. a) $f(1,0,0) = 1$, $f_x(1,0,0) = \frac{x}{\sqrt{x^2+y^2+z^2}}\Big|_{(1,0,0)} = 1$, $f_y(1,0,0) = \frac{y}{\sqrt{x^2+y^2+z^2}}\Big|_{(1,0,0)} = 0$,
$f_z(1,0,0) = \frac{z}{\sqrt{x^2+y^2+z^2}}\Big|_{(1,0,0)} = 0 \Rightarrow L(x,y,z) = x$
b) $f(1,1,0) = \sqrt{2}$, $f_x(1,1,0) = \frac{1}{\sqrt{2}}$, $f_y(1,1,0) = \frac{1}{\sqrt{2}}$, $f_z(1,1,0) = 0 \Rightarrow L(x,y,z) = \frac{1}{\sqrt{2}}x + \frac{1}{\sqrt{2}}y$
c) $f(1,2,2) = 3$, $f_x(1,2,2) = \frac{1}{3}$, $f_y(1,2,2) = \frac{2}{3}$, $f_z(1,2,2) = \frac{2}{3} \Rightarrow L(x,y,z) = \frac{1}{3}x + \frac{2}{3}y + \frac{2}{3}z$

27. a) $f(0,0,0) = 2$, $f_x(0,0,0) = e^x\,|_{(0,0,0)} = 1$, $f_y(0,0,0) = -\sin(y+z)|_{(0,0,0)} = 0$, $f_z(0,0,0) = -\sin(y+z)|_{(0,0,0)}$
$= 0 \Rightarrow L(x,y,z) = 2 + x$
b) $f\left(0,\frac{\pi}{2},0\right) = 1$, $f_x\left(0,\frac{\pi}{2},0\right) = 1$, $f_y\left(0,\frac{\pi}{2},0\right) = -1$, $f_z\left(0,\frac{\pi}{2},0\right) = -1 \Rightarrow L(x,y,z) = x - y - z + \frac{\pi}{2} + 1$
c) $f\left(0,\frac{\pi}{4},\frac{\pi}{4}\right) = 1$, $f_x\left(0,\frac{\pi}{4},\frac{\pi}{4}\right) = 1$, $f_y\left(0,\frac{\pi}{4},\frac{\pi}{4}\right) = -1$, $f_z\left(0,\frac{\pi}{4},\frac{\pi}{4}\right) = -1 \Rightarrow L(x,y,z) = x - y - z + \frac{\pi}{2} + 1$

29. $f(a,b,c,d) = \begin{vmatrix} a & b \\ c & d \end{vmatrix} = ad - bc \Rightarrow f_a = d$, $f_b = -c$, $f_c = -b$, $f_d = a \Rightarrow df = d\,da - c\,db - b\,dc + a\,dd$.
Since $|a|$ is much greater than $|b|$, $|c|$, and $|d|$, f is most sensitive to a change in d.

31. $V = lwh \Rightarrow V_l = wh$, $V_w = lh$, $V_h = lw \Rightarrow dV = wh\,dl + lh\,dw + lw\,dh \Rightarrow dV|_{(5,3,2)} = 6\,dl + 10\,dw + 15\,dh$
$dl = 1 \text{ in} = \frac{1}{12}$ ft, $dw = 1 \text{ in} = \frac{1}{12}$ ft, $dh = \frac{1}{2} \text{ in} = \frac{1}{24}$ ft $\Rightarrow dV = 6\left(\frac{1}{12}\right) + 10\left(\frac{1}{12}\right) + 15\left(\frac{1}{24}\right) = \frac{47}{24}\text{ ft}^3$

33. $u_x = e^y$, $u_y = xe^y + \sin z$, $u_z = y\cos z \Rightarrow du = e^y\,dx + (xe^y + \sin z)dy + (y\cos z)dz \Rightarrow$
$du|_{(2,\ln 3,\pi/2)} = 3\,dx + 7\,dy + 0\,dz = 3\,dx + 7\,dy \Rightarrow$ magnitude of the maximum possible error $\le$
$3(0.2) + 7(0.6) = 4.8$

35. If the first partial derivatives are continuous throughout on open region R, Then $f(x,y) = f(x_0,y_0) + f_x(x_0,y_0)\Delta x + f_y(x_0,y_0)\Delta y + \varepsilon_1\Delta x + \varepsilon_2\Delta y$ (Equation 3, Section 12.4) where $\varepsilon_1, \varepsilon_2 \to 0$ as $\Delta x, \Delta y \to 0$. Then as $(x,y) \to (x_0,y_0)$, $\Delta x \to 0$ and $\Delta y \to 0 \Rightarrow \lim_{(x,y)\to(x_0,y_0)} f(x,y) = f(x_0,y_0) \Rightarrow$ f is continuous at every (x_0,y_0) in R.

13.5 THE CHAIN RULE

1. a) $\frac{\partial w}{\partial x} = 2x, \frac{\partial w}{\partial y} = 2y, \frac{dx}{dt} = -\sin t, \frac{dy}{dt} = \cos t \Rightarrow \frac{dw}{dt} = -2x \sin t + 2y \cos t = -2 \cos t \sin t + 2 \sin t \cos t = 0$

$w = x^2 + y^2 = \cos^2 t + \sin^2 t = 1 \Rightarrow \frac{dw}{dt} = 0$

b) $\frac{dw}{dt}(\pi) = 0$

3. a) $\frac{\partial w}{\partial x} = \frac{1}{z}, \frac{\partial w}{\partial y} = \frac{1}{z}, \frac{\partial w}{\partial z} = \frac{-(x+y)}{z^2}, \frac{dx}{dt} = -2 \cos t \sin t, \frac{dy}{dt} = 2 \sin t \cos t, \frac{dz}{dt} = -\frac{1}{t^2} \Rightarrow$

$\frac{dw}{dt} = -\frac{2}{z} \cos t \sin t + \frac{2}{z} \sin t \cos t + \frac{x+y}{z^2t^2} = \frac{\cos^2 t + \sin^2 t}{\frac{1}{t^2}(t^2)} = 1.$ $w = \frac{x}{z} + \frac{y}{z} = \frac{\cos^2 t}{1/t} + \frac{\sin^2 t}{1/t} = t \Rightarrow \frac{dw}{dt} = 1$

b) $\frac{dw}{dt}(3) = 1$

5. a) $\frac{\partial w}{\partial x} = 2ye^x, \frac{\partial w}{\partial y} = 2e^x, \frac{\partial w}{\partial z} = -\frac{1}{z}, \frac{dx}{dt} = \frac{2t}{t^2+1}, \frac{dy}{dt} = \frac{1}{t^2+1}, \frac{dz}{dt} = e^t \Rightarrow \frac{dw}{dt} = \frac{4yte^x}{t^2+1} + \frac{2e^x}{t^2+1} - \frac{e^t}{z} =$

$\frac{4t \tan^{-1} t\, e^{\ln(t^2+1)}}{t^2+1} + \frac{2(t^2+1)}{t^2+1} - \frac{e^t}{e^t} = 4t \tan^{-1} t + 1.$ $w = 2ye^x - \ln z = 2 \tan^{-1} t\, e^{\ln(t^2+1)} - \ln e^t =$

$(2 \tan^{-1} t)(t^2+1) - t \Rightarrow \frac{dw}{dt} = \left(\frac{2}{t^2+1}\right)\left(t^2+1\right) + 2t\left(2 \tan^{-1} t\right)(2t) - 1 = 4t \tan^{-1} t + 1$

b) $\frac{dw}{dt}(1) = \pi + 1$

7. a) $z = 4e^x \ln y, x = \ln(r \cos\theta), y = r \sin\theta \Rightarrow \frac{\partial z}{\partial r} = 4e^x \ln y \left(\frac{1}{r\cos\theta}\right) \cos\theta + \frac{4e^x}{y}(\sin\theta) = \frac{4e^x \ln y}{r} + \frac{4e^x \sin\theta}{y}.$ $\frac{\partial z}{\partial \theta} = 4e^x \ln y \left(\frac{1}{r\cos\theta}\right)(-r\sin\theta) + \frac{4e^x}{y}(r\cos\theta) = -4e^x \ln y \tan\theta + \frac{4e^x r\cos\theta}{y}.$

As functions of r and θ only, $\frac{\partial z}{\partial r} = \frac{4e^{\ln(r\cos\theta)}\ln(r\sin\theta)}{r} + \frac{4e^{\ln(r\cos\theta)}\sin\theta}{r\sin\theta} = 4\cos\theta \ln(r\sin\theta) + 4\cos\theta$

$\frac{\partial z}{\partial \theta} = -4e^{\ln(r\cos\theta)}\ln(r\sin\theta)\tan\theta + \frac{4e^{\ln(r\cos\theta)} r\cos\theta}{r\sin\theta} = -4r\sin\theta \ln(r\sin\theta) + \frac{4r\cos^2\theta}{\sin\theta}.$ To find the partial derivatives directly, let $z = 4e^{\ln(r\cos\theta)}\ln(r\sin\theta) = 4r\cos\theta\ln(r\sin\theta)$. Then find the partial with respect to r and the partial with respect to θ. The answers will be the same.

b) At $\left(2, \frac{\pi}{4}\right)$, $\frac{\partial z}{\partial r} = 4\cos\frac{\pi}{4}\ln(2\sin\frac{\pi}{4}) + 4\cos\frac{\pi}{4} = 2\sqrt{2}\ln\sqrt{2} + 2\sqrt{2} = \sqrt{2}(\ln 2 + 2)$. $\frac{\partial z}{\partial \theta} = -4(2)\sin\frac{\pi}{4}\ln(2\sin\frac{\pi}{4}) + \frac{4(2)\cos^2\frac{\pi}{4}}{\sin\frac{\pi}{4}} = -4\sqrt{2}\ln\sqrt{2} + 4\sqrt{2} = -2\sqrt{2}\ln 2 + 4\sqrt{2}$

9. a) $w = xy + yz + xz,\ x = u + v,\ y = u - v,\ z = uv \Rightarrow \frac{\partial w}{\partial u} = (y + z)(1) + (x + z)(1) + (y + x)(v) = x + y + 2z +$ $v(y + x)$. As a function of u an v only, $\frac{\partial w}{\partial u} = u + v + u - v + 2uv + v(u - v + u + v) = 2u + 4uv$.
$\frac{\partial w}{\partial v} = (y + z)(1) + (x + z)(-1) + (y + x)(u) = y - x + (y + x)u$. As a function of u and v only, $\frac{\partial w}{\partial v} = u - v -$ $(u + v) + (u - v + u + v)u = -2v + 2u^2$. To find the partial derivatives directly, let $w = (u + v)(u - v) +$ $(u - v)uv + (u + v)uv = u^2 - v^2 + 2u^2v$. Then find the partial with respect to u and the partial with respect to v. The answers will be the same as above.

b) At $\left(\frac{1}{2}, 1\right)$, $\frac{\partial w}{\partial u} = 2\left(\frac{1}{2}\right) + 4\left(\frac{1}{2}\right)(1) = 3$, $\frac{\partial w}{\partial v} = -2(1) + 2\left(\frac{1}{2}\right)^2 = -\frac{3}{2}$.

11. a) $u = \frac{p - q}{q - r},\ p = x + y + z,\ q = x - y + z,\ r = x + y - z \Rightarrow \frac{\partial u}{\partial x} = \frac{1}{q - r} + \frac{r - p}{(q - r)^2} + \frac{p - q}{(q - r)^2} = 0$. $\frac{\partial u}{\partial y} = \frac{1}{q - r} -$ $\frac{r - p}{(q - r)^2} + \frac{p - q}{(q - r)^2} = \frac{2p - 2r}{(q - r)^2}$. As a function of x, y, and z only, $\frac{\partial u}{\partial y} = \frac{2(x + y + z) - 2(x + y - z)}{(2z - 2y)^2} = \frac{z}{(z - y)^2}$.
$\frac{\partial u}{\partial z} = \frac{1}{q - r} + \frac{r - p}{(q - r)^2} - \frac{p - q}{(q - r)^2} = \frac{-2p + 2q}{(q - r)^2}$. As a function of x, y, and z only, $\frac{\partial u}{\partial z} =$ $\frac{-2(x + y + z) + 2(x - y + z)}{(2z - 2y)^2} = \frac{-y}{(z - y)^2}$. To find the partial derivatives directly, let $u =$ $\frac{(x + y + z) - (x - y + z)}{(x - y + z) - (x + y - z)} = \frac{y}{z - y}$. Then find the partial with respect to x, the partial with respect to y, and the partial with respect to z. The answers will be the same as above.

b) At $\left(\sqrt{3}, 2, 1\right)$, $\frac{\partial u}{\partial x} = 0$, $\frac{\partial u}{\partial y} = \frac{1}{(1 - 2)^2} = 1$, $\frac{\partial u}{\partial z} = \frac{-2}{(1 - 2)^2} = -2$.

13. $\frac{dz}{dt} = \frac{\partial z}{\partial x}\frac{dx}{dt} + \frac{\partial z}{\partial y}\frac{dy}{dt}$

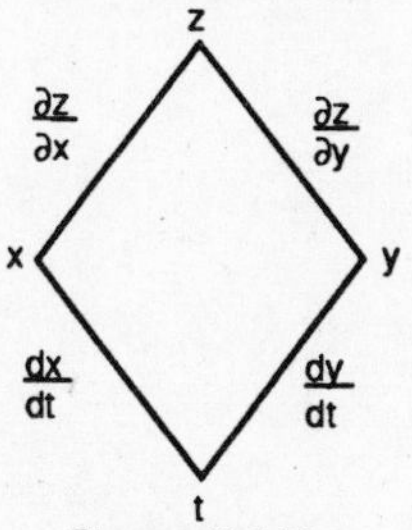

Diagram 13.5.13

15. $\frac{\partial w}{\partial u} = \frac{\partial w}{\partial x}\frac{\partial x}{\partial u} + \frac{\partial w}{\partial y}\frac{\partial y}{\partial u} + \frac{\partial w}{\partial z}\frac{\partial z}{\partial u}$ $\qquad \frac{\partial w}{\partial v} = \frac{\partial w}{\partial x}\frac{\partial x}{\partial v} + \frac{\partial w}{\partial y}\frac{\partial y}{\partial v} + \frac{\partial w}{\partial z}\frac{\partial z}{\partial v}$

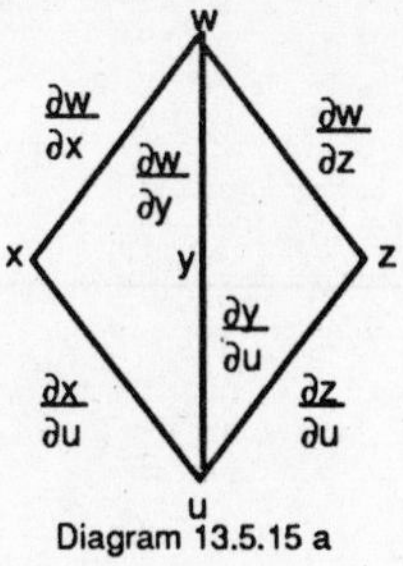

Diagram 13.5.15 a

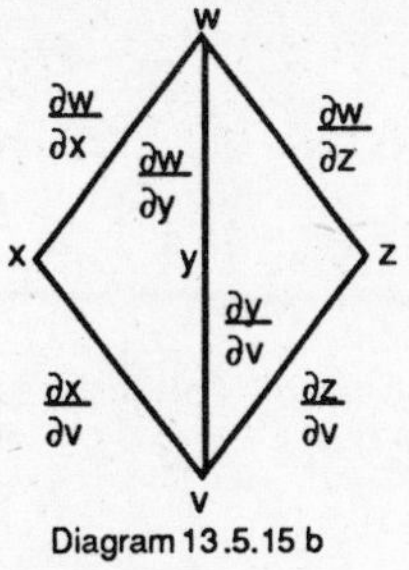

Diagram 13.5.15 b

17. $\dfrac{\partial w}{\partial u}=\dfrac{\partial w}{\partial x}\dfrac{\partial x}{\partial u}+\dfrac{\partial w}{\partial y}\dfrac{\partial y}{\partial u}$

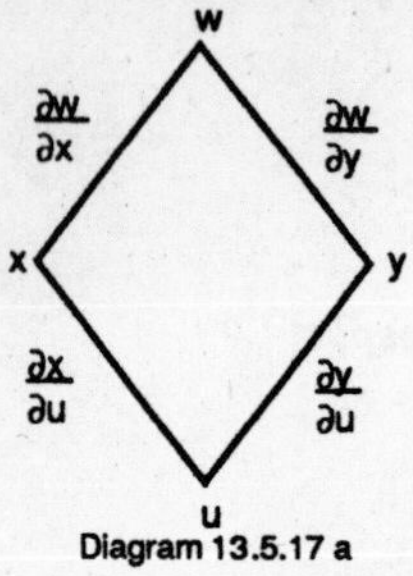

Diagram 13.5.17 a

$\dfrac{\partial w}{\partial v}=\dfrac{\partial w}{\partial x}\dfrac{\partial x}{\partial v}+\dfrac{\partial w}{\partial y}\dfrac{\partial y}{\partial v}$

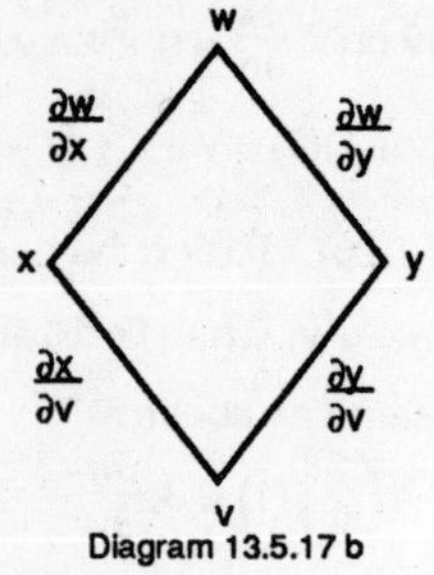

Diagram 13.5.17 b

19. $\dfrac{\partial z}{\partial t}=\dfrac{\partial z}{\partial x}\dfrac{\partial x}{\partial t}+\dfrac{\partial z}{\partial y}\dfrac{\partial y}{\partial t}$

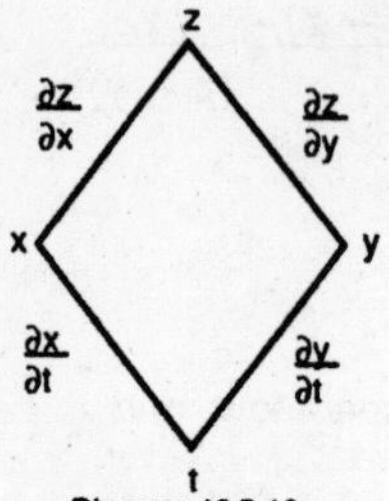

Diagram 13.5.19 a

$\dfrac{\partial z}{\partial s}=\dfrac{\partial z}{\partial x}\dfrac{\partial x}{\partial s}+\dfrac{\partial z}{\partial y}\dfrac{\partial y}{\partial s}$

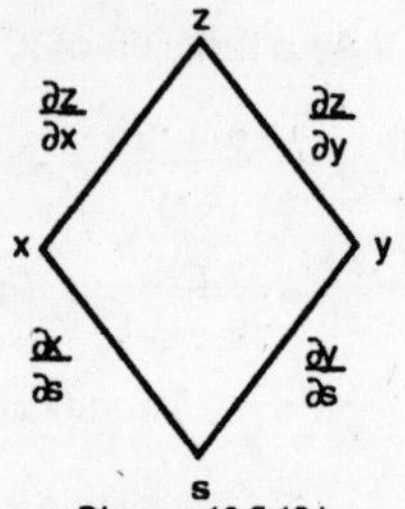

Diagram 13.5.19 b

21. $\dfrac{\partial w}{\partial s}=\dfrac{dw}{du}\dfrac{\partial u}{\partial s}$

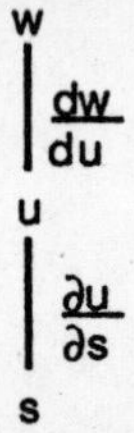

Diagram 13.5.21 a

$\dfrac{\partial w}{\partial t}=\dfrac{dw}{du}\dfrac{\partial u}{\partial t}$

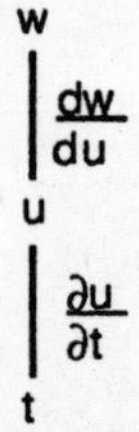

Diagram 13.5.21 b

23. $\dfrac{\partial w}{\partial r}=\dfrac{\partial w}{\partial x}\dfrac{dx}{dr}+\dfrac{\partial w}{\partial y}\dfrac{dy}{dr}=\dfrac{\partial w}{\partial x}\dfrac{dx}{dr}$

since $\dfrac{dy}{dr}=0$

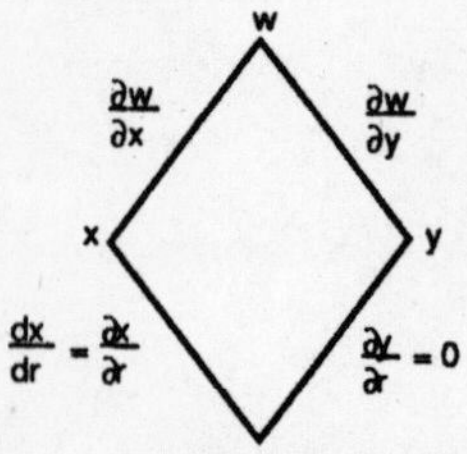

Diagram 13.5.23 a

$\dfrac{\partial w}{\partial s}=\dfrac{\partial w}{\partial x}\dfrac{dx}{ds}+\dfrac{\partial w}{\partial y}\dfrac{dy}{ds}=\dfrac{\partial w}{\partial y}\dfrac{dy}{ds}$

since $\dfrac{dx}{ds}=0$

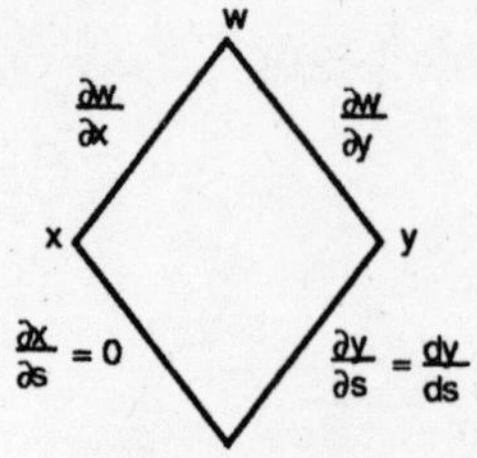

Diagram 13.5.23 b

25. Let $F(x,y) = x^3 - 2y^2 + xy = 0 \Rightarrow F_x(x,y) = 3x^2 + y,\ F_y(x,y) = -4y + x \Rightarrow \frac{dy}{dx} = -\frac{F_x}{F_y} = -\frac{3x^2 + y}{(-4y + x)} \Rightarrow$
$\frac{dy}{dx}(1,1) = \frac{4}{3}$

27. Let $F(x,y) = x^2 + xy + y^2 - 7 = 0 \Rightarrow F_x(x,y) = 2x + y,\ F_y(x,y) = x + 2y \Rightarrow \frac{dy}{dx} = -\frac{F_x}{F_y} = -\frac{2x + y}{x + 2y} \Rightarrow \frac{dy}{dx}(1,2) = -\frac{4}{5}$

29. Let $F(x,y,z)) = z^3 - xy + yz + y^3 - 2 = 0 \Rightarrow F_x(x,y,z) = -y,\ F_y(x,y,z) = -x + z + 3y^2,\ F_z(x,y,z) = 3z^2 + y \Rightarrow$
$\frac{\partial z}{\partial x} = -\frac{F_x}{F_z} = -\frac{-y}{3z^2 + y} = \frac{y}{3z^2 + y} \Rightarrow \frac{\partial z}{\partial x}(1,1,1) = \frac{1}{4}.\ \frac{\partial z}{\partial y} = -\frac{F_y}{F_z} = -\frac{-x + z + 3y^2}{3z^2 + y} = \frac{x - z - 3y^2}{3z^2 + y} \Rightarrow$
$\frac{\partial z}{\partial y}(1,1,1) = -\frac{3}{4}$

31. Let $F(x,y,z) = \sin(x + y) + \sin(y + z) + \sin(x + z) = 0 \Rightarrow F_x(x,y,z) = \cos(x + y) + \cos(x + z)$,
$F_y(x,y,z) = \cos(x + y) + \cos(y + z),\ F_z(x,y,z) = \cos(y + z) + \cos(x + z) \Rightarrow \frac{\partial z}{\partial x} = -\frac{F_x}{F_z} = -\frac{\cos(x + y) + \cos(x + z)}{\cos(y + z) + \cos(x + z)}$
$\Rightarrow \frac{\partial z}{\partial x}(\pi,\pi,\pi) = -1.\ \frac{\partial z}{\partial y} = -\frac{F_y}{F_z} = -\frac{\cos(x + y) + \cos(y + z)}{\cos(y + z) + \cos(x + z)} \Rightarrow \frac{\partial z}{\partial y}(\pi,\pi,\pi) = -1$

33. $\frac{\partial w}{\partial r} = \frac{\partial w}{\partial x}\frac{\partial x}{\partial r} + \frac{\partial w}{\partial y}\frac{\partial y}{\partial r} + \frac{\partial w}{\partial z}\frac{\partial z}{\partial r} = 2(x + y + z)(1) + 2(x + y + z)(-\sin(r + s)) + 2(x + y + z)(\cos(r + s)) =$
$2(x + y + z)(1 - \sin(r + s) + \cos(r + s)) = 2(r - s + \cos(r + s) + \sin(r + s))(1 - \sin(r + s) + \cos(r + s)) \Rightarrow$
$\left.\frac{\partial w}{\partial r}\right|_{r=1,s=-1} = 12$

35. $\frac{\partial w}{\partial v} = \frac{\partial w}{\partial x}\frac{\partial x}{\partial v} + \frac{\partial w}{\partial y}\frac{\partial y}{\partial v} = \left(2x - \frac{y}{x^2}\right)(-2) + \frac{1}{x}(1) = \left(2(u - 2v + 1) - \frac{2u + v - 2}{(u - 2v + 1)^2}\right)(-2) + \frac{1}{u - 2v + 1} \Rightarrow$
$\left.\frac{\partial w}{\partial v}\right|_{u=0,v=0} = -7$

37. $\frac{\partial z}{\partial u} = \frac{dz}{dx}\frac{\partial x}{\partial u} = \frac{5}{1 + x^2}\, e^u = \frac{5}{1 + \left(e^u + \ln v\right)^2}\, e^u \Rightarrow \left.\frac{\partial z}{\partial u}\right|_{u=\ln 2,v=1} = 2$
$\frac{\partial z}{\partial v} = \frac{dz}{dx}\frac{\partial x}{\partial v} = \frac{5}{1 + x^2}\left(\frac{1}{v}\right) = \frac{5}{1 + \left(e^u + \ln v\right)^2}\left(\frac{1}{v}\right) \Rightarrow \left.\frac{\partial z}{\partial v}\right|_{u=\ln 2,v=1} = 1$

39. $\frac{dV}{dt} = \frac{\partial V}{\partial I}\frac{dI}{dt} + \frac{\partial V}{\partial R}\frac{dR}{dt}.\ V = IR \Rightarrow \frac{\partial V}{\partial I} = R, \frac{\partial V}{\partial R} = I \Rightarrow \frac{dV}{dt} = R\frac{dI}{dt} + I\frac{dR}{dt} \Rightarrow -0.01$ volts/sec = (600 ohms)$\frac{dI}{dt}$ +
(0.04 amps)(0.5 ohms/sec) $\Rightarrow \frac{dI}{dt} = -0.00005$ amps/sec.

41. $f_x(x,y,z) = \cos t,\ f_y(x,y,z) = \sin t,\ f_z(x,y,z) = t^2 + t - 2.\ \frac{df}{dt} = \frac{\partial f}{\partial x}\frac{dx}{dt} + \frac{\partial f}{\partial y}\frac{dy}{dt} + \frac{\partial f}{\partial z}\frac{dz}{dt} = (\cos t)(-\sin t) +$
$(\sin t)(\cos t) + (t^2 + t - 2)(1) = t^2 + t - 2.\ \frac{df}{dt} = 0 \Rightarrow t^2 + t - 2 = 0 \Rightarrow t = -2$ or $t = 1$
$t = -2 \Rightarrow x = \cos(-2),\ y = \sin(-2),\ z = -2;\ t = 1 \Rightarrow x = \cos 1,\ y = \sin 1,\ z = 1$

43. a) $\frac{\partial T}{\partial x} = 8x - 4y, \frac{\partial T}{\partial y} = 8y - 4x. \frac{dT}{dt} = \frac{\partial T}{\partial x}\frac{dx}{dt} + \frac{\partial T}{\partial y}\frac{dy}{dt} = (8x - 4y)(-\sin t) + (8y - 4x)(\cos t) =$

$(8\cos t - 4\sin t)(-\sin t) + (8\sin t - 4\cos t)(\cos t) = 4\sin^2 t - 4\cos^2 t \Rightarrow \frac{d^2T}{dt^2} = 16\sin t\cos t$

$\frac{dT}{dt} = 0 \Rightarrow 4\sin^2 t - 4\cos^2 t = 0 \Rightarrow \sin^2 t = \cos^2 t \Rightarrow \sin t = \cos t$ or $\sin t = -\cos t \Rightarrow$

$t = \frac{\pi}{4}, \frac{5\pi}{4}$ or $\frac{3\pi}{4}, \frac{7\pi}{4}$ on the interval $0 \le t \le 2\pi$.

$\left.\frac{d^2T}{dt^2}\right|_{t=\pi/4} = 16\sin\frac{\pi}{4}\cos\frac{\pi}{4} > 0 \Rightarrow T$ has a minimum at $t = \frac{\pi}{4}$

$\left.\frac{d^2T}{dt^2}\right|_{t=3\pi/4} = 16\sin\frac{3\pi}{4}\cos\frac{3\pi}{4} < 0 \Rightarrow T$ has a maximum at $t = \frac{3\pi}{4}$

$\left.\frac{d^2T}{dt^2}\right|_{t=5\pi/4} = 16\sin\frac{5\pi}{4}\cos\frac{5\pi}{4} > 0 \Rightarrow T$ has a minimum at $t = \frac{5\pi}{4}$

$\left.\frac{d^2T}{dt^2}\right|_{t=7\pi/4} = 16\sin\frac{7\pi}{4}\cos\frac{7\pi}{4} < 0 \Rightarrow T$ has a maximum at $t = \frac{7\pi}{4}$

b) $T = 4x^2 - 4xy + 4y^2 \Rightarrow \frac{\partial T}{\partial x} = 8x - 4y, \frac{\partial T}{\partial y} = 8y - 4x$ (See part a above.)

$t = \frac{\pi}{4} \Rightarrow x = \cos\frac{\pi}{4} = \frac{\sqrt{2}}{2}, y = \sin\frac{\pi}{4} = \frac{\sqrt{2}}{2} \Rightarrow T\left(\frac{\pi}{4}\right) = 2$

$t = \frac{3\pi}{4} \Rightarrow x = \cos\frac{3\pi}{4} = -\frac{\sqrt{2}}{2}, y = \sin\frac{3\pi}{4} = \frac{\sqrt{2}}{2} \Rightarrow T\left(\frac{3\pi}{4}\right) = 6$

$t = \frac{5\pi}{4} \Rightarrow x = \cos\frac{5\pi}{4} = -\frac{\sqrt{2}}{2}, y = \sin\frac{5\pi}{4} = -\frac{\sqrt{2}}{2} \Rightarrow T\left(\frac{5\pi}{4}\right) = 2$

$t = \frac{7\pi}{4} \Rightarrow x = \cos\frac{7\pi}{4} = \frac{\sqrt{2}}{2}, y = \sin\frac{7\pi}{4} = -\frac{\sqrt{2}}{2} \Rightarrow T\left(\frac{7\pi}{4}\right) = 6$

$\therefore T_{max} = 6$ and $T_{min} = 2$.

13.6 DIRECTIONAL DERIVATIVES, GRADIENT VECTORS, AND TANGENT PLANES

1. $\frac{\partial f}{\partial x} = -1, \frac{\partial f}{\partial y} = 1 \Rightarrow \nabla f = -\mathbf{i} + \mathbf{j}$

$-1 = y - x$ is the level curve.

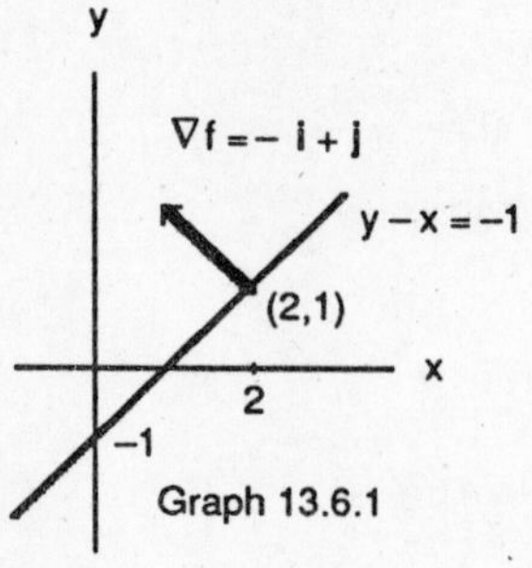

Graph 13.6.1

3. $\frac{\partial g}{\partial x} = -2x \Rightarrow \frac{\partial g}{\partial x}(-1,0) = 2. \frac{\partial g}{\partial y} = 1 \Rightarrow$

$\nabla g = 2\mathbf{i} + \mathbf{j}$. $-1 = y - x^2$ is the level curve.

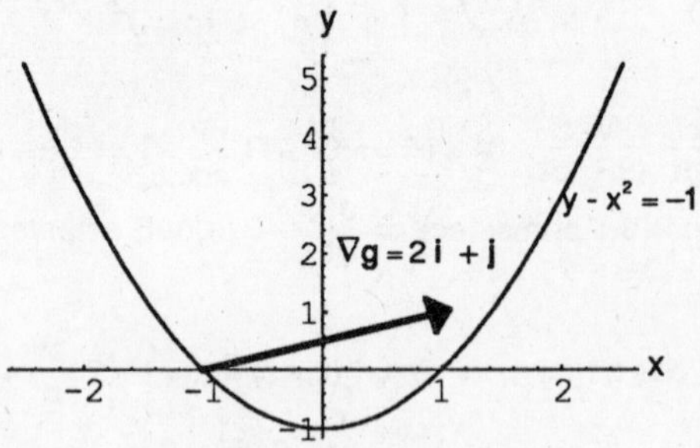

Graph 13.6.3

5. $\frac{\partial f}{\partial x} = 2x \Rightarrow \frac{\partial f}{\partial x}(1,1,1) = 2$. $\frac{\partial f}{\partial y} = 2y \Rightarrow \frac{\partial f}{\partial y}(1,1,1) = 2$. $\frac{\partial f}{\partial z} = -4z \Rightarrow \frac{\partial f}{\partial z}(1,1,1) = -4 \Rightarrow \nabla f = 2\,\mathbf{i} + 2\,\mathbf{j} - 4\,\mathbf{k}$

7. $\frac{\partial h}{\partial x} = -x(x^2+y^2+z^2)^{-3/2} \Rightarrow \frac{\partial h}{\partial x}(1,2,-2) = -\frac{1}{27}$. $\frac{\partial h}{\partial y} = -y(x^2+y^2+z^2)^{-3/2} \Rightarrow \frac{\partial h}{\partial y}(1,2,-2) = -\frac{2}{27}$.
$\frac{\partial h}{\partial z} = -z(x^2+y^2+z^2)^{-3/2} \Rightarrow \frac{\partial h}{\partial z}(1,2,-2) = \frac{2}{27} \Rightarrow \nabla h = -\frac{1}{27}\mathbf{i} - \frac{2}{27}\mathbf{j} + \frac{2}{27}\mathbf{k}$

9. $\mathbf{u} = \frac{\mathbf{A}}{|\mathbf{A}|} = \frac{12\,\mathbf{i} + 5\,\mathbf{j}}{\sqrt{12^2+5^2}} = \frac{12}{13}\mathbf{i} + \frac{5}{13}\mathbf{j}$. $f_x(x,y) = 1 + \frac{y^2}{x^2} \Rightarrow f_x(1,1) = 2$, $f_y(x,y) = -\frac{2y}{x} \Rightarrow f_y(1,1) = -2 \Rightarrow$
$\nabla f = 2\,\mathbf{i} - 2\,\mathbf{j} \Rightarrow (D_{\mathbf{u}}f)_{P_0} = \nabla f \cdot \mathbf{u} = \frac{24}{13} - \frac{10}{13} = \frac{14}{13}$

11. $\mathbf{u} = \frac{\mathbf{A}}{|\mathbf{A}|} = \frac{3\,\mathbf{i} + 4\,\mathbf{j}}{\sqrt{3^2+4^2}} = \frac{3}{5}\mathbf{i} + \frac{4}{5}\mathbf{j}$. $g_x(x,y) = 2x + 2y \Rightarrow g_x(1,1) = 4$, $g_y(x,y) = 2x = 6y \Rightarrow g_y(1,1) = -4 \Rightarrow$
$\nabla g = 4\,\mathbf{i} - 4\,\mathbf{j} \Rightarrow (D_{\mathbf{u}}g)_{P_0} = \nabla g \cdot \mathbf{u} = \frac{12}{5} - \frac{16}{5} = -\frac{4}{5}$

13. $\mathbf{u} = \frac{\mathbf{A}}{|\mathbf{A}|} = \frac{3\,\mathbf{i} + 6\,\mathbf{j} - 2\,\mathbf{k}}{\sqrt{3^2+6^2+(-2)^2}} = \frac{3}{7}\mathbf{i} + \frac{6}{7}\mathbf{j} - \frac{2}{7}\mathbf{k}$. $f_x(x,y,z) = y + z \Rightarrow f_x(1,-1,2) = 1$, $f_y(x,y,z) = x + z \Rightarrow$
$f_y(1,-1,2) = 3$, $f_z(x,y,z) = y + x \Rightarrow f_z(1,-1,2) = 0 \Rightarrow \nabla f = \mathbf{i} + 3\,\mathbf{j} \Rightarrow (D_{\mathbf{u}}f)_{P_0} = \nabla f \cdot \mathbf{u} = \frac{3}{7} + \frac{18}{7} = 3$

15. $\mathbf{u} = \frac{\mathbf{A}}{|\mathbf{A}|} = \frac{2\,\mathbf{i} + \mathbf{j} - 2\mathbf{k}}{\sqrt{2^2+1^2+(-2)^2}} = \frac{2}{3}\mathbf{i} + \frac{1}{3}\mathbf{j} - \frac{2}{3}\mathbf{k}$. $g_x(x,y,z) = 3e^x \cos yz \Rightarrow g_x(0,0,0) = 3$, $g_y(x,y,z) =$
$-3ze^x \sin yz \Rightarrow g_y(0,0,0) = 0$, $g_z(x,y,z) = -3ye^x \sin yz \Rightarrow g_z(0,0,0) = 0 \Rightarrow \nabla g = 3\,\mathbf{i} \Rightarrow (D_{\mathbf{u}}g)_{P_0} = \nabla g \cdot \mathbf{u} = 2$

17. $\nabla f = (2x + y)\,\mathbf{i} + (x + 2y)\,\mathbf{j} \Rightarrow \nabla f(-1,1) = -\mathbf{i} + \mathbf{j} \Rightarrow \mathbf{u} = \frac{\nabla f}{|\nabla f|} = \frac{-\,\mathbf{i} + \mathbf{j}}{\sqrt{(-1)^2+1^2}} = -\frac{1}{\sqrt{2}}\mathbf{i} + \frac{1}{\sqrt{2}}\mathbf{j}$. f increases most rapidly in the direction $\mathbf{u} = -\frac{1}{\sqrt{2}}\mathbf{i} + \frac{1}{\sqrt{2}}\mathbf{j}$ and decreases most rapidly in the direction $-\mathbf{u} = \frac{1}{\sqrt{2}}\mathbf{i} - \frac{1}{\sqrt{2}}\mathbf{j}$. $(D_{\mathbf{u}}f)_{P_0} = \nabla f \cdot \mathbf{u} = \sqrt{2}$, $(D_{-\mathbf{u}}f)_{P_0} = -\sqrt{2}$

19. $\nabla f = 2x\,\mathbf{i} + \cos y\,\mathbf{j} \Rightarrow \nabla f(1,0) = 2\,\mathbf{i} + \mathbf{j} \Rightarrow \mathbf{u} = \frac{\nabla f}{|\nabla f|} = \frac{2\,\mathbf{i} + \mathbf{j}}{\sqrt{2^2+1^2}} = \frac{2}{\sqrt{5}}\mathbf{i} + \frac{1}{\sqrt{5}}\mathbf{j}$. f increases most rapidly in the direction of $\mathbf{u} = \frac{2}{\sqrt{5}}\mathbf{i} + \frac{1}{\sqrt{5}}\mathbf{j}$; decreases most rapidly in the direction $-\mathbf{u} = -\frac{2}{\sqrt{5}}\mathbf{i} - \frac{1}{\sqrt{5}}\mathbf{j}$.
$(D_{\mathbf{u}}f)_{P_0} = \nabla f \cdot \mathbf{u} = \sqrt{5}$, $(D_{-\mathbf{u}}f)_{P_0} = -\sqrt{5}$.

21. $\nabla g = e^y\,\mathbf{i} + xe^y\,\mathbf{j} + 2z\,\mathbf{k} \Rightarrow \nabla g\left(1, \ln 2, \frac{1}{2}\right) = 2\,\mathbf{i} + 2\,\mathbf{j} + \mathbf{k} \Rightarrow \mathbf{u} = \frac{\nabla g}{|\nabla g|} = \frac{2\,\mathbf{i} + 2\,\mathbf{j} + \mathbf{k}}{\sqrt{2^2+2^2+1^2}} = \frac{2}{3}\mathbf{i} + \frac{2}{3}\mathbf{j} + \frac{1}{3}\mathbf{k}$.
g increases most rapidly in the direction $\mathbf{u} = \frac{2}{3}\mathbf{i} + \frac{2}{3}\mathbf{j} + \frac{1}{3}\mathbf{k}$; decreases most rapidly in the direction $-\mathbf{u} = -\frac{2}{3}\mathbf{i} - \frac{2}{3}\mathbf{j} - \frac{1}{3}\mathbf{k}$. $(D_{\mathbf{u}}g)_{P_0} = \nabla g \cdot \mathbf{u} = 3$; $(D_{-\mathbf{u}}g)_{P_0} = -3$

23. $\nabla f = \frac{x}{x^2+y^2+z^2}\mathbf{i} + \frac{y}{x^2+y^2+z^2}\mathbf{j} + \frac{z}{x^2+y^2+z^2}\mathbf{k} \Rightarrow \nabla f(3,4,12) = \frac{3}{169}\mathbf{i} + \frac{4}{169}\mathbf{j} + \frac{12}{169}\mathbf{k}$. $\mathbf{u} = \frac{\mathbf{A}}{|\mathbf{A}|} = \frac{3\mathbf{i}+6\mathbf{j}-2\mathbf{k}}{\sqrt{3^2+6^2+(-2)^2}} = \frac{3}{7}\mathbf{i} + \frac{6}{7}\mathbf{j} - \frac{2}{7}\mathbf{k} \Rightarrow \nabla f \cdot \mathbf{u} = \frac{9}{1183}$. $\therefore$ $df = (\nabla f \cdot \mathbf{u})ds = \frac{9}{1183}(0.1) \approx 0.000760$

25. $\mathbf{A} = \overrightarrow{P_0P_1} = -2\mathbf{i} + 2\mathbf{j} + 2\mathbf{k}$. $\nabla g = (1+\cos z)\mathbf{i} + (1-\sin z)\mathbf{j} + (-x\sin z - y\cos z)\mathbf{k} \Rightarrow \nabla g(2,-1,0) = 2\mathbf{i} + \mathbf{j} + \mathbf{k}$. $\mathbf{u} = \frac{\mathbf{A}}{|\mathbf{A}|} = \frac{-2\mathbf{i}+2\mathbf{j}+2\mathbf{k}}{\sqrt{(-2)^2+2^2+2^2}} = -\frac{1}{\sqrt{3}}\mathbf{i} + \frac{1}{\sqrt{3}}\mathbf{j} + \frac{1}{\sqrt{3}}\mathbf{k} \Rightarrow \nabla g \cdot \mathbf{u} = 0$. $\therefore$ $dg = (\nabla g \cdot \mathbf{u})ds = 0(0.2) = 0$

27. $\nabla f = 2x\mathbf{i} + 2y\mathbf{j} + 2z\mathbf{k} \Rightarrow \nabla f(1,1,1) = 2\mathbf{i} + 2\mathbf{j} + 2\mathbf{k} \Rightarrow$ Tangent plane: $2(x-1) + 2(y-1) + 2(z-1) = 0 \Rightarrow x + y + z = 3$; Normal line: $x = 1 + 2t$, $y = 1 + 2t$, $z = 1 + 2t$

29. $\nabla f = -2x\mathbf{i} + 2\mathbf{k} \Rightarrow \nabla f(2,0,2) = -4\mathbf{i} + 2\mathbf{k} \Rightarrow$ Tangent plane: $-4(x-2) + 2(z-2) = 0 \Rightarrow -4x + 2z + 4 = 0 \Rightarrow -2x + z + 2 = 0$; Normal line: $x = 2 - 4t$, $y = 0$, $z = 2 + 2t$

31. $\nabla f = \left(-\pi\sin\pi x - 2xy + ze^{xz}\right)\mathbf{i} + \left(-x^2 + z\right)\mathbf{j} + \left(xe^{xz} + y\right)\mathbf{k} \Rightarrow \nabla f(0,1,2) = 2\mathbf{i} + 2\mathbf{j} + \mathbf{k} \Rightarrow$ Tangent plane: $2(x-0) + 2(y-1) + 1(z-2) = 0 \Rightarrow 2x + 2y + z - 4 = 0$; Normal line: $x = 2t$, $y = 1 + 2t$, $z = 2 + t$

33. $\nabla f = \mathbf{i} + \mathbf{j} + \mathbf{k}$ for all points $\Rightarrow$ Tangent plane: $1(x-0) + 1(y-1) + 1(z-0) = 0 \Rightarrow x + y + z - 1 = 0$; Normal line: $x = t$, $y = 1 + t$, $z = t$

35. $z = f(x,y) = \ln(x^2+y^2) \Rightarrow f_x(x,y) = \frac{2x}{x^2+y^2}$, $f_y(x,y) = \frac{2y}{x^2+y^2} \Rightarrow f_x(1,0) = 2$, $f_y(1,0) = 0 \Rightarrow$ Tangent Plane at (1,0,0): $2(x-1) - z = 0$ or $2x - z = 2$.

37. $z = f(x,y) = \sqrt{y-x} \Rightarrow f_x(x,y) = -\frac{1}{2}(y-x)^{-1/2}$, $f_y(x,y) = \frac{1}{2}(y-x)^{-1/2} \Rightarrow f_x(1,2) = -\frac{1}{2}$, $f_y(1,2) = \frac{1}{2} \Rightarrow$ Tangent Plane at (1,2,1): $-\frac{1}{2}(x-1) + \frac{1}{2}(y-2) - (z-1) = 0 \Rightarrow -\frac{1}{2}x + \frac{1}{2}y - z + \frac{1}{2} = 0$ or $x - y + 2z - 1 = 0$.

39. $\nabla f = 2x\mathbf{i} + 2y\mathbf{j} \Rightarrow \nabla f(\sqrt{2},\sqrt{2}) = 2\sqrt{2}\mathbf{i} + 2\sqrt{2}\mathbf{j} \Rightarrow$ Tangent line: $2\sqrt{2}\left(x - \sqrt{2}\right) + 2\sqrt{2}\left(y - \sqrt{2}\right) = 0 \Rightarrow \sqrt{2}x + \sqrt{2}y = 4$

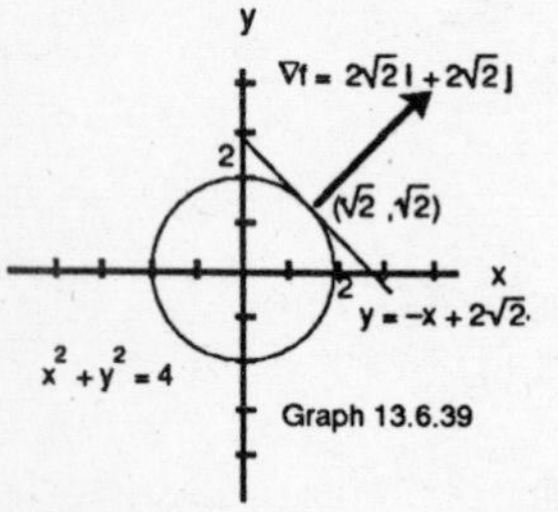

Graph 13.6.39

41. $\nabla f = y\mathbf{i} + x\mathbf{j} \Rightarrow \nabla f(2,-2) = -2\mathbf{i} + 2\mathbf{j} \Rightarrow$ Tangent line: $-2(x - 2) + 2(y + 2) = 0 \Rightarrow y = x - 4$

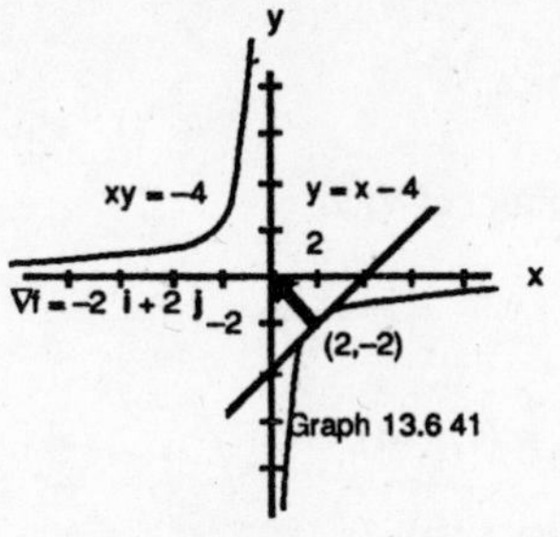

Graph 13.6 41

43. $\nabla f = \mathbf{i} + 2y\mathbf{j} + 2\mathbf{k} \Rightarrow \nabla f(1,1,1) = \mathbf{i} + 2\mathbf{j} + 2\mathbf{k}$. $\nabla g = \mathbf{i}$ for all points P. $\mathbf{v} = \nabla f \times \nabla g \Rightarrow$

$$\mathbf{v} = \begin{vmatrix} \mathbf{i} & \mathbf{j} & \mathbf{k} \\ 1 & 2 & 2 \\ 1 & 0 & 0 \end{vmatrix} = 2\mathbf{j} - 2\mathbf{k} \Rightarrow \text{Tangent line: } x = 1,\ y = 1 + 2t,\ z = 1 - 2t$$

45. $\nabla f = 2x\mathbf{i} + 2\mathbf{j} + 2\mathbf{k} \Rightarrow \nabla f(1,1,\frac{1}{2}) = 2\mathbf{i} + 2\mathbf{j} + 2\mathbf{k}$. $\nabla g = \mathbf{j}$ for all points P. $\mathbf{v} = \nabla f \times \nabla g \Rightarrow$

$$\mathbf{v} = \begin{vmatrix} \mathbf{i} & \mathbf{j} & \mathbf{k} \\ 2 & 2 & 2 \\ 0 & 1 & 0 \end{vmatrix} = -2\mathbf{i} + 2\mathbf{k} \Rightarrow \text{Tangent line: } x = 1 - 2t,\ y = 1,\ z = \frac{1}{2} + 2t$$

47. $\nabla f = \left(3x^2 + 6xy^2 + 4y\right)\mathbf{i} + \left(6x^2y + 3y^2 + 4x\right)\mathbf{j} - 2z\mathbf{k} \Rightarrow \nabla f(1,1,3) = 13\mathbf{i} + 13\mathbf{j} - 6\mathbf{k}$.

$$\nabla g = 2x\mathbf{i} + 2y\mathbf{j} + 2z\mathbf{k} \Rightarrow \nabla g(1,1,3) = 2\mathbf{i} + 2\mathbf{j} + 6\mathbf{k}.\ \mathbf{v} = \nabla f \times \nabla g \Rightarrow \mathbf{v} = \begin{vmatrix} \mathbf{i} & \mathbf{j} & \mathbf{k} \\ 13 & 13 & -6 \\ 2 & 2 & 6 \end{vmatrix} =$$

$90\mathbf{i} - 90\mathbf{j} \Rightarrow$ Tangent line: $x = 1 + 90t,\ y = 1 - 90t,\ z = 3$

49. $\nabla f = y\mathbf{i} + (x + 2y)\mathbf{j} \Rightarrow \nabla f(3,2) = 2\mathbf{i} + 7\mathbf{j}$. $\mathbf{A}$, orthogonal to ∇f, is $\mathbf{A} = 7\mathbf{i} - 2\mathbf{j} \Rightarrow \mathbf{u} = \frac{\mathbf{A}}{|\mathbf{A}|} = \frac{7\mathbf{i} - 2\mathbf{j}}{\sqrt{7^2 + (-2)^2}} =$

$\frac{7}{\sqrt{53}}\mathbf{i} - \frac{2}{\sqrt{53}}\mathbf{j} \Rightarrow -\mathbf{u} = -\frac{7}{\sqrt{53}}\mathbf{i} + \frac{2}{\sqrt{53}}\mathbf{j}$

51. $\nabla f = 2x\mathbf{i} + 2y\mathbf{j} + 2z\mathbf{k} = (2\cos t)\mathbf{i} + (2\sin t)\mathbf{j} + 2t\mathbf{k}$. $\mathbf{v} = (-\sin t)\mathbf{i} + (\cos t)\mathbf{j} + \mathbf{k} \Rightarrow \mathbf{T} = \frac{\mathbf{v}}{|\mathbf{v}|} =$

$\frac{(-\sin t)\mathbf{i} + (\cos t)\mathbf{j} + \mathbf{k}}{\sqrt{(\sin t)^2 + (\cos t)^2 + 1^2}} = \left(\frac{-\sin t}{\sqrt{2}}\right)\mathbf{i} + \left(\frac{\cos t}{\sqrt{2}}\right)\mathbf{j} + \frac{1}{\sqrt{2}}\mathbf{k}$.

$\left(D_{\mathbf{T}}f\right)_{P_0} = \nabla f \cdot \mathbf{T} = (2\cos t)\left(\frac{-\sin t}{\sqrt{2}}\right) + (2\sin t)\left(\frac{\cos t}{\sqrt{2}}\right) + 2t\left(\frac{1}{\sqrt{2}}\right) = \frac{2t}{\sqrt{2}} \Rightarrow \left(D_{\mathbf{T}}f\right)\left(\frac{-\pi}{4}\right) = \frac{-\pi}{2\sqrt{2}}$,

$\left(D_{\mathbf{T}}f\right)(0) = 0,\ \left(D_{\mathbf{T}}f\right)\left(\frac{\pi}{4}\right) = \frac{\pi}{2\sqrt{2}}$

53. $\nabla f = f_x(1,2)\mathbf{i} + f_y(1,2)\mathbf{j}$. $\mathbf{u}_1 = \frac{\mathbf{i} + \mathbf{j}}{\sqrt{1^2 + 1^2}} = \frac{1}{\sqrt{2}}\mathbf{i} + \frac{1}{\sqrt{2}}\mathbf{j}$. $\left(D_{\mathbf{u}_1}f\right)(1,2) =$

$f_x(1,2)\left(\frac{1}{\sqrt{2}}\right) + f_y(1,2)\left(\frac{1}{\sqrt{2}}\right) = 2\sqrt{2} \Rightarrow f_x(1,2) + f_y(1,2) = 4$. $\mathbf{u}_2 = -\mathbf{j}$. $\left(D_{\mathbf{u}_2}f\right)(1,2) = f_x(1,2)(0) + f_y(1,2)(-1)$

$= -3 \Rightarrow -f_y(1,2) = -3 \Rightarrow f_y(1,2) = 3$. $\therefore$ $f_x(1,2) + 3 = 4 \Rightarrow f_x(1,2) = 1$. Then $\nabla f(1,2) = \mathbf{i} + 3\mathbf{j}$.

53. (Continued)

$$\mathbf{u}=\frac{\mathbf{A}}{|\mathbf{A}|}=\frac{-\mathbf{i}-2\mathbf{j}}{\sqrt{(-1)^2+(-2)^2}}=-\frac{1}{\sqrt{5}}\mathbf{i}-\frac{2}{\sqrt{5}}\mathbf{j}\Rightarrow\left(D_{\mathbf{u}}f\right)_{P_0}=\nabla f\cdot\mathbf{u}=-\frac{1}{\sqrt{5}}-\frac{6}{\sqrt{5}}=-\frac{7}{\sqrt{5}}$$

55. The directional derivative is the scalar component. With ∇f evaluated at P_0, the scalar component of ∇f in the direction of **u** is $\nabla f\cdot\mathbf{u}=D_{\mathbf{u}}(f)_{P_0}$.

57. a) $\mathbf{r}=\sqrt{t}\,\mathbf{i}+\sqrt{t}\,\mathbf{j}-\frac{1}{4}(t+3)\,\mathbf{k}\Rightarrow\mathbf{v}=\frac{1}{2}t^{-1/2}\,\mathbf{i}+\frac{1}{2}t^{-1/2}\,\mathbf{j}-\frac{1}{4}\mathbf{k}$. $t=1\Rightarrow x=1, y=1, z=-1\Rightarrow P_0=(1,1,-1)$. Also $t=1\Rightarrow\mathbf{v}(1)=\frac{1}{2}\mathbf{i}+\frac{1}{2}\mathbf{j}-\frac{1}{4}\mathbf{k}$. $f(x,y,z)=x^2+y^2-z-3=0\Rightarrow\nabla f=2x\,\mathbf{i}+2y\,\mathbf{j}-\mathbf{k}\Rightarrow\nabla f(1,1,-1)=$ $2\mathbf{i}+2\mathbf{j}-\mathbf{k}$. $\therefore$ $\mathbf{v}=\frac{1}{4}(\nabla f)\Rightarrow$ The curve is normal to the surface.

b) $\mathbf{r}=\sqrt{t}\,\mathbf{i}+\sqrt{t}\,\mathbf{j}+(2t-1)\,\mathbf{k}\Rightarrow\mathbf{v}=\frac{1}{2}t^{-1/2}\,\mathbf{i}+\frac{1}{2}t^{-1/2}\,\mathbf{j}+2\mathbf{k}$. $t=1\Rightarrow x=1, y=1, z=1$. Also $t=1\Rightarrow$ $\mathbf{v}(1)=\frac{1}{2}\mathbf{i}+\frac{1}{2}\mathbf{j}+2\mathbf{k}$. $f(x,y,z)=x^2+y^2-z-1=0\Rightarrow\nabla f=2x\,\mathbf{i}+2y\,\mathbf{j}-\mathbf{k}\Rightarrow\nabla f(1,1,1)=2\mathbf{i}+2\mathbf{j}-\mathbf{k}\Rightarrow$ $\mathbf{v}\cdot\nabla f=\frac{1}{2}(2)+\frac{1}{2}(2)+2(-1)=0\Rightarrow$ The curve is tangent to the surface when $t=1$.

13.7 MAXIMA, MINIMA, AND SADDLE POINTS

1. $f_x(x,y)=2x+y+3=0$ and $f_y(x,y)=x+2y-3=0\Rightarrow x=-3, y=3\Rightarrow$ critical point is $(-3,3)$. $f_{xx}(-3,3)=2$, $f_{yy}(-3,3)=2$, $f_{xy}(-3,3)=1\Rightarrow f_{xx}f_{yy}-f_{xy}{}^2=3>0$ and $f_{xx}>0\Rightarrow$ local minimum. $f(-3,3)=-5$

3. $f_x(x,y)=2y-10x+4=0$ and $f_y(x,y)=2x-4y+4=0\Rightarrow x=\frac{2}{3}, y=\frac{4}{3}\Rightarrow$ critical point is $\left(\frac{2}{3},\frac{4}{3}\right)$. $f_{xx}\left(\frac{2}{3},\frac{4}{3}\right)=-10$, $f_{yy}\left(\frac{2}{3},\frac{4}{3}\right)=-4$, $f_{xy}\left(\frac{2}{3},\frac{4}{3}\right)=2\Rightarrow f_{xx}f_{yy}-f_{xy}{}^2=36>0$ and $f_{xx}<0\Rightarrow$ local maximum. $f\left(\frac{2}{3},\frac{4}{3}\right)=0$

5. $f_x(x,y)=2x+y+3=0$ and $f_y(x,y)=x+2=0\Rightarrow x=-2, y=1\Rightarrow$ critical point is $(-2,1)$. $f_{xx}(-2,1)=2$, $f_{yy}(-2,1)=0$, $f_{xy}(-2,1)=1\Rightarrow f_{xx}f_{yy}-f_{xy}{}^2=-1\Rightarrow$ saddle point.

7. $f_x(x,y)=5y-14x+3=0$ and $f_y(x,y)=5x-6=0\Rightarrow x=\frac{6}{5}, y=\frac{69}{25}\Rightarrow$ critical point is $\left(\frac{6}{5},\frac{69}{25}\right)$. $f_{xx}\left(\frac{6}{5},\frac{69}{25}\right)=-14$, $f_{yy}\left(\frac{6}{5},\frac{69}{25}\right)=0$, $f_{xy}\left(\frac{6}{5},\frac{69}{25}\right)=5\Rightarrow f_{xx}f_{yy}-f_{xy}{}^2=-25<0\Rightarrow$ saddle point.

9. $f_x(x,y)=2x-4y=0$ and $f_y(x,y)=-4x+2y+6=0\Rightarrow x=2, y=1\Rightarrow$ critical point is $(2,1)$. $f_{xx}(2,1)=2$, $f_{yy}(2,1)=2$, $f_{xy}(2,1)=-4\Rightarrow f_{xx}f_{yy}-f_{xy}{}^2=-12\Rightarrow$ saddle point.

11. $f_x(x,y)=4x+3y-5=0$ and $f_y(x,y)=3x+8y+2=0\Rightarrow x=2, y=-1\Rightarrow$ critical point is $(2,-1)$. $f_{xx}(2,-1)=4$, $f_{yy}(2,-1)=8$, $f_{xy}(2,-1)=3\Rightarrow f_{xx}f_{yy}-f_{xy}{}^2=23>0$ and $f_{xx}>0\Rightarrow$ local minimum. $f(2,-1)=-6$.

13. $f_x(x,y) = 2x - 2 = 0$ and $f_y(x,y) = -2y + 4 = 0 \Rightarrow x = 1, y = 2 \Rightarrow$ critical point is (1,2). $f_{xx}(1,2) = 2$, $f_{yy}(1,2) = -2$, $f_{xy}(1,2) = 0 \Rightarrow f_{xx}f_{yy} - f_{xy}^2 = -4 \Rightarrow$ saddle point.

15. $f_x(x,y) = 2x + 2y = 0$ and $f_y(x,y) = 2x = 0 \Rightarrow x = 0, y = 0 \Rightarrow$ critical point is (0,0). $f_{xx}(0,0) = 2$, $f_{yy}(0,0) = 0$, $f_{xy}(0,0) = 2 \Rightarrow f_{xx}f_{yy} - f_{xy}^2 = -4 \Rightarrow$ saddle point.

17. $f_x(x,y) = 3x^2 - 2y = 0$ and $f_y(x,y) = -3y^2 - 2x = 0 \Rightarrow x = 0, y = 0$ or $x = -\frac{2}{3}, y = \frac{2}{3} \Rightarrow$ critical points are (0,0) and $\left(-\frac{2}{3}, \frac{2}{3}\right)$. For (0,0): $f_{xx}(0,0) = 6x\big|_{(0,0)} = 0$, $f_{yy}(0,0) = -6y\big|_{(0,0)} = 0$, $f_{xy}(0,0) = -2 \Rightarrow f_{xx}f_{yy} - f_{xy}^2 = -4 \Rightarrow$ saddle point. For $\left(-\frac{2}{3}, \frac{2}{3}\right)$: $f_{xx}\left(-\frac{2}{3}, \frac{2}{3}\right) = -4$, $f_{yy}\left(-\frac{2}{3}, \frac{2}{3}\right) = -4$, $f_{xy}\left(-\frac{2}{3}, \frac{2}{3}\right) = -2$ $\Rightarrow f_{xx}f_{yy} - f_{xy}^2 = 12 > 0$ and $f_{xx} < 0 \Rightarrow$ local maximum. $f\left(-\frac{2}{3}, \frac{2}{3}\right) = \frac{170}{27}$

19. $f_x(x,y) = 12x - 6x^2 + 6y = 0$ and $f_y(x,y) = 6y + 6x = 0 \Rightarrow x = 0, y = 0$ or $x = 1, y = -1 \Rightarrow$ critical points are (0,0) and (1,−1). For (0,0): $f_{xx}(0,0) = 12 - 12x\big|_{(0,0)} = 12$, $f_{yy}(0,0) = 6$, $f_{xy}(0,0) = 6 \Rightarrow$ $f_{xx}f_{yy} - f_{xy}^2 = 36 > 0$ and $f_{xx} > 0 \Rightarrow$ local minimum. $f(0,0) = 0$. For (1,−1): $f_{xx}(1,-1) = 0$, $f_{yy}(1,-1) = 6$, $f_{xy}(1,-1) = 6 \Rightarrow f_{xx}f_{yy} - f_{xy}^2 = -36 \Rightarrow$ saddle point.

21. $f_x(x,y) = 27x^2 - 4y = 0$ and $f_y(x,y) = y^2 - 4x = 0 \Rightarrow x = 0, y = 0$ or $x = \frac{4}{9}, y = \frac{4}{3} \Rightarrow$ critical points are (0,0) and $\left(\frac{4}{9}, \frac{4}{3}\right)$. For (0,0): $f_{xx}(0,0) = 54x\big|_{(0,0)} = 0$, $f_{yy}(0,0) = 2y\big|_{(0,0)} = 0$, $f_{xy}(0,0) = -4 \Rightarrow$ $f_{xx}f_{yy} - f_{xy}^2 = -16 \Rightarrow$ saddle point. For $\left(\frac{4}{9}, \frac{4}{3}\right)$: $f_{xx}\left(\frac{4}{9}, \frac{4}{3}\right) = 24$, $f_{yy}\left(\frac{4}{9}, \frac{4}{3}\right) = \frac{8}{3}$, $f_{xy}\left(\frac{4}{9}, \frac{4}{3}\right) = -4 \Rightarrow$ $f_{xx}f_{yy} - f_{xy}^2 = 48 > 0$ and $f_{xx} > 0 \Rightarrow$ local minimum. $f\left(\frac{4}{9}, \frac{4}{3}\right) = -\frac{64}{81}$

23. $f_x(x,y) = 3x^2 + 6x = 0$ and $f_y(x,y) = 3y^2 - 6y = 0 \Rightarrow x = 0, y = 0$ or $x = 0, y = 2$ or $x = -2, y = 0$ or $x = -2, y = 2 \Rightarrow$ critical points are (0,0), (0,2), (−2,0), and (−2,2). For (0,0): $f_{xx}(0,0) =$ $6x + 6\big|_{(0,0)} = 6$, $f_{yy}(0,0) = 6y - 6\big|_{(0,0)}{-6}$, $f_{xy}(0,0) = 0 \Rightarrow f_{xx}f_{yy} - f_{xy}^2 = -36 \Rightarrow$ saddle point. For (0,2): $f_{xx}(0,2) = 6$, $f_{yy}(0,2) = 6$, $f_{xy}(0,2) = 0 \Rightarrow f_{xx}f_{yy} - f_{xy}^2 = 36 > 0$ and $f_{xx} > 0 \Rightarrow$ local minimum. $f(0,2) =$ -12. For (−2,0): $f_{xx}(-2,0) = -6$, $f_{yy}(-2,0) = -6$, $f_{xy}(-2,0) = 0 \Rightarrow f_{xx}f_{yy} - f_{xy}^2 = 36 > 0$ and $f_{xx} < 0 \Rightarrow$ local maximum. $f(-2,0) = -4$.
For (−2,2): $f_{xx}(-2,2) = -6$, $f_{yy}(-2,2) = 6$, $f_{xy}(-2,2) = 0 \Rightarrow f_{xx}f_{yy} - f_{xy}^2 = -36 \Rightarrow$ saddle point. No absolute extrema.

25. $f_x(x,y) = 4y - 4x^3 = 0$ and $f_y(x,y) = 4x - 4y^3 = 0 \Rightarrow x = 0, y = 0$ or $x = 1, y = 1$ or $x = -1, y = -1 \Rightarrow$ critical points are (0,0), (1,1), and (−1,−1). For (0,0): $f_{xx}(0,0) = -12x^2\big|_{(0,0)} = 0$, $f_{yy}(0,0) = -12y^2\big|_{(0,0)} = 0$, $f_{xy}(0,0) = 4 \Rightarrow$ $f_{xx}f_{yy} - f_{xy}^2 = -16 \Rightarrow$ saddle point. For (1,1): $f_{xx}(1,1) = -12$, $f_{yy}(1,1) = -12$, $f_{xy}(1,1) = 4 \Rightarrow$ $f_{xx}f_{yy} - f_{xy}^2 = 128 > 0$ and $f_{xx} < 0 \Rightarrow$ local maximum. $f(1,1) = 2$. For (−1,−1): $f_{xx}(-1,-1) = -12$, $f_{yy}(-1,-1) = -12$, $f_{xy}(-1,-1) = 4 \Rightarrow f_{xx}f_{yy} - f_{xy}^2 = 128 > 0$ and $f_{xx} < 0 \Rightarrow$ local maximum. $f(-1,-1) = 2$

27.

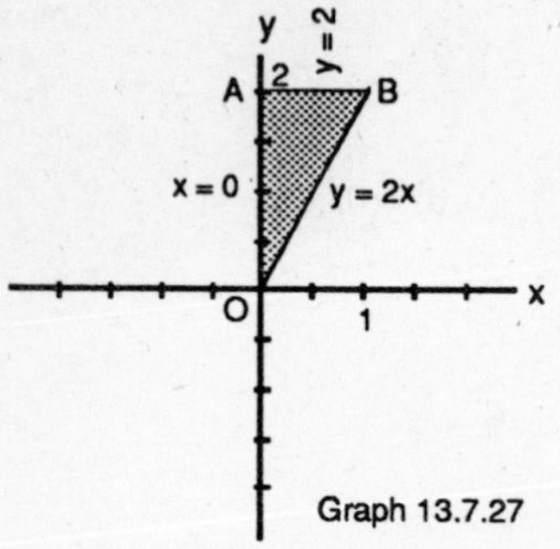

Graph 13.7.27

1. On OA, $f(x,y) = y^2 - 4y + 1 = f(0,y)$ on $0 \le y \le 2$. $y = 0 \Rightarrow f(0,0) = 1$. $y = 2 \Rightarrow f(0,2) = -3$. $f'(0,y) = 2y - 4 = 0 \Rightarrow y = 2 \Rightarrow f(0,2) = -3$.
2. On AB, $f(x,y) = 2x^2 - 4x - 3 = f(x,2)$ on $0 \le x \le 1$. $x = 0 \Rightarrow f(0,2) = -3$. $x = 1 \Rightarrow f(1,2) = -5$. $f'(x,2) = 4x - 4 = 0 \Rightarrow x = 1 \Rightarrow f(1,2) = -5$.
3. On OB, $f(x,y) = 6x^2 - 12x + 1 = f(x,2x)$ on $0 \le x \le 1$. Endpoint values have been found above. $f'(x,2x) = 12x - 12 = 0 \Rightarrow x = 1, y = 2 \Rightarrow (1,2)$, not an interior point of OB.
4. For interior points of the triangular region, $f_x(x,y) = 4x - 4 = 0$ and $f_y(x,y) = 2y - 4 = 0 \Rightarrow x = 1, y = 2 \Rightarrow (1,2)$, not an interior point of the region. $\therefore$ absolute maximum is 1 at (0,0); absolute minimum is -5 at (1,2)

29.

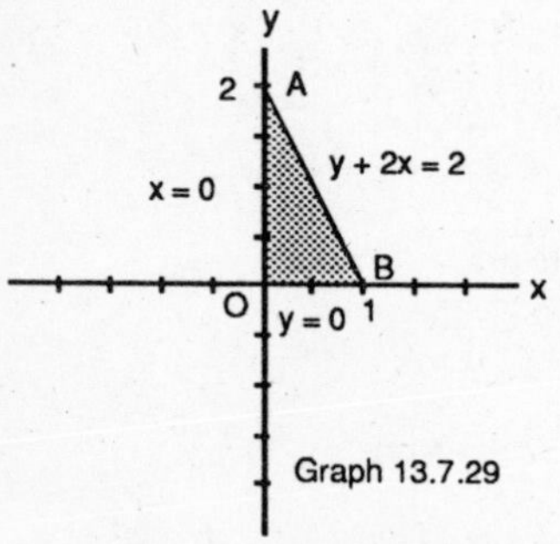

Graph 13.7.29

1. On OA, $f(x,y) = y^2 = f(0,y)$ on $0 \le y \le 2$. $f(0,0) = 0$. $f(0,2) = 4$. $f'(0,y) = 2y = 0 \Rightarrow y = 0, x = 0 \Rightarrow (0,0)$
2. On OB, $f(x,y) = x^2 = f(x,0)$ on $0 \le x \le 1$. $f(1,0) = 1$. $f'(x,0) = 2x = 0 \Rightarrow x = 0, y = 0 \Rightarrow (0,0)$
3. On AB, $f(x,y) = 5x^2 - 8x + 4 = f(x,-2x+2)$ on $0 \le x \le 1$. $f(0,2) = 4$. $f'(x,-2x+2) = 10x - 8 = 0 \Rightarrow x = \frac{4}{5}, y = \frac{2}{5}$. $f\left(\frac{4}{5}, \frac{2}{5}\right) = \frac{4}{5}$
4. For interior points of the triangular region, $f_x(x,y) = 2x = 0$ and $f_y(x,y) = 2y = 0 \Rightarrow x = 0, y = 0 \Rightarrow (0,0)$, not an interior point of the region. $\therefore$ absolute maximum is 4 at (0,2); absolute minimum is 0 at (0,0)

31.

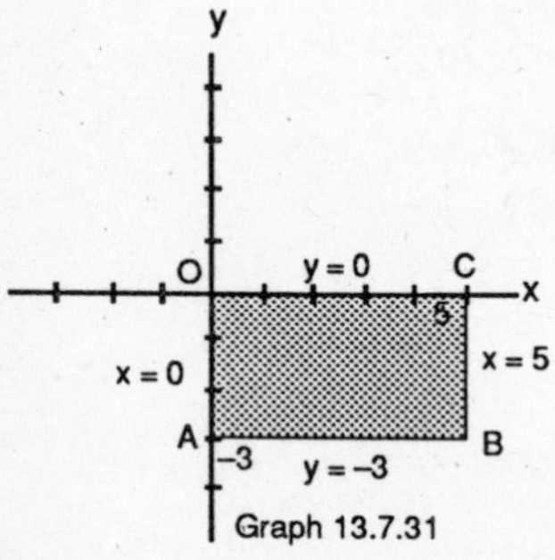

Graph 13.7.31

1. On OC, $T(x,y) = x^2 - 6x + 2 = T(x,0)$ on $0 \le x \le 5$. $T(0,0) = 2$. $T(5,0) = -3$. $T'(x,0) = 2x - 6 = 0 \Rightarrow x = 3, y = 0$. $T(3,0) = -7$
2. On CB, $T(x,y) = y^2 + 5y - 3 = T(5,y)$ on $-3 \le y \le 0$. $T(5,-3) = -9$ $T'(5,y) = 2y + 5 = 0 \Rightarrow y = -\frac{5}{2}, x = 5$. $T\left(5, -\frac{5}{2}\right) = -\frac{37}{4}$
3. On AB, $T(x,y) = x^2 - 9x + 11 = T(x,-3)$ on $0 \le x \le 5$. $T(0,-3) = 11$. $T'(x,-3) = 2x - 9 = 0 \Rightarrow x = \frac{9}{2}, y = -3$. $T\left(\frac{9}{2}, -3\right) = -\frac{37}{4}$
4. On AO, $T(x,y) = y^2 + 2 = T(0,y)$ on $-3 \le y \le 0$. $T'(0,y) = 2y = 0 \Rightarrow y = 0, x = 0$. (0,0) not an interior point of AO.
5. For interior points of the rectangular region, $T_x(x,y) = 2x + y - 6 = 0$ and $T_y(x,y) = x + 2y = 0 \Rightarrow x = 4, y = -2 \Rightarrow T(4,-2) = -10$. $\therefore$ absolute maximum is 11 at $(0,-3)$; absolute minimum is -10 at $(4,-2)$.

33.

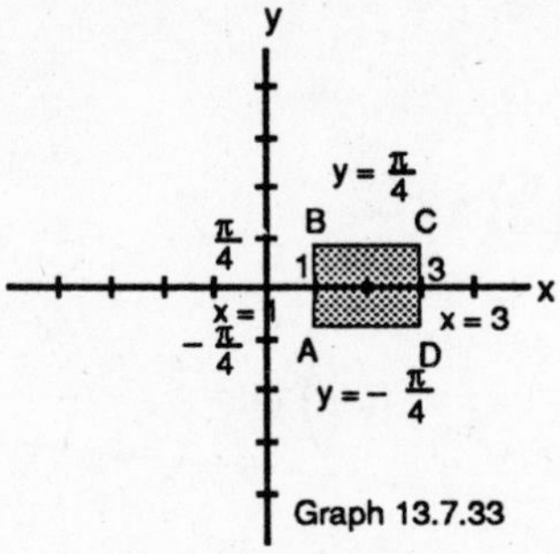

Graph 13.7.33

1. On AB, $f(x,y) = 3\cos y = f(1,y)$ on $-\frac{\pi}{4} \le y \le \frac{\pi}{4}$. $f\left(1,-\frac{\pi}{4}\right) = \frac{3\sqrt{2}}{2}$. $f\left(1,\frac{\pi}{4}\right) = \frac{3\sqrt{2}}{2}$. $f'(1,y) = -3\sin y = 0 \Rightarrow y = 0, x = 1 \Rightarrow f(1,0) = 3$.
2. On CD, $f(x,y) = 3\cos y = f(3,y)$ on $-\frac{\pi}{4} \le y \le \frac{\pi}{4}$. $f\left(3,-\frac{\pi}{4}\right) = \frac{3\sqrt{2}}{2}$. $f\left(3,\frac{\pi}{4}\right) = \frac{3\sqrt{2}}{2}$. $f'(3,y) = -3\sin y = 0 \Rightarrow y = 0, x = 3$. $f(3,0) = 3$.
3. On BC, $f(x,y) = \frac{\sqrt{2}}{2}(4x - x^2) = f\left(x,\frac{\pi}{4}\right)$ on $1 \le x \le 3$. $f'\left(x,\frac{\pi}{4}\right) = \sqrt{2}(2 - x) = 0 \Rightarrow x = 2, y = \frac{\pi}{4}$. $f\left(2,\frac{\pi}{4}\right) = 2\sqrt{2}$.
4. On AD, $f(x,y) = \frac{\sqrt{2}}{2}(4x - x^2) = f\left(x,-\frac{\pi}{4}\right)$ on $1 \le x \le 3$. $f'\left(x,-\frac{\pi}{4}\right) = \sqrt{2}(2 - x) = 0 \Rightarrow x = 2, y = -\frac{\pi}{4}$. $f\left(2,-\frac{\pi}{4}\right) = 2\sqrt{2}$.
5. For interior points of the region, $f_x(x,y) = (4 - 2x)\cos y = 0$ and $f_y(x,y) = -(4x - x^2)\sin y = 0 \Rightarrow x = 2$, $y = 0 \Rightarrow f(2,0) = 4$.

$\therefore$ the absolute maximum is 4 at (2,0) and the absolute minimum is $\frac{3\sqrt{2}}{2}$ at $\left(3,-\frac{\pi}{4}\right)$, $\left(3,\frac{\pi}{4}\right)$, $\left(1,-\frac{\pi}{4}\right)$, and $\left(1,\frac{\pi}{4}\right)$.

35. $T_x(x,y) = 2x - 1 = 0$ and $T_y(x,y) = 4y = 0 \Rightarrow x = \frac{1}{2}, y = 0 \Rightarrow T\left(\frac{1}{2},0\right) = -\frac{1}{4}$. On $x^2 + y^2 = 1$, $T(x,y) = -x^2 - x + 2$ on $-1 \le x \le 1$. $T(-1,0) = 2$, $T(1,0) = 0$. $T'(x,y) = -2x - 1 = 0 \Rightarrow x = -\frac{1}{2}, y = \pm\frac{\sqrt{3}}{2}$. $T\left(-\frac{1}{2},\frac{\sqrt{3}}{2}\right) = \frac{9}{4}$, $T\left(-\frac{1}{2},-\frac{\sqrt{3}}{2}\right) = \frac{9}{4}$. $\therefore$ hottest is $2\frac{1}{4}°$ at $\left(-\frac{1}{2},\frac{\sqrt{3}}{2}\right)$ and $\left(-\frac{1}{2},-\frac{\sqrt{3}}{2}\right)$; coldest is $-\frac{1}{4}°$ at $\left(\frac{1}{2},0\right)$.

37. a) $f_x(x,y) = 2x - 4y = 0$ and $f_y(x,y) = 2y - 4x = 0 \Rightarrow x = 0, y = 0$. $f_{xx}(0,0) = 2$, $f_{yy}(0,0) = 2$, $f_{xy}(0,0) = -4$ $\Rightarrow f_{xx}f_{yy} - f_{xy}^2 = -12 \Rightarrow$ saddle point.

b) $f_x(x,y) = 2x - 2 = 0$ and $f_y(x,y) = 2y - 4 = 0 \Rightarrow x = 1, y = 2$. $f_{xx}(1,2) = 2$, $f_{yy}(1,2) = 2$, $f_{xy}(1,2) = 0 \Rightarrow$ $f_{xx}f_{yy} - f_{xy}^2 = 4 > 0$ and $f_{xx} > 0 \Rightarrow$ local minimum at (1,2).

c) $f_x(x,y) = 9x^2 - 9 = 0$ and $f_y(x,y) = 2y + 4 = 0 \Rightarrow x = \pm 1, y = -2$. For (1,−2), $f_{xx}(1,-2) = 18x\big|_{(1,-2)} = 18$, $f_{yy}(1,-2) = 2$, $f_{xy}(1,-2) = 0 \Rightarrow f_{xx}f_{yy} - f_{xy}^2 = 36 > 0$ and $f_{xx} > 0 \Rightarrow$ local minimum at (1,−2). For (−1,−2), $f_{xx}(-1,-2) = -18$, $f_{yy}(-1,-2) = 2$, $f_{xy}(-1,-2) = 0 \Rightarrow f_{xx}f_{yy} - f_{xy}^2 = -36 \Rightarrow$ saddle point.

39. If $k = 0$, $f(x,y) = x^2 + y^2 \Rightarrow f_x(x,y) = 2x = 0 \Rightarrow x = 0$ and $f_y(x,y) = 2y = 0 \Rightarrow y = 0 \Rightarrow$ (0,0) is the critical point. If $k \ne 0$, $f_x(x,y) = 2x + ky = 0 \Rightarrow y = -\frac{2}{k}x$. $f_y(x,y) = kx + 2y = 0 \Rightarrow kx + 2\left(-\frac{2}{k}x\right) = 0 \Rightarrow kx - \frac{4x}{k} = 0 \Rightarrow$ $\left(k - \frac{4}{k}\right)x = 0 \Rightarrow x = 0 \Rightarrow y = -\frac{2}{k}(0) = 0 \Rightarrow$ (0,0) is the critical point.

41. No: for example $f(x,y) = xy$ has a saddle point at $(a,b) = (0,0)$ where $f_x = f_y = 0$.

43. a) $x = 2\cos t$, $y = 2\sin t \Rightarrow f(t) = 2\cos t + 2\sin t \Rightarrow \frac{df}{dt} = -2\sin t + 2\cos t$. $\frac{df}{dt} = 0 \Rightarrow \cos t = \sin t$

i) On the semicircle $x^2 + y^2 = 4$, $y \geq 0$, $0 \leq t \leq \pi$ and $f(0) = 2$, $f(\pi) = -2$. $\frac{df}{dt} = 0 \Rightarrow t = \frac{\pi}{4} \Rightarrow f\left(\frac{\pi}{4}\right) = 2\sqrt{2}$.

$\therefore$ absolute minimum is -2 at $t = \pi$; absolute maximum is $2\sqrt{2}$ at $t = \frac{\pi}{4}$.

ii) On the quarter circle $x^2 + y^2 = 4$, $x \geq 0$, $y \geq 0$, $0 \leq t \leq \frac{\pi}{2}$ and $f(0) = 2$, $f\left(\frac{\pi}{2}\right) = 2$. $\frac{df}{dt} = 0 \Rightarrow t = \frac{\pi}{4} \Rightarrow f\left(\frac{\pi}{4}\right) = 2\sqrt{2}$.

$\therefore$ absolute minimum is 2 at $t = 0, \frac{\pi}{2}$; absolute maximum is $2\sqrt{2}$ at $t = \frac{\pi}{4}$.

b) $x = 2\cos t$, $y = 2\sin t \Rightarrow g(t) = 4\cos t \sin t \Rightarrow \frac{dg}{dt} = y(-2\sin t) + x(2\cos t) = -4\sin^2 t + 4\cos^2 t$.

$\frac{dg}{dt} = 0 \Rightarrow 4\cos^2 t - 4\sin^2 t = 0 \Rightarrow \cos t = \sin t$ or $\cos t = -\sin t$

i) On the semicircle $x^2 + y^2 = 4$, $y \geq 0$, $0 \leq t \leq \pi$ and $\frac{dg}{dt} = 0 \Rightarrow t = \frac{\pi}{4}, \frac{3\pi}{4}$. $g(0) = 0$, $g\left(\frac{\pi}{4}\right) = 2$, $g\left(\frac{3\pi}{4}\right) = -2$, $g(\pi) = 0$

$\therefore$ absolute minimum is -2 at $t = \frac{3\pi}{4}$; absolute maximum is 2 at $t = \frac{\pi}{4}$.

ii) On the quarter circle $x^2 + y^2 = 4$, $x \geq 0$, $y \geq 0$, $0 \leq t \leq \frac{\pi}{2}$ and $\frac{dg}{dt} = 0$ at $t = \frac{\pi}{4}$. $g(0) = 0$, $g\left(\frac{\pi}{2}\right) = 0$, $g\left(\frac{\pi}{4}\right) = 2$

$\therefore$ absolute minimum is 0 at $t = 0, \frac{\pi}{2}$; absolute maximum is 2 at $t = \frac{\pi}{4}$.

c) $x = 2\cos t$, $y = 2\sin t \Rightarrow h(t) = 8\cos^2 t + 4\sin^2 t = 4\cos^2 t + 4 \Rightarrow \frac{dh}{dt} = -8\cos t \sin t$. $\frac{dh}{dt} = 0 \Rightarrow$

$\cos t = 0$ or $\sin t = 0$

i) On the semicircle $x^2 + y^2 = 4$, $y \geq 0$, $0 \leq t \leq \pi$ and $h(0) = 8$, $h(\pi) = 8$. $\frac{dh}{dt} = 0 \Rightarrow t = 0, \pi, \frac{\pi}{2}$. $h\left(\frac{\pi}{2}\right) = 4$.

$\therefore$ absolute minimum is 4 at $t = \frac{\pi}{2}$; absolute maximum is 8 at $t = 0, \pi$.

ii) On the quarter circle $x^2 + y^2 = 4$, $x \geq 0$, $y \geq 0$, $0 \leq t \leq \frac{\pi}{2}$ and $h(0) = 8$, $h\left(\frac{\pi}{2}\right) = 4$. $\frac{dh}{dt} = 0 \Rightarrow t = 0, \frac{\pi}{2}$.

$\therefore$ absolute minimum is 4 at $t = \frac{\pi}{2}$; absolute maximum is 8 at $t = 0$.

45. a) $x = 2t$, $y = t + 1 \Rightarrow f(t) = 2t^2 + 2t \Rightarrow \frac{df}{dt} = 4t + 2$. $\frac{df}{dt} = 0 \Rightarrow t = -\frac{1}{2}$. $f''(t) = 4 \Rightarrow f\left(-\frac{1}{2}\right)$ is a minimum.

$f\left(-\frac{1}{2}\right) = -\frac{1}{2}$ is an absolute minimum since $f(t)$ is an upward parabola. No absolute maximum.

b) $x = 2t$, $y = t + 1$ on $-1 \leq t \leq 0 \Rightarrow t = -\frac{1}{2}$ is a critical number on the interval (see part a) above).

$f\left(-\frac{1}{2}\right) = -\frac{1}{2}$. $f(0) = 0$, $f(-1) = 0 \Rightarrow$ absolute minimum is $-\frac{1}{2}$ at $t = -\frac{1}{2}$; absolute maximum is 0 at $t = 0, -1$.

c) $x = 2t$, $y = t + 1$ on $0 \leq t \leq 1 \Rightarrow$ no critical numbers on the interval (see part a) above). $f(0) = 0$,

$f(1) = 4 \Rightarrow$ absolute minimum is 0 at $t = 0$; absolute maximum is 4 at $t = 1$.

47.

k	x_k	y_k	x_k^2	$x_k y_k$
1	-1	2	1	-2
2	0	1	0	0
3	3	-4	9	-12
Σ	2	-1	10	-14

$m = \frac{2(-1) - 3(-14)}{2^2 - 3(10)} = -\frac{20}{13}$, $b = \frac{1}{3}\left(-1 - \left(-\frac{20}{13}\right)2\right) = \frac{9}{13}$

$\therefore y = -\frac{20}{13}x + \frac{9}{13}$. $y\big|_{x=4} = -\frac{71}{13}$

49.

k	x_k	y_k	x_k^2	x_ky_k
1	0	0	0	0
2	1	2	1	2
3	2	3	4	6
Σ	3	5	5	8

$m = \dfrac{3(5) - 3(8)}{3^2 - 3(5)} = 1.5$, $b = \dfrac{1}{3}(5 - 1.5(3)) = \dfrac{1}{6}$

$\therefore\ y = \dfrac{3}{2}x + \dfrac{1}{6}.\ y\Big|_{x=4} = \dfrac{37}{6}$

51.

k	x_k	y_k	x_k^2	x_ky_k
1	12	5.27	144	63.24
2	18	5.68	324	102.24
3	24	6.25	576	150
4	30	7.21	900	216.3
5	36	8.20	1296	295.2
6	42	8.71	1764	365.82
Σ	162	41.32	5004	1192.8

$m = \dfrac{162(41.32) - 6(1192.8)}{162^2 - 6(5004)} \approx 0.122$,

$b = \dfrac{1}{6}(41.32 - (0.122)(162)) \approx 3.59$

$\therefore\ y = 0.122\,x + 3.59$

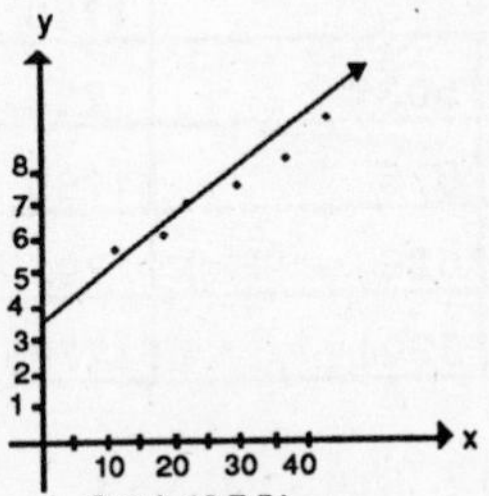

Graph 13.7.51

53. a)

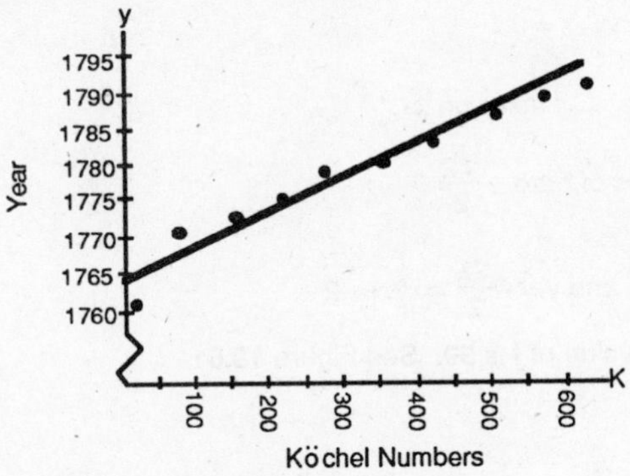

Graph 13.7.53 a

b) See the Table on the next page. $m = \dfrac{(3201)(17785) - 10(5710292)}{(3201)^2 - 10(1430389)} \approx 0.0427$.

$b = \dfrac{1}{10}(17785 - 0.0427(3201)) \approx 1764.8$. $\therefore\ y = 0.0427\,K + 1764.8$

c) $K = 364 \Rightarrow y = 0.0427(364) + 1764.8 \approx 1780$

53. b) The table for 53, part b.

k	K_k	y_k	K_k^2	$K_k y_k$
1	1	1761	1	1761
2	75	1771	5625	132825
3	155	1772	24025	274660
4	219	1775	47961	388725
5	271	1777	73441	481567
6	351	1780	123201	624780
7	425	1783	180625	757775
8	503	1786	253009	898358
9	575	1789	330625	1028675
10	626	1791	391876	1121166
Σ	3201	17785	1430389	5710292

13.8 LAGRANGE MULTIPLIERS

1. $\nabla f = y\mathbf{i} + x\mathbf{j}$, $\nabla g = 2x\mathbf{i} + 4y\mathbf{j}$. $\nabla f = \lambda \nabla g \Rightarrow y\mathbf{i} + x\mathbf{j} = \lambda(2x\mathbf{i} + 4y\mathbf{j}) \Rightarrow y = 2x\lambda$ and $x = 4y\lambda \Rightarrow$ $\lambda = \pm\frac{\sqrt{2}}{4}$ or $x = 0$. CASE 1: If $x = 0$, then $y = 0$ but $(0,0)$ not on the ellipse. $\therefore\ x \neq 0$.
CASE 2: $x \neq 0 \Rightarrow \lambda = \pm\frac{\sqrt{2}}{4} \Rightarrow x = \pm\sqrt{2}\,y \Rightarrow (\pm\sqrt{2}\,y)^2 + 2y^2 = 1 \Rightarrow y = \pm\frac{1}{2}$. $\therefore$ f takes on its extreme values at $\left(\pm\sqrt{2}, \frac{1}{2}\right)$ and $\left(\pm\sqrt{2}, -\frac{1}{2}\right) \Rightarrow$ the extreme values of f are $\pm\frac{\sqrt{2}}{2}$.

3. $\nabla f = -2x\mathbf{i} - 2y\mathbf{j}$, $\nabla g = \mathbf{i} + 3\mathbf{j}$. $\nabla f = \lambda\nabla g \Rightarrow -2x\mathbf{i} - 2y\mathbf{j} = \lambda(\mathbf{i} + 3\mathbf{j}) \Rightarrow x = -\frac{\lambda}{2}$ and $y = -\frac{3\lambda}{2} \Rightarrow \lambda = -2$ $\Rightarrow x = 1$ and $y = 3 \Rightarrow$ f takes on its extreme value at $(1,3) \Rightarrow$ the extreme value of f is 39. See Figure 13.61 page 864 in the text.

5. $\nabla f = 2xy\mathbf{i} + x^2\mathbf{j}$, $\nabla g = \mathbf{i} + \mathbf{j}$. $\nabla f = \lambda\nabla g \Rightarrow 2xy\mathbf{i} + x^2\mathbf{j} = \lambda(\mathbf{i} + \mathbf{j}) \Rightarrow 2xy = \lambda$ and $x^2 = \lambda \Rightarrow 2xy = x^2 \Rightarrow$ $x = 0, y = 3$ or $x = 2, y = 1$. $\therefore$ f takes on its extreme values at $(0,3)$ and $(2,1)$. $\therefore$ the extreme values of f are $f(0,3) = 0$ and $f(2,1) = 4$.

7. a) $\nabla f = \mathbf{i} + \mathbf{j}$, $\nabla g = y\mathbf{i} + x\mathbf{j}$. $\nabla f = \lambda\nabla g \Rightarrow \mathbf{i} + \mathbf{j} = \lambda(y\mathbf{i} + x\mathbf{j}) \Rightarrow 1 = \lambda y$ and $1 = \lambda x \Rightarrow y = \frac{1}{\lambda}$ and $x = \frac{1}{\lambda}$ $\Rightarrow \frac{1}{\lambda^2} = 16 \Rightarrow \lambda = \pm 4$. Use $\lambda = 4$ since $x > 0, y > 0$. Then $x = 4, y = 4 \Rightarrow$ the minimum value is 8 at $x = 4, y = 4$.

   $xy = 16, x > 0, y > 0$ is a branch of a hyperbola in the first quadrant with the x- and y-axes as asymptotes. $x + y = c$ is a family of parallel lines with $m = -1$. As these move away from the origin, c increases. $\therefore$ the minimum value of c will occur where $x + y = c$ is tangent to the hyperbola's branch.

   b) $\nabla f = y\mathbf{i} + x\mathbf{j}$, $\nabla g = \mathbf{i} + \mathbf{j}$. $\nabla f = \lambda\nabla g \Rightarrow y\mathbf{i} + x\mathbf{j} = \lambda(\mathbf{i} + \mathbf{j}) \Rightarrow y = \lambda = x \Rightarrow y = x \Rightarrow y + y = 16 \Rightarrow y = 8$ $\Rightarrow x = 8 \Rightarrow f(8,8) = 64$ is the maximum value. $xy = c > 0$ ($x > 0$ and $y > 0$ or $x < 0$ and $y < 0$ to get a maximum value) is a family of hyperbolas in the first and third quadrants with the x- and y-axes as asymptotes. The maximum value of c will occur where the hyperbola $xy = c$ is tangent to the line $x + y = 16$.

9. $V = \pi r^2 h \Rightarrow 16\pi = \pi r^2 h \Rightarrow 16 = r^2 h \Rightarrow g(r,h) = r^2 h - 16$. $S = 2\pi rh + 2\pi r^2 \Rightarrow \nabla S = (2\pi h + 4\pi r)\,\mathbf{i} + 2\pi r\,\mathbf{j}$, $\nabla g = 2rh\,\mathbf{i} + r^2\,\mathbf{j}$. $\nabla S = \lambda \nabla g \Rightarrow (2\pi rh + 4\pi r)\,\mathbf{i} + 2\pi r\,\mathbf{j} = \lambda(2rh\,\mathbf{i} + r^2\,\mathbf{j}) \Rightarrow 2\pi h + 4\pi r = 2rh\lambda$ and $2\pi r = \lambda r^2 \Rightarrow 0 = \lambda r^2 - 2\pi r \Rightarrow r = 0$ or $\lambda = \frac{2\pi}{r}$. Now $r \neq 0 \Rightarrow \lambda = \frac{2\pi}{r} \Rightarrow 2\pi h + 4\pi r = 2rh\left(\frac{2\pi}{r}\right) \Rightarrow 2r = h \Rightarrow 16 = r^2(2r) \Rightarrow r = 2 \Rightarrow h = 4$. $\therefore$ $r = 2$ cm, $h = 4$ cm give the only extreme surface area, $24\pi\ \text{cm}^2$. Since $r = 4$ cm, $h = 1$ cm $\Rightarrow V = 16\pi\ \text{cm}^3$ and $S = 40\pi\ \text{cm}^2$, a larger surface area, $24\pi\ \text{cm}^2$ must be the minumum surface area.

11. $\nabla T = (8x - 4y)\,\mathbf{i} + (-4x + 2y)\,\mathbf{j}$, $g(x,y) = x^2 + y^2 - 25 = 0 \Rightarrow \nabla g = 2x\,\mathbf{i} + 2y\,\mathbf{j}$. $\nabla T = \lambda \nabla g \Rightarrow (8x - 4y)\,\mathbf{i} + (-4x + 2y)\,\mathbf{j} = \lambda(2x\,\mathbf{i} + 2y\,\mathbf{j}) \Rightarrow 8x - 4y = 2\lambda x$ and $-4x + 2y = 2\lambda y \Rightarrow y = \frac{-2x}{\lambda - 1}$, $\lambda \neq 1 \Rightarrow 8x - 4\left(\frac{-2x}{\lambda - 1}\right) = 2\lambda x \Rightarrow x = 0$ or $\lambda = 0$ or $\lambda = 5$. $x = 0 \Rightarrow y = 0$. But $(0,0)$ not on $x^2 + y^2 = 25$. $\therefore$ $x \neq 0 \Rightarrow \lambda = 0$ or $\lambda = 5$. $\lambda = 0 \Rightarrow y = 2x \Rightarrow x^2 + (2x)^2 = 25 \Rightarrow x = \pm\sqrt{5} \Rightarrow y = \pm 2\sqrt{5}$. $\lambda = 5 \Rightarrow y = \frac{-2x}{4} = -\frac{1}{2}x \Rightarrow x^2 + \left(-\frac{1}{2}x\right)^2 = 25 \Rightarrow x = \pm 2\sqrt{5}$. $x = 2\sqrt{5} \Rightarrow y = -\sqrt{5}$, $x = -2\sqrt{5} \Rightarrow y = \sqrt{5}$. $T(\sqrt{5}, 2\sqrt{5}) = 0° = T(-\sqrt{5}, -2\sqrt{5})$, the minimum value; $T(2\sqrt{5}, -\sqrt{5}) = 125° = T(-2\sqrt{5}, \sqrt{5})$, the maximum value.

13. $\nabla f = 2x\,\mathbf{i} + 2y\,\mathbf{j}$, $\nabla g = (2x - 2)\,\mathbf{i} + (2y - 4)\,\mathbf{j}$. $\nabla f = \lambda \nabla g \Rightarrow 2x\,\mathbf{i} + 2y\,\mathbf{j} = \lambda((2x - 2)\,\mathbf{i} + (2y - 4)\,\mathbf{j}) \Rightarrow 2x = \lambda(2x - 2)$ and $2y = \lambda(2y - 4) \Rightarrow x = \frac{\lambda}{\lambda - 1}$ and $y = \frac{2\lambda}{\lambda - 1}$, $\lambda \neq 1 \Rightarrow y = 2x \Rightarrow x^2 - 2x + (2x)^2 - 4(2x) = 0 \Rightarrow x = 0, y = 0$ or $x = 2, y = 4$. $\therefore$ $f(0,0) = 0$ is the minimum value, $f(2,4) = 20$ is the maximum value.

15. $\nabla f = \mathbf{i} - 2\,\mathbf{j} + 5\,\mathbf{k}$, $\nabla g = 2x\,\mathbf{i} + 2y\,\mathbf{j} + 2z\,\mathbf{k}$. $\nabla f = \lambda \nabla g \Rightarrow \mathbf{i} - 2\,\mathbf{j} + 5\,\mathbf{k} = \lambda(2x\,\mathbf{i} + 2y\,\mathbf{j} + 2z\,\mathbf{k}) \Rightarrow 1 = 2x\lambda$, $-2 = 2y\lambda$, and $5 = 2z\lambda \Rightarrow x = \frac{1}{2\lambda}$, $y = -\frac{1}{\lambda} = -2x$, $z = \frac{5}{2\lambda} = 5x \Rightarrow x^2 + (-2x)^2 + (5x)^2 = 30 \Rightarrow x = \pm 1$. $x = 1 \Rightarrow y = -2, z = 5$. $x = -1 \Rightarrow y = 2, z = -5$. $f(1,-2,5) = 30$, the maximum value; $f(-1,2,-5) = -30$, the minimum value.

17. Let $f(x,y,z) = x^2 + y^2 + z^2$ be the square of the distance to the origin. Then $\nabla f = 2x\,\mathbf{i} + 2y\,\mathbf{j} + 2z\,\mathbf{k}$, $\nabla g = y\,\mathbf{i} + x\,\mathbf{j} - \mathbf{k}$. $\nabla f = \lambda \nabla g \Rightarrow 2x\,\mathbf{i} + 2y\,\mathbf{j} + 2z\,\mathbf{k} = \lambda(y\,\mathbf{i} + x\,\mathbf{j} - \mathbf{k}) \Rightarrow 2x = \lambda y$, $2y = \lambda x$, and $2z = -\lambda \Rightarrow x = \frac{\lambda y}{2} \Rightarrow 2y = \lambda\left(\frac{\lambda y}{2}\right) \Rightarrow y = 0$ or $\lambda = \pm 2$. $y = 0 \Rightarrow x = 0 \Rightarrow -z + 1 = 0 \Rightarrow z = 1$. $\lambda = 2 \Rightarrow x = y$, $z = -1 \Rightarrow x^2 - (-1) + 1 = 0 \Rightarrow x^2 + 2 = 0$, no solution. $\lambda = -2 \Rightarrow x = -y$, $z = 1 \Rightarrow (-y)y - 1 + 1 = 0 \Rightarrow y = 0$, again. $\therefore$ $(0,0,1)$ is the point on the surface closest to the origin since this point gives the only extreme value and there is no maximum distance.

19. $\nabla f = \mathbf{i} + 2\,\mathbf{j} + 3\,\mathbf{k}$, $\nabla g = 2x\,\mathbf{i} + 2y\,\mathbf{j} + 2z\,\mathbf{k}$. $\nabla f = \lambda \nabla g \Rightarrow \mathbf{i} + 2\,\mathbf{j} + 3\,\mathbf{k} = \lambda(2x\,\mathbf{i} + 2y\,\mathbf{j} + 2z\,\mathbf{k}) \Rightarrow 1 = 2x\lambda$, $2 = 2y\lambda$, and $3 = 2z\lambda \Rightarrow x = \frac{1}{2\lambda}$, $y = \frac{1}{\lambda}$, and $z = \frac{3}{2\lambda} \Rightarrow y = 2x$ and $z = 3x \Rightarrow x^2 + (2x)^2 + (3x)^2 = 25 \Rightarrow x = \pm\frac{5}{\sqrt{14}}$. $x = \frac{5}{\sqrt{14}} \Rightarrow y = \frac{10}{\sqrt{14}}$, $z = \frac{15}{\sqrt{14}}$. $x = -\frac{5}{\sqrt{14}} \Rightarrow y = -\frac{10}{\sqrt{14}}$, $z = -\frac{15}{\sqrt{14}}$. $f\left(\frac{5}{\sqrt{14}}, \frac{10}{\sqrt{14}}, \frac{15}{\sqrt{14}}\right) = 5\sqrt{14}$, the maximum value; $f\left(-\frac{5}{\sqrt{14}}, -\frac{10}{\sqrt{14}}, -\frac{15}{\sqrt{14}}\right) = -5\sqrt{14}$, the minimum value.

21. $f(x,y,z) = xyz$ and $g(x,y,z) = x + y + z^2 - 16 = 0 \Rightarrow \nabla f = yz\,\mathbf{i} + xz\,\mathbf{j} + xy\,\mathbf{k}$, $\nabla g = \mathbf{i} + \mathbf{j} + 2z\,\mathbf{k}$. $\nabla f = \lambda \nabla g \Rightarrow yz\,\mathbf{i} + xz\,\mathbf{j} + xy\,\mathbf{k} = \lambda(\mathbf{i} + \mathbf{j} + 2z\,\mathbf{k}) \Rightarrow yz = \lambda$, $xz = \lambda$, and $xy = 2z\lambda \Rightarrow yz = xz \Rightarrow z = 0$ or $y = x$. But $z > 0 \Rightarrow y = x \Rightarrow x^2 = 2z\lambda$ and $xz = \lambda$. Then $x^2 = 2z(xz) \Rightarrow x = 0$ or $x = 2z^2$. But $x > 0 \Rightarrow x = 2z^2 \Rightarrow y = 2z^2 \Rightarrow 2z^2 + 2z^2 + z^2 = 16 \Rightarrow z = \pm\frac{4}{\sqrt{5}}$. Use $z = \frac{4}{\sqrt{5}}$ since $z > 0 \Rightarrow x = \frac{32}{5}$, $y = \frac{32}{5}$. $f\left(\frac{32}{5}, \frac{32}{5}, \frac{4}{\sqrt{5}}\right) = \frac{4096}{25\sqrt{5}}$

23. $\nabla U = (y+2)\,\mathbf{i} + x\,\mathbf{j}$, $\nabla g = 2\,\mathbf{i} + \mathbf{j}$. $\nabla U = \lambda \nabla g \Rightarrow (y+2)\,\mathbf{i} + x\,\mathbf{j} = \lambda(2\,\mathbf{i} + \mathbf{j}) \Rightarrow y + 2 = 2\lambda$ and $x = \lambda \Rightarrow$ $y + 2 = 2x \Rightarrow y = 2x - 2 \Rightarrow 2x + 2x - 2 = 30 \Rightarrow x = 8 \Rightarrow y = 14$. $\therefore$ $U(8,14) = \$128$, the maximum value of U under the constraint.

25. $\nabla f = \mathbf{i} + \mathbf{j}$, $\nabla g = y\,\mathbf{i} + x\,\mathbf{j}$. $\nabla f = \lambda \nabla g \Rightarrow \mathbf{i} + \mathbf{j} = \lambda(y\,\mathbf{i} + x\,\mathbf{j}) \Rightarrow 1 = y\lambda$ and $1 = x\lambda \Rightarrow y = x \Rightarrow y^2 = 16 \Rightarrow$ $y = \pm 4 \Rightarrow x = \pm 4$. But as $x \to \infty$, $y \to 0$ and $f(x,y) \to \infty$; as $x \to -\infty$, $y \to 0$ and $f(x,y) \to -\infty$.

27. Let $f(x,y,z) = x^2 + y^2 + z^2$. Maximize f subject to $g_1(x,y,z) = y + 2z - 12 = 0$ and $g_2(x,y,z) = x + y - 6 = 0$
$\nabla f = 2x\,\mathbf{i} + 2y\,\mathbf{j} + 2z\,\mathbf{k}$, $\nabla g_1 = \mathbf{j} + 2\,\mathbf{k}$, $\nabla g_2 = \mathbf{i} + \mathbf{j}$. Then $2x\,\mathbf{i} + 2y\,\mathbf{j} + 2z\,\mathbf{k} = \lambda(\mathbf{j} + 2\,\mathbf{k}) + \mu(\mathbf{i} + \mathbf{j}) \Rightarrow$
$2x\,\mathbf{i} + 2y\,\mathbf{j} + 2z\,\mathbf{k} = \mu\,\mathbf{i} + (\lambda + \mu)\,\mathbf{j} + 2\lambda\,\mathbf{k} \Rightarrow 2x = \mu,\ 2y = \lambda + \mu,\ 2z = 2\lambda \Rightarrow x = \frac{\mu}{2},\ z = \lambda \Rightarrow$
$2x = 2y - z \Rightarrow x = \frac{2y - z}{2}$. Then $y + 2z - 12 = 0$ and $\frac{2y - z}{2} + y - 6 = 0 \Rightarrow 4y - z - 12 = 0 \Rightarrow z = 4 \Rightarrow$
$y = 4 \Rightarrow x = 2 \Rightarrow (2,4,4)$ is the point closest since this is the only extreme value and there is no maximum.

29. Let $g_1(x,y,z) = z - 1 = 0$ and $g_2(x,y,z) = x^2 + y^2 + z^2 - 10 = 0 \Rightarrow \nabla g_1 = \mathbf{k}$, $\nabla g_2 = 2x\,\mathbf{i} + 2y\,\mathbf{j} + 2z\,\mathbf{k}$.
$\nabla f = 2xyz\,\mathbf{i} + x^2z\,\mathbf{j} + x^2y\,\mathbf{k} \Rightarrow 2xyz\,\mathbf{i} + x^2z\,\mathbf{j} + x^2y\,\mathbf{k} = \lambda(\mathbf{k}) + \mu(2x\,\mathbf{i} + 2y\,\mathbf{j} + 2z\,\mathbf{k}) \Rightarrow 2xyz = 2x\mu$, $x^2z =$ $2y\mu$, $x^2y = 2z + \lambda \Rightarrow xyz = x\mu \Rightarrow x = 0$ or $yz = \mu \Rightarrow \mu = y$ since $z = 1$. $x = 0$ and $z = 1 \Rightarrow y^2 - 9 = 0$ (from g_2)
$\Rightarrow y = \pm 3 \Rightarrow (0, \pm 3, 1)$. $\mu = y \Rightarrow x^2z = 2y^2 \Rightarrow x^2 = 2y^2$ since $z = 1$
$\Rightarrow 2y^2 + y^2 + 1 - 10 = 0 \Rightarrow 3y^2 - 9 = 0 \Rightarrow y = \pm\sqrt{3} \Rightarrow x^2 = 2\left(\pm\sqrt{3}\right)^2 = 6 \Rightarrow x =$
$\pm\sqrt{6} \Rightarrow (\pm\sqrt{6}, \pm\sqrt{3}, 1)$. $f(0,\pm 3,1) = 1$. $f(\pm\sqrt{6},\pm\sqrt{3},1) = 6\left(\pm\sqrt{3}\right) + 1 = 1 \pm 6\sqrt{3} \Rightarrow$ Maximum of f is $1 + 6\sqrt{3}$ at $(\pm\sqrt{6},\sqrt{3},1)$; minimum of f is $1 - 6\sqrt{3}$ at $(\pm\sqrt{6},-\sqrt{3},1)$

31. Let $g_1(x,y,z) = y - x = 0 \Rightarrow x = y$. Let $g_2(x,y,z) = x^2 + y^2 + z^2 - 4 = 0 \Rightarrow \nabla g_2 = 2x\,\mathbf{i} + 2y\,\mathbf{j} + 2z\,\mathbf{k}$. $\nabla g_1 =$ $-\mathbf{i} + \mathbf{j}$; $\nabla f = y\,\mathbf{i} + x\,\mathbf{j} + 2z\,\mathbf{k} \Rightarrow y\,\mathbf{i} + x\,\mathbf{j} + 2z\,\mathbf{k} = \lambda(-\mathbf{i} + \mathbf{j}) + \mu(2x\,\mathbf{i} + 2y\,\mathbf{j} + 2z\,\mathbf{k}) \Rightarrow y = -\lambda + 2x\mu$, $x =$ $\lambda + 2y\mu$, $2z = 2z\mu \Rightarrow z = z\mu \Rightarrow z = 0$ or $\mu = 1$. $z = 0 \Rightarrow x^2 + y^2 - 4 = 0 \Rightarrow 2x^2 - 4 = 0$ (since $x = y$) $\Rightarrow$ $x^2 = 2 \Rightarrow x = \pm\sqrt{2} \Rightarrow y = \pm\sqrt{2} \Rightarrow (\pm\sqrt{2}, \pm\sqrt{2}, 0)$. $\mu = 1 \Rightarrow y = -\lambda + 2x$ and $x = \lambda + 2y \Rightarrow x + y =$ $2(x + y) \Rightarrow 2x = 2(2x)$ since $x = y \Rightarrow x = 0 \Rightarrow y = 0 \Rightarrow z^2 - 4 = 0 \Rightarrow z = \pm 2 \Rightarrow (0,0,\pm 2)$.
$f(0,0,\pm 2) = 4$, $f(\pm\sqrt{2}, \pm\sqrt{2}, 0) = 2 \Rightarrow$ Maximum value of f is 4 at $(0,0,\pm 2)$; minimum value of f is 2 at $(\pm\sqrt{2},\pm\sqrt{2},0)$

13.P PRACTICE EXERCISES

1.

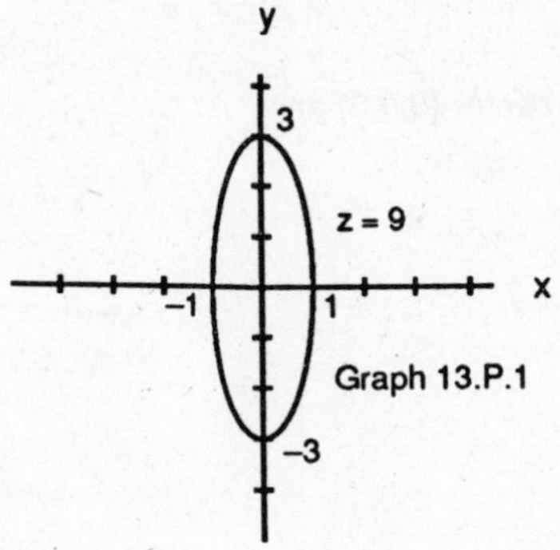

Graph 13.P.1

Domain: All points in the xy–plane

Range: $f(x,y) \geq 0$

Level curves are ellipses with major axis along the y–axis and minor axis along the x–axis.

3.

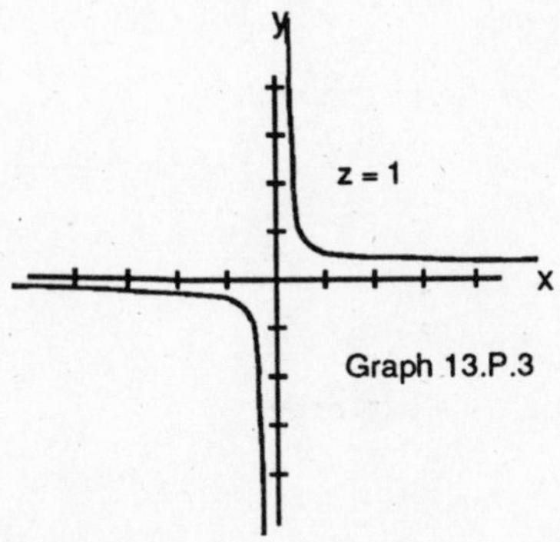

Graph 13.P.3

Domain: All (x,y) such that $x \neq 0$ and $y \neq 0$

Range: $f(x,y) \neq 0$

Level curves are hyperbolas with the x and y axes as asymptotes.

5.

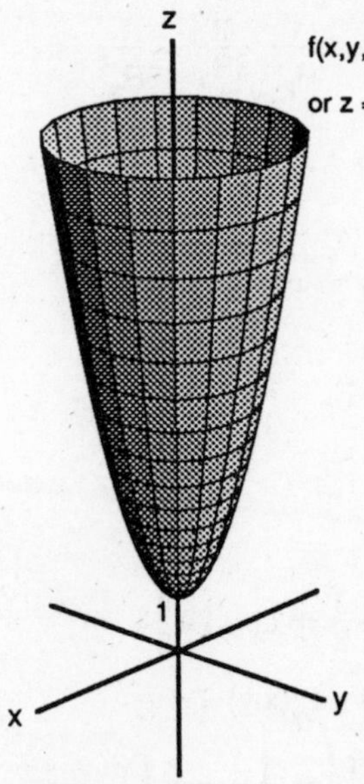

Graph 13.P.5

$f(x,y,z) = x^2 + y^2 - z = -1$

or $z = x^2 + y^2 + 1$

Domain: All (x,y,z)

Range: All Real Numbers

Level surfaces are paraboloids of revolution with the z–axis as axis.

7.

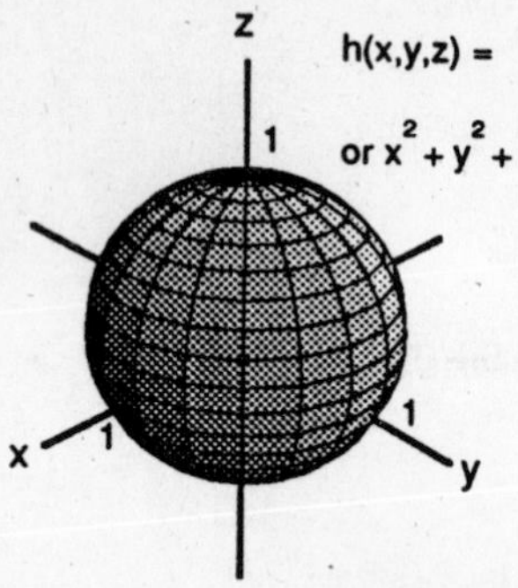

$h(x,y,z) = \dfrac{1}{x^2+y^2+z^2+1} = 1$

or $x^2+y^2+z^2=1$

Domain: All (x,y,z) such that $(x,y,z) \neq (0,0,0)$

Range: $f(x,y,z) > 0$

Level surfaces are spheres with center (0,0,0) and radius $r > 0$.

Graph 13.P.7

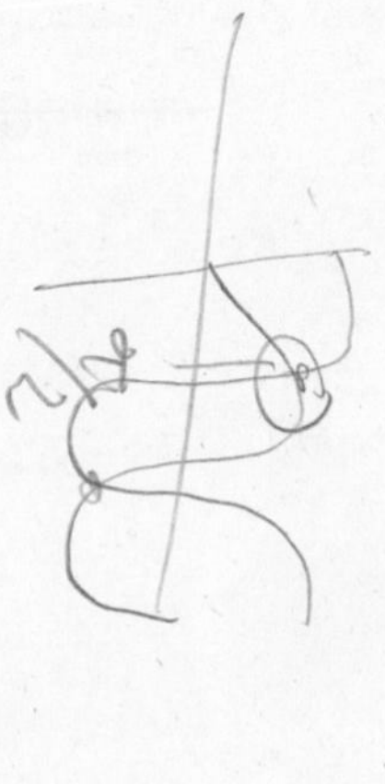

9. $\displaystyle\lim_{(x,y)\to(\pi,\ln 2)} e^y \cos x = e^{\ln 2} \cos \pi = -2$

11. $\displaystyle\lim_{\substack{(x,y)\to(1,1)\\ x\neq\pm y}} \frac{x-y}{x^2-y^2} = \lim_{\substack{(x,y)\to(1,1)\\ x\neq\pm y}} \frac{x-y}{(x-y)(x+y)} = \lim_{(x,y)\to(1,1)} \frac{1}{x+y} = \frac{1}{2}$

13. $\displaystyle\lim_{P\to(1,-1,e)} \ln|x+y+z| = \ln|1+(-1)+e| = 1$

15. Let $y = kx^2$, $k \neq 1$. Then $\displaystyle\lim_{\substack{(x,y)\to(0,0)\\ y\neq x^2}} \frac{y}{x^2-y} = \lim_{(x,kx^2)\to(0,0)} \frac{kx^2}{x^2-kx^2} = \frac{k}{1-k^2}$ which gives different limits for different values of k. ∴ the limit doesn't exist.

17. $\dfrac{\partial g}{\partial r} = \cos\theta + \sin\theta, \dfrac{\partial g}{\partial \theta} = -r\sin\theta + r\cos\theta$

19. $\dfrac{\partial f}{\partial R_1} = -\dfrac{1}{R_1^2}, \dfrac{\partial f}{\partial R_2} = -\dfrac{1}{R_2^2}, \dfrac{\partial f}{\partial R_3} = -\dfrac{1}{R_3^2}$

21. $\dfrac{\partial P}{\partial n} = \dfrac{RT}{V}, \dfrac{\partial P}{\partial R} = \dfrac{nT}{V}, \dfrac{\partial P}{\partial T} = \dfrac{nR}{V}, \dfrac{\partial P}{\partial V} = -\dfrac{nRT}{V^2}$

23. $\dfrac{\partial f}{\partial x} = \dfrac{1}{y}, \dfrac{\partial f}{\partial y} = 1 - \dfrac{x}{y^2} \Rightarrow \dfrac{\partial^2 f}{\partial x^2} = 0, \dfrac{\partial^2 f}{\partial y^2} = \dfrac{2x}{y^3}, \dfrac{\partial^2 f}{\partial y \partial x} = \dfrac{\partial^2 f}{\partial x \partial y} = -\dfrac{1}{y^2}$

25. $\dfrac{\partial f}{\partial x} = 1 + y - 15x^2 + \dfrac{2x}{x^2+1}, \dfrac{\partial f}{\partial y} = x \Rightarrow \dfrac{\partial^2 f}{\partial x^2} = -30x + \dfrac{2-2x^2}{(x^2+1)^2}, \dfrac{\partial^2 f}{\partial y^2} = 0, \dfrac{\partial^2 f}{\partial y \partial x} = \dfrac{\partial^2 f}{\partial x \partial y} = 1$

27. $f\left(\frac{\pi}{4},\frac{\pi}{4}\right) = \frac{1}{2}, f_x\left(\frac{\pi}{4},\frac{\pi}{4}\right) = \cos x \cos y\Big|_{(\pi/4,\pi/4)} = \frac{1}{2}, f_y\left(\frac{\pi}{4},\frac{\pi}{4}\right) = -\sin x \sin y\Big|_{(\pi/4,\pi/4)} = -\frac{1}{2} \Rightarrow L(x,y) = \frac{1}{2} + \frac{1}{2}\left(x - \frac{\pi}{4}\right) - \frac{1}{2}\left(y - \frac{\pi}{4}\right) = \frac{1}{2} + \frac{1}{2}x - \frac{1}{2}y$. $f_{xx}(x,y) = -\sin x \cos y$, $f_{yy}(x,y) = -\sin x \cos y$, $f_{xy}(x,y) = -\cos x \sin y$. ∴ maximum of $|f_{xx}|$, $|f_{yy}|$, and $|f_{xy}|$ is 1 $\Rightarrow M = 1 \Rightarrow |E(x,y)| \leq \frac{1}{2}(1)\left(\left|x - \frac{\pi}{4}\right| + \left|y - \frac{\pi}{4}\right|\right)^2 \leq 0.02$.

29. a) $f(1,0,0) = 0, f_x(1,0,0) = y - 3z\Big|_{(1,0,0)} = 0, f_y(1,0,0) = x + 2z\Big|_{(1,0,0)} = 1, f_z(1,0,0) = 2y - 3x\Big|_{(1,0,0)} =$
$-3 \Rightarrow L(x,y,z) = 0(x-1) + (y-0) - 3(z-0) = y - 3z.$

b) $f(1,1,0) = 1, f_x(1,1,0) = 1, f_y(1,1,0) = 1, f_z(1,1,0) = -1 \Rightarrow L(x,y,z) = 1 + (x-1) + (y-1) - 1(z-0) =$
$x + y - z - 1$

31. $dV = 2\pi rh\, dr + \pi r^2\, dh \Rightarrow dV\Big|_{(1.5,5280)} = 2\pi(1.5)(5280)\, dr + \pi(1.5)^2\, dh = 15840\pi\, dr + 2.25\pi\, dh.$ Be more careful with the diameter since it has a greater effect on dV.

33. $dI = \frac{1}{R}dV - \frac{V}{R^2}dR \Rightarrow dI\Big|_{(24,100)} = \frac{1}{100}dV - \frac{24}{100^2}dR \Rightarrow dI\Big|_{dV=-1,dR=-20} = 0.038$ or increases by 0.038 amps.
% change in V $= (100)\left(-\frac{1}{24}\right) \approx -4.17\%$; % change in R $= -\frac{20}{100}(100) = -20\%$. $I = \frac{24}{100} = 0.24 \Rightarrow$ Estimated % change in I $=$
$\frac{dI}{I} \times 100 = \frac{0.038}{0.24} \times 100 \approx 15.83\%$

35. $\frac{\partial w}{\partial x} = y\cos(xy+\pi), \frac{\partial w}{\partial y} = x\cos(xy+\pi), \frac{dx}{dt} = e^t, \frac{dy}{dt} = \frac{1}{t+1} \Rightarrow \frac{dw}{dt} = y\cos(xy+\pi)\,e^t + x\cos(xy+\pi)\left(\frac{1}{t+1}\right)$
$= e^t\ln(t+1)\cos(e^t\ln(t+1)+\pi) + \frac{e^t}{t+1}\cos(e^t\ln(t+1)+\pi) \Rightarrow \frac{dw}{dt}\Big|_{t=0} = -1$

37. $\frac{\partial w}{\partial x} = 2\cos(2x-y), \frac{\partial w}{\partial y} = -\cos(2x-y), \frac{\partial x}{\partial r} = 1, \frac{\partial x}{\partial s} = \cos s, \frac{\partial y}{\partial r} = s, \frac{\partial y}{\partial s} = r \Rightarrow \frac{\partial w}{\partial r} = 2\cos(2x-y)(1) +$
$(-\cos(2x-y)(s)) = 2\cos(2r + 2\sin s - rs) - s\cos(2r + 2\sin s - rs) \Rightarrow \frac{\partial w}{\partial r}\Big|_{r=\pi,s=0} = 2.$
$\frac{\partial w}{\partial s} = 2\cos(2x-y)(\cos s) + (-\cos(2x-y)(r)) = 2\cos(2r + 2\sin s - rs)(\cos s) - r\cos(2r + 2\sin s - rs) \Rightarrow$
$\frac{\partial w}{\partial s}\Big|_{r=\pi,s=0} = 2 - \pi$

39. $F_x = -1 - y\cos xy, F_y = -2y - x\cos xy.\ \frac{dy}{dx} = -\frac{F_x}{F_y} = -\frac{-1 - y\cos xy}{-2y - x\cos xy} = \frac{1 + y\cos xy}{-2y - x\cos xy} \Rightarrow$
$\frac{dy}{dx}\Big|_{(x,y)=(0,1)} = -1$

41. $\frac{\partial f}{\partial x} = y + z, \frac{\partial f}{\partial y} = x + z, \frac{\partial f}{\partial z} = y + x, \frac{dx}{dt} = -\sin t, \frac{dy}{dt} = \cos t, \frac{dz}{dt} = -2\sin 2t \Rightarrow \frac{df}{dt} = -(y+z)\sin t + (x+z)\cos t$
$- 2(y+x)\sin 2t = -(\sin t + \cos 2t)\sin t + (\cos t + \cos 2t)\cos t - 2(\sin t + \cos t)\sin 2t \Rightarrow \frac{df}{dt}\Big|_{t=1} =$
$-(\sin 1 + \cos 2)\sin 1 + (\cos 1 + \cos 2)\cos 1 - 2(\sin 1 + \cos 1)\sin 2$

43. $\nabla f = (-\sin x\cos y)\,\mathbf{i} - (\cos x\sin y)\,\mathbf{j} \Rightarrow \nabla f\Big|_{(\pi/4,\pi/4)} = -\frac{1}{2}\mathbf{i} - \frac{1}{2}\mathbf{j} \Rightarrow |\nabla f| = \sqrt{\left(-\frac{1}{2}\right)^2 + \left(-\frac{1}{2}\right)^2} = \frac{1}{\sqrt{2}}$
$\mathbf{u} = \frac{\nabla f}{|\nabla f|} = \frac{-\frac{1}{2}\mathbf{i} - \frac{1}{2}\mathbf{j}}{\frac{1}{\sqrt{2}}} = -\frac{\sqrt{2}}{2}\mathbf{i} - \frac{\sqrt{2}}{2}\mathbf{j}$. f increases most rapidly in the direction $\mathbf{u} = -\frac{\sqrt{2}}{2}\mathbf{i} - \frac{\sqrt{2}}{2}\mathbf{j}$;
decreases most rapidly in the direction $-\mathbf{u} = \frac{\sqrt{2}}{2}\mathbf{i} + \frac{\sqrt{2}}{2}\mathbf{j}$. $\left(D_{\mathbf{u}}f\right)_{P_0} = \frac{\sqrt{2}}{2}, \left(D_{-\mathbf{u}}f\right)_{P_0} = -\frac{\sqrt{2}}{2}.$
$\mathbf{u}_1 = \frac{\mathbf{A}}{|\mathbf{A}|} = \frac{3\mathbf{i} + 4\mathbf{j}}{\sqrt{3^2 + 4^2}} = \frac{3}{5}\mathbf{i} + \frac{4}{5}\mathbf{j}.\ \left(D_{\mathbf{u}_1}f\right)_{P_0} = \nabla f \cdot \mathbf{u}_1 = -\frac{7}{10}.$

45. $\nabla f = \left(\frac{2}{2x+3y+6z}\right)\mathbf{i} + \left(\frac{3}{2x+3y+6z}\right)\mathbf{j} + \left(\frac{6}{2x+3y+6z}\right)\mathbf{k} \Rightarrow \nabla f\big|_{(-1,-1,1)} = 2\mathbf{i}+3\mathbf{j}+6\mathbf{k}$.
$\mathbf{u} = \frac{\nabla f}{|\nabla f|} = \frac{2\mathbf{i}+3\mathbf{j}+6\mathbf{k}}{\sqrt{2^2+3^2+6^2}} = \frac{2}{7}\mathbf{i}+\frac{3}{7}\mathbf{j}+\frac{6}{7}\mathbf{k}$. f increases most rapidly in the direction $\mathbf{u} = \frac{2}{7}\mathbf{i}+\frac{3}{7}\mathbf{j}+\frac{6}{7}\mathbf{k}$; decreases most rapidly in the direction $-\mathbf{u} = -\frac{2}{7}\mathbf{i}-\frac{3}{7}\mathbf{j}-\frac{6}{7}\mathbf{k}$. $\left(D_{\mathbf{u}}f\right)_{P_0} = \nabla f \cdot \mathbf{u} = 7$,
$\left(D_{-\mathbf{u}}f\right)_{P_0} = -7$. $\mathbf{u}_1 = \frac{\mathbf{A}}{|\mathbf{A}|} = \frac{2}{7}\mathbf{i}+\frac{3}{7}\mathbf{j}+\frac{6}{7}\mathbf{k}$ since $\mathbf{A} = \nabla f$. $\Rightarrow \left(D_{\mathbf{u}_1}f\right)_{P_0} = 7$.

47.

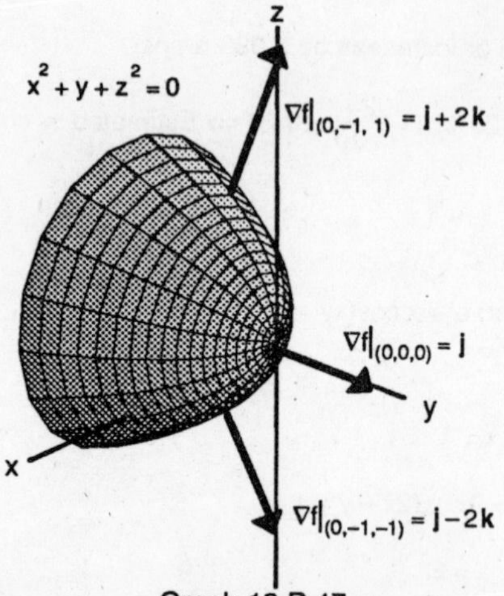

Graph 13.P.47

$\nabla f = 2x\mathbf{i}+\mathbf{j}+2z\mathbf{k} \Rightarrow$
$\nabla f\big|_{(0,-1,-1)} = \mathbf{j}-2\mathbf{k}$,
$\nabla f\big|_{(0,0,0)} = \mathbf{j}$,
$\nabla f\big|_{(0,-1,1)} = \mathbf{j}+2\mathbf{k}$

49. $\nabla f = 2x\mathbf{i}-\mathbf{j}-5\mathbf{k} \Rightarrow \nabla f\big|_{(2,-1,1)} = 4\mathbf{i}-\mathbf{j}-5\mathbf{k} \Rightarrow$ Tangent Plane: $4(x-2)-(y+1)-5(z-1) = 0 \Rightarrow$ $4x-y-5z = 4$; Normal Line: $x = 2+4t$, $y = -1-t$, $z = 1-5t$

51. $\frac{\partial z}{\partial x} = \frac{2x}{x^2+y^2} \Rightarrow \frac{\partial z}{\partial x}\Big|_{(0,1,0)} = 0$. $\frac{\partial z}{\partial y} = \frac{2y}{x^2+y^2} \Rightarrow \frac{\partial z}{\partial y}\Big|_{(0,1,0)} = 2$. $\therefore$ the tangent plane is $2(y-1)-(z-0) = 0$ or $2y-z-2 = 0$.

53.

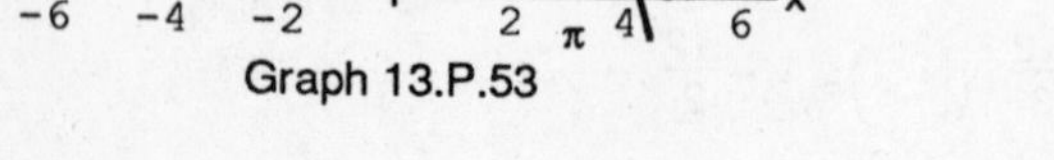

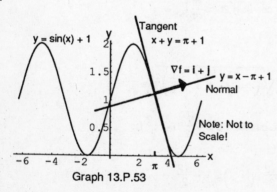

Graph 13.P.53

$\nabla f = (-\cos x)\mathbf{i}+\mathbf{j} \Rightarrow \nabla f\big|_{(\pi,1)} = \mathbf{i}+\mathbf{j} \Rightarrow$ Tangent Line: $(x-\pi)+(y-1) = 0 \Rightarrow x+y = \pi+1$;
Normal Line: $y-1 = 1(x-\pi) \Rightarrow y = x-\pi+1$

55. Let $f(x,y,z) = x^2 + 2y + 2z - 4$ and $g(x,y,z) = y - 1$ at $P_0\left(1,1,\frac{1}{2}\right)$. $\nabla f = 2x\,\mathbf{i} + 2\,\mathbf{j} + 2\,\mathbf{k}\Big|_{\left(1,1,\frac{1}{2}\right)} = 2\,\mathbf{i} + 2\,\mathbf{j} +$ $2\,\mathbf{k}$. $\nabla g = \mathbf{j} \Rightarrow \nabla f \times \nabla g = \begin{vmatrix} \mathbf{i} & \mathbf{j} & \mathbf{k} \\ 2 & 2 & 2 \\ 0 & 1 & 0 \end{vmatrix} = -2\,\mathbf{i} + 2\,\mathbf{k} \Rightarrow$ the line is $x = 1 - 2t$, $y = 1$, $z = \frac{1}{2} + 2t$.

57. $f(x,y,z) = xyz \Rightarrow \nabla f = yz\,\mathbf{i} + xz\,\mathbf{j} + xy\,\mathbf{k}$. At (1,1,1), $\nabla f = \mathbf{i} + \mathbf{j} + \mathbf{k} \Rightarrow$ the maximum value of $D_u f\Big|_{(1,1,1)} = |\nabla f| = \sqrt{3}$.

59. $f_x(x,y) = 2x - y + 2 = 0$ and $f_y(x,y) = -x + 2y + 2 = 0 \Rightarrow x = -2, y = -2 \Rightarrow (-2,-2)$ is the critical point. $f_{xx}(-2,-2) = 2$, $f_{yy}(-2,-2) = 2$, $f_{xy}(-2,-2) = -1 \Rightarrow f_{xx}f_{yy} - f_{xy}^2 = 3 > 0$ and $f_{xx} > 0 \Rightarrow$ Minimum $f(-2,-2) = -8$

61. $f_x(x,y) = 6x^2 + 3y = 0$ and $f_y(x,y) = 3x + 6y^2 = 0 \Rightarrow x = 0, y = 0$ or $x = -\frac{1}{2}, y = -\frac{1}{2} \Rightarrow$ critical points are (0,0) and $\left(-\frac{1}{2},-\frac{1}{2}\right)$ For (0,0): $f_{xx}(0,0) = 12x\Big|_{(0,0)} = 0$, $f_{yy}(,0) = 6y\Big|_{(0,0)} = 0$, $f_{xy}(0,0) = 3 \Rightarrow$ $f_{xx}f_{yy} - f_{xy}^2 = -9 \Rightarrow$ Saddle Point. $f(0,0) = 0$. For $\left(-\frac{1}{2},-\frac{1}{2}\right)$, $f_{xx} = -6$, $f_{yy} = -6$, $f_{xy} = 3 \Rightarrow f_{xx}f_{yy} - f_{xy}^2 = 27 > 0$ and $f_{xx} < 0 \Rightarrow$ Maximum. $f\left(-\frac{1}{2},-\frac{1}{2}\right) = \frac{1}{4}$

63. $f_x(x,y) = 3x^2 + 6x = 0$ and $f_y(x,y) = 3y^2 - 6y = 0 \Rightarrow x = 0, y = 0$; $x = 0, y = 2$; $x = -2, y = 0$; or $x = -2, y = 2 \Rightarrow$ critical points are (0,0), (0,2), (−2,0), and (−2,2). For (0,0): $f_{xx}(0,0) = 6x + 6\Big|_{(0,0)} = 6$, $f_{yy}(0,0) = 6y - 6\Big|_{(0,0)} = -6$, $f_{xy}(0,0) = 0 \Rightarrow f_{xx}f_{yy} - f_{xy}^2 = -36 \Rightarrow$ Saddle Point. $f(0,0) = 0$. For (0,2): $f_{xx}(0,2) = 6$, $f_{yy}(0,2) = 6$, $f_{xy}(0,2) = 0 \Rightarrow f_{xx}f_{yy} - f_{xy}^2 = 36 > 0$ and $f_{xx} > 0 \Rightarrow$ Minimum (local). $f(0,2) = -4$. For (−2,0): $f_{xx}(-2,0) = -6$, $f_{yy}(-2,0) = -6$, $f_{xy}(-2,0) = 0 \Rightarrow f_{xx}f_{yy} - f_{xy}^2 = 36 > 0$ and $f_{xx} < 0 \Rightarrow$ Maximum (local). $f(-2,0) = 4$. For (−2,2): $f_{xx}(-2,2) = -6$, $f_{yy}(-2,2) = 6$, $f_{xy}(-2,2) = 0 \Rightarrow$ $f_{xx}f_{yy} - f_{xy}^2 = -36 \Rightarrow$ Saddle Point. $f(-2,2) = 0$

65.

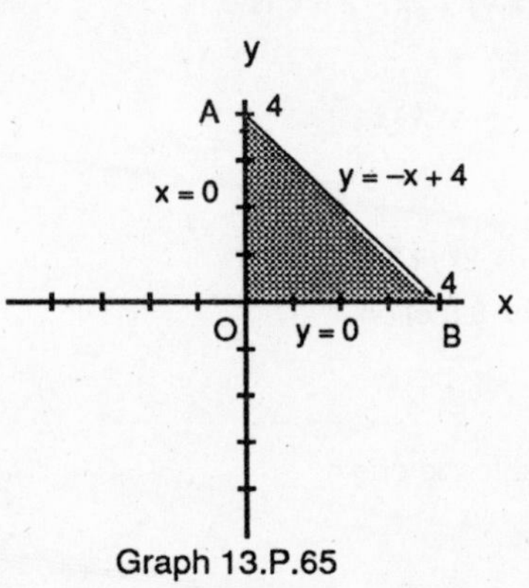

Graph 13.P.65

1. On OA, $f(x,y) = y^2 + 3y = f(0,y)$ for $0 \le y \le 4$. $f(0,0) = 0$, $f(0,4) = 28$. $f'(0,y) = 2y + 3 = 0 \Rightarrow y = -\frac{3}{2}$. But $\left(0,-\frac{3}{2}\right)$ is not in the region.
2. On AB, $f(x,y) = x^2 - 10x + 28 = f(x,-x+4)$ for $0 \le x \le 4$. $f(4,0) = 4$. $f'(x,-x+4) = 2x - 10 = 0 \Rightarrow x = 5, y = -1$. But (5,−1) not in the region.
3. On OB, $f(x,y) = x^2 - 3x = f(x,0)$ for $0 \le x \le 4$. $f'(x,0) = 2x - 3 \Rightarrow$ $x = \frac{3}{2}, y = 0 \Rightarrow \left(\frac{3}{2},0\right)$ is a critical point. $f\left(\frac{3}{2},0\right) = -\frac{9}{4}$

65. (Continued)

4. For the interior of the triangular region, $f_x(x,y) = 2x + y - 3 = 0$ and $f_y(x,y) = x + 2y + 3 = 0 \Rightarrow x = 3, y = -3$. But (3,−3) is not in the region.

65. (Continued)

$\therefore$ the absolute maximum is 28 at (0,4); the absolute minimum is $-\frac{9}{4}$ at $\left(\frac{3}{2}, 0\right)$.

67.

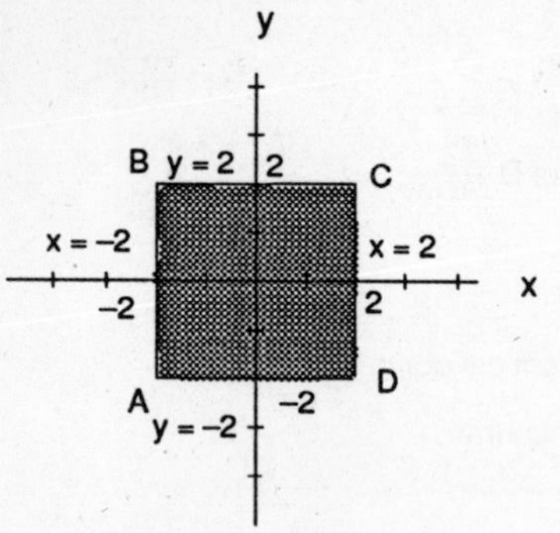

Graph 13.P.67

1. On AB, $f(x,y) = y^2 - y - 4 = f(-2,y)$ for $-2 \le y \le 2$. $f(-2,-2) = 2$, $f(-2,2) = -2$. $f'(-2,y) = 2y - 1 \Rightarrow y = \frac{1}{2}, x = -2 \Rightarrow \left(-2, \frac{1}{2}\right)$ is a critical point. $f\left(-2, \frac{1}{2}\right) = -\frac{17}{4}$.
2. On BC, $f(x,y) = -2 = f(x,2)$ for $-2 \le x \le 2$. $f(2,2) = -2$. $f'(x,2) = 0 \Rightarrow$ no critical points in the interior of BC.
3. On CD, $f(x,y) = y^2 - 5y + 4 = f(2,y)$ for $-2 \le y \le 2$. $f(2,-2) = 18$. $f'(2,y) = 2y - 5 = 0 \Rightarrow y = \frac{5}{2}, x = 2 \Rightarrow \left(2, \frac{5}{2}\right)$ which is not in the region.
4. On AD, $f(x,y) = 4x + 10 = f(x,-2)$ for $-2 \le x \le 2$. $f'(x,-2) = 4 \Rightarrow$ no critical points in the interior of AD.
5. For the interior of the square, $f_x(x,y) = -y + 2 = 0$ and $f_y(x,y) = 2y - x - 3 = 0 \Rightarrow x = 1, y = 2 \Rightarrow (1,2)$ is a critical point. $f(1,2) = -2$

$\therefore$ the absolute maximum is 18 at (2,−2); the absolute minimum is $-\frac{17}{4}$ at $\left(-2, \frac{1}{2}\right)$.

69.

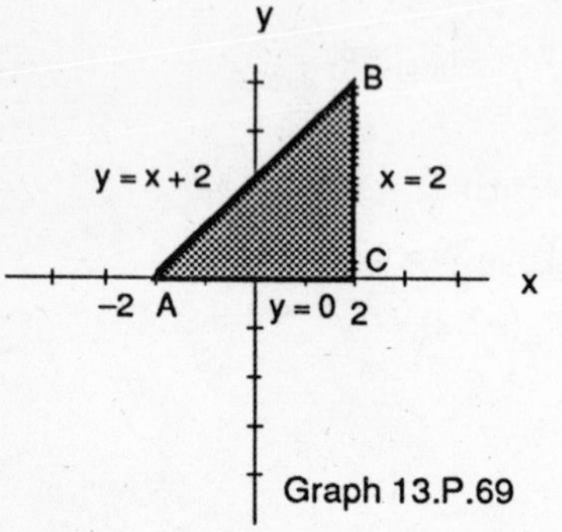

Graph 13.P.69

1. On AB, $f(x,y) = -2x + 4 = f(x,x+2)$ for $-2 \le x \le 2$. $f(-2,0) = 8$, $f(2,4) = 0$. $f'(x,x+2) = -2 \Rightarrow$ no critical points in the interior of AB.
2. On BC, $f(x,y) = -y^2 + 4y = f(2,y)$ for $0 \le y \le 4$. $f(2,0) = 0$. $f'(2,y) = -2y + 4 = 0 \Rightarrow y = 2, x = 2 \Rightarrow (2,2)$ is a critical point. $f(2,2) = 4$.
3. On AC, $f(x,y) = x^2 - 2x = f(x,0)$ for $-2 \le x \le 2$. $f'(x,0) = 2x - 2 = 0 \Rightarrow x = 1, y = 0 \Rightarrow (1,0)$ is a critical point. $f(1,0) = -1$.
4. For the interior of the triangular region, $f_x(x,y) = 2x - 2 = 0$ and $f_y(x,y) = -2y + 4 = 0 \Rightarrow x = 1, y = 2 \Rightarrow (1,2)$ is a critical point $f(1,2) = 3$.

$\therefore$ the absolute maximum is 8 at (−2,0); the absolute minimum is −1 at (1,0).

71. Let $f(x,y) = x^2 + y^2$ be the square of the distance to the origin. $\nabla f = 2x\,\mathbf{i} + 2y\,\mathbf{j}$, $\nabla g = y^2\,\mathbf{i} + 2xy\,\mathbf{j}$. $\nabla f = \lambda \nabla g \Rightarrow 2x\,\mathbf{i} + 2y\,\mathbf{j} = \lambda(y^2\,\mathbf{i} + 2xy\,\mathbf{j}) \Rightarrow 2x = \lambda y^2$ and $2y = 2xy\lambda \Rightarrow 2y = \lambda y^2(y\lambda) \Rightarrow y = 0$ (not on $xy^2 = 54$) or $\lambda^2 y^2 - 2 = 0 \Rightarrow y^2 = \frac{2}{\lambda^2}$. $2y = 2xy\lambda \Rightarrow 1 = x\lambda$ since $y \ne 0 \Rightarrow x = \frac{1}{\lambda}$.
$\therefore \frac{1}{\lambda}\left(\frac{2}{\lambda^2}\right) = 54 \Rightarrow \lambda^3 = \frac{1}{27} \Rightarrow \lambda = \frac{1}{3} \Rightarrow x = 3, y^2 = 18 \Rightarrow y = \pm 3\sqrt{2} \Rightarrow$ the points nearest to the origin are $(3, \pm 3\sqrt{2})$.

73. $\nabla T = 400yz^2\,\mathbf{i} + 400xz^2\,\mathbf{j} + 800xyz\,\mathbf{k}$, $\nabla g = 2x\,\mathbf{i} + 2y\,\mathbf{j} + 2z\,\mathbf{k}$. $\nabla T = \lambda\,\nabla g \Rightarrow 400yz^2\,\mathbf{i} + 400xz^2\,\mathbf{j} + 800xyz\,\mathbf{k} = \lambda(2x\,\mathbf{i} + 2y\,\mathbf{j} + 2z\,\mathbf{k}) \Rightarrow 400yz^2 = 2x\lambda$, $400xz^2 = 2y\lambda$, and $800xyz = 2z\lambda$. Solving this system yields the following points: $(0,\pm1,0)$, $(\pm1,0,0)$, $\left(\pm\frac{1}{2},\pm\frac{1}{2},\pm\frac{\sqrt{2}}{2}\right)$. $T(0,\pm1,0) = 0$, $T(\pm1,0,0) = 0$, $T\left(\pm\frac{1}{2},\pm\frac{1}{2},\pm\frac{\sqrt{2}}{2}\right) = \pm50$. $\therefore$ 50 is the maximum at $\left(\frac{1}{2},\frac{1}{2},\pm\frac{\sqrt{2}}{2}\right)$ and $\left(-\frac{1}{2},-\frac{1}{2},\pm\frac{\sqrt{2}}{2}\right)$; -50 is the minimum at $\left(\frac{1}{2},-\frac{1}{2},\pm\frac{\sqrt{2}}{2}\right)$ and $\left(-\frac{1}{2},\frac{1}{2},\pm\frac{\sqrt{2}}{2}\right)$.

75. Let $f(x,y,z) = x^2 + y^2 + z^2$ (the square of the distance to the origin) and $g(x,y,z) = xyz - 1$. Then $\nabla f = \lambda\nabla g$

$$\Rightarrow \begin{cases} 2x = \lambda yz \\ 2y = \lambda xz \\ 2z = \lambda xy \end{cases} \Rightarrow \begin{cases} 2x^2 = \lambda xyz \\ 2y^2 = \lambda xyz \end{cases} \Rightarrow 2x^2 = 2y^2 \Rightarrow y = \pm x.$$ Similarly, $z = \pm x$. $\therefore x(\pm x)(\pm x) = 1 \Rightarrow x^3 = \pm1 \Rightarrow x = \pm1 \Rightarrow$ the points are $(1,1,1)$, $(1,-1,-1)$, $(-1,-1,1)$, and $(-1,1,-1)$.

CHAPTER 14

MULTIPLE INTEGRALS

14.1 DOUBLE INTEGRALS

1. $\int_0^3 \int_0^2 \left(4 - y^2\right) dy\, dx = \int_0^3 \left[4y - \frac{y^3}{3}\right]_0^2 dx = \frac{16}{3}\int_0^3 dx = 16$

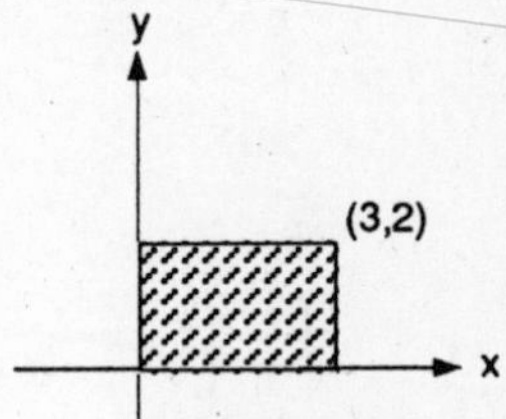

Graph 14.1.1

3. $\int_{-1}^0 \int_{-1}^1 (x + y + 1)\, dx\, dy = \int_{-1}^0 \left[\frac{x^2}{2} + yx + x\right]_{-1}^{1} dy =$
$\int_{-1}^0 (2y + 2)\, dy = 1$

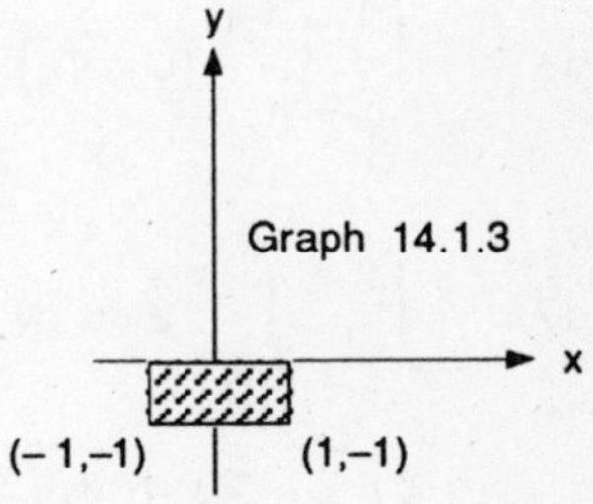

Graph 14.1.3

5. $\int_0^\pi \int_0^x (x \sin y)\, dy\, dx = \int_0^\pi [-x \cos y]_0^x dx = \int_0^\pi (x - x\cos x)\, dx = \frac{\pi^2}{2} + 2$

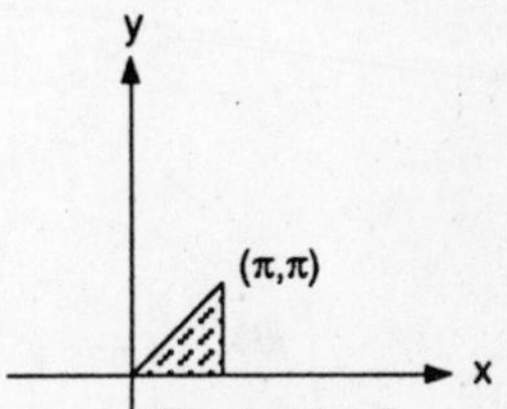

Graph 14.1.5

7. $\int_1^{\ln 8} \int_0^{\ln y} e^{x+y}\, dx\, dy = \int_1^{\ln 8} \left[e^{x+y}\right]_0^{\ln y} dy = \int_1^{\ln 8} \left(y e^y - e^y\right) dy =$
$\left[(y-1)e^y - e^y\right]_1^{\ln 8} = 8\ln(8) - 16 + e$

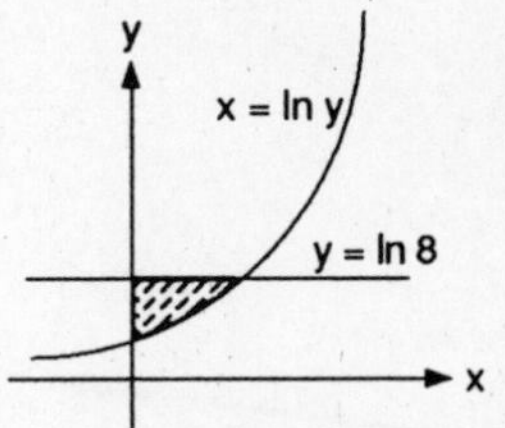

Graph 14.1.7

9. $$\int_0^1 \int_0^{y^2} 3y^3e^{xy}\,dx\,dy = \int_0^1 \left[3y^2e^{xy}\right]_0^{y^2} dy =$$
$$\int_0^1 \left(3y^2e^{y^3} - 3y^2\right) dy = \left[e^{y^3} - y^3\right]_0^1 = e - 2$$

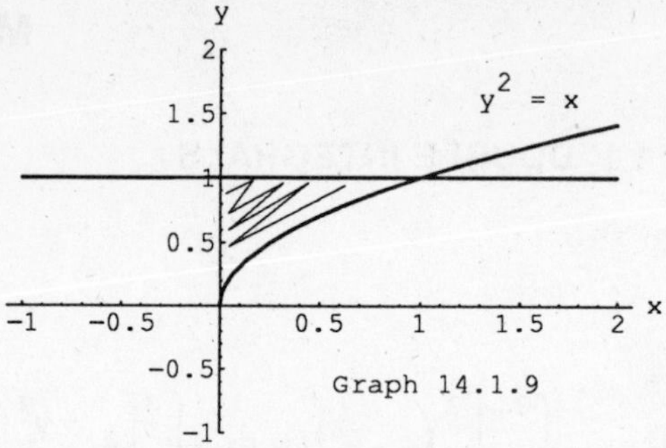

Graph 14.1.9

11. $$\int_1^2 \int_x^{2x} \frac{x}{y}\,dy\,dx = \int_1^2 [x \ln y]_x^{2x}\,dx = \ln(2)\int_1^2 x\,dx = (3/2)\ln 2$$

13. $$\int_0^1 \int_0^{1-y} \left(x^2 + y^2\right) dx\,dy = \int_0^1 \left[\frac{x^3}{3} + y^2x\right]_0^{1-y} dy = \int_0^1 \left[\frac{(1-y)^3}{3} + y^2 - y^3\right] dy = \frac{1}{6}$$

15. $$\int_0^1 \int_0^{1-u} \left(v - \sqrt{u}\right) dv\,du = \int_0^1 \left[\frac{v^2}{2} - v\sqrt{u}\right]_0^{1-u} du = \int_0^1 \left[\frac{1 - 2u + u^2}{2} - \sqrt{u}(1-u)\right] du = -\frac{1}{10}$$

17. $$\int_{-2}^0 \int_v^{-v} 2\,dp\,dv = \int_{-2}^0 [p]_v^{-v}\,dv =$$
$$2\int_{-2}^0 -2\,v\,dv = -2\left[2\,v^2\right]_{-2}^0 = 8$$

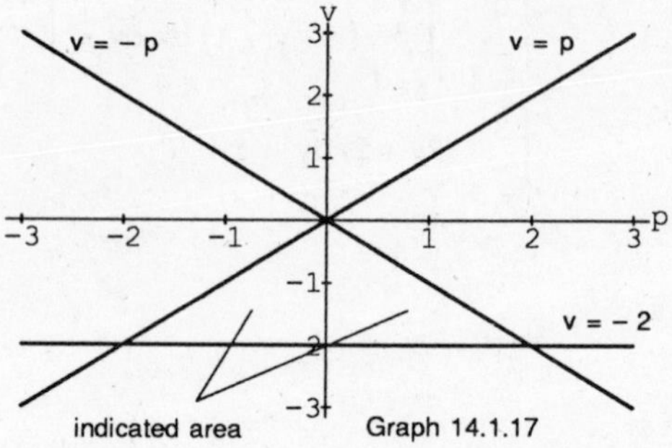

Graph 14.1.17

19. $$\int_{-\pi/3}^{\pi/3} \int_0^{\sec t} 3\cos t\,du\,dt =$$
$$\int_{-\pi/3}^{\pi/3} 3\cos t[u]_0^{\sec t}\,dt = 3\int_{-\pi/3}^{\pi/3} dt = 2\pi$$

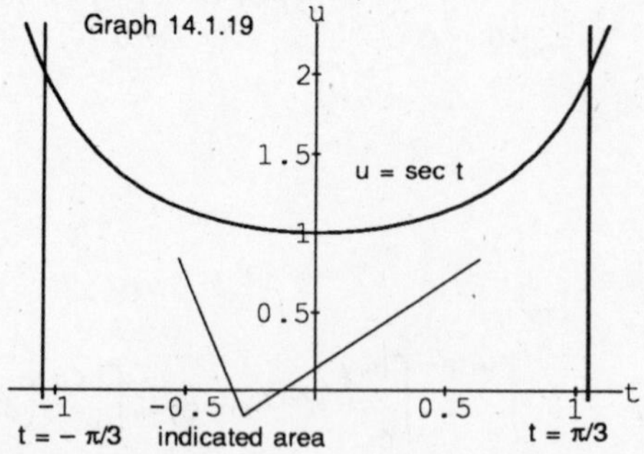

Graph 14.1.19

21. $$\int_2^4 \int_0^{(4-y)/2} dx\,dy$$

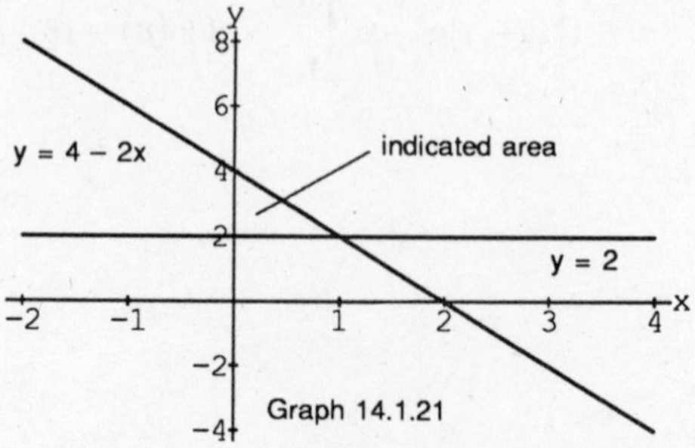

Graph 14.1.21

23. $\int_0^1 \int_{x^2}^{x} dy\, dx$

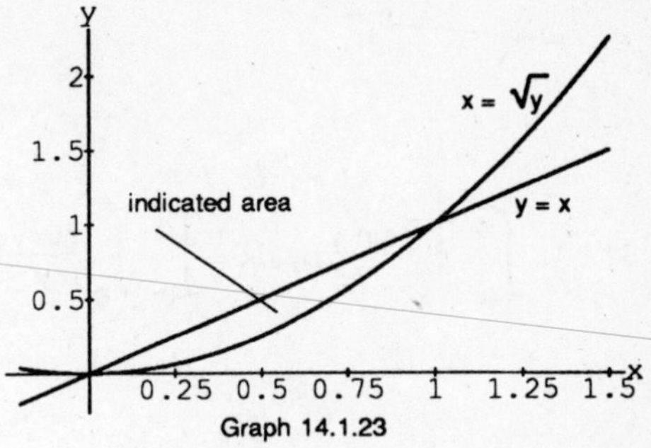

Graph 14.1.23

25. $\int_1^e \int_{\ln y}^{1} dx\, dy$

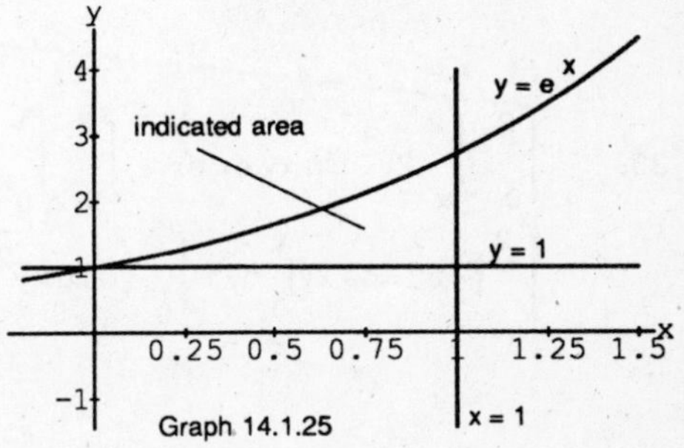

Graph 14.1.25

27. $\int_0^9 \int_0^{(\sqrt{9-y})/2} 16x\, dx\, dy$

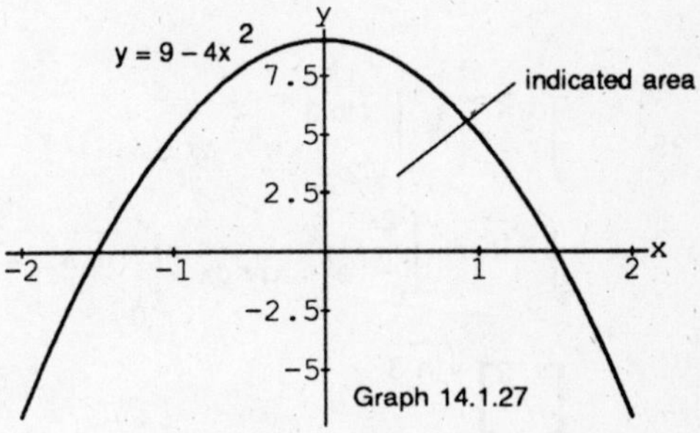

Graph 14.1.27

29. $\int_{-1}^1 \int_0^{\sqrt{1-x^2}} 3y\, dy\, dx$

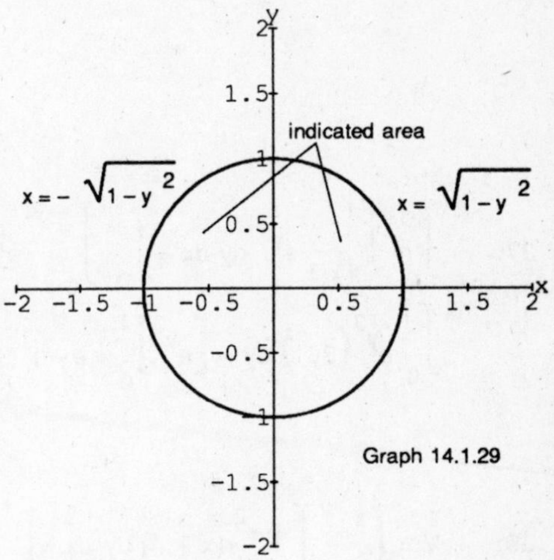

Graph 14.1.29

31. $\int_0^{\pi} \int_x^{\pi} \frac{\sin y}{y} \, dy \, dx = \int_0^{\pi} \int_0^{y} \frac{\sin y}{y} \, dx \, dy = \int_0^{\pi} \sin y \, dy = 2$

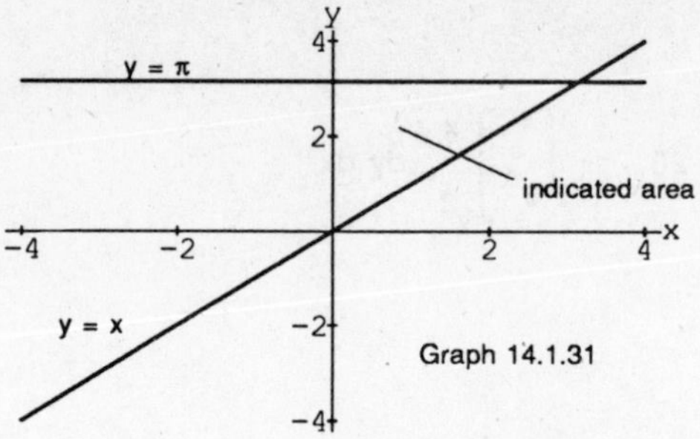

Graph 14.1.31

33. $\int_0^2 \int_x^2 2y^2 \sin xy \, dy \, dx = \int_0^2 \int_0^y 2y^2 \sin xy \, dx \, dy =$

$\int_0^2 \left[-2y \cos xy\right]_0^y dy = \int_0^2 \left(-\cos y^2(2y) + 2y\right) dy = 4 - \sin 4$

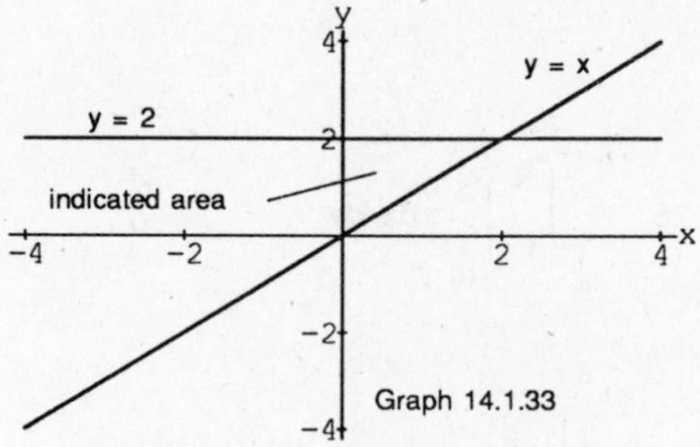

Graph 14.1.33

35. $\int_0^{2\sqrt{\ln 3}} \int_{y/2}^{\sqrt{\ln 3}} e^{x^2} \, dx \, dy =$

$\int_0^{\sqrt{\ln 3}} \int_0^{2x} e^{x^2} \, dy \, dx = \int_0^{\sqrt{\ln 3}} e^{x^2} \, 2x \, dx =$

$\left[e^{x^2}\right]_0^{\sqrt{\ln 3}} = 2$

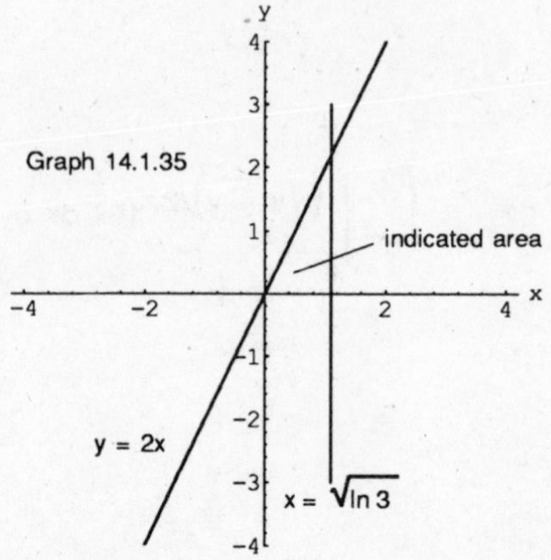

37. $\int_0^3 \int_{\sqrt{x/3}}^1 e^{y^3} \, dy \, dx = \int_0^1 \int_0^{3y^2} e^{y^3} \, dx \, dy =$

$\int_0^1 e^{y^3}\left(3y^2\right) dy = \left[e^{y^3}\right]_0^1 = e - 1$

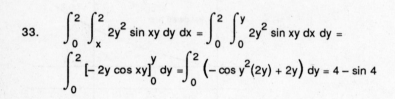

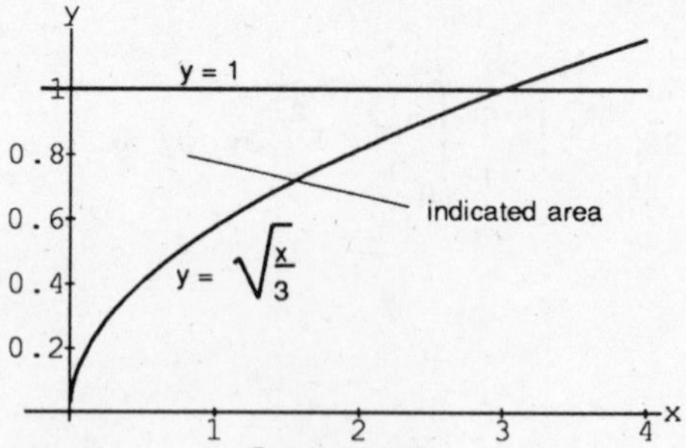

Graph 14.1.37

39. $V = \int_{-4}^{1} \int_{3x}^{4-x^2} (x + 4) \, dy \, dx = \int_{-4}^{1} \left[(xy + 4y)\right]_{3x}^{4-x^2} dx = \int_{-4}^{1} \left(-x^3 - 7x^2 - 8x + 16\right) dx = \frac{625}{12}$

41. $V = \int_0^2 \int_0^3 \left(4 - y^2\right) dx \, dy = \int_0^2 \left[4x - y^2 x\right]_0^3 dy = \int_0^2 \left(12 - 3y^2\right) dy = 16$

43. $V = \int_0^1 \int_x^{2-x} \left(x^2 + y^2\right) dy\, dx = \int_0^1 \left[x^2 y + \frac{y^3}{3}\right]_x^{2-x} dx = \int_0^1 2x^2 - \frac{7x^3}{3} + \frac{(2-x)^3}{3}\, dx =$

$\left[\frac{2x^3}{3} - \frac{7x^4}{12} - \frac{(2-x)^4}{12}\right]_0^1 = \frac{4}{3}$

45. To maximize the integral, we want the domain to include all points where the integrand is positive and to exclude all points where the integrand is negative. These criteria are met by the set of points (x, y) such that $4 - x^2 - 2y^2 \geq 0$ or $x^2 + 2y^2 \leq 4$. It consists of the ellipse $x^2 + 2y^2 = 4$ and its interior.

47. No, it is not all right. By Fubini's theorem, the two orders of integration must give the same result.

49. $\int_1^3 \int_1^x \frac{1}{xy}\, dy\, dx = 0.603$

51. $\int_0^1 \int_0^1 \tan^{-1} xy\, dy\, dx = 0.233$

14.2 AREAS, MOMENTS, AND CENTERS OF MASS

1. $\int_0^2 \int_0^{2-x} dy\, dx = \int_0^2 (2-x)\, dx = 2$ or

$\int_0^2 \int_0^{2-y} dx\, dy = \int_0^2 (2-y)\, dx = 2$

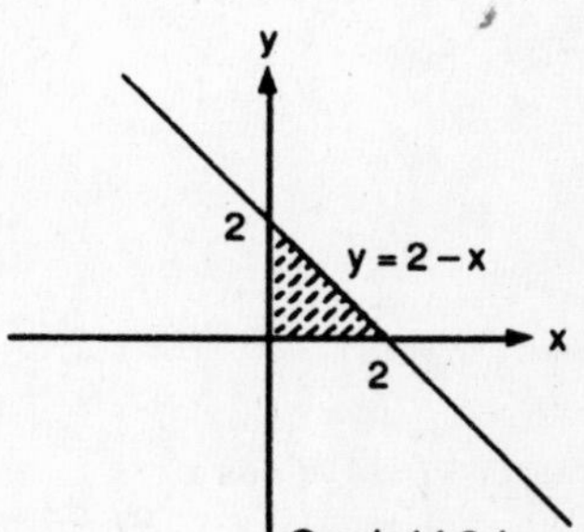

Graph 14.2.1

3. $\int_{-2}^{1} \int_{y-2}^{-y^2} dx\, dy = \int_{-2}^{1} (-y^2 - y + 2)\, dy = \frac{9}{2}$

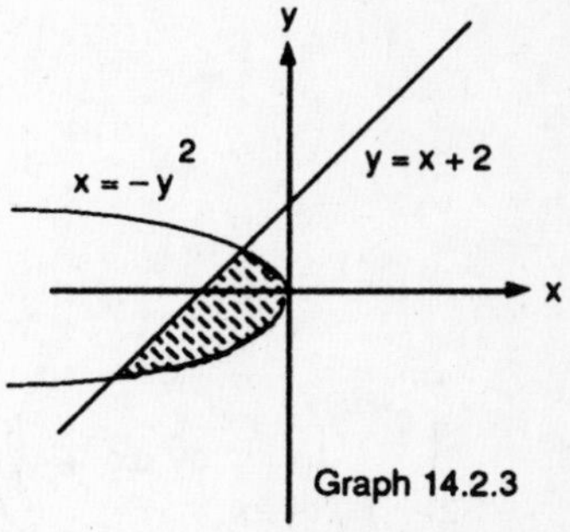

Graph 14.2.3

5. $\int_0^{\ln 2} \int_0^{e^x} dy\, dx = \int_0^{\ln 2} e^x\, dx = 1$

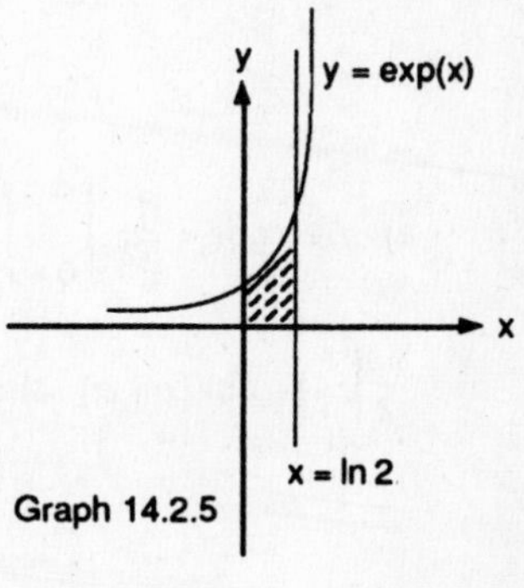

Graph 14.2.5

7. $\int_0^1 \int_{y^2}^{2y-y^2} dx\, dy = \int_0^1 \left(2y - 2y^2\right) dy = \frac{1}{3}$

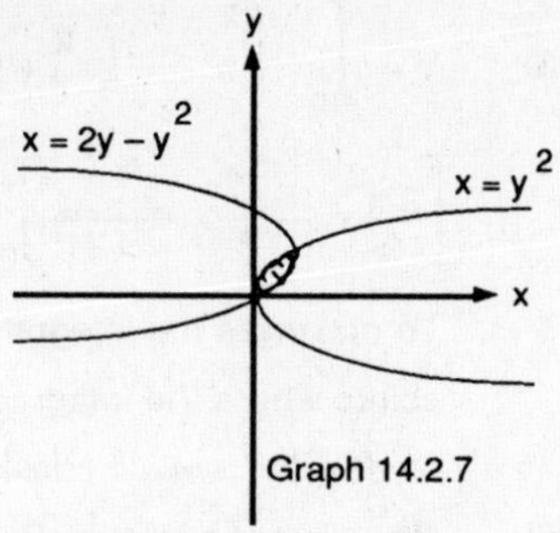

Graph 14.2.7

9. $\int_0^6 \int_{y^2/3}^{2y} dx\, dy = \int_0^6 \left(2y - y^2/3\right) dy = 12$

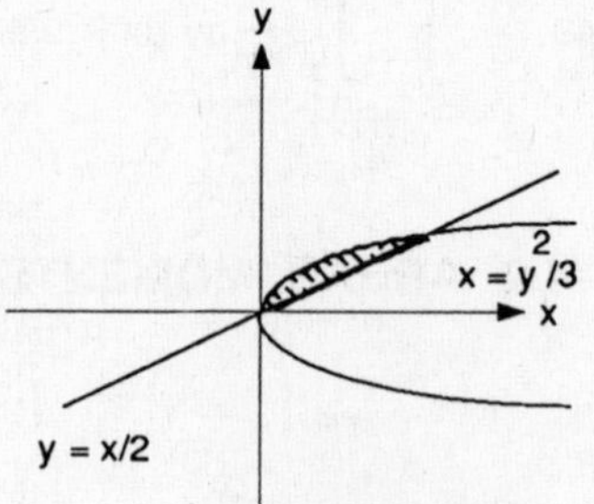

Graph 14.2.9

11. $\int_0^{\pi/4} \int_{\sin x}^{\cos x} dy\, dx = \int_0^{\pi/4} (\cos x - \sin x)\, dx = \sqrt{2} - 1$

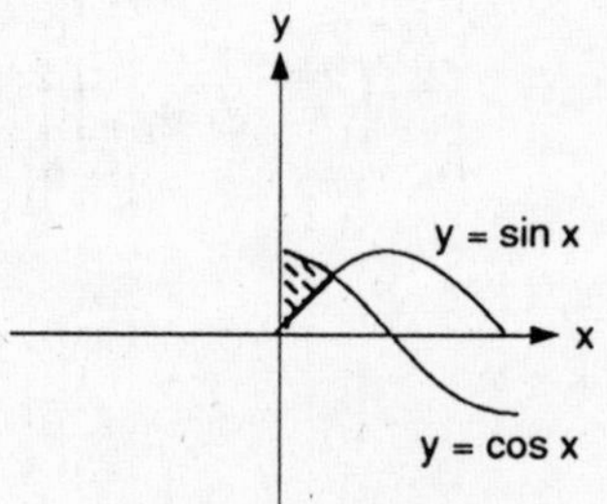

Graph 14.2.11

13. $\int_{-1}^{0} \int_{-2x}^{1-x} dy\, dx + \int_0^2 \int_{-x/2}^{1-x} dy\, dx = \int_{-1}^{0} (1 + x)\, dx + \int_0^2 (1 - x/2)\, dx = \frac{3}{2}$

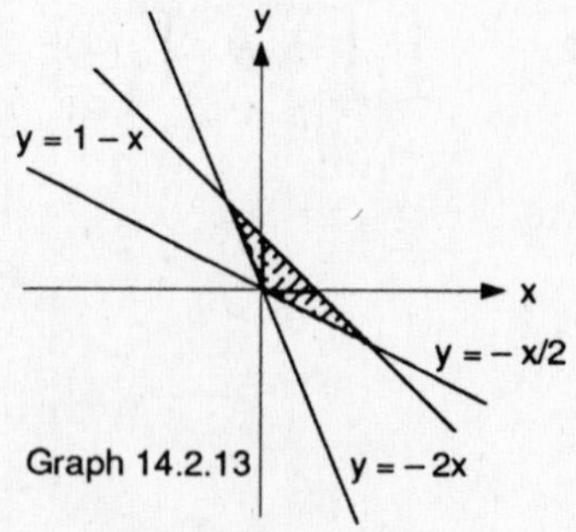

Graph 14.2.13

15. a) Average $= \frac{1}{\pi^2} \int_0^{\pi} \int_0^{\pi} \sin(x + y)\, dy\, dx = \frac{1}{\pi^2} \int_0^{\pi} \left[-\cos(x + y)\right]_0^{\pi} dx =$

$\frac{1}{\pi^2} \int_0^{\pi} (-\cos(x + \pi) + \cos x)\, dx = \frac{1}{\pi^2} \left[-\sin(x + \pi) + \sin x\right]_0^{\pi} = 0$

b) $\text{Average} = \frac{1}{\pi^2/2}\int_0^{\pi}\int_0^{\pi/2}\sin(x+y)\,dy\,dx = \frac{2}{\pi^2}\int_0^{\pi}\left[-\cos(x+y)\right]_0^{\pi/2}dx =$

$\frac{2}{\pi^2}\int_0^{\pi}\left(-\cos(x+\pi/2)+\cos x\right)dx = \frac{2}{\pi^2}\left[-\sin(x+\pi/2)+\sin x\right]_0^{\pi} = \frac{4}{\pi^2}$

17. $\text{Average height} = \frac{1}{4}\int_0^2\int_0^2\left(x^2+y^2\right)dy\,dx = \frac{1}{4}\int_0^2\left[x^2y+\frac{y^3}{3}\right]_0^2dx = \frac{1}{4}\int_0^2\left(2x^2+\frac{8}{3}\right)dx = \frac{1}{2}\left[\frac{x^3}{3}+\frac{4x}{3}\right]_0^2 = \frac{8}{3}$

19. $M = \int_0^1\int_x^{2-x^2}3\,dy\,dx = 3\int_0^1\left(2-x^2-x\right)dx = \frac{7}{2}$

$M_y = \int_0^1\int_x^{2-x^2}3x\,dy\,dx = 3\int_0^1\left[xy\right]_x^{2-x^2}dx = 3\int_0^1\left(2x-x^3-x^2\right)dx = \frac{5}{4}$

$M_x = \int_0^1\int_x^{2-x^2}3y\,dy\,dx = \frac{3}{2}\int_0^1\left[y^2\right]_x^{2-x^2}dx = \frac{3}{2}\int_0^1\left(4-5x^2+x^4\right)dx = \frac{19}{5} \Rightarrow \overline{x} = \frac{5}{14},\ \overline{y} = \frac{38}{35}$

21. $M = \int_0^2\int_{y^2/2}^{4-y}dx\,dy = \int_0^2\left(4-y-\frac{y^2}{2}\right)dy = \frac{14}{3}$

$M_y = \int_0^2\int_{y^2/2}^{4-y}x\,dx\,dy = \frac{1}{2}\int_0^2\left[x^2\right]_{y^2/2}^{4-y}dx = \frac{1}{2}\int_0^2\left(16-8y+y^2-\frac{y^4}{4}\right)dy = \frac{128}{15}$

$M_x = \int_0^2\int_{y^2/2}^{4-y}y\,dx\,dy = \int_0^2 y\left(4-y-\frac{y^2}{2}\right)dy = \frac{10}{3} \Rightarrow \overline{x} = \frac{64}{35},\ \overline{y} = \frac{5}{7}$

23. $M = 2\int_0^1\int_0^{\sqrt{1-x^2}}dy\,dx = 2\int_0^1\sqrt{1-x^2}\,dx = \frac{\pi}{2}$

$M_x = 2\int_0^1\int_0^{\sqrt{1-x^2}}y\,dy\,dx = \int_0^1\left[y^2\right]_0^{\sqrt{1-x^2}}dx = \int_0^1\left(1-x^2\right)dx = \frac{2}{3} \Rightarrow \overline{x} = 0$, by symmetry; $\overline{y} = \frac{4}{3\pi}$

25. $M = \int_0^a\int_0^{\sqrt{a^2-x^2}}dy\,dx = \frac{\pi a^2}{4}$; $M_y = \int_0^a\int_0^{\sqrt{a^2-x^2}}x\,dy\,dx =$

$\int_0^a\left[xy\right]_0^{\sqrt{a^2-x^2}}dx = -\frac{1}{2}\int_0^a\sqrt{a^2-y^2}(-2x)\,dx = \frac{a^3}{3} \Rightarrow \overline{x} = \overline{y} = \frac{4a}{3\pi}$, by symmetry

27. $M = \int_0^{\pi}\int_0^{\sin x}dy\,dx = \int_0^{\pi}\sin x\,dx = 2$; $M_x = \int_0^{\pi}\int_0^{\sin x}y\,dy\,dx = \frac{1}{2}\int_0^{\pi}\left[y^2\right]_0^{\sin x}dx =$

$\frac{1}{4}\int_0^{\pi}(1-\cos 2x)\,dx = \frac{\pi}{4} \Rightarrow \overline{x} = \frac{\pi}{2},\ \overline{y} = \frac{\pi}{8}$

29. $M = \int_{-\infty}^{0}\int_{0}^{\exp(x)} dy\,dx = \int_{-\infty}^{0} e^x\,dx = \lim_{t \to -\infty}\int_{t}^{0} e^x\,dx = 1$

$M_y = \int_{-\infty}^{0}\int_{0}^{\exp(x)} x\,dy\,dx = \int_{-\infty}^{0} x\,e^x\,dx = \lim_{t \to -\infty}\int_{t}^{0} x\,e^x\,dx = -1$

$M_x = \int_{-\infty}^{0}\int_{0}^{\exp(x)} y\,dy\,dx = \frac{1}{2}\int_{-\infty}^{0} e^{2x}\,dx = \frac{1}{2}\lim_{t \to -\infty}\int_{t}^{0} e^{2x}\,dx = \frac{1}{4} \Rightarrow \overline{x} = -1,\ \overline{y} = \frac{1}{4}$

31. $M = \int_{0}^{2}\int_{-y}^{y-y^2} (x + y)\,dx\,dy = \int_{0}^{2}\left[\frac{x^2}{2} + xy\right]_{-y}^{y-y^2} dy = \int_{0}^{2}\left(\frac{y^4}{2} - 2y^3 + 2y^2\right)dy = \frac{8}{15};$

$I_x = \int_{0}^{2}\int_{-y}^{y-y^2} y^2(x + y)\,dx\,dy = \int_{0}^{2}\left[\frac{x^2y^2}{2} + xy^3\right]_{-y}^{y-y^2} dy = \int_{0}^{2}\left(\frac{y^6}{2} - 2y^5 + 2y^4\right)dy = \frac{64}{105};\ R_x = \sqrt{\frac{I_x}{M}} = \sqrt{\frac{8}{7}}$

33. $M = \int_{0}^{1}\int_{x}^{2-x} (6x + 3y + 3)\,dy\,dx = \int_{0}^{1}\left(-12x^2 + 12\right)dx = 8;\ M_y = \int_{0}^{1}\int_{x}^{2-x} x(6x + 3y + 3)\,dy\,dx =$

$\int_{0}^{1}\left(12x - 12x^3\right)dx = 3;\ M_x = \int_{0}^{1}\int_{x}^{2-x} y(6x + 3y + 3)\,dy\,dx = \int_{0}^{1}\left(14 - 6x - 6x^2 - 2x^3\right)dx = \frac{17}{2} \Rightarrow$

$\overline{x} = \frac{3}{8},\ \overline{y} = \frac{17}{16}$

35. $M = \int_{0}^{1}\int_{0}^{6} (x + y + 1)\,dx\,dy = \int_{0}^{1} (6y + 24)\,dy = 27,\ M_x = \int_{0}^{1}\int_{0}^{6} y(x + y + 1)\,dx\,dy = \int_{0}^{1} y(6y + 24)\,dy = 14$

$M_y = \int_{0}^{1}\int_{0}^{6} x(x + y + 1)\,dx\,dy = \int_{0}^{1} (18y + 90)\,dy = 99 \Rightarrow \overline{x} = \frac{11}{3},\ \overline{y} = \frac{14}{27}$

$I_y = \int_{0}^{1}\int_{0}^{6} x^2(x + y + 1)\,dx\,dy = 216\int_{0}^{1}\left(\frac{y}{3} + \frac{11}{6}\right)dy = 432 \Rightarrow R_y = \sqrt{\frac{I_y}{M}} = 4$

37. $M = \int_{-1}^{1}\int_{0}^{x^2} (7y + 1)\,dy\,dx = \int_{-1}^{1}\left(\frac{7x^4}{2} + x^2\right)dx = \frac{31}{15}$

$M_x = \int_{-1}^{1}\int_{0}^{x^2} y(7y + 1)\,dy\,dx = \int_{-1}^{1}\left(\frac{7x^6}{3} + \frac{x^4}{2}\right)dx = \frac{13}{15},\ M_y = \int_{-1}^{1}\int_{0}^{x^2} x(7y + 1)\,dy\,dx = \int_{-1}^{1}\left(\frac{7x^5}{2} + x^3\right)dx = 0$

$\therefore \overline{x} = 0,\ \overline{y} = \frac{13}{31},\ I_y = \int_{-1}^{1}\int_{0}^{x^2} x^2(7y + 1)\,dy\,dx = \int_{-1}^{1} x^2\left(\frac{7x^4}{2} + x^2\right)dx = \frac{7}{5} \Rightarrow R_y = \sqrt{\frac{I_y}{M}} = \sqrt{\frac{21}{31}}$

39. $M = \int_{0}^{1}\int_{-y}^{y} (1 + y)\,dx\,dy = \int_{0}^{1}\left(2y^2 + 2y\right)dy = \frac{5}{3},\ M_x = \int_{0}^{1}\int_{-y}^{y} y(1 + y)\,dx\,dy = 2\int_{0}^{1}\left(y^3 + y^2\right)dy = \frac{7}{6}$

$M_y = \int_{0}^{1}\int_{-y}^{y} x(1 + y)\,dx\,dy = \int_{0}^{1} 0\,dy = 0 \Rightarrow \overline{x} = 0,\ \overline{y} = \frac{7}{10}$

$I_x = \int_{0}^{1}\int_{-y}^{y} y^2(1 + y)\,dx\,dy = \int_{0}^{1} y^2\left(2y^2 + 2y\right)dy = \frac{9}{10} \Rightarrow R_x = \sqrt{\frac{I_x}{M}} = \frac{3\sqrt{6}}{10}$

$I_y = \int_{0}^{1}\int_{-y}^{y} x^2(1 + y)\,dx\,dy = \frac{1}{3}\int_{0}^{1} y^2\left(2y^2 + 2y\right)dy = \frac{3}{10} \Rightarrow R_y = \sqrt{\frac{I_y}{M}} = \frac{3\sqrt{2}}{10}$

$I_o = I_x + I_y = \frac{6}{5}$ and $R_o = \sqrt{\frac{I_o}{M}} = \frac{3\sqrt{2}}{5}$

41. $M = \int_{-1}^{1}\int_{0}^{a(1-x^2)} dy\,dx = 2a\int_{0}^{1}\left(1-x^2\right)dx = 2a\left[x-\frac{x^3}{3}\right]_0^1 = \frac{4a}{3}$, $M_x = \int_{-1}^{1}\int_{0}^{a(1-x^2)} y\,dy\,dx =$

$\frac{2a^2}{2}\int_0^1\left(1-2x^2+x^4\right)dx = a^2\left[x-\frac{2x^3}{3}+\frac{x^5}{5}\right]_0^1 = \frac{8a^2}{15}$ $\therefore$ $\overline{y} = \frac{M_x}{M} = \frac{8a^2/15}{4a/3} = \frac{2a}{5}$; The angle formed by the c. m., the fulcrum and origin plus 45° must remain less than 90°. i. e. $\tan^{-1}\left(\frac{2a}{5}\right) + \frac{\pi}{4} < \frac{\pi}{2} \Rightarrow 0 < a < \frac{5}{2}$.

14.3 DOUBLE INTEGRALS IN POLAR FORM

1. $\int_{-1}^{1}\int_{0}^{\sqrt{1-x^2}} dy\,dx = \int_0^{\pi}\int_0^1 r\,dr\,d\theta = \frac{1}{2}\int_0^{\pi} d\theta = \frac{\pi}{2}$

3. $\int_0^1\int_0^{\sqrt{1-y^2}}\left(x^2+y^2\right)dx\,dy = \int_0^{\pi/2}\int_0^1 r^3\,dr\,d\theta = \frac{1}{4}\int_0^{\pi/2} d\theta = \frac{\pi}{8}$

5. $\int_{-a}^{a}\int_{-\sqrt{a^2-x^2}}^{\sqrt{a^2-x^2}} dy\,dx = \int_0^{2\pi}\int_0^a r\,dr\,d\theta = \frac{a^2}{2}\int_0^{2\pi} d\theta = \pi a^2$

7. $\int_0^6\int_0^y x\,dx\,dy = \int_{\pi/4}^{\pi/2}\int_0^{6\csc\theta} r^2\cos\theta\,dr\,d\theta = 72\int_{\pi/4}^{\pi/2}\cot\theta\csc^2\theta d\theta = -36\left[\cot^2\theta\right]_{\pi/4}^{\pi/2} = 36$

9. $\int_{-1}^{0}\int_{-\sqrt{1-x^2}}^{0}\frac{2}{1+\sqrt{x^2+y^2}}dy\,dx = \int_{\pi}^{3\pi/2}\int_0^1\frac{2r}{1+r}dr\,d\theta = 2\int_{\pi}^{3\pi/2}(1-\ln 2)\,d\theta = (1-\ln 2)\pi$

11. $\int_0^{\ln 2}\int_0^{\sqrt{\ln^2 2-y^2}} e^{\sqrt{x^2+y^2}}dx\,dy = \int_0^{\pi/2}\int_0^{\ln 2} e^r r\,dr\,d\theta = \int_0^{\pi/2}(\ln(4)-1)\,d\theta = \frac{\pi(\ln(4)-1)}{2}$

13. $\int_0^2\int_0^{\sqrt{1-(x-1)^2}} 3xy\,dy\,dx = \int_0^{\pi/2}\int_0^{2\cos\theta} 3(r\cos\theta)(r\sin\theta)r\,dr\,d\theta = 12\int_0^{\pi/2}\cos^5\theta\sin\theta\,d\theta =$

$-2\left[\frac{\cos^6\theta}{6}\right]_0^{\pi/2} = 2$

15. $\int_{-1}^{1}\int_{-\sqrt{1-y^2}}^{\sqrt{1-y^2}}\ln(x^2+y^2+1)\,dx\,dy = 4\int_0^{\pi/2}\int_0^1\ln(r^2+1)\,r\,dr\,d\theta = 2\int_0^{\pi/2}(\ln(4)-1)\,d\theta = \pi(\ln(4)-1)$

17. $\int_0^{\pi/2}\int_0^{2\sqrt{2-\sin 2\theta}} r\,dr\,d\theta = 2\int_0^{\pi/2}\left(2-\sin 2\theta\right)d\theta = 2(\pi-1)$

19. $A = 2\int_0^{\pi/6}\int_0^{12\cos 3\theta} r\,dr\,d\theta = 144\int_0^{\pi/6}\cos^2 3\theta\,d\theta = 12\pi$

21. $A = \int_0^{\pi/2}\int_0^{1+\sin\theta} r\,dr\,d\theta = \frac{1}{2}\int_0^{\pi/2}\left(\frac{3}{2}+2\sin\theta-\frac{\cos 2\theta}{2}\right)d\theta = \frac{3\pi+8}{8}$

23. $\int_0^{2\pi}\int_0^{\sqrt{3}/2}\frac{1}{1-r^2}r\,dr\,d\theta = \ln(2)\int_0^{2\pi} d\theta = \pi\ln 4$

25. $M_x = \int_0^{\pi}\int_0^{1-\cos\theta} 3r^2 \sin\theta \; dr\, d\theta = \int_0^{\pi} \left(1-\cos\theta\right)^3 \sin\theta\, d\theta = 4$

27. $M = 2\int_{\pi/6}^{\pi/2}\int_3^{6\sin\theta} dr\,d\theta = 2\int_{\pi/6}^{\pi/2}(6\sin\theta - 3)\,d\theta = 6[-2\cos\theta - \theta]_{\pi/6}^{\pi/2} = 6\sqrt{3} - 2\pi$

29. $M = 2\int_0^{\pi}\int_0^{1+\cos\theta} r\,dr\,d\theta = \int_0^{\pi} \left(1+\cos\theta\right)^2 d\theta = \frac{3\pi}{2}$, $M_y = \int_0^{2\pi}\int_0^{1+\cos\theta}\cos\theta\, r^2\, dr\,d\theta =$

$\int_0^{2\pi}\left(\frac{4\cos\theta}{3}+\frac{15}{24}+\cos 2\theta - \sin^2\theta\cos\theta + \frac{\cos 4\theta}{4}\right)d\theta = \frac{5\pi}{4} \Rightarrow \overline{x} = \frac{5}{6}$, $\overline{y} = 0$ by symmetry

31. Average $= \frac{4}{\pi a^2}\int_0^{\pi/2}\int_0^a \sqrt{a^2-r^2}\, r\,dr\,d\theta = -\frac{4}{3\pi a^2}\int_0^{\pi/2} -a^3\, d\theta = \frac{2a}{3}$

33. Average $= \frac{1}{\pi a^2}\int_{-a}^{a}\int_{-\sqrt{a^2-x^2}}^{\sqrt{a^2-x^2}}\sqrt{x^2+y^2}\,dy\,dx = \frac{1}{\pi a^2}\int_0^{2\pi}\int_0^a r^2\,dr\,d\theta = \frac{a}{3\pi}\int_0^{2\pi} d\theta = \frac{2a}{3}$

14.4 TRIPLE INTEGRALS IN RECTANGULAR COORDINATES

1. $\int_0^1\int_0^{1-z}\int_0^2 dx\,dy\,dz = 2\int_0^1\int_0^{1-z} dy\,dz = 2\int_0^1 1 - z\,dz = 1$

3. $\int_0^1\int_0^{2-2x}\int_0^{3-3x-3y/2} dz\,dy\,dx = \int_0^1\int_0^{2-2x}\left(3-3x-\frac{3}{2}y\right)dy\,dx = \int_0^1\left(3-6x+3x^2\right)dx = 1,$

$\int_0^2\int_0^{1-y/2}\int_0^{3-3x-3y/2} dz\,dx\,dy,\ \int_0^1\int_0^{3-3x}\int_0^{2-2x-2z/3} dy\,dz\,dx,$

$\int_0^3\int_0^{1-z/3}\int_0^{2-2x-2z/3} dy\,dx\,dz,\ \int_0^2\int_0^{3-3y/2}\int_0^{1-y/2-z/3} dx\,dz\,dy,\ \int_0^3\int_0^{2-2z/3}\int_0^{1-y/2-z/3} dx\,dy\,dz$

5. $\int_0^1\int_0^1\int_0^1\left(x^2+y^2+z^2\right)dz\,dy\,dx = \int_0^1\int_0^1\left(x^2+y^2+\frac{1}{3}\right)dy\,dx = \int_0^1\left(x^2+\frac{2}{3}\right)dx = 1$

7. $\int_1^e\int_1^e\int_1^e \frac{1}{xyz}\,dx\,dy\,dz = \int_1^e\int_1^e\left[\frac{\ln x}{yz}\right]_1^e dy\,dz = \int_1^e \frac{1}{z}\,dz = 1$

9. $\int_0^1\int_0^{\pi}\int_0^{\pi} y\sin z\,dx\,dy\,dz = \int_0^1\int_0^{\pi}\pi y\sin z\,dy\,dz = \frac{\pi^3}{2}\int_0^1 \sin z\,dz = \frac{\pi^3}{2}(1-\cos 1)$

11. $\int_0^3\int_0^{\sqrt{9-x^2}}\int_0^{\sqrt{9-x^2}} dz\,dy\,dx = \int_0^3\int_0^{\sqrt{9-x^2}}\sqrt{9-x^2}\,dy\,dx = \int_0^3\left(9-x^2\right)dx = 18$

13. $\int_0^1\int_0^{2-x}\int_0^{2-x-y} dz\,dy\,dx = \int_0^1\int_0^{2-x}(2-x-y)\,dy\,dx = \int_0^1\left(\frac{x^2}{2}-2x+2\right)dx = \frac{7}{6}$

15. $\int_0^{\pi}\int_0^{\pi}\int_0^{\pi}\cos(u+v+w)\,du\,dv\,dw = \int_0^{\pi}\int_0^{\pi}\left[\sin(w+v+\pi)-\sin(w+v)\right]dv\,dw =$

$\int_0^{\pi}\left(-\cos(w+2\pi)+\cos(w+\pi)+\cos(w)-\cos(w+\pi)\right)dw = 0$

17. $\int_0^{\pi/4}\int_0^{\ln\sec v}\int_{-\infty}^{2t} e^x\,dx\,dt\,dv = \int_0^{\pi/4}\int_0^{\ln\sec v} e^{2t}\,dt\,dv = \int_0^{\pi/4}\frac{\sec^2 v}{2}\,dv = \left[\frac{\tan v}{2}\right]_0^{\pi/4} = \frac{1}{2} - \frac{\pi}{8} = \frac{4-\pi}{8}$

19. a) $\int_{-1}^{1}\int_0^{1-x^2}\int_{x^2}^{1-z} dy\,dz\,dx$ b) $\int_0^1\int_{-\sqrt{1-z}}^{\sqrt{1-z}}\int_{x^2}^{1-z} dy\,dx\,dz$

c) $\int_0^1\int_0^{1-z}\int_{-\sqrt{y}}^{\sqrt{y}} dx\,dy\,dz$ d) $\int_0^1\int_0^{1-y}\int_{-\sqrt{y}}^{\sqrt{y}} dx\,dz\,dy$

e) $\int_0^1\int_{-\sqrt{y}}^{\sqrt{y}}\int_0^{1-y} dz\,dx\,dy$

21. $V = \int_0^1\int_{-1}^1\int_0^{y^2} dz\,dy\,dx = \int_0^1\int_{-1}^1 y^2\,dy\,dx = \frac{2}{3}\int_0^1 dx = \frac{2}{3}$

23. $V = \int_0^4\int_0^{\sqrt{4-x}}\int_0^{2-y} dz\,dy\,dx = \int_0^4\int_0^{\sqrt{4-x}}(2-y)\,dy\,dx = \int_0^4\left[2\sqrt{4-x} - \left(\frac{4-x}{2}\right)\right]dx = \frac{20}{3}$

25. $V = \int_0^1\int_0^{2-2x}\int_0^{3-3x-3y/2} dz\,dy\,dx = \int_0^1\int_0^{2-2x}\left(3 - 3x - \frac{3}{2}y\right)dy\,dx = \int_0^1\left(3 - 6x + 3x^2\right)dx = 1$

27. $V = 8\int_0^1\int_0^{\sqrt{1-x^2}}\int_0^{\sqrt{1-x^2}} dz\,dy\,dx = 8\int_0^1\int_0^{\sqrt{1-x^2}}\sqrt{1-x^2}\,dy\,dx = 8\int_0^1\left(1 - x^2\right)dx = \frac{16}{3}$

29. $V = \int_0^4\int_0^{\left(\sqrt{16-y^2}\right)/2}\int_0^{4-y} dx\,dz\,dy = \int_0^4\int_0^{\left(\sqrt{16-y^2}\right)/2}(4-y)\,dz\,dy = \int_0^4 \frac{\sqrt{16-y^2}}{2}(4-y)\,dy = 8\pi - \frac{32}{3}$

31. $\text{average} = \frac{1}{8}\int_0^2\int_0^2\int_0^2\left(x^2+9\right)dz\,dy\,dx = \frac{1}{8}\int_0^2\int_0^2\left(2x^2+18\right)dy\,dx = \frac{1}{8}\int_0^2\left(4x^2+36\right)dx = \frac{31}{3}$

33. $\text{average} = \int_0^1\int_0^1\int_0^1\left(x^2+y^2+z^2\right)dz\,dy\,dx = \int_0^1\int_0^1\left(x^2+y^2+\frac{1}{3}\right)dy\,dx = \int_0^1\left(x^2+\frac{2}{3}\right)dx = 1$

35. To minimize the integral, we want the domain to include all points where the integrand is negative and to exclude all points where it is positive. These criteria are met by the set of points (x,y,z) such that $4x^2 + 4y^2 + z^2 - 4 \le 0$ or $4x^2 + 4y^2 + z^2 \le 4$. It is the solid ellipsoid $4x^2 + 4y^2 + z^2 \le 4$.

14.5 MASSES AND MOMENTS IN THREE DIMENSIONS

1. $I_x = \int_{-c/2}^{c/2}\int_{-b/2}^{b/2}\int_{-a/2}^{a/2}\left(y^2+z^2\right)dx\,dy\,dz = 4a\int_0^{c/2}\int_0^{b/2}\left(y^2+z^2\right)dy\,dz = 4a\int_0^{c/2}\left(\frac{b^3}{24}+\frac{z^2 b}{2}\right)dz =$

$\frac{abc}{12}\left(b^2+c^2\right) \Rightarrow I_x = \frac{M}{12}\left(b^2+c^2\right);\ R_x = \sqrt{\frac{b^2+c^2}{12}},\ R_y = \sqrt{\frac{a^2+c^2}{12}},\ R_z = \sqrt{\frac{a^2+b^2}{12}}$

3. $I_x = \int_0^a\int_0^b\int_0^c\left(y^2+z^2\right)dz\,dy\,dx = \int_0^a\int_0^b cy^2 + \frac{c^3}{3}\,dy\,dx = \int_0^a \frac{cb^3}{3} + \frac{c^3 b}{3}\,dx = \frac{abc\left(b^2+c^2\right)}{3} =$

$\frac{M}{3}\left(b^2+c^2\right);\ I_y = \frac{M}{3}\left(a^2+c^2\right)$ and $I_z = \frac{M}{3}\left(a^2+b^2\right)$ by symmetry, where $M = abc$

5. $M = 4\int_0^1 \int_0^1 \int_{4y^2}^4 dz\,dy\,dx = 4\int_0^1 \int_0^1 \left(4-4y^2\right) dy\,dx = 16\int_0^1 \frac{2}{3}\,dx = \frac{32}{3}$

$M_{xy} = 4\int_0^1 \int_0^1 \int_{4y^2}^4 z\,dz\,dy\,dx = 2\int_0^1 \int_0^1 \left(16-16y^4\right) dy\,dx = \frac{128}{5}\int_0^1 dx = \frac{128}{5} \Rightarrow \overline{z} = \frac{12}{5}$, $\overline{x} = \overline{y} = 0$ by symmetry; $I_z = 4\int_0^1 \int_0^1 \int_{4y^2}^4 \left(x^2+y^2\right) dz\,dy\,dx = 16\int_0^1 \int_0^1 \left(x^2 - x^2y^2 + y^2 - y^4\right) dy\,dx = 16\int_0^1 \left(\frac{2x^2}{3} + \frac{2}{15}\right) dx =$ $\frac{256}{45}$; $I_x = 4\int_0^1 \int_0^1 \int_{4y^2}^4 \left(y^2+z^2\right) dz\,dy\,dx = 4\int_0^1 \int_0^1 \left[\left(4y^2 + \frac{64}{3}\right) - \left(4y^4 + \frac{64y^6}{3}\right)\right] dy\,dx = 4\int_0^1 \frac{1976}{105}\,dx =$ $\frac{7904}{105}$; $I_y = 4\int_0^1 \int_0^1 \int_{4y^2}^4 \left(x^2+z^2\right) dz\,dy\,dx = 4\int_0^1 \int_0^1 \left[\left(4x^2 + \frac{64}{3}\right) - \left(4x^2y^2 + \frac{64y^6}{3}\right)\right] dy\,dx =$ $4\int_0^1 \left(\frac{8}{3}x^2 + \frac{128}{7}\right) dx = \frac{4832}{63}$

7. a) $M = 4\int_0^2 \int_0^{\sqrt{4-x^2}} \int_{x^2+y^2}^4 dz\,dy\,dx = 4\int_0^{\pi/2} \int_0^2 \int_{r^2}^4 r\,dz\,dr\,d\theta =$ $4\int_0^{\pi/2} \int_0^2 \left(4r - r^3\right) dr\,d\theta = 4\int_0^{\pi/2} 4\,d\theta = 8\pi$; $M_{xy} = \int_0^{2\pi} \int_0^2 \int_{r^2}^4 z\,r\,dz\,dr\,d\theta =$ $\int_0^{2\pi} \int_0^2 \frac{r}{2}\left(16 - r^4\right) dr\,d\theta = \frac{32}{3}\int_0^{2\pi} d\theta = \frac{64\pi}{3} \Rightarrow \overline{z} = \frac{8}{3}$, $\overline{x} = \overline{y} = 0$ by symmetry

b) $M = 8\pi$; $4\pi = \int_0^{2\pi} \int_0^{\sqrt{c}} \int_{r^2}^c r\,dz\,dr\,d\theta = \int_0^{2\pi} \int_0^{\sqrt{c}} \left(cr - r^3\right) dr\,d\theta = \int_0^{2\pi} \left(\frac{c^2}{4}\right) d\theta = \frac{c^2\pi}{2} \Rightarrow$ $c^2 = 8 \Rightarrow c = 2\sqrt{2}$, since $c > 0$

9. $I_L = \int_{-2}^2 \int_{-2}^4 \int_{-1}^{1-y/2} \left((y-6)^2 + z^2\right) dz\,dy\,dx = \int_{-2}^2 \int_{-2}^4 \left(\frac{(y-6)^2(4-y)}{2} + \frac{(2-y)^3}{24} + \frac{1}{3}\right) dy\,dx =$ $4\int_{-2}^4 \left[\frac{13t^3}{24} + 5t^2 + 16t + \frac{49}{3}\right] dt = 1386$, where $t = 2 - y$; $M = 36$, $R_L = \sqrt{\frac{I_L}{M}} = \sqrt{\frac{77}{2}}$

11. $M = 8$, $I_L = \int_0^4 \int_0^2 \int_0^1 \left(z^2 + (y-2)^2\right) dz\,dy\,dx = \int_0^4 \int_0^2 \left(y^2 - 4y + \frac{13}{3}\right) dy\,dx = \frac{10}{3}\int_0^4 dx = \frac{40}{3} \Rightarrow$ $R_L = \sqrt{\frac{I_L}{M}} = \sqrt{\frac{5}{3}}$

13. $M = \int_0^2 \int_0^{2-x} \int_0^{2-x-y} 2x\,dz\,dy\,dx = \int_0^2 \int_0^{2-x} \left(4x - 2x^2 - 2xy\right) dy\,dx = \int_0^2 \left(x^3 - 4x^2 + 4x\right) dx = \frac{4}{3}$

$M_{xy} = \int_0^2 \int_0^{2-x} \int_0^{2-x-y} 2xz\,dz\,dy\,dx = \int_0^2 \int_0^{2-x} x(2-x-y)^2\,dy\,dx = \int_0^2 \frac{x(2-x)^3}{3}\,dx = \frac{8}{15}$;

$M_{xz} = \frac{8}{15}$ by symmetry; $M_{yz} = \int_0^2 \int_0^{2-x} \int_0^{2-x-y} 2x^2\,dz\,dy\,dx = \int_0^2 \int_0^{2-x} 2x^2(2-x-y)\,dy\,dx =$ $\int_0^2 \left(2x - x^2\right)^2 dx = \frac{16}{15} \Rightarrow \overline{x} = \frac{4}{5}$, $\overline{y} = \overline{z} = \frac{2}{5}$

15. $M = \int_0^1\int_0^1\int_0^1 (x+y+z+1)\,dz\,dy\,dx = \int_0^1\int_0^1 \left(x+y+\frac{3}{2}\right) dy\,dx = \int_0^1 (x+2)\,dx = \frac{5}{2}$

$M_{xy} = \int_0^1\int_0^1\int_0^1 (x+y+z+1)z\,dz\,dy\,dx = \frac{1}{2}\int_0^1\int_0^1 \left(x+y+\frac{5}{3}\right) dy\,dx =$

$\frac{1}{2}\int_0^1 \left(x+\frac{13}{6}\right) dx = \frac{4}{3} \Rightarrow M_{xy} = M_{yz} = M_{xz} = \frac{4}{3}$ by symmetry $\therefore\ \overline{x} = \overline{y} = \overline{z} = \frac{8}{15}$

$I_z = \int_0^1\int_0^1\int_0^1 (x+y+z+1)\left(x^2+y^2\right) dz\,dy\,dx = \int_0^1\int_0^1 \left(x+y+\frac{3}{2}\right)\left(x^2+y^2\right) dy\,dx =$

$\int_0^1 \left(x^3+2x^2+\frac{1}{3}x+\frac{3}{4}\right) dx = \frac{11}{6} \Rightarrow I_x = I_y = I_z = \frac{11}{6}$ by symmetry $\Rightarrow R_x = R_y = R_z = \sqrt{\frac{I_z}{M}} = \sqrt{\frac{11}{15}}$

14.6 TRIPLE INTEGRALS IN CYLINDRICAL AND SPHERICAL COORDINATES

1. $\int_0^{2\pi}\int_0^1\int_r^{\sqrt{2-r^2}} r\,dz\,dr\,d\theta = \int_0^{2\pi}\int_0^1 \left(\left(2-r^2\right)^{1/2} r - r^2\right) dr\,d\theta = \int_0^{2\pi}\left(\frac{2^{3/2}}{3}-\frac{2}{3}\right)d\theta = \frac{4\pi\left(\sqrt{2}-1\right)}{3}$

3. $\int_0^{2\pi}\int_0^{\theta/2\pi}\int_0^{3+24r^2} r\,dz\,dr\,d\theta = \int_0^{2\pi}\int_0^{\theta/2\pi}\left(3+24r^2\right) r\,dr\,d\theta = \frac{3}{2}\int_0^{2\pi}\left(\frac{\theta^2}{4\pi^2}+\frac{4\theta^4}{16\pi^4}\right)d\theta = \frac{17\pi}{5}$

5. $\int_0^{2\pi}\int_0^1\int_r^{\left(2-r^2\right)^{-1/2}} 3\,r\,dz\,dr\,d\theta = 3\int_0^{2\pi}\int_0^1 \left(\left(2-r^2\right)^{-1/2} r - r^2\right) dr\,d\theta = 3\int_0^{2\pi}\left(\sqrt{2}-\frac{4}{3}\right) d\theta = \pi\left(6\sqrt{2}-8\right)$

7. $\int_0^{\pi}\int_0^{\pi}\int_0^{2\sin\phi} \rho^2 \sin\phi\,d\rho\,d\phi\,d\theta = \frac{8}{3}\int_0^{\pi}\int_0^{\pi} \sin^4\phi\,d\phi\,d\theta = \frac{2}{3}\int_0^{\pi}\frac{3\pi}{2}\,d\theta = \pi^2$

9. $\int_0^{2\pi}\int_0^{\pi}\int_0^{(1-\cos\phi)/2} \rho^2 \sin\phi\,d\rho\,d\phi\,d\theta = \frac{1}{24}\int_0^{2\pi}\int_0^{\pi}\left(1-\cos\phi\right)^3 \sin\phi\,d\phi\,d\theta = \frac{1}{6}\int_0^{2\pi} d\theta = \frac{\pi}{3}$

11. $\int_0^{2\pi}\int_0^{\pi/3}\int_{\sec\phi}^{2} 3\rho^2 \sin\phi\,d\rho\,d\phi\,d\theta = \int_0^{2\pi}\int_0^{\pi/3}\left(8-\sec^3\phi\right)\sin\phi\,d\phi\,d\theta = \frac{5}{2}\int_0^{2\pi} d\theta = 5\pi$

13. $\int_0^{2\pi}\int_0^{3}\int_0^{z/3} r^3\,dr\,dz\,d\theta = \int_0^{2\pi}\int_0^{3}\frac{z^4}{324}\,dz\,d\theta = \int_0^{2\pi}\frac{3}{20}\,d\theta = \frac{3\pi}{10}$

15. $\int_0^1\int_0^{\sqrt{z}}\int_0^{2\pi} \left(r^2\cos^2\theta + z^2\right) r\,d\theta\,dr\,dz = \int_0^1\int_0^{\sqrt{z}}\left(\pi r^3 + 2\pi r z^2\right) dr\,dz = \int_0^1\left(\frac{\pi z^2}{4}+\pi z^3\right)dz = \left[\frac{\pi z^3}{12}+\frac{\pi z^4}{4}\right]_0^1 = \frac{\pi}{3}$

17. $\int_0^2\int_{-\pi}^{0}\int_{\pi/4}^{\pi/2} \rho^3 \sin 2\phi\,d\phi\,d\theta\,d\rho = \int_0^2\int_{-\pi}^{0}\frac{\rho^3}{2}\,d\theta\,d\rho = \int_0^2 \frac{\rho^3\pi}{2}\,d\rho = \left[\frac{\pi\rho^4}{8}\right]_0^2 = 2\pi$

19. $\int_0^1\int_0^{\pi}\int_0^{\pi/4} 12\rho\sin^3\phi\,d\phi\,d\theta\,d\rho = \int_0^1\int_0^{\pi}\left(8\rho-\frac{10\rho}{\sqrt{2}}\right) d\theta\,d\rho = \pi\int_0^1\left(8\rho-\frac{10\rho}{\sqrt{2}}\right) d\rho =$

$\pi\left[4\rho^2-\frac{5\rho^2}{\sqrt{2}}\right]_0^1 = \frac{(4\sqrt{2}-5)\pi}{\sqrt{2}}$

21. a) $8\int_0^{\pi/2}\int_0^{\pi/2}\int_0^2 \rho^2 \sin\phi \, d\rho \, d\phi \, d\theta$ b) $8\int_0^{\pi/2}\int_0^2\int_0^{\sqrt{4-r^2}} r \, dz \, dr \, d\theta$

c) $8\int_0^2\int_0^{\sqrt{4-x^2}}\int_0^{\sqrt{4-x^2-y^2}} dz \, dy \, dx$

23. a) $V = \int_0^{2\pi}\int_0^{\pi/3}\int_{\sec\phi}^2 \rho^2 \sin\phi \, d\rho \, d\phi \, d\theta$ b) $V = \int_0^{2\pi}\int_0^{\sqrt{3}}\int_1^{\sqrt{4-r^2}} r \, dz \, dr \, d\theta$

c) $V = \int_{-\sqrt{3}}^{\sqrt{3}}\int_{-\sqrt{3-x^2}}^{\sqrt{3-x^2}}\int_1^{\sqrt{4-x^2-y^2}} dz \, dy \, dx$

d) $V = \int_0^{2\pi}\int_0^{\sqrt{3}}\left((4-r^2)^{1/2} r - r\right) dr \, d\theta = -\int_0^{2\pi}\left[\frac{(4-r^2)^{3/2}}{3} + \frac{r^2}{2}\right]_0^{\sqrt{3}} d\theta = \frac{5}{6}\int_0^{2\pi} d\theta = \frac{5\pi}{3}$

25. $\int_0^{\pi}\int_0^{2\sin\theta}\int_0^{4 - r\sin\theta} f(r,\theta,z) \, dz \, r \, dr \, d\theta$

27. $\int_{-\pi/2}^{\pi/2}\int_1^{1+\cos\theta}\int_0^4 f(r,\theta,z) \, dz \, r \, dr \, d\theta$

29. $\int_0^{\pi/4}\int_0^{\sec\theta}\int_0^{2 - r\sin\theta} f(r,\theta,z) \, dz \, r \, dr \, d\theta$

31. $V = 4\int_0^{\pi/2}\int_0^1\int_{r^4-1}^{4-4r^2} dz \, r \, dr \, d\theta = 4\int_0^{\pi/2}\int_0^1 \left(-r^5 - 4r^3 + 5r\right) dr \, d\theta = 4\int_0^{\pi/2} \frac{4}{3} \, d\theta = \frac{8\pi}{3}$

33. $V = \int_{-\pi/2}^0\int_0^{3\cos\theta}\int_0^{-r\sin\theta} dz \, r \, dr \, d\theta = \int_{-\pi/2}^0\int_0^{3\cos\theta} (-r\sin\theta) \, r \, dr \, d\theta = 9\int_{-\pi/2}^0 \cos^3\theta \sin\theta \, d\theta = \frac{9}{4}$

35. $V = \int_0^{\pi/2}\int_0^{\sin\theta}\int_0^{\sqrt{1-r^2}} dz \, r \, dr \, d\theta = \int_0^{\pi/2}\int_0^{\sin\theta}\left(\sqrt{1-r^2}\right) r \, dr \, d\theta = -\frac{1}{3}\int_0^{\pi/2}\left(\cos^3\theta - 1\right) d\theta =$

$-\frac{1}{3}\int_0^{\pi/2}\left(\cos\theta - \sin^2\theta\cos\theta - 1\right) d\theta = \frac{3\pi - 4}{18}$

37. $V = 4\int_0^{\pi/2}\int_0^1\int_0^{r^2} r \, dz \, dr \, d\theta = 4\int_0^{\pi/2}\int_0^1 r^3 \, dr \, d\theta = \int_0^{\pi/2} d\theta = \frac{\pi}{2}$

39. $V = \int_0^{2\pi}\int_0^2\int_0^{4 - r\sin\theta} r \, dz \, dr \, d\theta = \int_0^{2\pi}\int_0^2 \left(4 - r\sin\theta\right) r \, dr \, d\theta = 8\int_0^{2\pi}\left(1 - \frac{\sin\theta}{3}\right) d\theta = 16\pi$

41. $V = 4\int_0^{\pi/2}\int_0^1\int_{4r^2}^{5-r^2} r \, dz \, dr \, d\theta = 4\int_0^{\pi/2}\int_0^1 \left(5 - 5r^2\right) r \, dr \, d\theta = 5\int_0^{\pi/2} d\theta = \frac{5\pi}{2}$

43. $V = 8\int_0^{\pi/2}\int_0^1\int_0^{\sqrt{4-r^2}} r \, dz \, dr \, d\theta = 8\int_0^{\pi/2}\int_0^1 \left(4 - r^2\right)^{1/2} r \, dr \, d\theta = -\frac{8}{3}\int_0^{\pi/2}\left(3^{3/2} - 8\right) d\theta = \frac{4\pi\left(8 - 3\sqrt{3}\right)}{3}$

45. $\text{average} = \frac{1}{2\pi}\int_0^{2\pi}\int_0^1\int_{-1}^1 r^2 \, dz \, dr \, d\theta = \frac{1}{2\pi}\int_0^{2\pi}\int_0^1 2r^2 \, dr \, d\theta = \frac{1}{3\pi}\int_0^{2\pi} d\theta = \frac{2}{3}$

47. $M = 4\int_0^{\pi/2}\int_0^1\int_0^r r \, dz \, dr \, d\theta = 4\int_0^{\pi/2}\int_0^1 r^2 \, dr \, d\theta = \frac{4}{3}\int_0^{\pi/2} d\theta = \frac{2\pi}{3}$, $M_{xy} = \int_0^{2\pi}\int_0^1\int_0^r r z \, dz \, dr \, d\theta =$

$\frac{1}{2}\int_0^{2\pi}\int_0^1 r^3 \, dr \, d\theta = \frac{1}{8}\int_0^{2\pi} d\theta = \frac{\pi}{4} \Rightarrow \overline{z} = \frac{3}{8}$, $\overline{x} = \overline{y} = 0$ by symmetry

49. $M = \int_0^{2\pi}\int_0^4\int_0^{\sqrt{r}} dz\ r\ dr\ d\theta = \int_0^{2\pi}\int_0^4 r^{3/2} dr\ d\theta = \frac{64}{5}\int_0^{2\pi} d\theta = \frac{128\pi}{5}$, $M_{xy} = \int_0^{2\pi}\int_0^4\int_0^{\sqrt{r}} z\ dz\ r\ dr\ d\theta =$ $\frac{1}{2}\int_0^{2\pi}\int_0^4 r^2\ dr\ d\theta = \frac{32}{3}\int_0^{2\pi} d\theta = \frac{64\pi}{3}$, $\overline{z} = \frac{M_{xy}}{M} = \frac{5}{6}$, $\overline{x} = \overline{y} = 0$, by symmetry

51. a) $I_z = \int_0^{2\pi}\int_0^1\int_{-1}^1 r^3\ dz\ dr\ d\theta = 2\int_0^{2\pi}\int_0^1 r^3\ dr\ d\theta = \frac{1}{2}\int_0^{2\pi} d\theta = \pi$

b) $I_x = \int_0^{2\pi}\int_0^1\int_{-1}^1 \left(r^2\sin^2\theta + z^2\right) r\ dz\ dr\ d\theta = \int_0^{2\pi}\int_0^1 \left(2r^2\sin^2\theta + \frac{2}{3}\right) r\ dr\ d\theta = \int_0^{2\pi}\left(\frac{\sin^2\theta}{2} + \frac{1}{3}\right) d\theta = \frac{7\pi}{6}$

53. $I_z = \int_0^{2\pi}\int_0^a\int_{-\sqrt{a^2-r^2}}^{\sqrt{a^2-r^2}} r^3\ dz\ dr\ d\theta = -\int_0^{2\pi}\int_0^a r^2\left(a^2-r^2\right)^{1/2}(-2r)\ dr\ d\theta = \int_0^{2\pi}\frac{4a^5}{15} d\theta = \frac{8\pi a^5}{15}$

55. a) $4\int_0^{\pi/2}\int_0^{\pi/2}\int_{\cos\phi}^2 \rho^2\sin\phi\ d\rho\ d\phi\ d\theta$ b) $4\int_0^{\pi/2}\int_0^{\pi/2}\int_{\cos\phi}^2 \rho^2\sin\phi\ d\rho\ d\phi\ d\theta =$ $\frac{4}{3}\int_0^{\pi/2}\int_0^{\pi/2}\left(8-\cos^3\phi\right)\sin\phi\ d\phi\ d\theta = \frac{31}{3}\int_0^{\pi/2} d\theta = \frac{31\pi}{6}$

57. $V = 4\int_0^{\pi/2}\int_0^{\pi}\int_0^{1-\cos\phi} \rho^2\sin\phi\ d\rho\ d\phi\ d\theta = \frac{4}{3}\int_0^{\pi/2}\int_0^{\pi}\left(1-\cos\phi\right)^3\sin\phi\ d\phi\ d\theta = \frac{16}{3}\int_0^{\pi/2} d\theta = \frac{8\pi}{3}$

59. $V = 4\int_0^{\pi/2}\int_{\pi/4}^{\pi/2}\int_0^{2\cos\phi} \rho^2\sin\phi\ d\rho\ d\phi\ d\theta = \frac{32}{3}\int_0^{\pi/2}\int_{\pi/4}^{\pi/2}\cos^3\phi\sin\phi\ d\phi\ d\theta = \frac{2}{3}\int_0^{\pi/2} d\theta = \frac{\pi}{3}$

61. $\int_0^{2\pi}\int_{\pi/3}^{2\pi/3}\frac{a^3}{3}\sin\phi\ d\phi\ d\theta = \frac{a^3}{3}\int_0^{2\pi} d\theta = \frac{2\pi a^3}{3} \Rightarrow V = \frac{4}{3}\pi a^3 - \frac{2\pi a^3}{3} = \frac{2\pi a^3}{3}$

63. $V = \int_0^{2\pi}\int_0^{\pi/3}\int_{\sec\phi}^2 \rho^2\sin\phi\ d\rho\ d\phi\ d\theta = \frac{1}{3}\int_0^{2\pi}\int_0^{\pi/3}\left(8\sin\phi - \tan\phi\sec^2\phi\right) d\phi\ d\theta = \frac{5}{6}\int_0^{2\pi} d\theta = \frac{5\pi}{3}$

65. $V = 8\int_0^{\pi/2}\int_1^{\sqrt{2}}\int_0^r r\ dz\ dr\ d\theta = 8\int_0^{\pi/2}\int_1^{\sqrt{2}} r^2\ dr\ d\theta = \frac{2\sqrt{2}-1}{3}\int_0^{\pi/2} d\theta = \frac{4\pi(2\sqrt{2}-1)}{3}$

67. $\text{average} = \frac{3}{4\pi}\int_0^{2\pi}\int_0^{\pi}\int_0^1 \rho^3\sin\phi\ d\rho\ d\phi\ d\theta = \frac{3}{16\pi}\int_0^{2\pi}\int_0^{\pi}\sin\phi\ d\phi\ d\theta = \frac{3}{8\pi}\int_0^{2\pi} d\theta = \frac{3}{4}$

69. $M = \int_0^{2\pi}\int_0^{\pi/4}\int_0^a \rho^2\sin\phi\ d\rho\ d\phi\ d\theta = \frac{a^3}{3}\int_0^{2\pi}\int_0^{\pi/4}\sin\phi\ d\phi\ d\theta = \frac{a^3}{3}\int_0^{2\pi}\frac{\sqrt{2}-1}{\sqrt{2}} d\theta = \frac{\pi a^3\left(2-\sqrt{2}\right)}{3}$

$M_{xy} = \int_0^{2\pi}\int_0^{\pi/4}\int_0^a \rho^3\sin\phi\cos\phi\ d\rho\ d\phi\ d\theta = \frac{a^3}{4}\int_0^{2\pi}\int_0^{\pi/4}\sin\phi\cos\phi\ d\phi\ d\theta = \frac{a^4}{16}\int_0^{2\pi} d\theta = \frac{\pi a^4}{8} \Rightarrow$

$\overline{z} = \frac{3\left(2+\sqrt{2}\right)a}{16}$, $\overline{x} = \overline{y} = 0$ by symmetry

14.7 SUBSTITUTIONS IN MULTIPLE INTEGRALS

1. $\int_0^4 \int_{y/2}^{1+y/2} \frac{2x-y}{2}\,dx\,dy = \int_0^4 \left[\frac{x^2}{2} - \frac{xy}{2}\right]_{y/2}^{1+y/2} dy = \frac{1}{2}\int_0^4 dy = 2$

3. a) $x = \frac{u+v}{3}$, $y = \frac{v-2u}{3}$, $J(u,v) = \begin{vmatrix} 1/3 & 1/3 \\ -2/3 & 1/3 \end{vmatrix} = \frac{1}{9} + \frac{2}{9} = \frac{1}{3}$ b) $\int_4^7 \int_{-1}^2 \frac{vu}{3}\,dv\,du = \frac{1}{2}\int_4^7 u\,du = \frac{33}{4}$

5. $J(u,v) = \begin{vmatrix} v^{-1} & -uv^{-2} \\ v & u \end{vmatrix} = v^{-1}u + v^{-1}u = \frac{2u}{v}$; $\int_1^3 \int_1^2 (v+u)\left(\frac{2u}{v}\right) dv\,du = \int_1^3 \left[2u + (\ln 4)\,u^2\right] du = 8 + \frac{26 \ln 4}{3}$

7. $J(r,\theta) = \begin{vmatrix} a\cos\theta & -ar\sin\theta \\ b\sin\theta & br\cos\theta \end{vmatrix} = abr\cos^2\theta + abr\sin^2\theta = abr$, $I_o = \int_{-a}^{a} \int_{(-b/a)\sqrt{a^2-x^2}}^{(b/a)\sqrt{a^2-x^2}} \left(x^2+y^2\right) dy\,dx =$

$\int_0^{2\pi} \int_0^1 r^2\left(a^2\cos^2\theta + b^2\sin^2\theta\right) abr\,dr\,d\theta = \frac{ab}{4}\int_0^{2\pi} \left[a^2\cos^2\theta + b^2\sin^2\theta\right] d\theta = \frac{\pi ab\left(a^2+b^2\right)}{4}$

9. $\begin{vmatrix} \sin\phi\cos\theta & \rho\cos\phi\cos\theta & -\rho\sin\phi\sin\theta \\ \sin\phi\sin\theta & \rho\cos\phi\sin\theta & \rho\sin\phi\cos\theta \\ \cos\phi & -\rho\sin\phi & 0 \end{vmatrix} = \cos\phi \begin{vmatrix} \rho\cos\phi\cos\theta & -\rho\sin\phi\sin\theta \\ \rho\cos\phi\sin\theta & \rho\sin\phi\cos\theta \end{vmatrix} +$

$\rho\sin\phi \begin{vmatrix} \sin\phi\cos\theta & -\rho\sin\phi\sin\theta \\ \sin\phi\sin\theta & \rho\sin\phi\cos\theta \end{vmatrix} = \rho^2\cos\phi\left(\sin\phi\cos\phi\cos^2\theta + \sin\phi\cos\phi\sin^2\theta\right) +$

$\rho^2\sin\phi\left(\sin^2\phi\cos^2\theta + \sin^2\phi\sin^2\theta\right) = \rho^2\sin\phi\cos^2\phi + \rho^2\sin^3\phi = \rho^2\sin\phi\left(\cos^2\phi + \sin^2\phi\right) = \rho^2\sin\phi$

11. $J(u,v,w) = \begin{vmatrix} a & 0 & 0 \\ 0 & b & 0 \\ 0 & 0 & c \end{vmatrix} = abc$. The transformation takes the $\frac{x^2}{a^2} + \frac{y^2}{b^2} + \frac{z^2}{c^2} = 1$ region in the xyz–space into the $u^2 + v^2 + w^2 = 1$ region in the uvw–space which is a unit sphere with $V = \frac{4}{3}\pi$. $\therefore\ V = \int\int_R\int dx\,dy\,dz =$

$\int\int_G\int abc\,du\,dv\,dw = \frac{4\pi abc}{3}$

13. $J(u,v,w) = \begin{vmatrix} 1 & 0 & 0 \\ -v/u^2 & 1/u & 0 \\ 0 & 0 & 1/3 \end{vmatrix} = \frac{1}{3u}$; $\int\int_R\int \left(x^2y + 3xyz\right) dx\,dy\,dz =$

$\int\int_G\int \left(u^2\left(\frac{v}{u}\right) + 3u\,\frac{v}{u}\,\frac{w}{3}\right) J(u,v,w)\,du\,dv\,dw = \frac{1}{3}\int_0^3 \int_0^2 \int_1^2 \left(v + \frac{vw}{u}\right) du\,dv\,dw =$

$\frac{1}{3}\int_0^3 \int_0^2 (v + vw\ln 2)\,dv\,dw = \frac{1}{3}\int_0^3 \left(2 + (\ln 4)w\right) dw = 2 + \ln 8$

PRACTICE EXERCISES

1. $\int_1^{10}\int_0^{1/y} ye^{xy}\,dx\,dy = \int_1^{10}\left[e^{xy}\right]_0^{1/y} dy =$

$\int_1^{10} (e-1)\,dy = 9e - 9$

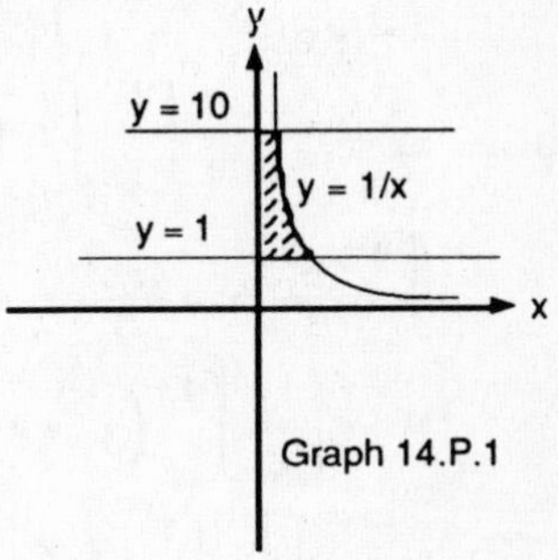

Graph 14.P.1

3. $\int_0^1 \int_y^{\sqrt{y}} f(x,y)\,dx\,dy$

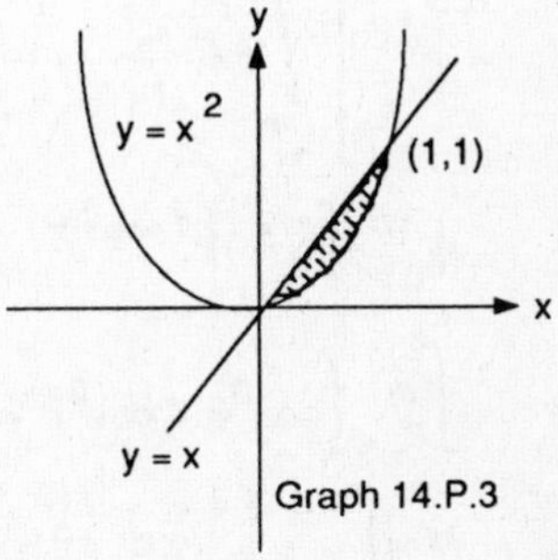

Graph 14.P.3

5. a) $\int_0^{3/2}\int_{-\sqrt{9-4y^2}}^{\sqrt{9-4y^2}} y\,dx\,dy = \int_0^{3/2}\left[\,yx\,\right]_{-\sqrt{9-4y^2}}^{\sqrt{9-4y^2}} dy =$

$\int_0^{3/2} 2y\sqrt{9-4y^2}\,dy = \frac{9}{2}$

b) $\int_{-3}^{3}\int_0^{(9-x^2)^{1/2}/2} y\,dy\,dx = \frac{1}{2}\int_{-3}^{3}\left[y^2\right]_0^{(9-x^2)^{1/2}/2} dx = \frac{1}{8}\int_{-3}^{3}\left(9-x^2\right) dx = \frac{9}{2}$

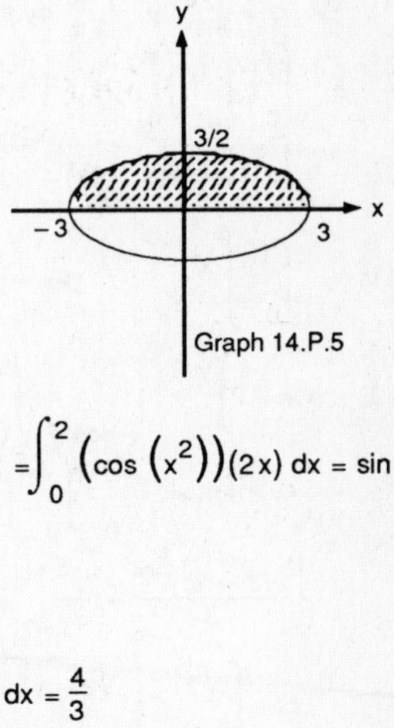

Graph 14.P.5

7. $\int_0^1\int_{2y}^{2} 4\cos\left(x^2\right) dx\,dy = \int_0^2\int_0^{x/2} 4\cos\left(x^2\right) dy\,dx = \int_0^2 \left[\left(4\cos\left(x^2\right)\right)y\right]_0^{x/2} dx = \int_0^2 \left(\cos\left(x^2\right)\right)(2x)\,dx = \sin 4$

9. $\int_0^8\int_{\sqrt[3]{x}}^{2} \frac{1}{y^4+1}\,dy\,dx = \int_0^2\int_0^{y^3} \frac{1}{y^4+1}\,dx\,dy = \frac{1}{4}\int_0^2 \frac{4y^3}{y^4+1}\,dy = \frac{\ln 17}{4}$

11. $V = \int_0^1\int_x^{2-x} \left(x^2+y^2\right) dy\,dx = \int_0^1\left[x^2y + \frac{y^3}{3}\right]_x^{2-x} dx = \int_0^1\left(-\frac{8}{3}x^3 + 4x^2 - 4x + \frac{8}{3}\right) dx = \frac{4}{3}$

13. $A = \int_{-2}^{0}\int_{2x+4}^{4-x^2} dy\,dx = \int_{-2}^{0}\left(-x^2-2x\right) dx = \frac{4}{3}$

15. average value $= \int_0^1\int_0^1 xy\,dy\,dx = \int_0^1\left[\frac{xy^2}{2}\right]_0^1 dx = \int_0^1 \frac{x}{2}\,dx = \frac{1}{4}$

17. $M = \int_1^2 \int_{2/x}^2 dy\,dx = \int_1^2 \left(2 - \frac{2}{x}\right) dx = 2 - \ln 4$, $M_y = \int_1^2 \int_{2/x}^2 x\,dy\,dx = \int_1^2 x\left[2 - \frac{2}{x}\right] dx = 1$, $M_x =$ $\int_1^2 \int_{2/x}^2 y\,dy\,dx = \int_1^2 \left(2 - \frac{2}{x^2}\right) dx = 1 \Rightarrow \overline{x} = \frac{1}{2 - \ln 4}$, $\overline{y} = \frac{1}{2 - \ln 4}$

19. a) $I_o = \int_{-2}^2 \int_{-1}^1 (x^2 + y^2)\,dy\,dx = \int_{-2}^2 \left(2x^2 + \frac{2}{3}\right) dx = \frac{40}{3}$ b) $I_x = \int_{-a}^a \int_{-b}^b y^2\,dy\,dx =$ $\int_{-a}^a \frac{2b^3}{3} dx = \frac{4ab^3}{3}$, $I_y = \int_{-b}^b \int_{-a}^a x^2\,dx\,dy = \int_{-b}^b \frac{2a^3}{3} dy = \frac{4a^3b}{3} \Rightarrow I_o = I_x + I_y = \frac{4ab^3}{3} + \frac{4a^3b}{3} = \frac{4ab\left(b^2 + a^2\right)}{3}$

21. $M = \int_{-1}^1 \int_{-1}^1 \left(x^2 + y^2 + \frac{1}{3}\right) dy\,dx = \int_{-1}^1 \left(2x^2 + \frac{4}{3}\right) dx = 4$, $M_x = \int_{-1}^1 \int_{-1}^1 y\left(x^2 + y^2 + \frac{1}{3}\right) dy\,dx = \int_{-1}^1 0\,dx = 0$
$M_y = \int_{-1}^1 \int_{-1}^1 x\left(x^2 + y^2 + \frac{1}{3}\right) dy\,dx = \int_{-1}^1 x\left(2x^2 + \frac{4}{3}\right) dx = 0$

23. $\int_{-1}^1 \int_{-\sqrt{1-x^2}}^{\sqrt{1-x^2}} \frac{2}{(1 + x^2 + y^2)^2}\,dy\,dx = \int_0^{2\pi} \int_0^1 \frac{2r}{(1 + r^2)^2}\,dr\,d\theta = \frac{1}{2}\int_0^{2\pi} d\theta = \pi$

25. $M = 2\int_0^{\pi/2} \int_1^{1+\cos\theta} r\,dr\,d\theta = \int_0^{\pi/2} \left(2\cos\theta + \frac{1 + \cos 2\theta}{2}\right) d\theta = \frac{8 + \pi}{4}$, $M_y = \int_{-\pi/2}^{\pi/2} \int_1^{1+\cos\theta} \cos\theta\, r^2\,dr\,d\theta =$ $\int_{-\pi/2}^{\pi/2} \left(\cos^2\theta + \cos^3\theta + \frac{\cos^4\theta}{3}\right) d\theta = \frac{32 + 15\pi}{24} \Rightarrow \overline{x} = \frac{15\pi + 32}{6\pi + 48}$, $\overline{y} = 0$ by symmetry

27. $M = \int_0^{\pi/2} \int_1^3 r\,dr\,d\theta = 4\int_0^{\pi/2} d\theta = 2\pi$, $M_y = \int_0^{\pi/2} \int_1^3 r^2\cos\theta\,dr\,d\theta = \frac{26}{3}\int_0^{\pi/2} \cos\theta\,d\theta = \frac{26}{3} \Rightarrow$ $\overline{x} = \frac{13}{3\pi}$, $\overline{y} = \frac{13}{3\pi}$ by symmetry

29. $\int_0^\pi \int_0^\pi \int_0^\pi \cos(x + y + z)\,dx\,dy\,dz = \int_0^\pi \int_0^\pi \left[\sin(z + y + \pi) - \sin(z + y)\right] dy\,dz =$ $\int_0^\pi \left(-\cos(z + 2\pi) + \cos(z + \pi) + \cos(z) - \cos(z + \pi)\right) dz = 0$

31. $\int_0^1 \int_0^{x^2} \int_0^{x+y} (2x - y - z)\,dz\,dy\,dx = \int_0^1 \int_0^{x^2} \left(\frac{3x^2}{2} - \frac{3y^2}{2}\right) dy\,dx = \int_0^1 \left(\frac{3x^4}{2} - \frac{x^6}{2}\right) dx = \frac{8}{35}$

33. $V = 2\int_0^{\pi/2} \int_{-\cos y}^0 \int_0^{-2x} dz\,dx\,dy = -4\int_0^{\pi/2} \int_{-\cos y}^0 x\,dx\,dy = 2\int_0^{\pi/2} \cos^2 y\,dy = \frac{\pi}{2}$

35. Average $= \frac{1}{3}\int_0^1 \int_0^3 \int_0^1 30xz\sqrt{x^2 + y}\,dz\,dy\,dx = \int_0^1 \int_0^3 5x\sqrt{x^2 + y}\,dy\,dx = \frac{1}{3}\int_0^1 \left(10x(3 + x^2)^{3/2} - 10x^4\right) dx =$ $\left[\frac{2(3 + x^2)^{5/2}}{3} - \frac{3x^5}{3}\right]_0^1 = \frac{2(31 - 3^{5/2})}{3}$

37. a) $\int_{-\sqrt{2}}^{\sqrt{2}} \int_{-\sqrt{2 - y^2}}^{\sqrt{2 - y^2}} \int_{\sqrt{x^2 + y^2}}^{\sqrt{4 - \left(x^2 + y^2\right)}} 3\,dz\,dx\,dy$

b) $\int_0^{2\pi} \int_0^{\pi/4} \int_0^2 3\rho^2 \sin\phi\,d\rho\,d\phi\,d\theta$

c) $\int_0^{2\pi} \int_0^{\sqrt{2}} \int_r^{\sqrt{4 - r^2}} 3\,dz\,r\,dr\,d\theta = 3\int_0^{2\pi} \int_0^{\sqrt{2}} \left(\left(4 - r^2\right)^{1/2} - r\right) r\,dr\,d\theta = \int_0^{2\pi} \left(8 - 2^{5/2}\right) d\theta = 2\pi\left(8 - 2^{5/2}\right)$

39. a) $4\int_0^{\pi/2}\int_0^{\pi/4}\int_0^{\sec\phi}\rho^2\sin\phi\,d\rho\,d\phi\,d\theta$

b) $4\int_0^{\pi/2}\int_0^{\pi/4}\int_0^{\sec\phi}\rho^2\sin\phi\,d\rho\,d\phi\,d\theta = \frac{4}{3}\int_0^{\pi/2}\int_0^{\pi/4}(\sec\phi)(\sec\phi\tan\phi)\,d\phi\,d\theta = \frac{2}{3}\int_0^{\pi/2}d\theta = \frac{\pi}{3}$

41. $\int_0^1\int_{\sqrt{1-x^2}}^{\sqrt{3-x^2}}\int_1^{\sqrt{4-x^2-y^2}} yxz^2\,dz\,dy\,dx + \int_1^{\sqrt{3}}\int_0^{\sqrt{3-x^2}}\int_1^{\sqrt{4-x^2-y^2}} yxz^2\,dz\,dy\,dx$

43. a) $\int_{-\sqrt{3}}^{\sqrt{3}}\int_{-\sqrt{3-x^2}}^{\sqrt{3-x^2}}\int_1^{\sqrt{4-x^2-y^2}} dz\,dy\,dx$ b) $\int_0^{2\pi}\int_0^{\sqrt{3}}\int_1^{\sqrt{4-r^2}} r\,dz\,dr\,d\theta$

c) $\int_0^{2\pi}\int_0^{\pi/3}\int_{\sec\phi}^{2}\rho^2\sin\phi\,d\rho\,d\phi\,d\theta$

45. $V = \int_0^{\pi/2}\int_1^2\int_0^{r^2\sin\theta\cos\theta} r\,dz\,dr\,d\theta = \int_0^{\pi/2}\int_1^2 r^3\sin\theta\cos\theta\,dr\,d\theta = \frac{15}{4}\int_0^{\pi/2}\sin\theta\cos\theta\,d\theta = \frac{15}{8}$

47. $V = 4\int_0^1\int_0^{\sqrt{1-x^2}}\int_{2x^2+2y^2}^{3-x^2-y^2} dz\,dy\,dx = 4\int_0^{\pi/2}\int_0^1\int_{2r^2}^{3-r^2} r\,dz\,dr\,d\theta = 4\int_0^{\pi/2}\int_0^1\left(3r-3r^3\right)dr\,d\theta =$

$3\int_0^{\pi/2}d\theta = \frac{3\pi}{2}$

49. a) $M = 4\int_0^{\pi/2}\int_0^1\int_0^{r^2} z\,r\,dz\,dr\,d\theta = 2\int_0^{\pi/2}\int_0^1 r^5\,dr\,d\theta = \frac{1}{3}\int_0^{\pi/2}d\theta = \frac{\pi}{6}$

$M_{xy} = \int_0^{2\pi}\int_0^1\int_0^{r^2} z^2\,r\,dz\,dr\,d\theta = \frac{1}{3}\int_0^{2\pi}\int_0^1 r^7\,dr\,d\theta = \frac{1}{24}\int_0^{2\pi}d\theta = \frac{\pi}{12} \Rightarrow \bar{z} = \frac{1}{2}$, $\bar{x} = \bar{y} = 0$ by symmetry;

$I_z = \int_0^{2\pi}\int_0^1\int_0^{r^2} z\,r^3\,dz\,dr\,d\theta = \frac{1}{2}\int_0^{2\pi}\int_0^1 r^7\,dr\,d\theta = \frac{1}{16}\int_0^{2\pi}d\theta = \frac{\pi}{8} \Rightarrow R_z = \sqrt{\frac{I_z}{M}} = \frac{\sqrt{3}}{2}$

b) $M = 4\int_0^{\pi/2}\int_0^1\int_0^{r^2} r^2\,dz\,dr\,d\theta = 4\int_0^{\pi/2}\int_0^1 r^4\,dr\,d\theta = \frac{4}{5}\int_0^{\pi/2}d\theta = \frac{2\pi}{5}$

$M_{xy} = \int_0^{2\pi}\int_0^1\int_0^{r^2} z\,r^2\,dz\,dr\,d\theta = \frac{1}{2}\int_0^{2\pi}\int_0^1 r^6\,dr\,d\theta = \frac{1}{14}\int_0^{2\pi}d\theta = \frac{\pi}{7} \Rightarrow \bar{z} = \frac{5}{14}$, $\bar{x} = \bar{y} = 0$ by symmetry;

$I_z = \int_0^{2\pi}\int_0^1\int_0^{r^2} r^4\,dz\,dr\,d\theta = \int_0^{2\pi}\int_0^1 r^6\,dr\,d\theta = \frac{1}{7}\int_0^{2\pi}d\theta = \frac{2\pi}{7} \Rightarrow R_z = \sqrt{\frac{I_z}{M}} = \sqrt{\frac{5}{7}}$

51. $V = 8\int_0^{\pi/2}\int_0^{\pi/2}\int_0^{2\sin\phi}\rho^2\sin\phi\,d\rho\,d\phi\,d\theta = \frac{64}{3}\int_0^{\pi/2}\int_0^{\pi/2}\sin^4\phi\,d\phi\,d\theta = 4\pi\int_0^{\pi/2}d\theta = 2\pi^2$

53. $M = \frac{2}{3}\pi a^3$; $M_{xy} = \int_0^{2\pi}\int_0^{\pi/2}\int_0^a \rho^3\sin\phi\cos\phi\,d\rho\,d\phi\,d\theta = \frac{a^4}{4}\int_0^{2\pi}\int_0^{\pi/2}\sin\phi\cos\phi\,d\phi\,d\theta =$

$\frac{a^4}{8}\int_0^{2\pi}d\theta = \frac{a^4\pi}{4} \Rightarrow \bar{z} = \frac{3a}{8}$, $\bar{x} = \bar{y} = 0$ by symmetry

CHAPTER 15

INTEGRATION IN VECTOR FIELDS

15.1 LINE INTEGRALS

1. $\mathbf{r} = t\mathbf{i} + (1-t)\mathbf{j} \Rightarrow x = t, y = 1 - t \Rightarrow y = 1 - x \Rightarrow c$

3. $\mathbf{r} = (2\cos t)\mathbf{i} + (2\sin t)\mathbf{j} \Rightarrow x = 2\cos t, y = 2\sin t \Rightarrow x^2 + y^2 = 4 \Rightarrow g$

5. $\mathbf{r} = t\mathbf{i} + t\mathbf{j} + t\mathbf{k} \Rightarrow x = t, y = t, z = t \Rightarrow d$

7. $\mathbf{r} = (t^2 - 1)\mathbf{j} + 2t\mathbf{k} \Rightarrow y = t^2 - 1, z = 2t \Rightarrow y = \frac{z^2}{4} - 1 \Rightarrow f$

9. $\mathbf{r} = t\mathbf{i} + (1-t)\mathbf{j} \Rightarrow x = t, y = 1 - t, z = 0 \Rightarrow f(g(t),h(t),k(t)) = 1.\ \frac{dx}{dt} = 1, \frac{dy}{dt} = -1, \frac{dz}{dt} = 0 \Rightarrow$

$$\sqrt{\left(\frac{dx}{dt}\right)^2 + \left(\frac{dy}{dt}\right)^2 + \left(\frac{dz}{dt}\right)^2}\ dt = \sqrt{2}\,dt \Rightarrow \int_C f(x,y,z)\,ds = \int_0^1 \sqrt{2}\,dt = \sqrt{2}$$

11. $\mathbf{r} = 2t\mathbf{i} + t\mathbf{j} + (2 - 2t)\mathbf{k} \Rightarrow x = 2t, y = t, z = 2 - 2t \Rightarrow f(g(t),h(t),k(t)) = 2t^2 - t + 2.\ \frac{dx}{dt} = 2, \frac{dy}{dt} = 1, \frac{dz}{dt} = -2$

$$\Rightarrow \sqrt{\left(\frac{dx}{dt}\right)^2 + \left(\frac{dy}{dt}\right)^2 + \left(\frac{dz}{dt}\right)^2}\ dt = 3\,dt \Rightarrow \int_C f(x,y,z)\,ds = \int_0^1 (2t^2 - t + 2)3\,dt = \frac{13}{2}$$

13. $\mathbf{r} = \mathbf{i} + \mathbf{j} + t\mathbf{k} \Rightarrow x = 1, y = 1, z = t \Rightarrow f(g(t),h(t),k(t)) = 3t\sqrt{4 + t^2}\ \ \frac{dx}{dt} = 0, \frac{dy}{dt} = 0, \frac{dz}{dt} = 1 \Rightarrow$

$$\sqrt{\left(\frac{dx}{dt}\right)^2 + \left(\frac{dy}{dt}\right)^2 + \left(\frac{dz}{dt}\right)^2}\ dt = 1\,dt = dt \Rightarrow \int_C f(x,y,z)\,ds = \int_{-1}^1 3t\sqrt{4 + t^2}\ dt = 0$$

15. C_1: $\mathbf{r} = t\mathbf{i} + t^2\mathbf{j} \Rightarrow x = t, y = t^2, z = 0 \Rightarrow f(g(t),h(t),k(t)) = t + \sqrt{t^2} = 2t$ since $0 \le t \le 1$. $\frac{dx}{dt} = 1, \frac{dy}{dt} = 2t,$

$$\frac{dz}{dt} = 0 \Rightarrow \sqrt{\left(\frac{dx}{dt}\right)^2 + \left(\frac{dy}{dt}\right)^2 + \left(\frac{dz}{dt}\right)^2}\ dt = \sqrt{1 + 4t^2}\ dt \Rightarrow \int_{C_1} f(x,y,z)\,ds = \int_0^1 2t\sqrt{1 + 4t^2}\ dt =$$

$\frac{1}{6}\left(5^{3/2}\right) - \frac{1}{6} = \frac{5}{6}\sqrt{5} - \frac{1}{6}$. C_2: $\mathbf{r} = \mathbf{i} + \mathbf{j} + t\mathbf{k} \Rightarrow x = 1, y = 1, z = t \Rightarrow f(g(t),h(t),k(t)) = 2 - t^2.\ \frac{dx}{dt} = 0,$

$$\frac{dy}{dt} = 0, \frac{dz}{dt} = 1 \Rightarrow \sqrt{\left(\frac{dx}{dt}\right)^2 + \left(\frac{dy}{dt}\right)^2 + \left(\frac{dz}{dt}\right)^2}\ dt = 1\,dt = dt \Rightarrow \int_{C_2} f(x,y,z)\,ds = \int_0^1 (2 - t^2)\,dt =$$

$$\frac{5}{3}.\ \therefore \int_C f(x,y,z)\,ds = \int_{C_1} f(x,y,z)\,ds + \int_{C_2} f(x,y,z)\,ds = \frac{5}{6}\sqrt{5} + \frac{3}{2}.$$

17. $\mathbf{r} = t\mathbf{i} + t\mathbf{j} + t\mathbf{k} \Rightarrow x = t, y = t, z = t \Rightarrow f(g(t),h(t),k(t)) = \frac{t + t + t}{t^2 + t^2 + t^2} = \frac{1}{t} \cdot \frac{dx}{dt} = 1, \frac{dy}{dt} = 1, \frac{dz}{dt} = 1 \Rightarrow$

$$\sqrt{\left(\frac{dx}{dt}\right)^2 + \left(\frac{dy}{dt}\right)^2 + \left(\frac{dz}{dt}\right)^2}\ dt = \sqrt{3}\,dt \Rightarrow \int_C f(x,y,z)\,ds = \int_a^b \frac{1}{t}\left(\sqrt{3}\,dt\right) = \sqrt{3}\ln|b| - \sqrt{3}\ln|a|$$

$$= \sqrt{3}\ln\left|\frac{b}{a}\right|$$

19. $\delta(x,y,z) = 2 - z$, $\mathbf{r} = (\cos t)\,\mathbf{j} + (\sin t)\,\mathbf{k}$, $0 \le t \le \pi$, $ds = dt$, $x = 0$, $y = \cos t$, $z = \sin t$, and $M = 2\pi - 2$ are all given or found in Example 3 in the text, page 935. $I_x = \int_C (y^2 + z^2)\delta\,ds$

$$= \int_0^\pi (\cos^2 t + \sin^2 t)(2 - \sin t)\,dt = \int_0^\pi (2 - \sin t)\,dt = 2\pi - 2 \Rightarrow R_x = \sqrt{\frac{I_x}{M}} = \sqrt{\frac{2\pi - 2}{2\pi - 2}} = 1$$

21. Let δ be constant. Let $x = a\cos t$, $y = a\sin t$. Then $\frac{dx}{dt} = -a\sin t, \frac{dy}{dt} = a\cos t$, $0 \le t \le 2\pi$, $\frac{dz}{dt} = 0 \Rightarrow$

$$\sqrt{\left(\frac{dx}{dt}\right)^2 + \left(\frac{dy}{dt}\right)^2 + \left(\frac{dz}{dt}\right)^2}\ dt = a\,dt. \ \therefore\ I_z = \int_C (x^2 + y^2)\delta\,ds = \int_0^{2\pi} (a^2\sin^2 t + a^2\cos^2 t)a\delta\,dt =$$

$$\int_0^{2\pi} a^3\delta\,dt = 2\pi a^3\delta.\ \ M = \int_C \delta(x,y,z)\,ds = \int_0^{2\pi} \delta a\,dt = 2\pi\delta a.\ \ R_z = \sqrt{\frac{I_z}{M}} = \sqrt{\frac{2\pi a^3\delta}{2\pi a\delta}} = a.$$

23. a) $\mathbf{r} = (\cos t)\,\mathbf{i} + (\sin t)\,\mathbf{j} + t\,\mathbf{k} \Rightarrow x = \cos t, y = \sin t, z = t \Rightarrow \frac{dx}{dt} = -\sin t, \frac{dy}{dt} = \cos t, \frac{dz}{dt} = 1 \Rightarrow$

$$\sqrt{\left(\frac{dx}{dt}\right)^2 + \left(\frac{dy}{dt}\right)^2 + \left(\frac{dz}{dt}\right)^2}\ dt = \sqrt{2}\,dt.\ M = \int_C \delta(x,y,z)\,ds = \int_0^{2\pi} \delta\sqrt{2}\,dt = 2\pi\delta\sqrt{2}.$$

$$I_z = \int_C (x^2 + y^2)\delta\,ds = \int_0^{2\pi} (\cos^2 t + \sin^2 t)\delta\sqrt{2}\,dt = \int_0^{2\pi} \delta\sqrt{2}\,dt = 2\pi\delta\sqrt{2}.\ R_z = \sqrt{\frac{I_z}{M}}$$

$$= \sqrt{\frac{2\pi\delta\sqrt{2}}{2\pi\delta\sqrt{2}}} = 1$$

b) $M = \int_C \delta(x,y,z)\,ds = \int_0^{4\pi} \delta\sqrt{2}\,dt = 4\pi\delta\sqrt{2}$. $I_z = \int_C (x^2 + y^2)\delta\,ds = \int_0^{4\pi} \delta\sqrt{2}\,dt = 4\pi\delta\sqrt{2}$.

$$R_z = \sqrt{\frac{I_z}{M}} = \sqrt{\frac{4\pi\delta\sqrt{2}}{4\pi\delta\sqrt{2}}} = 1$$

25. $\mathbf{r} = (1 - t)\mathbf{i} + (1 - t)\mathbf{j} + (1 - t)\mathbf{k}$, $0 \le t \le 1 \Rightarrow \mathbf{v}(t) = -\mathbf{i} - \mathbf{j} - \mathbf{k} \Rightarrow |\mathbf{v}(t)| = \sqrt{(-1)^2 + (-1)^2 + (-1)^2}$

$= \sqrt{3}$. The integral of f over C is $\int_C f(x,y,z)\,ds = \int_0^1 f(1 - t, 1 - t, 1 - t)\left(\sqrt{3}\right)dt =$

$$\int_0^1 \left((1 - t) - 3(1 - t)^2 + (1 - t)\right)\sqrt{3}\,dt = \sqrt{3}\int_0^1 \left(-1 + 4t - 3t^2\right)dt = \sqrt{3}\left[-t + 2t^2 - t^3\right]_0^1 = 0$$

27. No, parametrizations of the two curves may be different over different domains.

15.2 VECTOR FIELDS, WORK, CIRCULATION, AND FLUX

1. $f(x,y,z) = \left(x^2 + y^2 + z^2\right)^{-1/2} \Rightarrow \frac{\partial f}{\partial x} = -\frac{1}{2}\left(x^2 + y^2 + z^2\right)^{-3/2}(2x) = -x\left(x^2 + y^2 + z^2\right)^{-3/2}$. Similarly, $\frac{\partial f}{\partial y} = -y\left(x^2 + y^2 + z^2\right)^{-3/2}, \frac{\partial f}{\partial z} = -z\left(x^2 + y^2 + z^2\right)^{-3/2}$. $\therefore \nabla f = \frac{-x\,\mathbf{i} - y\,\mathbf{j} - z\,\mathbf{k}}{\left(x^2 + y^2 + z^2\right)^{3/2}}$.

3. $g(x,y,z) = e^z - \ln(x^2 + y^2) \Rightarrow \frac{\partial g}{\partial x} = -\frac{2x}{x^2 + y^2}, \frac{\partial g}{\partial y} = -\frac{2y}{x^2 + y^2}, \frac{\partial g}{\partial z} = e^z \Rightarrow \nabla g = -\frac{2x\,\mathbf{i}}{x^2 + y^2} - \frac{2y\,\mathbf{j}}{x^2 + y^2} + e^z\,\mathbf{k}$.

5. a) $\mathbf{F} = 3t\,\mathbf{i} + 2t\,\mathbf{j} + 4t\,\mathbf{k}, \frac{d\mathbf{r}}{dt} = \mathbf{i} + \mathbf{j} + \mathbf{k} \Rightarrow \mathbf{F}\cdot\frac{d\mathbf{r}}{dt} = 9t \Rightarrow W = \int_0^1 9t\,dt = \frac{9}{2}$

b) $\mathbf{F} = 3t^2\,\mathbf{i} + 2t\,\mathbf{j} + 4t^4\,\mathbf{k}, \frac{d\mathbf{r}}{dt} = \mathbf{i} + 2t\,\mathbf{j} + 4t^3\,\mathbf{k} \Rightarrow \mathbf{F}\cdot\frac{d\mathbf{r}}{dt} = 7t^2 + 16t^7 \Rightarrow W = \int_0^1 (7t^2 + 16t^7)\,dt = \frac{13}{3}$

c) $\mathbf{F}_1 = 3t\,\mathbf{i} + 2t\,\mathbf{j}, \frac{d\mathbf{r}_1}{dt} = \mathbf{i} + \mathbf{j} \Rightarrow \mathbf{F}_1\cdot\frac{d\mathbf{r}_1}{dt} = 5t \Rightarrow W_1 = \int_0^1 5t\,dt = \frac{5}{2}$. $\mathbf{F}_2 = 3\,\mathbf{i} + 2\,\mathbf{j} + 4t\,\mathbf{k}, \frac{d\mathbf{r}_2}{dt} = \mathbf{k} \Rightarrow$ $\mathbf{F}_2\cdot\frac{d\mathbf{r}_2}{dt} = 4t \Rightarrow W_2 = \int_0^1 4t\,dt = 2$. $\therefore\ W = W_1 + W_2 = \frac{9}{2}$

7. a) $\mathbf{F} = \sqrt{t}\,\mathbf{i} - 2t\,\mathbf{j} + \sqrt{t}\,\mathbf{k}, \frac{d\mathbf{r}}{dt} = \mathbf{i} + \mathbf{j} + \mathbf{k} \Rightarrow \mathbf{F}\cdot\frac{d\mathbf{r}}{dt} = 2\sqrt{t} - 2t \Rightarrow W = \int_0^1 (2\sqrt{t} - 2t)\,dt = \frac{1}{3}$

b) $\mathbf{F} = t^2\,\mathbf{i} - 2t\,\mathbf{j} + t\,\mathbf{k}, \frac{d\mathbf{r}}{dt} = \mathbf{i} + 2t\,\mathbf{j} + 4t^3\,\mathbf{k} \Rightarrow \mathbf{F}\cdot\frac{d\mathbf{r}}{dt} = 4t^4 - 3t^2 \Rightarrow W = \int_0^1 (4t^4 - 3t^2)\,dt = -\frac{1}{5}$

c) $\mathbf{F}_1 = -2t\,\mathbf{j} + \sqrt{t}\,\mathbf{k}, \frac{d\mathbf{r}_1}{dt} = \mathbf{i} + \mathbf{j} \Rightarrow \mathbf{F}_1\cdot\frac{d\mathbf{r}_1}{dt} = -2t \Rightarrow W_1 = \int_0^1 -2t\,dt = -1$. $\mathbf{F}_2 = \sqrt{t}\,\mathbf{i} - 2\,\mathbf{j} + \mathbf{k}$, $\frac{d\mathbf{r}_2}{dt} = \mathbf{k} \Rightarrow \mathbf{F}_2\cdot\frac{d\mathbf{r}_2}{dt} = 1 \Rightarrow W_2 = \int_0^1 dt = 1$. $\therefore\ W = W_1 + W_2 = 0$

9. a) $\mathbf{F} = (3t^2 - 3t)\,\mathbf{i} + 3t\,\mathbf{j} + \mathbf{k}, \frac{d\mathbf{r}}{dt} = \mathbf{i} + \mathbf{j} + \mathbf{k} \Rightarrow \mathbf{F}\cdot\frac{d\mathbf{r}}{dt} = 3t^2 + 1 \Rightarrow W = \int_0^1 (3t^2 + 1)\,dt = 2$

b) $\mathbf{F} = (3t^2 - 3t)\,\mathbf{i} + 3t^4\,\mathbf{j} + \mathbf{k}, \frac{d\mathbf{r}}{dt} = \mathbf{i} + 2t\,\mathbf{j} + 4t^3\,\mathbf{k} \Rightarrow \mathbf{F}\cdot\frac{d\mathbf{r}}{dt} = 6t^5 + 4t^3 + 3t^2 - 3t \Rightarrow$ $W = \int_0^1 \left(6t^5 + 4t^3 + 3t^2 - 3t\right)dt = \frac{3}{2}$

9. c) $\mathbf{F}_1 = (3t^2 - 3t)\,\mathbf{i} + \mathbf{k},\ \frac{d\mathbf{r}_1}{dt} = \mathbf{i} + \mathbf{j} \Rightarrow \mathbf{F}_1 \cdot \frac{d\mathbf{r}_1}{dt} = 3t^2 - 3t \Rightarrow W_1 = \int_0^1 (3t^2 - 3t)\,dt = -\frac{1}{2}$

$\mathbf{F}_2 = 3t\,\mathbf{j} + \mathbf{k},\ \frac{d\mathbf{r}_2}{dt} = \mathbf{k} \Rightarrow \mathbf{F}_2 \cdot \frac{d\mathbf{r}_2}{dt} = 1 \Rightarrow W_2 = \int_0^1 dt = 1.\ \therefore W = W_1 + W_2 = \frac{1}{2}$

11. $\mathbf{F} = t^3\,\mathbf{i} + t^2\,\mathbf{j} - t^3\,\mathbf{k},\ \frac{d\mathbf{r}}{dt} = \mathbf{i} + 2t\,\mathbf{j} + \mathbf{k} \Rightarrow \mathbf{F} \cdot \frac{d\mathbf{r}}{dt} = 2t^3 \Rightarrow W = \int_0^1 2t^3\,dt = \frac{1}{2}$

13. $\mathbf{F} = t\,\mathbf{i} + (\sin t)\,\mathbf{j} + (\cos t)\,\mathbf{k},\ \frac{d\mathbf{r}}{dt} = (\cos t)\,\mathbf{i} - (\sin t)\,\mathbf{j} + \mathbf{k} \Rightarrow \mathbf{F} \cdot \frac{d\mathbf{r}}{dt} = t\cos t - \sin^2 t + \cos t \Rightarrow$

$W = \int_0^{2\pi} (t\cos t - \sin^2 t + \cos t)\,dt = -\pi$

15. $\mathbf{F} = -4t^3\,\mathbf{i} + 8t^2\,\mathbf{j} + 2\,\mathbf{k},\ \frac{d\mathbf{r}}{dt} = \mathbf{i} + 2t\,\mathbf{j} \Rightarrow \mathbf{F} \cdot \frac{d\mathbf{r}}{dt} = 12t^3 \Rightarrow \text{Flow} = \int_0^2 12t^3\,dt = 48$

17. $\mathbf{F} = (\cos t - \sin t)\,\mathbf{i} + (\cos t)\,\mathbf{k},\ \frac{d\mathbf{r}}{dt} = (-\sin t)\,\mathbf{i} + (\cos t)\,\mathbf{k} \Rightarrow \mathbf{F} \cdot \frac{d\mathbf{r}}{dt} = -\sin t\cos t + 1 \Rightarrow$

$\text{Flow} = \int_0^{\pi} (-\sin t\cos t + 1)\,dt = \pi$

19. a) $\mathbf{F}_1 = (\cos t)\,\mathbf{i} + (\sin t)\,\mathbf{j},\ \frac{d\mathbf{r}}{dt} = (-\sin t)\,\mathbf{i} + (\cos t)\,\mathbf{j} \Rightarrow \mathbf{F}_1 \cdot \frac{d\mathbf{r}}{dt} = 0 \Rightarrow \text{Circulation} = 0.$

$M = \cos t,\ N = \sin t,\ dx = -\sin t\,dt,\ dy = \cos t\,dt \Rightarrow \text{Flux} = \int_C M\,dy - N\,dx =$

$\int_0^{2\pi} (\cos^2 t + \sin^2 t)\,dt = \int_0^{2\pi} dt = 2\pi$

$\mathbf{F}_2 = -\sin t\,\mathbf{i} + \cos t\,\mathbf{j},\ \frac{d\mathbf{r}}{dt} = (-\sin t)\,\mathbf{i} + (\cos t)\,\mathbf{j} \Rightarrow \mathbf{F}_2 \cdot \frac{d\mathbf{r}}{dt} = 1 \Rightarrow \text{Circulation} = \int_0^{2\pi} \mathbf{F}_2 \cdot \frac{d\mathbf{r}}{dt}\,dt = \int_0^{2\pi} dt = 2\pi$

$M = \cos t,\ N = \sin t,\ dx = -\cos t,\ dy = -\sin t \Rightarrow \text{Flux} = \int_C M\,dy - N\,dx = \int_0^{2\pi} 0\,dt = 0$

b) $\mathbf{F}_1 = (\cos t)\,\mathbf{i} + (4\sin t)\,\mathbf{j},\ \frac{d\mathbf{r}}{dt} = (-\sin t)\,\mathbf{i} + (4\cos t)\,\mathbf{j} \Rightarrow \mathbf{F}_1 \cdot \frac{d\mathbf{r}}{dt} = 15\sin t\cos t \Rightarrow \text{Circ} =$

$\int_0^{2\pi} 15\sin t\cos t\,dt = 0.$ $M = \cos t,\ N = 4\sin t,\ dx = -\sin t,\ dy = 4\cos t \Rightarrow \text{Flux} =$

$\int_C M\,dy - N\,dx = \int_0^{2\pi} (4\cos^2 t + 4\sin^2 t)\,dt = 8\pi$

19. (Continued)

$F_2 = -4 \sin t\, \mathbf{i} + \cos t\, \mathbf{j}, \frac{dr}{dt} = (-\sin t)\, \mathbf{i} + (4 \cos t)\, \mathbf{j} \Rightarrow F_2 \cdot \frac{dr}{dt} = 4 \sin^2 t + 4 \cos^2 t = 4 \Rightarrow$

$\text{Circ} = \int_0^{2\pi} 4\, dt = 8\pi$. $M = \cos t$, $N = 4 \sin t$, $dx = -4 \cos t$, $dy = -\sin t \Rightarrow \text{Flux} =$

$\int_C M dy - N dx = \int_0^{2\pi} 0\, dt = 0$

21. $F_1 = (a \cos t)\, \mathbf{i} + (a \sin t)\, \mathbf{j}, \frac{dr_1}{dt} = (-a \sin t)\, \mathbf{i} + (a \cos t)\, \mathbf{j} \Rightarrow F_1 \cdot \frac{dr_1}{dt} = 0 \Rightarrow \text{Circ}_1 = 0$. $M_1 = a \cos t$, $N_1 =$

$a \sin t$, $dx = -a \sin t\, dt$, $dy = a \cos t\, dt \Rightarrow \text{Flux}_1 = \int_C M_1 dy - N_1 dx = \int_0^{\pi} (a^2 \cos^2 t + a^2 \sin^2 t)\, dt =$

$\int_0^{\pi} a^2\, dt = a^2 \pi$.

$F_2 = t\, \mathbf{i}, \frac{dr_2}{dt} = \mathbf{i} \Rightarrow F_2 \cdot \frac{dr_2}{dt} = t \Rightarrow \text{Circ}_2 = \int_{-a}^{a} t\, dt = 0$. $M_2 = t$, $N_2 = 0$, $dx = dt$, $dy = 0 \Rightarrow \text{Flux}_2 =$

$\int_C M_2 dy - N_2 dx = \int_{-a}^{a} 0\, dt = 0$. $\therefore$ $\text{Circ} = \text{Circ}_1 + \text{Circ}_2 = 0$, $\text{Flux} = \text{Flux}_1 + \text{Flux}_2 = a^2 \pi$

23. $F_1 = (-a \sin t)\, \mathbf{i} + (a \cos t)\, \mathbf{j}, \frac{dr_1}{dt} = (-a \sin t)\, \mathbf{i} + (a \cos t)\, \mathbf{j} \Rightarrow F_1 \cdot \frac{dr_1}{dt} = a^2 \sin^2 t + a^2 \cos^2 t = a^2 \Rightarrow$

$\text{Circ}_1 = \int_0^{\pi} a^2\, dt = a^2 \pi$.

$F_2 = t\, \mathbf{j}, \frac{dr_2}{dt} = \mathbf{i} \Rightarrow F_2 \cdot \frac{dr_2}{dt} = 0 \Rightarrow \text{Circ}_2 = 0$. $\therefore$ $\text{Circ} = \text{Circ}_1 + \text{Circ}_2 = a^2 \pi$

$M_1 = -a \sin t$, $N_1 = a \cos t$, $dx = -a \sin t$, $dy = a \cos t \Rightarrow \text{Flux}_1 = \int_C M_1 dy - N_1 dx =$

$\int_0^{\pi} (-a^2 \sin t \cos t + a^2 \sin t \cos t)\, dt = 0$. $M_2 = 0$, $N_2 = t$, $dx = dt$, $dy = 0 \Rightarrow \text{Flux}_2 =$

$\int_C M_2 dy - N_2 dx = \int_{-a}^{a} -t\, dt = 0$. $\therefore$ $\text{Flux} = \text{Flux}_1 + \text{Flux}_2 = 0$

25.

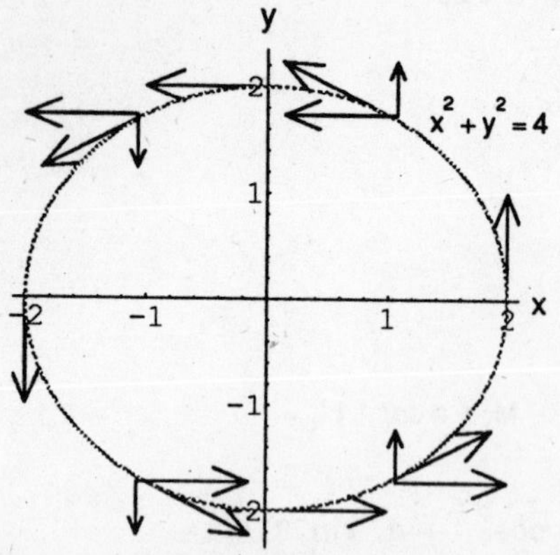

Graph 15.2.25

$\mathbf{F} = -\dfrac{y}{\sqrt{x^2+y^2}}\,\mathbf{i} + \dfrac{x}{\sqrt{x^2+y^2}}\,\mathbf{j}$ on $x^2+y^2=4$. At (2,0), $\mathbf{F} = \mathbf{j}$, at (0,2), $\mathbf{F} = -\mathbf{i}$, at (–2,0), $\mathbf{F} = -\mathbf{j}$, at (0,–2), $\mathbf{F} = \mathbf{i}$. At $(1,\sqrt{3})$, $\mathbf{F} = -\frac{\sqrt{3}}{2}\mathbf{i} + \frac{1}{2}\mathbf{j}$, at $(1,-\sqrt{3})$, $\mathbf{F} = \frac{\sqrt{3}}{2}\mathbf{i} + \frac{1}{2}\mathbf{j}$, at $(-1,\sqrt{3})$, $\mathbf{F} = -\frac{\sqrt{3}}{2}\mathbf{i} - \frac{1}{2}\mathbf{j}$, at $(-1,-\sqrt{3})$, $\mathbf{F} = \frac{\sqrt{3}}{2}\mathbf{i} - \frac{1}{2}\mathbf{j}$

27. a) $\mathbf{G} = P(x,y)\,\mathbf{i} + Q(x,y)\,\mathbf{j}$ is to have a magnitude $\sqrt{a^2+b^2}$ and to be tangent to $x^2+y^2 = a^2+b^2$ in a counterclockwise direction. $x^2+y^2 = a^2+b^2 \Rightarrow 2x + 2y\,y' = 0 \Rightarrow y' = -\frac{x}{y}$ is the slope of the tangent line at any point on the circle. $(a,b) \Rightarrow y' = -\frac{a}{b}$. Let $\mathbf{v} = -b\,\mathbf{i} + a\,\mathbf{j} \Rightarrow |\mathbf{v}| = \sqrt{a^2+b^2}$ and $\mathbf{v}$ is in a counterclockwise direction. $\therefore$ let $P(x,y) = -y$, $Q(x,y)\ x \Rightarrow \mathbf{G} = -y\,\mathbf{i} + x\,\mathbf{j}$. Then if (a,b) is on $x^2+y^2 = a^2+b^2$, $\mathbf{G} = -b\,\mathbf{i} + a\,\mathbf{j} \Rightarrow |\mathbf{G}| = \sqrt{a^2+b^2}$, $\mathbf{G}$ is tangent to the circle in a counterclockwise direction.

b) $\mathbf{G} = \sqrt{x^2+y^2}\ \mathbf{F} = \sqrt{a^2+b^2}\ \mathbf{F}$ since $x^2+y^2 = a^2+b^2$.

29. The slope of a line through (x,y) and the origin is $\frac{y}{x} \Rightarrow \mathbf{v} = x\,\mathbf{i} + y\,\mathbf{j}$ is a vector on that line. But $\mathbf{v}$ points away from the origin. $\therefore\ \mathbf{F} = -\dfrac{x\,\mathbf{i} + y\,\mathbf{j}}{\sqrt{x^2+y^2}}$ is the unit vector we want.

31. a) $x^2+y^2 = 1 \Rightarrow \mathbf{r} = (\cos t)\,\mathbf{i} + (\sin t)\,\mathbf{j},\ 0 \le t \le \pi \Rightarrow x = \cos t,\ y = \sin t \Rightarrow \mathbf{F} = (\cos t + \sin t)\,\mathbf{i} - \mathbf{j}$ and $\frac{d\mathbf{r}}{dt} = (-\sin t)\,\mathbf{i} + (\cos t)\,\mathbf{j} \Rightarrow \mathbf{F}\cdot\frac{d\mathbf{r}}{dt} = -\sin t\cos t - \sin^2 t - \cos t \Rightarrow$ Flow =

$$\int_0^{\pi} (-\sin t\cos t - \sin^2 t - \cos t)\,dt = -\frac{1}{2}\pi$$

b) $\mathbf{r} = -t\,\mathbf{i},\ -1 \le t \le 1 \Rightarrow x = -t,\ y = 0 \Rightarrow \mathbf{F} = -t\,\mathbf{i} - t^2\,\mathbf{j}$ and $\frac{d\mathbf{r}}{dt} = -\mathbf{i} \Rightarrow \mathbf{F}\cdot\frac{d\mathbf{r}}{dt} = t \Rightarrow$ Flow $= \displaystyle\int_{-1}^{1} t\,dt = 0$

c) $\mathbf{r}_1 = (1-t)\,\mathbf{i} - t\,\mathbf{j},\ 0 \le t \le 1 \Rightarrow \mathbf{F}_1 = (1-2t)\,\mathbf{i} - (1 - 2t + 2t^2)\,\mathbf{j}$ and $\frac{d\mathbf{r}_1}{dt} = -\mathbf{i} - \mathbf{j} \Rightarrow \mathbf{F}_1\cdot\frac{d\mathbf{r}_1}{dt} = 2t^2 \Rightarrow$

Flow$_1$ $= \displaystyle\int_0^1 2t^2\,dt = \frac{2}{3}$. $\mathbf{r}_2 = -t\,\mathbf{i} + (t-1)\,\mathbf{j},\ 0 \le t \le 1 \Rightarrow \mathbf{F}_2 = -\mathbf{i} - (2t^2 - 2t + 1)\,\mathbf{j}$ and $\frac{d\mathbf{r}_2}{dt} = -\mathbf{i} + \mathbf{j} \Rightarrow \mathbf{F}_2\cdot\frac{d\mathbf{r}_2}{dt} = 2t - 2t^2 \Rightarrow$ Flow$_2$ $= \displaystyle\int_0^1 (2t - 2t^2)\,dt = \frac{1}{3}$. $\therefore$ Flow = Flow$_1$ + Flow$_2$ = 1

33. C_1: $\mathbf{r}(t) = \cos t\,\mathbf{i} + \sin t\,\mathbf{j} + t\,\mathbf{k},\ 0 \le t \le \frac{\pi}{2} \Rightarrow \mathbf{F} = 2\cos t\,\mathbf{i} + 2t\,\mathbf{j} + 2\sin t\,\mathbf{k}.\ \frac{d\mathbf{r}}{dt} = -\sin t\,\mathbf{i} + \cos t\,\mathbf{j} + \mathbf{k} \Rightarrow \mathbf{F}\cdot\frac{d\mathbf{r}}{dt} =$ $-2\cos t \sin t + 2t\cos t + 2\sin t = -\sin 2t + 2t\cos t + 2\sin t$. Then $\text{Flow}_1 =$

$$\int_0^{\pi/2} (-\sin 2t + 2t\cos t + 2\sin t)\,dt = \left[\frac{1}{2}\cos 2t + 2t\sin t + 2\cos t - 2\cos t\right]_0^{\pi/2}$$

(do the middle term of the integral by Parts) $= -1 + \pi$.

C_2: $\mathbf{r}(t) = \mathbf{j} + \frac{\pi}{2}(1-t)\,\mathbf{k},\ 0 \le t \le 1 \Rightarrow \mathbf{F} = \pi(1-t)\,\mathbf{j} + 2\,\mathbf{k}.\ \frac{d\mathbf{r}}{dt} = -\frac{\pi}{2}\mathbf{k} \Rightarrow \mathbf{F}\cdot\frac{d\mathbf{r}}{dt} = -\pi.\ \therefore \text{Flow}_2 =$

$$\int_0^1 -\pi\,dt = \left[-\pi t\right]_0^1 = -\pi.$$

C_3: $\mathbf{r}(t) = t\,\mathbf{i} + (1-t)\,\mathbf{j},\ 0 \le t \le 1 \Rightarrow \mathbf{F} = 2t\,\mathbf{i} + 2(1-t)\,\mathbf{k}.\ \frac{d\mathbf{r}}{dt} = \mathbf{i} - \mathbf{j} \Rightarrow \mathbf{F}\cdot\frac{d\mathbf{r}}{dt} = 2t.\ \therefore \text{Flow}_3 =$

$$\int_0^1 2t\,dt = \left[t^2\right]_0^1 = 1.\ \therefore \text{Circulation} = (-1+\pi) - \pi + 1 = 0.$$

15.3 PATH INDEPENDENCE, POTENTIAL FUNCTIONS, AND CONSERVATIVE FIELDS

1. $\frac{\partial P}{\partial y} = x = \frac{\partial N}{\partial z},\ \frac{\partial M}{\partial z} = y = \frac{\partial P}{\partial x},\ \frac{\partial N}{\partial x} = z = \frac{\partial M}{\partial y} \Rightarrow$ Conservative

3. $\frac{\partial P}{\partial y} = -1 \ne \frac{\partial N}{\partial z} \Rightarrow$ Not Conservative

5. $\frac{\partial N}{\partial x} = 0 \ne \frac{\partial M}{\partial y} \Rightarrow$ Not Conservative

7. $\frac{\partial f}{\partial x} = 2x \Rightarrow f(x,y,z) = x^2 + g(y,z).\ \frac{\partial f}{\partial y} = \frac{\partial g}{\partial y} = 3y \Rightarrow g(y,z) = \frac{3y^2}{2} + h(z) \Rightarrow f(x,y,z) = x^2 + \frac{3y^2}{2} + h(z).\ \frac{\partial f}{\partial z} = h'(z)$ $= 4z. \Rightarrow h(z) = 2z^2 + C \Rightarrow f(x,y,z) = x^2 + \frac{3y^2}{2} + 2z^2 + C$

9. $\frac{\partial f}{\partial x} = e^{y+2z} \Rightarrow f(x,y,z) = x\,e^{y+2z} + g(y,z).\ \frac{\partial f}{\partial y} = x\,e^{y+2z} + \frac{\partial g}{\partial y} = x\,e^{y+2z} \Rightarrow \frac{\partial g}{\partial y} = 0.$ Then $f(x,y,z) = x\,e^{y+2z} +$ $h(z).\ \frac{\partial f}{\partial z} = 2x\,e^{y+2z} + h'(z) = 2x\,e^{y+2z} \Rightarrow h'(z) = 0 \Rightarrow h(z) = C.\ \therefore\ f(x,y,z) = x\,e^{y+2z} + C$

11. Let $\mathbf{F}(x,y,z) = 2x\,\mathbf{i} + 2y\,\mathbf{j} + 2z\,\mathbf{k} \Rightarrow \frac{\partial P}{\partial y} = 0 = \frac{\partial N}{\partial z},\ \frac{\partial M}{\partial z} = 0 = \frac{\partial P}{\partial x},\ \frac{\partial N}{\partial x} = 0 = \frac{\partial M}{\partial y} \Rightarrow M\,dx + N\,dy + P\,dz$ is exact. $\frac{\partial f}{\partial x} = 2x \Rightarrow f(x,y,z) = x^2 + g(y,z).\ \frac{\partial f}{\partial y} = \frac{\partial g}{\partial y} = 2y \Rightarrow g(y,z) = y^2 + h(z) \Rightarrow f(x,y,z) = x^2 + y^2 + h(z).$

$\frac{\partial f}{\partial z} = h'(z) = 2z \Rightarrow h(z) = z^2 + C.\ \therefore\ f(x,y,z) = x^2 + y^2 + z^2 + C \Rightarrow \int_{(0,0,0)}^{(2,3,-6)} 2x\,dx + 2y\,dy + 2z\,dz =$

$f(2,3,-6) - f(0,0,0) = 49$

13. Let $\mathbf{F}(x,y,z) = 2xy\,\mathbf{i} + (x^2 - z^2)\,\mathbf{j} - 2yz\,\mathbf{k} \Rightarrow \frac{\partial P}{\partial y} = -2z = \frac{\partial N}{\partial z}, \frac{\partial M}{\partial z} = 0 = \frac{\partial P}{\partial x}, \frac{\partial N}{\partial x} = 2x = \frac{\partial M}{\partial y} \Rightarrow M\,dx + N\,dy +$ $P\,dz$ is exact. $\frac{\partial f}{\partial x} = 2xy \Rightarrow f(x,y,z) = x^2y + g(y,z)$. $\frac{\partial f}{\partial y} = x^2 + \frac{\partial g}{\partial y} = x^2 - z^2 \Rightarrow \frac{\partial g}{\partial y} = -z^2 \Rightarrow g(y,z) = -yz^2 +$ $h(z) \Rightarrow f(x,y,z) = x^2y - yz^2 + h(z)$. $\frac{\partial f}{\partial z} = -2yz + h'(z) = -2yz \Rightarrow h'(z) = 0 \Rightarrow h(z) = C \Rightarrow f(x,y,z) =$

$$x^2y - yz^2 + C \Rightarrow \int_{(0,0,0)}^{(1,2,3)} 2xy\,dx + (x^2 - z^2)\,dy - 2yz\,dz = f(1,2,3) - f(0,0,0) = -16$$

15. Let $\mathbf{F}(x,y,z) = (\sin y \cos x)\,\mathbf{i} + (\cos y \sin x)\,\mathbf{j} + \mathbf{k} \Rightarrow \frac{\partial P}{\partial y} = 0 = \frac{\partial N}{\partial z}, \frac{\partial M}{\partial z} = 0 = \frac{\partial P}{\partial x}, \frac{\partial N}{\partial x} = \cos y \cos x = \frac{\partial M}{\partial y} \Rightarrow$ $M\,dx + N\,dy + P\,dz$ is exact. $\frac{\partial f}{\partial x} = \sin y \cos x \Rightarrow f(x,y,z) = \sin y \sin x + g(y,z)$. $\frac{\partial f}{\partial y} = \cos y \sin x + \frac{\partial g}{\partial y} =$ $\cos y \sin x \Rightarrow \frac{\partial g}{\partial y} = 0 \Rightarrow g(y,z) = h(z) \Rightarrow f(x,y,z) = \sin y \sin x + h(z)$. $\frac{\partial f}{\partial z} = h'(z) = 1 \Rightarrow h(z) = z + C \Rightarrow$

$$f(x,y,z) = \sin y \sin x + z + C \Rightarrow \int_{(1,0,0)}^{(0,1,1)} \sin y \cos x\,dx + \cos y \sin x\,dy + dz = f(0,1,1) - f(1,0,0) = 1$$

17. Let $\mathbf{F}(x,y,z) = (2\cos y)\,\mathbf{i} + \left(\frac{1}{y} - 2x \sin y\right)\mathbf{j} + \frac{1}{z}\mathbf{k} \Rightarrow \frac{\partial P}{\partial y} = 0 = \frac{\partial N}{\partial z}, \frac{\partial M}{\partial z} = 0 = \frac{\partial P}{\partial x}, \frac{\partial N}{\partial x} = -2\sin y = \frac{\partial M}{\partial y} \Rightarrow$ $M\,dx + N\,dy + P\,dz$ is exact. $\frac{\partial f}{\partial x} = 2\cos y \Rightarrow f(x,y,z) = 2x\cos y + g(y,z)$. $\frac{\partial f}{\partial y} = -2x\sin y + \frac{\partial g}{\partial y} = \frac{1}{y} - 2x\sin y$ $\Rightarrow \frac{\partial g}{\partial y} = \frac{1}{y} \Rightarrow g(y,z) = \ln y + h(z) \Rightarrow f(x,y,z) = 2x\cos y + \ln y + h(z)$. $\frac{\partial f}{\partial z} = h'(z) = \frac{1}{z} \Rightarrow h(z) = \ln z + C \Rightarrow$

$$f(x,y,z) = 2x\cos y + \ln y + \ln z + C \Rightarrow \int_{(0,2,1)}^{(1,\pi/2,2)} 2\cos y\,dx + \left(\frac{1}{y} - 2x\sin y\right)dy + \frac{1}{z}dz =$$

$$f\left(1,\frac{\pi}{2},2\right) - f(0,2,1) = \ln\frac{\pi}{2}$$

19. Let $\mathbf{F}(x,y,z) = \frac{1}{y}\mathbf{i} + \left(\frac{1}{z} - \frac{x}{y^2}\right)\mathbf{j} - \frac{y}{z^2}\mathbf{k} \Rightarrow \frac{\partial P}{\partial y} = -\frac{1}{z^2} = \frac{\partial N}{\partial z}, \frac{\partial M}{\partial z} = 0 = \frac{\partial P}{\partial x}, \frac{\partial N}{\partial x} = -\frac{1}{y^2} = \frac{\partial M}{\partial y} \Rightarrow$ $M\,dx + N\,dy + P\,dz$ is exact. $\frac{\partial f}{\partial x} = \frac{1}{y} \Rightarrow f(x,y,z) = \frac{x}{y} + g(y,z)$. $\frac{\partial f}{\partial y} = -\frac{x}{y^2} + \frac{\partial g}{\partial y} = \frac{1}{z} - \frac{x}{y^2} \Rightarrow \frac{\partial g}{\partial y} = \frac{1}{z} \Rightarrow$ $g(y,z) = \frac{y}{z} + h(z) \Rightarrow f(x,y,z) = \frac{x}{y} + \frac{y}{z} + h(z)$. $\frac{\partial f}{\partial z} = -\frac{y}{z^2} + h'(z) = -\frac{y}{z^2} \Rightarrow h'(z) = 0 \Rightarrow h(z) = C \Rightarrow$

$$f(x,y,z) = \frac{x}{y} + \frac{y}{z} + C \Rightarrow \int_{(1,1,1)}^{(2,2,2)} \frac{1}{y}\,dx + \left(\frac{1}{z} - \frac{x}{y^2}\right)dy - \frac{y}{z^2}dz = f(2,2,2) - f(1,1,1) = 0$$

21. Let $x - 1 = t$, $y - 1 = 2t$, $z - 1 = -2t$, $0 \le t \le 1 \Rightarrow dx = dt$, $dy = 2\,dt$, $dz = -2\,dt \Rightarrow \int_{(1,1,1)}^{(2,3,-1)} y\,dx + x\,dy + 4\,dz$

$$= \int_0^1 (2t + 1)dt + (t + 1)2\,dt + 4(-2)dt = \int_0^1 (4t - 5)\,dt = -3$$

23. $\frac{\partial P}{\partial y} = 0 = \frac{\partial N}{\partial z}, \frac{\partial M}{\partial z} = 2z = \frac{\partial P}{\partial x}, \frac{\partial N}{\partial x} = 0 = \frac{\partial M}{\partial y} \Rightarrow M\,dx + N\,dy + P\,dz$ is exact $\Rightarrow$ **F** is conservative $\Rightarrow$ path independence.

25. $\frac{\partial P}{\partial y} = 0 = \frac{\partial N}{\partial z}, \frac{\partial M}{\partial z} = 0 = \frac{\partial P}{\partial x}, \frac{\partial N}{\partial x} = -\frac{2x}{y^2} = \frac{\partial M}{\partial y} \Rightarrow$ **F** is conservative $\Rightarrow$ there exists an f so that $\mathbf{F} = \nabla f$.
$\frac{\partial f}{\partial x} = \frac{2x}{y} \Rightarrow f(x,y) = \frac{x^2}{y} + g(y) \Rightarrow \frac{\partial f}{\partial y} = -\frac{x^2}{y} + g'(y) = \frac{1 - x^2}{y^2} \Rightarrow g'(y) = \frac{1}{y^2} \Rightarrow g(y) = -\frac{1}{y} + C \Rightarrow f(x,y) = \frac{x^2}{y} - \frac{1}{y} + C$. Let C = 0. Then $f(x,y) = \frac{x^2 - 1}{y}$. $\therefore \mathbf{F} = \nabla\left(\frac{x^2 - 1}{y}\right)$.

27. $\mathbf{F} = (x^2 + y)\,\mathbf{i} + (y^2 + x)\,\mathbf{j} + z e^z\,\mathbf{k} \Rightarrow \frac{\partial P}{\partial y} = 0 = \frac{\partial N}{\partial z}, \frac{\partial M}{\partial z} = 0 = \frac{\partial P}{\partial x}, \frac{\partial N}{\partial x} = 1 = \frac{\partial M}{\partial y} \Rightarrow$ **F** is conservative. $\frac{\partial f}{\partial x} = x^2 + y \Rightarrow f(x,y,z) = \frac{x^3}{3} + xy + g(y,z) \Rightarrow \frac{\partial f}{\partial y} = x + \frac{\partial g}{\partial y} = y^2 + x \Rightarrow \frac{\partial g}{\partial y} = y^2 \Rightarrow g(y,z) = \frac{y^3}{3} + h(z)$. Then $f(x,y,z) = \frac{x^3}{3} + xy + \frac{y^3}{3} + h(z) \Rightarrow \frac{\partial f}{\partial z} = h'(z) = z e^z \Rightarrow h(z) = z e^z - e^z + C$ (let C = 0). $\therefore f(x,y,z) = \frac{x^3}{3} + xy + \frac{y^3}{3} + z e^z - e^z$ is a potential function for **F**. Thus work $= \int_{t=a}^{t=b} \mathbf{F} \cdot d\mathbf{R} = \int_{t=a}^{t=b} M\,dx + N\,dy + P\,dz = f(1,0,1) - f(1,0,0) = 1$ regardless of the path taken.

29. a) $\mathbf{F} = \nabla(x^3y^2) \Rightarrow \mathbf{F} = 3x^2y^2\,\mathbf{i} + 2x^3y\,\mathbf{j}$. Let C_1 be the path from (−1,1) to (0,0) $\Rightarrow x = t - 1, y = -t + 1$, $0 \le t \le 1 \Rightarrow \mathbf{F} = 3(t-1)^2(-t+1)^2\,\mathbf{i} + 2(t-1)^3(-t+1)\,\mathbf{j} = 3(t-1)^4\,\mathbf{i} - 2(t-1)^4\,\mathbf{j}$ and $\mathbf{R}_1 = (t-1)\,\mathbf{i} + (-t+1)\,\mathbf{j} \Rightarrow d\mathbf{R}_1 = dt\,\mathbf{i} - dt\,\mathbf{j}$. $\therefore \int_C \mathbf{F} \cdot d\mathbf{R}_1 = \int_0^1 \left(3(t-1)^4 + 2(t-1)^4\right) dt = \int_0^1 5(t-1)^4\,dt = 1$

Let C_2 be the path from (0,0) to (1,1) y $x = t, y = t, 0 \le t \le 1 \Rightarrow \mathbf{F} = 3t^4\,\mathbf{i} + 2t^4\,\mathbf{j}$ and $\mathbf{R}_2 = t\,\mathbf{i} + t\,\mathbf{j} \Rightarrow d\mathbf{R}_2 = dt\,\mathbf{i} + dt\,\mathbf{j}$. $\therefore \int_{C_2} \mathbf{F} \cdot d\mathbf{R}_2 = \int_0^1 (3t^4 + 2t^4)\,dt = \int_0^1 5t^4\,dt = 1$. Then $\int_C \mathbf{F} \cdot d\mathbf{R} = \int_{C_1} \mathbf{F} \cdot d\mathbf{R}_1 + \int_{C_2} \mathbf{F} \cdot d\mathbf{R}_2 = 2$

b) Since $f(x,y) = x^3y^2$ is a potential function for **F**, $\int_{(-1,1)}^{(1,1)} \mathbf{F} \cdot d\mathbf{R} = f(1,1) - f(-1,1) = 2$.

31. The path will not matter. The work along any path will be the same because the field is conservative.

15.4 GREEN'S THEOREM IN THE PLANE

1. Equation 15: $M = -y = -a \sin t$, $N = x = a \cos t$, $dx = -a \sin t\, dt$, $dy = a \cos t\, dt \Rightarrow \frac{\partial M}{\partial x} = 0, \frac{\partial M}{\partial y} = -1,$

$\frac{\partial N}{\partial x} = 1, \frac{\partial N}{\partial y} = 0 \Rightarrow \oint_C M dy - N dx = \int_0^{2\pi} ((-a \sin t)(a \cos t)\, dt - (a \cos t)(-a \sin t))dt =$

$0. \iint_R \left(\frac{\partial M}{\partial x} + \frac{\partial N}{\partial y}\right) dx\, dy = \iint_R 0\, dx\, dy = 0$

Equation 16: $\oint_C M dx + N dy = \int_0^{2\pi} ((-a \sin t)(-a \sin t) + (a \cos t)(a \cos t))\, dt = 2\pi a^2$

$\iint_R \left(\frac{\partial N}{\partial x} - \frac{\partial M}{\partial y}\right) dx\, dy = \int_{-a}^{a} \int_{-\sqrt{a^2 - x^2}}^{\sqrt{a^2 - x^2}} 2\, dy\, dx \quad \int_{-a}^{a} 4\sqrt{a^2 - x^2}\, dx = 2a^2\pi$

3. $M = 2x = 2a \cos t$, $N = -3y = -3a \sin t$, $dx = -a \sin t$, $dy = a \cos t \Rightarrow \frac{\partial M}{\partial x} = 2, \frac{\partial M}{\partial y} = 0, \frac{\partial N}{\partial x} = 0, \frac{\partial N}{\partial y} = -3$

Equation 15: $\oint_C M dy - N dx = \int_0^{2\pi} (2a \cos t(a \cos t) + 3a \sin t(-a \sin t))dt =$

$\int_0^{2\pi} \left(2a^2 \cos^2 t - 3a^2 \sin^2 t\right) dt = -\pi a^2. \iint_R \left(\frac{\partial M}{\partial x} + \frac{\partial N}{\partial y}\right) = \iint_R -1\, dx\, dy =$

$\int_{-a}^{a} \int_{-\sqrt{a^2 - x^2}}^{\sqrt{a^2 - x^2}} -1\, dy\, dx = -\pi a^2$

Equation 16: $\oint_C M dx + N dy = \int_0^{2\pi} (2a \cos t(-a \sin t) + (-3a \sin t)(a \cos t))\, dt =$

$\int_0^{2\pi} \left(-2a^2 \sin t \cos t - 3a^2 \sin t \cos t\right) dt = 0. \iint_R 0\, dx\, dy = 0$

5. $M = x - y$, $N = y - x \Rightarrow \frac{\partial M}{\partial x} = 1, \frac{\partial M}{\partial y} = -1, \frac{\partial N}{\partial x} = -1, \frac{\partial N}{\partial y} = 1 \Rightarrow \text{Flux} = \iint_R 2\, dx\, dy = \int_0^1 \int_0^1 2\, dx\, dy$

$= 2.\ \text{Circ} = \iint_R (-1 - (-1))\, dx\, dy = 0$

7. $M = y^2 - x^2,\ N = x^2 + y^2 \Rightarrow \frac{\partial M}{\partial x} = -2x,\ \frac{\partial M}{\partial y} = 2y,\ \frac{\partial N}{\partial x} = 2x,\ \frac{\partial N}{\partial y} = 2y \Rightarrow \text{Flux} = \iint_R (-2x + 2y)\,dx\,dy$

$= \int_0^3 \int_0^x (-2x + 2y)\,dy\,dx = \int_0^3 (-2x^2 + x^2)\,dx = -9.$ Circ $= \iint_R (2x - 2y)\,dx\,dy =$

$\int_0^3 \int_0^x (2x - 2y)\,dy\,dx = \int_0^3 x^2\,dx = 9$

9. $M = x + e^x \sin y,\ N = x + e^x \cos y \Rightarrow \frac{\partial M}{\partial x} = 1 + e^x \sin y,\ \frac{\partial M}{\partial y} = e^x \cos y,\ \frac{\partial N}{\partial x} = 1 + e^x \cos y,\ \frac{\partial N}{\partial y} = e^x \sin y$

$\Rightarrow \text{Flux} = \iint_R dx\,dy = \int_{-\pi/4}^{\pi/4} \int_0^{\sqrt{\cos 2\theta}} r\,dr\,d\theta = \int_{-\pi/4}^{\pi/4} \left(\frac{1}{2}\cos 2\theta\right) d\theta = \frac{1}{2}.$

Circ $= \iint_R (1 + e^x \cos y - e^x \cos y)\,dx\,dy = \iint_R dx\,dy =$

$\int_{-\pi/4}^{\pi/4} \int_0^{\sqrt{\cos 2\theta}} r\,dr\,d\theta = \int_{-\pi/4}^{\pi/4} \frac{1}{2}\cos 2\theta\,d\theta = \frac{1}{2}$

11. $M = xy,\ N = y^2 \Rightarrow \frac{\partial M}{\partial x} = y,\ \frac{\partial M}{\partial y} = x,\ \frac{\partial N}{\partial x} = 0,\ \frac{\partial N}{\partial y} = 2y \Rightarrow \text{Flux} = \iint_R (y + 2y)\,dy\,dx = \int_0^1 \int_{x^2}^{x} 3y\,dy\,dx$

$= \int_0^1 \left(\frac{3x^2}{2} - \frac{3x^4}{2}\right) dx = \frac{1}{5}.$ Circ $= \iint_R -x\,dy\,dx = \int_0^1 \int_{x^2}^{x} -x\,dy\,dx = \int_0^1 (-x^2 + x^3)\,dx = -\frac{1}{12}$

13. $M = 3xy - \frac{x}{1 + y^2},\ N = e^x + \tan^{-1} y \Rightarrow \frac{\partial M}{\partial x} = 3y - \frac{1}{1 + y^2},\ \frac{\partial N}{\partial y} = \frac{1}{1 + y^2} \Rightarrow \text{Flux} =$

$\iint_R \left(3y - \frac{1}{1 + y^2} + \frac{1}{1 + y^2}\right) dx\,dy = \iint_R 3y\,dx\,dy = \int_0^{2\pi} \int_0^{a(1+\cos\theta)} 3r \sin\theta\,r\,dr\,d\theta = \int_0^{2\pi} a^3(1 + \cos\theta)^3 \sin\theta\,d\theta$

$= -4a^3 - (-4a^3) = 0$

15. $M = y^2, N = x^2 \Rightarrow \frac{\partial M}{\partial y} = 2y, \frac{\partial N}{\partial x} = 2x \Rightarrow \oint_C y^2\,dx + x^2\,dy = \iint_R (2x - 2y)\,dy\,dx =$

$$\int_0^1 \int_0^{-x+1} (2x - 2y)\,dy\,dx = \int_0^1 (-3x^2 + 4x - 1)\,dx = 0.$$

17. $M = 6y + x, N = y + 2x \Rightarrow \frac{\partial M}{\partial y} = 6, \frac{\partial N}{\partial x} = 2 \Rightarrow \oint_C (6y + x)dx + (y + 2x)dy = \iint_R (2 - 6)\,dy\,dx =$

-4(Area of the circle) $= -16\pi$

19. $M = 2xy^3, N = 4x^2y^2 \Rightarrow \frac{\partial M}{\partial y} = 6xy^2, \frac{\partial N}{\partial x} = 8xy^2 \Rightarrow \oint_C 2xy^3\,dx + 4x^2y^2\,dy = \iint_R (8xy^2 - 6xy^2)\,dx\,dy$

$$\int_0^1 \int_0^{x^3} 2xy^2\,dy\,dx = \int_0^1 \frac{2}{3}x^{10}\,dx = \frac{2}{33}$$

21. a) $M = f(x), N = g(y) \Rightarrow \frac{\partial M}{\partial y} = 0, \frac{\partial N}{\partial x} = 0 \Rightarrow \oint_C f(x)\,dx + g(y)\,dy = \iint_R 0\,dy\,dx = 0$

b) $M = ky, N = hx \Rightarrow \frac{\partial M}{\partial y} = k, \frac{\partial N}{\partial x} = h \Rightarrow \oint_C ky\,dx + hx\,dy = \iint_R (h - k)\,dx\,dy =$

$(h - k)$(Area of the region)

23. $M = 4x^3y, N = x^4 \Rightarrow \frac{\partial M}{\partial y} = 4x^3, \frac{\partial N}{\partial x} = 4x^3 \Rightarrow \oint_C 4x^3y\,dx + x^4\,dy = \iint_R (4x^3 - 4x^3)\,dx\,dy = 0$

25. If a two-dimensional vector field is conservative, then $\frac{\partial N}{\partial x} = \frac{\partial M}{\partial y} \Rightarrow \frac{\partial N}{\partial x} - \frac{\partial M}{\partial y} = \text{curl } \mathbf{F} = 0$. A two-dimensional field $\mathbf{F} = M\,\mathbf{i} + N\,\mathbf{j}$ can be considered to be the restriction to the xy-plane of a three-dimensional field whose **k** component is zero, and whose **i** and **j** components are independent of z.

27. Area $= \frac{1}{2}\oint_C x\,dy - y\,dx$. $M = x = a\cos t, N = y = a\sin t \Rightarrow dx = -a\sin t\,dt, dy = a\cos t\,dt \Rightarrow$ Area $=$

$$\frac{1}{2}\int_0^{2\pi} \left(a^2\cos^2 t + a^2\sin^2 t\right) dt = \frac{1}{2}\int_0^{2\pi} a^2\,dt = \pi a^2$$

29. Area $= \frac{1}{2}\oint_C x\,dy - y\,dx$. $M = x = \cos^3 t$, $N = y = \sin^3 t \Rightarrow dx = -3\cos^2 t \sin t\,dt$, $dy = 3\sin^2 t \cos t\,dt$

$$\Rightarrow \text{Area} = \frac{1}{2}\int_0^{2\pi} (3\sin^2 t\cos^2 t(\cos^2 t + \sin^2 t))\,dt = \frac{1}{2}\int_0^{2\pi} (3\sin^2 t\cos^2 t)\,dt = \frac{3\pi}{8}$$

15.5 SURFACE AREA AND SURFACE INTEGRALS

1. $\mathbf{p} = \mathbf{k}$, $\nabla f = 2x\,\mathbf{i} + 2y\,\mathbf{j} - \mathbf{k} \Rightarrow |\nabla f| = \sqrt{(2x)^2 + (2y)^2 + (-1)^2} = \sqrt{4x^2 + 4y^2 + 1}$. $|\nabla f\cdot\mathbf{p}| = 1 \Rightarrow$

$$S = \iint_R \frac{|\nabla f|}{|\nabla f\cdot \mathbf{p}|}\,dA = \iint_R \sqrt{4x^2 + 4y^2 + 1}\,dx\,dy =$$

$$\iint_R \sqrt{4r^2\cos^2\theta + 4r^2\sin^2\theta + 1}\;r\,dr\,d\theta = \int_0^{2\pi}\int_0^{\sqrt{2}} \sqrt{4r^2 + 1}\;r\,dr\,d\theta = \frac{13}{3}\pi$$

3. $\mathbf{p} = \mathbf{k}$. $\nabla f = \mathbf{i} + 2\,\mathbf{j} + 2\,\mathbf{k} \Rightarrow |\nabla f| = 3$. $|\nabla f\cdot\mathbf{p}| = 2 \Rightarrow S = \iint_R \frac{|\nabla f|}{|\nabla f\cdot\mathbf{p}|}\,dA = \iint_R \frac{3}{2}\,dx\,dy =$

$$\int_{-1}^{1}\int_{y^2}^{2-y^2} \frac{3}{2}\,dx\,dy = \int_{-1}^{1} (3 - 3y^2)\,dy = 4$$

5. $\mathbf{p} = \mathbf{k}$. $\nabla f = 2x\,\mathbf{i} - 2\,\mathbf{j} - 2\,\mathbf{k} \Rightarrow |\nabla f| = \sqrt{(2x)^2 + (-2)^2 + (-2)^2} = \sqrt{4x^2 + 8}$. $|\nabla f\cdot\mathbf{p}| = 2 \Rightarrow S =$

$$\iint_R \frac{|\nabla f|}{|\nabla f\cdot\mathbf{p}|}\,dA = \iint_R \frac{\sqrt{4x^2 + 8}}{2}\,dx\,dy = \int_0^2\int_0^{3x} \sqrt{x^2 + 2}\,dy\,dx = \int_0^2 3x\sqrt{x^2 + 2}\,dx = 6\sqrt{6} - 2\sqrt{2}$$

7. $\mathbf{p} = \mathbf{k}$. $\nabla f = c\,\mathbf{i} - \mathbf{k} \Rightarrow |\nabla f| = \sqrt{c^2 + 1}$. $|\nabla f\cdot\mathbf{p}| = 1 \Rightarrow S = \iint_R \frac{|\nabla f|}{|\nabla f\cdot\mathbf{p}|}\,dA = \iint_R \sqrt{c^2 + 1}\,dx\,dy =$

$$\int_0^{2\pi}\int_0^1 \sqrt{c^2 + 1}\;r\,dr\,d\theta = \int_0^{2\pi} \frac{\sqrt{c^2 + 1}}{2}\,d\theta = \pi\sqrt{c^2 + 1}$$

9. $\mathbf{p} = \mathbf{i}$, $\nabla f = \mathbf{i} + 2y\,\mathbf{j} + 2z\,\mathbf{k} \Rightarrow |\nabla f| = \sqrt{1^2 + (2y)^2 + (2z)^2} = \sqrt{1 + 4y^2 + 4z^2}$. $|\nabla f \cdot \mathbf{p}| = 1 \Rightarrow$

$$S = \iint_R \frac{|\nabla f|}{|\nabla f \cdot \mathbf{p}|}\,dA = \iint_R \sqrt{1 + 4y^2 + 4z^2}\,dy\,dz = \int_0^{2\pi}\int_1^2 \sqrt{1 + 4r^2\cos^2\theta + 4r^2\sin^2\theta}\;r\,dr\,d\theta$$

$$= \int_0^{2\pi}\int_1^2 \sqrt{1 + 4r^2}\;r\,dr\,d\theta = \frac{17\pi\sqrt{17} - 5\pi\sqrt{5}}{6}$$

11. The bottom face of the cube is in the xy–plane $\Rightarrow z = 0 \Rightarrow g(x,y,0) = x + y$ and $f(x,y,z) = z = 0 \Rightarrow$

$\nabla f = \mathbf{k} \Rightarrow |\nabla f| = 1$. $\mathbf{p} = \mathbf{k} \Rightarrow |\nabla f \cdot \mathbf{p}| = 1 \Rightarrow d\sigma = dx\,dy \Rightarrow \displaystyle\iint_{z=0} (x + y)\,dx\,dy = \int_0^a\int_0^a (x + y)\,dx\,dy =$

a^3. Because of symmetry, you get a^3 over the face of the cube in the xz–plane and a^3 over the face of the cube in the yz–plane.

In the top of the cube, $g(x,y,z) = g(x,y,a) = x + y + a$ and $f(x,y,z) = z = a \Rightarrow \nabla f = \mathbf{k} \Rightarrow |\nabla f| = 1$. $\mathbf{p} = \mathbf{k} \Rightarrow$

$|\nabla f \cdot \mathbf{p}| = 1 \Rightarrow d\sigma = dx\,dy \Rightarrow \displaystyle\iint_{z=a} (x + y + a)\,dx\,dy = \int_0^a\int_0^a (x + y + a)\,dx\,dy = 2a^3.$

Because of symmetry, the integral is $2a^3$ over each of the other two faces. $\therefore \displaystyle\iint_{\text{cube}} (x + y + z)\,d\sigma = 9a^3.$

13. On the faces in the coordinate planes, $g(x,y,z) = 0 \Rightarrow$ the integral over these faces is 0.

On the face, $x = a$, $f(x,y,z) = x = a$ and $g(x,y,z) = g(a,y,z) = ayz \Rightarrow \nabla f = \mathbf{i} \Rightarrow |\nabla f| = 1$. $\mathbf{p} = \mathbf{i} \Rightarrow |\nabla f \cdot \mathbf{p}| = 1$

$\Rightarrow d\sigma = dy\,dz \Rightarrow \displaystyle\iint_{x=a} xyz\,d\sigma = \int_0^c\int_0^b ayz\,dy\,dz = \frac{ab^2c^2}{4}$

On the face, $y = b$, $f(x,y,z) = y = b$ and $g(x,y,z) = g(x,b,z) = bxz \Rightarrow \nabla f = \mathbf{j} \Rightarrow |\nabla f| = 1$. $\mathbf{p} = \mathbf{j} \Rightarrow |\nabla f \cdot \mathbf{p}| = 1$

$\Rightarrow d\sigma = dx\,dz \Rightarrow \displaystyle\iint_{y=b} xyz\,dx\,dz = \int_0^c\int_0^a bxz\,dz\,dx = \frac{a^2bc^2}{4}$

On the face, $z = c$, $f(x,y,z) = z = c$ and $g(x,y,z) = g(x,y,c) = cxy \Rightarrow \nabla f = \mathbf{k} \Rightarrow |\nabla f| = 1$. $\mathbf{p} = \mathbf{k} \Rightarrow |\nabla f \cdot \mathbf{p}| = 1$

$\Rightarrow d\sigma = dy\,dx \Rightarrow \displaystyle\iint_{z=c} xyz\,d\sigma = \int_0^b\int_0^a cxy\,dx\,dy = \frac{a^2b^2c}{4}$

$$\therefore \iint_S g(x,y,z)\,d\sigma = \frac{abc(ab + ac + bc)}{4}$$

15. $\nabla f = 2\mathbf{i} + 2\mathbf{j} + \mathbf{k}$ and $g(x,y,z) = x + y + (2 - 2x - 2y) = 2 - x - y \Rightarrow |\nabla f| = 3$. $\mathbf{p} = \mathbf{k} \Rightarrow |\nabla f \cdot \mathbf{p}| = 1 \Rightarrow$

$$d\sigma = 3\, dy\, dx \Rightarrow \iint_S (x + y + z)\, d\sigma = 3 \int_0^1 \int_0^{1-x} (2 - x - y)\, dy\, dx = 2$$

17. $\nabla G = 2x\mathbf{i} + 2y\mathbf{j} + 2z\mathbf{k} \Rightarrow |\nabla G| = \sqrt{4x^2 + 4y^2 + 4z^2} = 2a.$ $\mathbf{n} = \dfrac{2x\,\mathbf{i} + 2y\,\mathbf{j} + 2z\,\mathbf{k}}{2\sqrt{x^2 + y^2 + z^2}} = \dfrac{x\,\mathbf{i} + y\,\mathbf{j} + z\,\mathbf{k}}{a} \Rightarrow$

$$\mathbf{F} \cdot \mathbf{n} = \frac{z^2}{a}.\ |\nabla G \cdot \mathbf{k}| = 2z \Rightarrow d\sigma = \frac{2a}{2z}\, dA = \frac{a}{z}\, dA.\ \therefore \text{Flux} = \iint_R \frac{z^2}{a}\left(\frac{a}{z}\right) dA = \iint_R z\, dA =$$

$$\iint_R \sqrt{a^2 - (x^2 + y^2)}\, dx\, dy = \int_0^{\pi/2} \int_0^a \sqrt{a^2 - r^2}\, r\, dr\, d\theta = \frac{a^3 \pi}{6}$$

19. $\mathbf{n} = \dfrac{x\,\mathbf{i} + y\,\mathbf{j} + z\,\mathbf{k}}{a}$, $d\sigma = \dfrac{a}{z}\, dA$ (See Exercise 15) and $\mathbf{F} \cdot \mathbf{n} = \dfrac{xy}{a} - \dfrac{xy}{a} + \dfrac{z}{a} = \dfrac{z}{a}$.

$$\therefore \text{Flux} = \iint_R \frac{z}{a}\left(\frac{a}{z}\right) dA = \iint_R 1\, dA = \frac{\pi a^2}{4}$$

21. $\mathbf{n} = \dfrac{x\,\mathbf{i} + y\,\mathbf{j} + z\,\mathbf{k}}{a}$, $d\sigma = \dfrac{a}{z}\, dA$ (See Exercise 15) and $\mathbf{F} \cdot \mathbf{n} = \dfrac{x^2}{a} + \dfrac{y^2}{a} + \dfrac{z^2}{a} = a \Rightarrow \text{Flux} = \displaystyle\iint_R a\left(\frac{a}{z}\right) dA$

$$= \iint_R \frac{a^2}{z}\, dA = \iint_R \frac{a^2}{\sqrt{a^2 - (x^2 + y^2)}}\, dA = \int_0^{\pi/2} \int_0^a \frac{a^2}{\sqrt{a^2 - r^2}}\, r\, dr\, d\theta = \frac{a^3 \pi}{2}$$

23. $\nabla G = 2y\,\mathbf{j} + \mathbf{k} \Rightarrow |\nabla G| = \sqrt{4y^2 + 1} \Rightarrow \mathbf{n} = \dfrac{2y\,\mathbf{j} + \mathbf{k}}{\sqrt{4y^2 + 1}} \Rightarrow \mathbf{F} \cdot \mathbf{n} = \dfrac{2xy - 3z}{\sqrt{4y^2 + 1}}$. $\mathbf{p} = \mathbf{k} \Rightarrow |\nabla G \cdot \mathbf{k}| = 1 \Rightarrow$

$$d\sigma = \sqrt{4y^2 + 1}\, dA \Rightarrow \text{Flux} = \iint_R \left(\frac{2xy - 3z}{\sqrt{4y^2 + 1}}\right)\sqrt{4y^2 + 1}\, dA = \iint_R (2xy - 3z)\, dA =$$

$$\iint_R (2xy - 3(4 - y^2))\, dA = \int_0^1 \int_{-2}^2 (2xy - 12 + 3y^2)\, dy\, dx = -32$$

25. $\nabla G = -e^x\mathbf{i} + \mathbf{j} \Rightarrow |\nabla G| = \sqrt{e^{2x} + 1}$. $\mathbf{p} = \mathbf{i} \Rightarrow |\nabla G \cdot \mathbf{i}| = e^x$. $\mathbf{n} = \dfrac{e^x\,\mathbf{i} - \mathbf{j}}{\sqrt{e^{2x} + 1}} \Rightarrow \mathbf{F} \cdot \mathbf{n} = \dfrac{-2e^x - 2y}{\sqrt{e^{2x} + 1}}$.

$$d\sigma = \frac{\sqrt{e^{2x} + 1}}{e^x}\, dA \Rightarrow \text{Flux} = \iint_R \frac{-2e^x - 2y}{\sqrt{e^{2x} + 1}}\left(\frac{\sqrt{e^{2x} + 1}}{e^x}\right) dA = \iint_R -4\, dA = \int_0^1 \int_1^2 -4\, dy\, dz$$

$= -4$

27. On the face, $z = a$: $G(x,y,z) = G(x,y,a) = z \Rightarrow \nabla G = \mathbf{k} \Rightarrow |\nabla G| = 1$. $\mathbf{n} = \mathbf{k} \Rightarrow \mathbf{F}\cdot\mathbf{n} = 2xz = 2ax$ since $z = a$

$d\sigma = dA \Rightarrow \text{Flux} = \iint_R 2ax\,dx\,dy = \int_0^a\int_0^a 2ax\,dx\,dy = a^4$. On the face, $z = 0$: $G(x,y,z) = G(x,y,0) =$

$z \Rightarrow \nabla G = \mathbf{k} \Rightarrow |\nabla G| = 1$. $\mathbf{n} = -\mathbf{k} \Rightarrow \mathbf{F}\cdot\mathbf{n} = -2xz = 0$ since $z = 0 \Rightarrow \text{Flux} = \iint_R 0\,dx\,dy = 0$

On the face, $x = a$: $G(x,y,z) = G(a,y,z) = x \Rightarrow \nabla G = \mathbf{i} \Rightarrow |\nabla G| = 1$. $\mathbf{n} = \mathbf{i} \Rightarrow \mathbf{F}\cdot\mathbf{n} = 2xy = 2ay$ since $x = a$

$\text{Flux} = \int_0^a\int_0^a 2ay\,dy\,dz = a^4$ On the face, $x = 0$: $G(x,y,z) = G(0,y,z) = x \Rightarrow \nabla G = \mathbf{i} \Rightarrow |\nabla G| = 1$. $\mathbf{n} = -\mathbf{i}$

$\Rightarrow \mathbf{F}\cdot\mathbf{n} = -2xy = 0$ since $x = 0 \Rightarrow \text{Flux} = 0$. On the face, $y = a$: $G(x,y,z) = G(x,a,z) = y \Rightarrow \nabla G = \mathbf{j} \Rightarrow |\nabla G|$

$= 1$. $\mathbf{n} = \mathbf{j} \Rightarrow \mathbf{F}\cdot\mathbf{n} = 2yz = 2az$ since $y = a \Rightarrow \text{Flux} = \int_0^a\int_0^a 2az\,dz\,dx = a^4$. On the face, $y = 0$: $G(x,y,z) =$

$G(z,0,z) = y \Rightarrow \nabla G = \mathbf{j} \Rightarrow |\nabla G| = 1$. $\mathbf{n} = -\mathbf{j} \Rightarrow \mathbf{F}\cdot\mathbf{n} = -2yz = 0$ since $y = 0 \Rightarrow \text{Flux} = 0$.

$\therefore$ Total Flux $= 3a^4$

29. $\nabla F = 2x\,\mathbf{i} + 2y\,\mathbf{j} + 2z\,\mathbf{k} \Rightarrow |\nabla F| = \sqrt{4x^2 + 4y^2 + 4z^2} = 2a,\ a > 0$ $\mathbf{p} = \mathbf{k} \Rightarrow |\nabla F\cdot\mathbf{k}| = 2z$ since $z \ge 0 \Rightarrow d\sigma =$

$\frac{2a}{2z}\,dA = \frac{a}{z}\,dA$. $\therefore\ M = \iint_S \delta\,d\sigma = \frac{\pi a^2}{2}\delta$. $M_{xy} = \iint_S z\delta\,d\sigma = \delta\iint_S z\left(\frac{a}{z}\right)dA =$

$a\delta\int_0^a\int_0^{\sqrt{a^2-x^2}} dy\,dx = \frac{\pi a^3}{4}\delta$. $\therefore\ \bar{z} = \dfrac{\frac{\pi a^3}{4}\delta}{\frac{\pi a^2}{2}\delta} = \frac{a}{2}$. Because of symmetry, $\bar{x} = \bar{y} = \frac{a}{2}$. $\therefore$ Centroid $= \left(\frac{a}{2}, \frac{a}{2}, \frac{a}{2}\right)$

31. Because of symmetry, $\bar{x} = \bar{y} = 0$. $M = \iint_S \delta\,d\sigma = \delta\iint_S d\sigma = \delta(\text{Area of S}) = 3\pi\sqrt{2}\,\delta$.

$\nabla F = 2x\,\mathbf{i} + 2y\,\mathbf{j} - 2z\,\mathbf{k} \Rightarrow |\nabla F| = \sqrt{4x^2 + 4y^2 + 4z^2} = 2\sqrt{x^2 + y^2 + z^2}$. $\mathbf{p} = \mathbf{k} \Rightarrow |\nabla F\cdot\mathbf{k}| = 2z \Rightarrow$

$d\sigma = \frac{2\sqrt{x^2 + y^2 + z^2}}{2z}\,dA = \frac{\sqrt{x^2 + y^2 + z^2}}{z}\,dA = \frac{\sqrt{x^2 + y^2 + (x^2 + y^2)}}{z}\,dA = \frac{\sqrt{2}\sqrt{x^2 + y^2}}{z}\,dA$.

$\therefore\ M_{xy} = \delta\iint_S z\left(\frac{\sqrt{2}\sqrt{x^2 + y^2}}{z}\right)dA = \delta\iint_S \sqrt{2}\sqrt{x^2 + y^2}\,dA = \delta\int_0^{2\pi}\int_1^2 \sqrt{2}\,r^2\,dr\,d\theta =$

$\frac{14\pi\sqrt{2}}{3}\delta$. $\bar{z} = \dfrac{\frac{14\pi\sqrt{2}}{3}\delta}{3\pi\sqrt{2}\,\delta} = \frac{14}{9}$. $\therefore\ (\bar{x}, \bar{y}, \bar{z}) = \left(0, 0, \frac{14}{9}\right)$. $I_z = \iint_S (x^2 + y^2)\,\delta\,d\sigma =$

31. (Continued)

$$\int\!\!\int_S (x^2+y^2)\left(\frac{\sqrt{2}\sqrt{x^2+y^2}}{z}\right)\delta\, dA = \delta\sqrt{2}\int\!\!\int_S (x^2+y^2)\, dA = \delta\sqrt{2}\int_0^{2\pi}\int_1^2 r^3\, dr\, d\theta = \frac{15\pi\sqrt{2}}{2}\delta.$$

$$R_z = \sqrt{I_z/M} = \frac{\sqrt{10}}{2}$$

33. $f_x(x,y) = 2x,\ f_y(x,y) = 2y \Rightarrow \sqrt{f_x^2 + f_y^2 + 1} = \sqrt{4x^2 + 4y^2 + 1} \Rightarrow$

$$\text{Area} = \int\!\!\int_R \sqrt{4x^2 + 4y^2 + 1}\ dx\, dy = \int_0^{2\pi}\int_0^{\sqrt{3}} \sqrt{4r^2+1}\ r\, dr\, d\theta = \frac{\pi}{6}\left(13\sqrt{13} - 1\right)$$

35. $f_z(y,z) = -2y,\ f_y(y,z) = -2z \Rightarrow \sqrt{f_y^2 + f_z^2 + 1} = \sqrt{4y^2 + 4z^2 + 1} \Rightarrow$

$$\text{Area} = \int\!\!\int_R \sqrt{4y^2 + 4z^2 + 1}\ dy\, dz = \int_0^{2\pi}\int_0^{1} \sqrt{4r^2+1}\ r\, dr\, d\theta = \frac{\pi}{6}\left(5\sqrt{5} - 1\right)$$

37. $y = \frac{2}{3}z^{3/2} \Rightarrow f_x(x,z) = 0,\ f_z(x,z) = z^{1/2} \Rightarrow \sqrt{f_x^2 + f_z^2 + 1} = \sqrt{z+1} \Rightarrow \text{Area} = \int_0^4\int_0^1 \sqrt{z+1}\ dx\, dz =$

$$\int_0^4 \sqrt{z+1}\ dz = \frac{2}{3}\left(5\sqrt{5} - 1\right)$$

15.6 PARAMETRIZED SURFACES

1. In cylindrical coordinates, let $x = r\cos\theta$, $y = r\sin\theta$, $z = \left(\sqrt{x^2+y^2}\right)^2 = r^2$. Then $\mathbf{r}(r,\theta) = (r\cos\theta)\,\mathbf{i} + (r\sin\theta)\,\mathbf{j} + r^2\,\mathbf{k}$, $0 \le r \le 2$, $0 \le \theta \le 2\pi$.

3. In cylindrical coordinates, let $x = r\cos\theta$, $y = r\sin\theta$, $z = \frac{\sqrt{x^2+y^2}}{2} \Rightarrow z = \frac{r}{2}$. Then $\mathbf{r}(r,\theta) = (r\cos\theta)\,\mathbf{i} + (r\sin\theta)\,\mathbf{j} + \frac{r}{2}\mathbf{k}$. For $0 \le z \le 3$, $0 \le \frac{r}{2} \le 3 \Rightarrow 0 \le r \le 6$. To get only the first octant, let $0 \le \theta \le \frac{\pi}{2}$.

5. In cylindrical coordinates, let $x = r\cos\theta$, $y = r\sin\theta$, $z = \sqrt{x^2+y^2} = r$. Then $\mathbf{r}(r,\theta) = (r\cos\theta)\,\mathbf{i} + (r\sin\theta)\,\mathbf{j} + \sqrt{9-r^2}\ \mathbf{k}$ since $x^2+y^2+z^2 = 9 \Rightarrow z^2 = 9-(x^2+y^2) = 9-r^2 \Rightarrow z = \sqrt{9-r^2}$ $\left(z \ge 0 \text{ since } z = \sqrt{x^2+y^2}\right)$. Let $0 \le \theta \le 2\pi$. For the domain of r: $z = \sqrt{x^2+y^2}$ and $x^2+y^2+z^2 = 9 \Rightarrow x^2+y^2+\left(\sqrt{x^2+y^2}\right)^2 = 9 \Rightarrow 2\left(x^2+y^2\right) = 9 \Rightarrow 2r^2 = 9 \Rightarrow r = \sqrt{9/2}\ (r \ge 0) = \frac{3}{\sqrt{2}}$. So, $0 \le r \le \frac{3}{\sqrt{2}}$.

7. In spherical coordinates, $x = \rho \sin\phi \cos\theta$, $y = \rho \sin\phi \sin\theta$, $\rho = \sqrt{x^2 + y^2 + z^2} \Rightarrow \rho^2 = 3 \Rightarrow \rho = \sqrt{3}$ $(\rho \geq 0)$. $x^2 + y^2 + z^2 = 3 \Rightarrow z^2 = 3 - (x^2 + y^2) \Rightarrow z^2 = 3 - (3 \sin^2\phi \cos^2\theta + 3 \sin^2\phi \sin^2\theta) = 3 - 3\sin^2\phi = 3(1 - \sin^2\phi) = 3\cos^2\phi \Rightarrow z = \sqrt{3}\cos\phi$. $z = \frac{\sqrt{3}}{2} \Rightarrow \frac{\sqrt{3}}{2} = \sqrt{3}\cos\phi \Rightarrow \cos\phi = \frac{1}{2} \Rightarrow \phi = \frac{\pi}{3}$. $z = -\frac{\sqrt{3}}{2} \Rightarrow -\frac{\sqrt{3}}{2} = \sqrt{3}\cos\phi \Rightarrow \cos\phi = -\frac{1}{2} \Rightarrow \phi = \frac{2\pi}{3}$. $\therefore$ let $\frac{\pi}{3} \leq \phi \leq \frac{2\pi}{3}$. Let $0 \leq \theta \leq 2\pi$. Then $\mathbf{r}(r,\theta) = (\sqrt{3}\sin\phi\cos\theta)\mathbf{i} + (\sqrt{3}\sin\phi\sin\theta)\mathbf{j} + (\sqrt{3}\cos\phi)\mathbf{k}$.

9. Since $z = 4 - y^2$, we can let $\mathbf{r}$ be a function of x and y $\Rightarrow \mathbf{r}(x,y) = x\,\mathbf{i} + y\,\mathbf{j} + (4 - y^2)\,\mathbf{k}$. Let $0 \leq x \leq 2$. $z = 0 \Rightarrow 0 = 4 - y^2 \Rightarrow y = \pm 2$. $\therefore$ let $-2 \leq y \leq 2$.

11. When $x = 0$, let $y^2 + z^2 = 9$ be the circular section in the yz-plane. For a point on that circle (0,y,z), let θ be the angle with vertex (0,0,0), one side through (0,y,z), the other side along the positive y axis. Then $y = 3\cos\theta$, $z = 3\sin\theta$. Thus let $\mathbf{r}$ be a function of x and $\theta \Rightarrow \mathbf{r}(x,\theta) = x\,\mathbf{i} + 3\cos\theta\,\mathbf{j} + 3\sin\theta\,\mathbf{k}$ where $0 \leq x \leq 3$, $0 \leq \theta \leq 2\pi$.

13. a) $x + y + z = 1 \Rightarrow z = 1 - x - y$. In cylindrical coordinates, let $x = r\cos\theta$, $y = r\sin\theta$, $z = \sqrt{x^2 + y^2}$. Then $\mathbf{r}(r,\theta) = r\cos\theta\,\mathbf{i} + r\sin\theta\,\mathbf{j} + (1 - r\cos\theta - r\sin\theta)\,\mathbf{k}$, $0 \leq \theta \leq 2\pi$, $0 \leq r \leq 3$.

b) In a fashion similar to cylindrical coordinates, but working in the yz-plane instead of the xy-plane, let $y = u\cos v$, $z = u\sin v$ where $u = \sqrt{y^2 + z^2}$ and v is the angle formed by (x,y,z), (x,0,0), and (x,y,0) with (x,0,0) as vertex. Then $\mathbf{r}$ is a function of u and v $\Rightarrow \mathbf{r}(u,v) = (1 - u\cos v - u\sin v)\,\mathbf{i} + u\cos v\,\mathbf{j} + u\sin v\,\mathbf{k}$ (since $x + y + z = 1 \Rightarrow x = 1 - y - z$), $0 \leq u \leq 3$, $0 \leq v \leq 2\pi$.

15. Let $x = w\cos v$, $z = w\sin v$. Then $(x - 2)^2 + z^2 = 4 \Rightarrow (w\cos v - 2)^2 + (w\sin v)^2 = 4 \Rightarrow w^2\cos^2 v - 4w\cos v + 4 + w^2\sin^2 v = 4 \Rightarrow w^2 - 4w\cos v = 0 \Rightarrow w = 0$ or $w - 4\cos v = 0 \Rightarrow w = 0$ or $w = 4\cos v$. $w = 0 \Rightarrow x = 0, y = 0$, a line not a cylinder. $\therefore$ let $w = 4\cos v \Rightarrow x = (4\cos v)(\cos v) = 4\cos^2 v$ and $z = 4\cos v\sin v$. Now, let $y = u$. Then $\mathbf{r}(v,u) = 4\cos^2 v\,\mathbf{i} + u\,\mathbf{j} + 4\cos v\sin v\,\mathbf{k}$, $-\frac{\pi}{2} \leq v \leq \frac{\pi}{2}$, $0 \leq u \leq 3$.

17. Let $x = r\cos\theta$, $y = r\sin\theta$, $z = r^2$. Then $\mathbf{r}(r,\theta) = r\cos\theta\,\mathbf{i} + r\sin\theta\,\mathbf{j} + \left(\frac{2 - r\sin\theta}{2}\right)\mathbf{k}$, $0 \leq r \leq 1$, $0 \leq \theta \leq 2\pi$.
$\mathbf{r}_r = \cos\theta\,\mathbf{i} + \sin\theta\,\mathbf{j} - \frac{\sin\theta}{2}\mathbf{k}$. $\mathbf{r}_\theta = -r\sin\theta\,\mathbf{i} + r\cos\theta\,\mathbf{j} - \frac{r\cos\theta}{2}\mathbf{k} \Rightarrow$

$$\mathbf{r}_r \times \mathbf{r}_\theta = \begin{vmatrix} \mathbf{i} & \mathbf{j} & \mathbf{k} \\ \cos\theta & \sin\theta & -\frac{\sin\theta}{2} \\ -r\sin\theta & r\cos\theta & -\frac{r\cos\theta}{2} \end{vmatrix} = \left(\frac{-r\sin\theta\cos\theta}{2} - \frac{-\sin\theta\,(r\cos\theta)}{2}\right)\mathbf{i} +$$

$$\left(\frac{r\sin^2\theta}{2} - \left(\frac{-r\cos^2\theta}{2}\right)\right)\mathbf{j} + \left(r\cos^2\theta - (-r\sin^2\theta)\right)\mathbf{k} = \frac{r}{2}\mathbf{j} + r\mathbf{k} \Rightarrow |\mathbf{r}_r \times \mathbf{r}_\theta| = \sqrt{\frac{r^2}{4} + r^2} = \frac{\sqrt{5}\,r}{2}.$$

$$\therefore A = \int_0^{2\pi}\int_0^1 \frac{\sqrt{5}\,r}{2}\,dr\,d\theta = \int_0^{2\pi} \left[\frac{\sqrt{5}\,r^2}{4}\right]_0^1 d\theta = \int_0^{2\pi} \frac{\sqrt{5}}{4}\,d\theta = \left[\frac{\sqrt{5}}{4}\theta\right]_0^{2\pi} = \frac{\pi\sqrt{5}}{2}.$$

19. Let $x = r\cos\theta$, $y = r\sin\theta$, $z = 2\sqrt{x^2+y^2} = 2r$, $1 \le r \le 3$, $0 \le \theta \le 2\pi$. Then $\mathbf{r}(r,\theta) = r\cos\theta\,\mathbf{i} + r\sin\theta\,\mathbf{j} + 2r\,\mathbf{k} \Rightarrow$ $\mathbf{r}_r = \cos\theta\,\mathbf{i} + \sin\theta\,\mathbf{j} + 2\,\mathbf{k}$ and $\mathbf{r}_\theta = -r\sin\theta\,\mathbf{i} + r\cos\theta\,\mathbf{j} \Rightarrow \mathbf{r}_r \times \mathbf{r}_\theta = \begin{vmatrix} \mathbf{i} & \mathbf{j} & \mathbf{k} \\ \cos\theta & \sin\theta & 2 \\ -r\sin\theta & r\cos\theta & 0 \end{vmatrix} = -2r\cos\theta\,\mathbf{i} -$ $2r\sin\theta\,\mathbf{j} + \left(r\cos^2\theta + r\sin^2\theta\right)\mathbf{k} = -2r\cos\theta\,\mathbf{i} - 2r\sin\theta\,\mathbf{j} + r\,\mathbf{k} \Rightarrow |\mathbf{r}_r \times \mathbf{r}_\theta| = \sqrt{4r^2\cos^2\theta + 4r^2\sin^2\theta + r^2}$ $= \sqrt{5r^2} = r\sqrt{5}$. $\therefore A = \int_0^{2\pi}\int_1^3 r\sqrt{5}\,dr\,d\theta = \int_0^{2\pi}\left[\frac{r^2\sqrt{5}}{2}\right]_1^3 d\theta = \int_0^{2\pi} 4\sqrt{5}\,d\theta = \left[4\sqrt{5}\,\theta\right]_0^{2\pi} = 8\pi\sqrt{5}$.

21. Let $x = r\cos\theta$, $y = r\sin\theta \Rightarrow x^2 + y^2 = r^2 = 1 \Rightarrow r = 1$. Then $\mathbf{r}(z,\theta) = \cos\theta\,\mathbf{i} + \sin\theta\,\mathbf{j} + 2\mathbf{k}$ $1 \le z \le 4$, $0 \le \theta \le 2\pi$. $\mathbf{r}_z = \mathbf{k}$ and $\mathbf{r}_\theta = -\sin\theta\,\mathbf{i} + \cos\theta\,\mathbf{j} \Rightarrow \mathbf{r}_\theta \times \mathbf{r}_z = \begin{vmatrix} \mathbf{i} & \mathbf{j} & \mathbf{k} \\ -\sin\theta & \cos\theta & 0 \\ 0 & 0 & 1 \end{vmatrix}$ $= \cos\theta\,\mathbf{i} + \sin\theta\,\mathbf{j} \Rightarrow |\mathbf{r}_\theta \times \mathbf{r}_z| = \sqrt{\cos^2\theta + \sin^2\theta} = 1$. $\therefore A = \int_0^{2\pi}\int_1^4 1\,dr\,d\theta = \int_0^{2\pi} 3\,d\theta = 6\pi$.

23. $z = 2 - x^2 - y^2$ and $z = \sqrt{x^2+y^2} \Rightarrow z = 2 - z^2 \Rightarrow z^2 + z - 2 = 0 \Rightarrow z = -2$ or $z = 1$. Since $z = \sqrt{x^2+y^2} \ge 0$, $z = 1$ is where the cone intersects the paraboloid. When $x = 0$ and $y = 0$, $z = 2 \Rightarrow$ The vertex of the paraboloid is $(0,0,2)$. $\therefore$ z ranges from 1 to 2 on the "cap" $\Rightarrow$ r ranges from 1 (when $x^2 + y^2 = 1$) to 0 (when $x = 0$ and $y = 0$ at the vertex). Let $x = r\cos\theta$, $y = r\sin\theta$, $z = 2 - r^2$. Then $\mathbf{r}(r,\theta) = r\cos\theta\,\mathbf{i} + r\sin\theta\,\mathbf{j} + (2 - r^2)\,\mathbf{k}$, $0 \le r \le 1$, $0 \le \theta \le 2\pi$. $\mathbf{r}_r = \cos\theta\,\mathbf{i} + \sin\theta\,\mathbf{j} - 2r\,\mathbf{k}$ and $\mathbf{r}_\theta = -r\sin\theta\,\mathbf{i} + r\cos\theta\,\mathbf{j} \Rightarrow \mathbf{r}_r \times \mathbf{r}_\theta = \begin{vmatrix} \mathbf{i} & \mathbf{j} & \mathbf{k} \\ \cos\theta & \sin\theta & -2r \\ -r\sin\theta & r\cos\theta & 0 \end{vmatrix} =$ $2r^2\cos\theta\,\mathbf{i} + 2r^2\sin\theta\,\mathbf{j} + r\,\mathbf{k} \Rightarrow |\mathbf{r}_r \times \mathbf{r}_\theta| = \sqrt{4r^4\cos^2\theta + 4r^4\sin^2\theta + r^2} = r\sqrt{4r^2 + 1}$

$\therefore A = \int_0^{2\pi}\int_0^1 r\sqrt{4r^2 + 1}\,dr\,d\theta = \int_0^{2\pi}\left[\frac{1}{12}(4r^2+1)^{3/2}\right]_0^1 d\theta = \int_0^{2\pi}\left(\frac{5\sqrt{5} - 1}{12}\right)d\theta = \left[\left(\frac{5\sqrt{5} - 1}{12}\right)\theta\right]_0^{2\pi}$ $= \frac{(5\sqrt{5} - 1)\pi}{6}$

25. Let $x = \rho\sin\phi\cos\theta$, $y = \rho\sin\phi\sin\theta$, $z = \rho\cos\phi$, $\rho = \sqrt{x^2 + y^2 + z^2} \Rightarrow \rho = \sqrt{2}$ on the sphere. $x^2 + y^2 + z^2 = 2$ and $z = \sqrt{x^2+y^2} \Rightarrow z^2 + z^2 = 2 \Rightarrow z^2 = 1 \Rightarrow z = 1$ $\left(z = \sqrt{x^2 + y^2} \Rightarrow z \ge 0\right)$. $z = 1 \Rightarrow \phi = \frac{\pi}{4}$. For the lower portion of the sphere cut by the cone, $z = -\sqrt{2}$ when $x = 0$ and $y = 0 \Rightarrow$ $\phi = \pi$. Then $\mathbf{r}(\phi,\theta) = \sqrt{2}\sin\phi\cos\theta\,\mathbf{i} + \sqrt{2}\sin\phi\sin\theta\,\mathbf{j} + \sqrt{2}\cos\phi\,\mathbf{k}$, $\frac{\pi}{4} \le \phi \le \pi$, $0 \le \theta \le 2\pi$. $\mathbf{r}_\phi = \sqrt{2}\cos\phi\cos\theta\,\mathbf{i} + \sqrt{2}\cos\phi\sin\theta\,\mathbf{j} - \sqrt{2}\sin\phi\,\mathbf{k}$ and $\mathbf{r}_\theta = -\sqrt{2}\sin\phi\sin\theta\,\mathbf{i} + \sqrt{2}\sin\phi\cos\theta\,\mathbf{j} \Rightarrow$ $\mathbf{r}_\phi \times \mathbf{r}_\theta = \begin{vmatrix} \mathbf{i} & \mathbf{j} & \mathbf{k} \\ \sqrt{2}\cos\phi\cos\theta & \sqrt{2}\cos\phi\sin\theta & -\sqrt{2}\sin\phi \\ -\sqrt{2}\sin\phi\sin\theta & \sqrt{2}\sin\phi\cos\theta & 0 \end{vmatrix} = 2\sin^2\phi\cos\theta\,\mathbf{i} + 2\sin^2\phi\sin\theta\,\mathbf{j} +$ $2\sin\phi\cos\phi\,\mathbf{k} \Rightarrow |\mathbf{r}_\phi \times \mathbf{r}_\theta| = \sqrt{4\sin^4\phi\cos^2\theta + 4\sin^4\phi\sin^2\theta + 4\sin^2\phi\cos^2\phi} = \sqrt{4\sin^2\phi} = 2\sin\phi$.

25. (Continued)

$$\therefore A = \int_0^{2\pi}\int_{\pi/4}^{\pi} 2\sin\phi\, d\phi\, d\theta = \int_0^{2\pi}\left(2+\sqrt{2}\right) d\theta = \left(4+2\sqrt{2}\right)\pi.$$

27. Let the parametrization be $\mathbf{r}(x,z) = x\,\mathbf{i} + x^2\,\mathbf{j} + z\,\mathbf{k} \Rightarrow \mathbf{r}_x = \mathbf{i} + 2x\,\mathbf{j}$ and $\mathbf{r}_z = \mathbf{k} \Rightarrow \mathbf{r}_x \times \mathbf{r}_z = \begin{vmatrix} \mathbf{i} & \mathbf{j} & \mathbf{k} \\ 1 & 2x & 0 \\ 0 & 0 & 1 \end{vmatrix} = 2x\,\mathbf{i} + \mathbf{j} \Rightarrow$

$|\mathbf{r}_x \times \mathbf{r}_z| = \sqrt{4x^2+1}$. $\therefore \iint_S G(x,y,z)\, d\sigma = \int_0^3\int_0^2 x\sqrt{4x^2+1}\, dx\, dz = \frac{17\sqrt{17}-1}{4}$.

29. Let the parametrization be $\mathbf{r}(\phi,\theta) = \sin\phi\cos\theta\,\mathbf{i} + \sin\phi\sin\theta\,\mathbf{j} + \cos\phi\,\mathbf{k}$ (spherical coordinates with $\rho = 1$ on the sphere), $0 \le \phi \le \pi$, $0 \le \theta \le 2\pi$. $\mathbf{r}_\phi = \cos\phi\cos\theta\,\mathbf{i} + \cos\phi\sin\theta\,\mathbf{j} - \sin\phi\,\mathbf{k}$ and $\mathbf{r}_\theta = -\sin\phi\sin\theta\,\mathbf{i} + \sin\phi\cos\theta\,\mathbf{j} \Rightarrow$

$\mathbf{r}_\phi \times \mathbf{r}_\theta = \begin{vmatrix} \mathbf{i} & \mathbf{j} & \mathbf{k} \\ \cos\phi\cos\theta & \cos\phi\sin\theta & -\sin\phi \\ -\sin\phi\sin\theta & \sin\phi\cos\theta & 0 \end{vmatrix} = \sin^2\phi\cos\theta\,\mathbf{i} + \sin^2\phi\sin\theta\,\mathbf{j} + \sin\phi\cos\phi\,\mathbf{k} \Rightarrow |\mathbf{r}_\phi \times \mathbf{r}_\theta|$

$= \sqrt{\sin^4\phi\cos^2\theta + \sin^4\phi\sin^2\theta + \sin^2\phi\cos^2\phi} = \sin\phi$. $\therefore$ since $x = \sin\phi\cos\theta \Rightarrow G(x,y,z) = G(\phi,\theta) =$

$\cos^2\theta\sin^2\phi$, $\iint_S G(x,y,z)\, d\sigma = \int_0^{2\pi}\int_0^{\pi} (\cos^2\theta\sin^2\phi)(\sin\phi)\, d\phi\, d\theta = \frac{4\pi}{3}$. (Hint: Write $\sin^2\phi$ as $1 - \cos^2\phi$ for the first integration.)

31. Let $x = x$, $y = y$, $z = 4 - x - y$, $0 \le x \le 1$, $0 \le y \le 1 \Rightarrow \mathbf{r}(x,y) = x\,\mathbf{i} + y\,\mathbf{j} + (4-x-y)\,\mathbf{k} \Rightarrow \mathbf{r}_x = \mathbf{i} - \mathbf{k}$ and $\mathbf{r}_y = \mathbf{j} - \mathbf{k}$

$\Rightarrow \mathbf{r}_x \times \mathbf{r}_y = \begin{vmatrix} \mathbf{i} & \mathbf{j} & \mathbf{k} \\ 1 & 0 & -1 \\ 0 & 1 & -1 \end{vmatrix} = \mathbf{i} + \mathbf{j} + \mathbf{k} \Rightarrow |\mathbf{r}_x \times \mathbf{r}_y| = \sqrt{3} \Rightarrow \iint_S G(x,y,z)\, d\sigma = \int_0^1\int_0^1 (4 - x - y)\sqrt{3}\, dy\, dx$

$= 3\sqrt{3}$

33. Let $x = r\cos\theta$, $y = r\sin\theta$, $z = 1 - \left(\sqrt{x^2+y^2}\right)^2 = 1 - r^2$ (cylindrical coordinates), $0 \le r \le 1$ since $0 \le z \le 1$, $0 \le \theta \le 2\pi \Rightarrow$ $\mathbf{r}(r,\theta) = r\cos\theta\,\mathbf{i} + r\sin\theta\,\mathbf{j} + (1-r^2)\,\mathbf{k} \Rightarrow \mathbf{r}_r = \cos\theta\,\mathbf{i} + \sin\theta\,\mathbf{j} - 2r\,\mathbf{k}$ and $\mathbf{r}_\theta = -r\sin\theta\,\mathbf{i} + r\cos\theta\,\mathbf{j} \Rightarrow \mathbf{r}_r \times \mathbf{r}_\theta =$

$\begin{vmatrix} \mathbf{i} & \mathbf{j} & \mathbf{k} \\ \cos\theta & \sin\theta & -2r \\ -r\sin\theta & r\cos\theta & 0 \end{vmatrix} = 2r^2\cos\theta\,\mathbf{i} + 2r^2\sin\theta\,\mathbf{j} + r\,\mathbf{k} \Rightarrow |\mathbf{r}_r \times \mathbf{r}_\theta| = \sqrt{(2r^2\cos\theta)^2 + (2r^2\sin\theta)^2 + (r)^2}$

$= r\sqrt{1+4r^2}$. Since $z = 1 - r^2$ and $x = r\cos\theta$, $H(x,y,z) = H(r,\theta) = r^2\cos^2\theta\sqrt{1+4r^2} \Rightarrow \iint_S H(x,y,z)\, d\sigma =$

$$\int_0^{2\pi}\int_0^1 \left(r^2\cos^2\theta\sqrt{1+4r^2}\right)\left(r\sqrt{1+4r^2}\right) dr\, d\theta = \int_0^{2\pi}\int_0^1 \cos^2\theta\, r^3(1+4r^2)\, dr\, d\theta = \frac{11\pi}{12}$$

35. Let $x = x$, $y = y$, $z = 4 - y^2$, $0 \le x \le 1$, $-2 \le y \le 2$ since $z = 0 \Rightarrow 0 = 4 - y^2 \Rightarrow y = \pm 2 \Rightarrow \mathbf{r}(x,y) = x\,\mathbf{i} + y\,\mathbf{j} + (4 - y^2)\,\mathbf{j}$

$\Rightarrow \mathbf{r}_x = \mathbf{i}$ and $\mathbf{r}_y = \mathbf{j} - 2y\,\mathbf{k} \Rightarrow \mathbf{r}_x \times \mathbf{r}_y = \begin{vmatrix} \mathbf{i} & \mathbf{j} & \mathbf{k} \\ 1 & 0 & 0 \\ 0 & 1 & -2y \end{vmatrix} = 2y\,\mathbf{j} + \mathbf{k} \Rightarrow \mathbf{F} \cdot \mathbf{n}\, d\sigma = \mathbf{F} \cdot \dfrac{\mathbf{r}_x \times \mathbf{r}_y}{|\mathbf{r}_x \times \mathbf{r}_y|} |\mathbf{r}_x \times \mathbf{r}_y|\, dy\, dx =$

$\left(2xy - 3(4 - y^2)\right) dy\, dx$ since $\mathbf{F} = (4 - y^2)^2\,\mathbf{i} + x\,\mathbf{j} - 3(4 - y^2)\,\mathbf{k} \Rightarrow \displaystyle\iint_S \mathbf{F} \cdot \mathbf{n}\, d\sigma$

$$= \int_0^1 \int_{-2}^2 \left(2xy - 3(4 - y^2)\right) dy\, dx = -32$$

37. Let the parametrization be $\mathbf{r}(\phi,\theta) = a \sin\phi \cos\theta\,\mathbf{i} + a \sin\phi \sin\theta\,\mathbf{j} + a \cos\phi\,\mathbf{k}$ (spherical coordinates with $\rho = a$, $a \ge 0$, on the sphere), $0 \le \phi \le \frac{\pi}{2}$, $0 \le \theta \le \frac{\pi}{2}$ (for the first octant). $\mathbf{r}_\phi = a \cos\phi \cos\theta\,\mathbf{i} + a \cos\phi \sin\theta\,\mathbf{j} - a \sin\phi\,\mathbf{k}$ and

$\mathbf{r}_\theta = -a \sin\phi \sin\theta\,\mathbf{i} + a \sin\phi \cos\theta\,\mathbf{j} \Rightarrow \mathbf{r}_\phi \times \mathbf{r}_\theta = \begin{vmatrix} \mathbf{i} & \mathbf{j} & \mathbf{k} \\ a\cos\phi\cos\theta & a\cos\phi\sin\theta & -a\sin\phi \\ -a\sin\phi\sin\theta & a\sin\phi\cos\theta & 0 \end{vmatrix} =$

$a^2 \sin^2\phi \cos\theta\,\mathbf{i} + a^2 \sin^2\phi \sin\theta\,\mathbf{j} + a^2 \sin\phi \cos\phi\,\mathbf{k} \Rightarrow \mathbf{F} \cdot \mathbf{n}\, d\sigma = \mathbf{F} \cdot \dfrac{\mathbf{r}_\phi \times \mathbf{r}_\theta}{|\mathbf{r}_\phi \times \mathbf{r}_\theta|} |\mathbf{r}_\phi \times \mathbf{r}_\theta|\, d\theta\, d\phi = a^3 \sin\phi \cos^2\phi$

since $\mathbf{F} = z\,\mathbf{k} = a \cos\phi\,\mathbf{k} \Rightarrow \displaystyle\iint_S \mathbf{F} \cdot \mathbf{n}\, d\sigma = \int_0^{\pi/2}\int_0^{\pi/2} a^3 \sin\phi \cos^2\phi\, d\phi\, d\theta = \frac{a^3\pi}{6}$

39. Let the parametrization be $\mathbf{r}(x,y) = x\,\mathbf{i} + y\,\mathbf{j} + (2a - x - y)\,\mathbf{k}$, $0 \le x \le a$, $0 \le y \le a \Rightarrow \mathbf{r}_x = \mathbf{i} - \mathbf{k}$ and $\mathbf{r}_y = \mathbf{j} - \mathbf{k} \Rightarrow$

$\mathbf{r}_x \times \mathbf{r}_y = \begin{vmatrix} \mathbf{i} & \mathbf{j} & \mathbf{k} \\ 1 & 0 & -1 \\ 0 & 1 & -1 \end{vmatrix} = \mathbf{i} + \mathbf{j} + \mathbf{k} \Rightarrow \mathbf{F} \cdot \mathbf{n}\, d\sigma = \mathbf{F} \cdot \dfrac{\mathbf{r}_x \times \mathbf{r}_y}{|\mathbf{r}_x \times \mathbf{r}_y|} |\mathbf{r}_x \times \mathbf{r}_y|\, dy\, dx$

$= (2xy + 2y(2a - x - y) + 2x(2a - x - y))\, dy\, dx$ since $\mathbf{F} = 2xy\,\mathbf{i} + 2yz\,\mathbf{j} + 2xz\,\mathbf{k} = 2xy\,\mathbf{i} + 2y(2a - x - y)\,\mathbf{j}$

$+ 2x(2a - x - y)\,\mathbf{k} \Rightarrow \displaystyle\iint_S \mathbf{F} \cdot \mathbf{n}\, d\sigma = \int_0^a\int_0^a (2xy + 2y(2a - x - y) + 2x(2a - x - y))\, dy\, dx = \frac{13a^4}{6}$

41. Let $x = r\cos\theta$, $y = r\sin\theta$, $z = \sqrt{x^2 + y^2} = r$ (cylindrical coordinates), $0 \le r \le 1$ since $0 \le z \le 1$, $0 \le \theta \le 2\pi \Rightarrow$ $\mathbf{r}(r,\theta) = r\cos\theta\,\mathbf{i} + r\sin\theta\,\mathbf{j} + r\,\mathbf{k} \Rightarrow \mathbf{r}_r = \cos\theta\,\mathbf{i} + \sin\theta\,\mathbf{j} + \mathbf{k}$ and $\mathbf{r}_\theta = -r\sin\theta\,\mathbf{i} + r\cos\theta\,\mathbf{j} \Rightarrow \mathbf{r}_\theta \times \mathbf{r}_r =$

$\begin{vmatrix} \mathbf{i} & \mathbf{j} & \mathbf{k} \\ -r\sin\theta & r\cos\theta & 0 \\ \cos\theta & \sin\theta & 1 \end{vmatrix} = r\cos\theta\,\mathbf{i} + r\sin\theta\,\mathbf{j} - r\,\mathbf{k} \Rightarrow \mathbf{F} \cdot \mathbf{n}\, d\sigma\, \mathbf{F} \cdot \dfrac{\mathbf{r}_\theta \times \mathbf{r}_r}{|\mathbf{r}_\theta \times \mathbf{r}_r|} |\mathbf{r}_\theta \times \mathbf{r}_r|\, d\theta\, dr =$

$\left(r^3 \sin\theta \cos^2\theta + r^2\right) d\theta\, dr$ since $\mathbf{F} = r^2 \sin\theta\cos\theta\,\mathbf{i} - r\,\mathbf{k} \Rightarrow \displaystyle\iint_S \mathbf{F} \cdot \mathbf{n}\, d\sigma = \int_0^{2\pi}\int_0^1 (r^3 \sin\theta \cos^2\theta + r^2)\, dr\, d\theta$

$= \frac{2\pi}{3}$

43. Let $x = r\cos\theta$, $y = r\sin\theta$, $z = \sqrt{x^2+y^2} = r$ (cylindrical coordinates), $1 \le r \le 2$ since $1 \le z \le 2$, $0 \le \theta \le 2\pi \Rightarrow$ $\mathbf{r}(r,\theta) = r\cos\theta\,\mathbf{i} + r\sin\theta\,\mathbf{j} + r\,\mathbf{k} \Rightarrow \mathbf{r}_r = \cos\theta\,\mathbf{i} + \sin\theta\,\mathbf{j} + \mathbf{k}$ and $\mathbf{r}_\theta = -r\sin\theta\,\mathbf{i} + r\cos\theta\,\mathbf{j} \Rightarrow \mathbf{r}_\theta \times \mathbf{r}_r =$

$$\begin{vmatrix} \mathbf{i} & \mathbf{j} & \mathbf{k} \\ -r\sin\theta & r\cos\theta & 0 \\ \cos\theta & \sin\theta & 1 \end{vmatrix} = r\cos\theta\,\mathbf{i} + r\sin\theta\,\mathbf{j} - r\,\mathbf{k} \Rightarrow \mathbf{F}\cdot\mathbf{n}\,d\sigma\ \mathbf{F}\cdot\frac{\mathbf{r}_\theta \times \mathbf{r}_r}{|\mathbf{r}_\theta \times \mathbf{r}_r|}|\mathbf{r}_\theta \times \mathbf{r}_r|\,d\theta\,dr =$$

$$\left(-r^2\cos^2\theta - r^2\sin^2\theta - r^3\right)d\theta\,dr = \left(-r^2 - r^3\right)d\theta\,dr$$ since $\mathbf{F} = -r\cos\theta\,\mathbf{i} - r\sin\theta\,\mathbf{j} + r^2\,\mathbf{k} \Rightarrow$

$$\int\!\!\int_S \mathbf{F}\cdot\mathbf{n}\,d\sigma = \int_0^{2\pi}\!\!\int_1^2 \left(-r^2 - r^3\right)dr\,d\theta = -\frac{73\pi}{6}$$

45. Let the parametrization be $\mathbf{r}(\phi,\theta) = a\sin\phi\cos\theta\,\mathbf{i} + a\sin\phi\sin\theta\,\mathbf{j} + a\cos\phi\,\mathbf{k}$, $0 \le \phi \le \frac{\pi}{2}$, $0 \le \theta \le \frac{\pi}{2}$, $\Rightarrow \mathbf{r}_\phi =$ $a\cos\phi\cos\theta\,\mathbf{i} + a\cos\phi\sin\theta\,\mathbf{j} - a\sin\phi\,\mathbf{k}$ and $\mathbf{r}_\theta = -a\sin\phi\sin\theta\,\mathbf{i} + a\sin\phi\cos\theta\,\mathbf{j} \Rightarrow \mathbf{r}_\phi \times \mathbf{r}_\theta =$

$$\begin{vmatrix} \mathbf{i} & \mathbf{j} & \mathbf{k} \\ a\cos\phi\cos\theta & a\cos\phi\sin\theta & -a\sin\phi \\ -a\sin\phi\sin\theta & a\sin\phi\cos\theta & 0 \end{vmatrix} = a^2\sin^2\phi\cos\theta\,\mathbf{i} + a^2\sin^2\phi\sin\theta\,\mathbf{j} + a^2\sin\phi\cos\phi\,\mathbf{k} \Rightarrow$$

$|\mathbf{r}_\phi \times \mathbf{r}_\theta| = \sqrt{a^4\sin^4\phi\cos^2\theta + a^4\sin^4\phi\sin^2\theta + a^4\sin^2\phi\cos^2} = \sqrt{a^4\sin^2\phi} = a^2\sin\phi$. The mass

$M = \int\!\!\int_S d\sigma = \int_0^{\pi/2}\!\!\int_0^{\pi/2} a^2\sin\phi\,d\phi\,d\theta = \frac{a^2\pi}{2}$. The first moment $M_{yz} = \int\!\!\int_S x\,d\sigma =$

$\int_0^{\pi/2}\!\!\int_0^{\pi/2} (a\sin\phi\cos\theta)(a^2\sin\phi)\,d\phi\,d\theta = \frac{a^3\pi}{4} \Rightarrow \bar{x} = \frac{a^3\pi/4}{a^2\pi/2} = \frac{a}{2}$. The centroid is located at $\left(\frac{a}{2}, \frac{a}{2}, \frac{a}{2}\right)$

by symmetry.

47. Let the parametrization be $\mathbf{r}(\phi,\theta) = a\sin\phi\cos\theta\,\mathbf{i} + a\sin\phi\sin\theta\,\mathbf{j} + a\cos\phi\,\mathbf{k}$, $0 \le \phi \le \pi$, $0 \le \theta \le 2\pi$, $\Rightarrow \mathbf{r}_\phi =$ $a\cos\phi\cos\theta\,\mathbf{i} + a\cos\phi\sin\theta\,\mathbf{j} - a\sin\phi\,\mathbf{k}$ and $\mathbf{r}_\theta = -a\sin\phi\sin\theta\,\mathbf{i} + a\sin\phi\cos\theta\,\mathbf{j} \Rightarrow \mathbf{r}_\phi \times \mathbf{r}_\theta =$

$$\begin{vmatrix} \mathbf{i} & \mathbf{j} & \mathbf{k} \\ a\cos\phi\cos\theta & a\cos\phi\sin\theta & -a\sin\phi \\ -a\sin\phi\sin\theta & a\sin\phi\cos\theta & 0 \end{vmatrix} = a^2\sin^2\phi\cos\theta\,\mathbf{i} + a^2\sin^2\phi\sin\theta\,\mathbf{j} + a^2\sin\phi\cos\phi\,\mathbf{k} \Rightarrow$$

$|\mathbf{r}_\phi \times \mathbf{r}_\theta| = \sqrt{a^4\sin^4\phi\cos^2\theta + a^4\sin^4\phi\sin^2\theta + a^4\sin^2\phi\cos^2} = \sqrt{a^4\sin^2\phi} = a^2\sin\phi$. The

moment of inertia, $I_z = \int\!\!\int_S \delta\left(x^2+y^2\right)d\sigma = \int_0^{2\pi}\!\!\int_0^{\pi} \delta\left((a\sin\phi\cos\theta)^2 + (a\sin\phi\sin\theta)^2\right)\left(a^2\sin\phi\right)d\phi\,d\theta$

$$= \int_0^{2\pi}\!\!\int_0^{\pi} \delta\left(a^2\sin^2\phi\right)\left(a^2\sin\phi\right)d\phi\,d\theta = \int_0^{2\pi}\!\!\int_0^{\pi} \delta\,a^4\sin^3\phi\,d\phi\,d\theta = \frac{8a^4\pi\delta}{3}.$$

49.

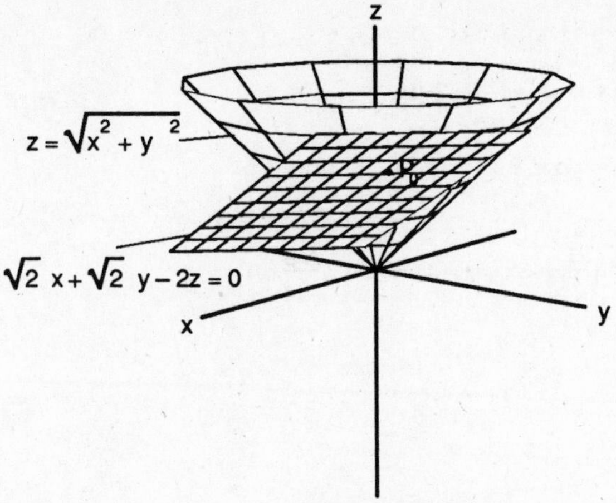

Graph 15.6.49

The parametrization $\mathbf{r}(r,\theta)$ at $P_0=\left(\sqrt{2}\,,\ \sqrt{2}\,,\ 2\right)$ $\Rightarrow \theta=\frac{\pi}{4}$, $r=2 \Rightarrow \mathbf{r}_r=\cos\theta\,\mathbf{i}+\sin\theta\,\mathbf{j}+\mathbf{k}=$ $\frac{\sqrt{2}}{2}\mathbf{i}+\frac{\sqrt{2}}{2}\mathbf{j}+\mathbf{k}$ and $\mathbf{r}_\theta=-r\sin\theta\,\mathbf{i}+r\cos\theta\,\mathbf{j}=-\sqrt{2}\,\mathbf{i}$ $+\sqrt{2}\,\mathbf{j} \Rightarrow \mathbf{r}_r \times \mathbf{r}_\theta=\begin{vmatrix}\mathbf{i}&\mathbf{j}&\mathbf{k}\\ \sqrt{2}/2&\sqrt{2}/2&1\\ -\sqrt{2}&\sqrt{2}&0\end{vmatrix}=$ $-\sqrt{2}\,\mathbf{i}-\sqrt{2}\,\mathbf{j}+2\mathbf{k} \Rightarrow$ the tangent plane is $\left(-\sqrt{2}\ \mathbf{i}-\sqrt{2}\ \mathbf{j}+2\mathbf{k}\right)\cdot\left(\left(x-\sqrt{2}\right)\mathbf{i}+\left(y-\sqrt{2}\right)\mathbf{j}+(z-2)\mathbf{k}\right)=\sqrt{2}\ x+\sqrt{2}\ y-2z=0.$ The parametrization $\mathbf{r}(r,\theta) \Rightarrow x=r\cos\theta$, $y=r\sin\theta$ and $z=r \Rightarrow x^2+y^2=r^2=z^2 \Rightarrow$ the surface $z=\sqrt{x^2+y^2}$

51.

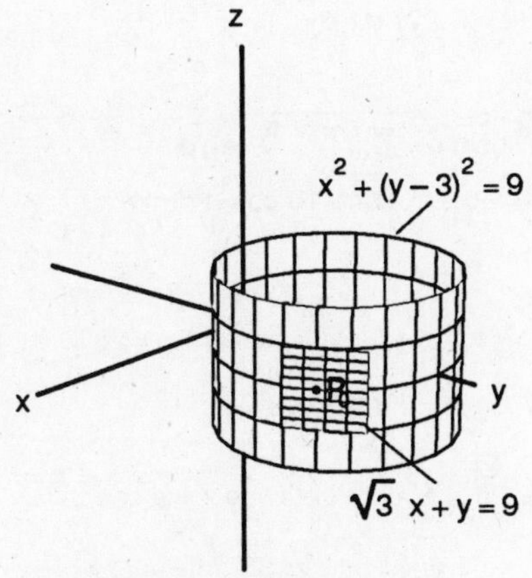

Graph 15.6.51

The parametrization $\mathbf{r}(\theta,z)$ at $P_0=\left(\frac{3\sqrt{3}}{2}\,,\ \frac{9}{2}\,,0\right)$ where $\theta=\frac{\pi}{3}$, $z=0 \Rightarrow \mathbf{r}_\theta=6\cos 2\theta\,\mathbf{i}+12\sin\theta\cos\theta\,\mathbf{j}=$ $-3\,\mathbf{i}+3\sqrt{3}\ \mathbf{j}$ and $\mathbf{r}_z=\mathbf{k} \Rightarrow \mathbf{r}_\theta \times \mathbf{r}_z=\begin{vmatrix}\mathbf{i}&\mathbf{j}&\mathbf{k}\\ -3&3\sqrt{3}&0\\ 0&0&1\end{vmatrix}$ $=3\sqrt{3}\,\mathbf{i}+3\mathbf{j} \Rightarrow$ the tangent plane is $\left(3\sqrt{3}\ \mathbf{i}+3\,\mathbf{j}\right)\cdot\left(\left(x-3\sqrt{3}/2\right)\mathbf{i}+(y-9/2)\,\mathbf{j}+(z-0)\,\mathbf{k}\right)$ $=0 \Rightarrow \sqrt{3}\ x+y=9.$ The parametrization $\Rightarrow x=$ $3\sin 2\theta$ and $y=6\sin^2\theta \Rightarrow x^2+y^2=9\sin^2 2\theta+$ $\left(6\sin^2\theta\right)^2=9(4\sin^2\theta\cos^2\theta)+36\sin^4\theta=$ $6(6\sin^2\theta)=6y \Rightarrow x^2+y^2=6y \Rightarrow x^2+y^2-6y+9=9$ $\Rightarrow x^2+(y-3)^2=9$ is the surface.

15.7 STOKE'S THEOREM

1. curl $\mathbf{F} = \nabla \times \mathbf{F} = 2\,\mathbf{k}$, $\mathbf{n} = \mathbf{k} \Rightarrow$ curl $\mathbf{F} \cdot \mathbf{n} = 2 \Rightarrow d\sigma = dx\,dy \Rightarrow \oint_C \mathbf{F} \cdot d\mathbf{r} = \iint_R 2\,dA =$ 2(Area of the ellipse) $= 4\pi$

3. curl $\mathbf{F} = \nabla \times \mathbf{F} = -x\,\mathbf{i} - 2x\,\mathbf{j} + (z-1)\,\mathbf{k}$, $\mathbf{n} = \dfrac{\mathbf{i}+\mathbf{j}+\mathbf{k}}{\sqrt{3}} \Rightarrow$ curl $\mathbf{F} \cdot \mathbf{n} = \dfrac{1}{\sqrt{3}}(-3x + z - 1) \Rightarrow d\sigma = \dfrac{\sqrt{3}}{1}\,dA \Rightarrow$

$$\oint_C \mathbf{F} \cdot d\mathbf{r} = \iint_R \frac{1}{\sqrt{3}}(-3x + z - 1)\sqrt{3}\,dA = \int_0^1 \int_0^{1-x} (-3x + z - 1)\,dy\,dx =$$

$$\int_0^1 \int_0^{1-x} (-3x + (1 - x - y) - 1)\,dy\,dx = \int_0^1 \int_0^{1-x} (-4x - y)\,dy\,dx = -\frac{5}{6}$$

5. curl $\mathbf{F} = \nabla \times \mathbf{F} = (2y - 0)\,\mathbf{i} + (2z - 2x)\,\mathbf{j} + (2x - 2y)\,\mathbf{k} = 2y\,\mathbf{i} + (2z - 2x)\,\mathbf{j} + (2x - 2y)\,\mathbf{k}$,

$\mathbf{n} = \mathbf{k} \Rightarrow$ curl $\mathbf{F} \cdot \mathbf{n} = 2x - 2y \Rightarrow d\sigma = dx\,dy \Rightarrow \oint_C \mathbf{F} \cdot d\mathbf{r} = \int_{-1}^{1} \int_{-1}^{1} (2x - 2y)\,dx\,dy = 0$

7. $x = 3\cos t$, $y = 2\sin t \Rightarrow \mathbf{F} = (2\sin t)\,\mathbf{i} + (9\cos^2 t)\,\mathbf{j} + (9\cos^2 t + 16\sin^4 t)\sin e^{\sqrt{6\sin t\cos t\,(0)}}$ and $\mathbf{r} = (3\cos t)\,\mathbf{i} + (2\sin t)\,\mathbf{j} \Rightarrow d\mathbf{r} = (-3\sin t)\,dt\,\mathbf{i} + (2\cos t)\,dt\,\mathbf{j} \Rightarrow \mathbf{F} \cdot d\mathbf{r} = -6\sin^2 t\,dt + 18\cos^3 t\,dt \Rightarrow$

$$\iint_S \nabla \times \mathbf{F} \cdot \mathbf{n}\,d\sigma = \int_0^{2\pi} (-6\sin^2 t + 18\cos^3 t)\,dt = -6\pi$$

9. curl $\mathbf{F} = \nabla \times \mathbf{F} = -2x\,\mathbf{j} + 2\,\mathbf{k}$. Flux of $\nabla \times \mathbf{F} = \iint_S \nabla \times \mathbf{F} \cdot \mathbf{n}\,d\sigma = \oint_C \mathbf{F} \cdot d\mathbf{r}$. Let C be $x = a\cos t$, $y = a\sin t$. Then $\mathbf{r} = (a\cos t)\,\mathbf{i} + (a\sin t)\,\mathbf{j} \Rightarrow d\mathbf{r} = (-a\sin t)\,dt\,\mathbf{i} + (a\cos t)\,dt\,\mathbf{j}$. Then $\mathbf{F} \cdot d\mathbf{r} = ay\sin t\,dt + ax\cos t\,dt = a^2\sin^2 t\,dt + a^2\cos^2 t\,dt = a^2 dt$. $\therefore$ Flux of $\nabla \times \mathbf{F} = \oint_C \mathbf{F} \cdot d\mathbf{r} = \int_0^{2\pi} a^2\,dt = 2\pi a^2$

11. Let S_1 and S_2 be oriented surfaces that span C and that induce the same positive direction on C.

Then $\displaystyle\iint_{S_1} \nabla \times \mathbf{F} \cdot \mathbf{n}_1\,d\sigma_1 = \int_C \mathbf{F} \cdot d\mathbf{r} = \iint_{S_2} \nabla \times \mathbf{F} \cdot \mathbf{n}_2\,d\sigma_2$

13. a) $\mathbf{F} = 2x\,\mathbf{i} + 2y\,\mathbf{j} + 2z\,\mathbf{k} \Rightarrow \text{curl } \mathbf{F} = \mathbf{0}$ $\therefore \oint_C \mathbf{F}\cdot d\mathbf{r} = \iint_S \nabla \times \mathbf{F}\cdot\mathbf{n}\, d\sigma = \iint_S d\sigma = 0$

b) $\mathbf{F} = \nabla(x^2y^2z^3)$. Let $f(x,y,z) = x^2y^2z^3 \Rightarrow \nabla\mathbf{F} = \nabla \times \nabla f = \mathbf{0} \Rightarrow \text{curl } \mathbf{F} = \mathbf{0}$

$\therefore \oint_C \mathbf{F}\cdot d\mathbf{r} = \iint_S \nabla \times \mathbf{F}\cdot\mathbf{n}\, d\sigma = \iint_S 0\, d\sigma = 0$

c) $\mathbf{F} = \nabla \times (x\,\mathbf{i} + y\,\mathbf{j} + z\,\mathbf{k}) = \mathbf{0} \Rightarrow \nabla \times \mathbf{F} = \mathbf{0}$. $\therefore \oint_C \mathbf{F}\cdot d\mathbf{r} = \iint_S \nabla \times \mathbf{F}\cdot\mathbf{n}\, d\sigma = \iint_S 0\, d\sigma = 0$

d) $\mathbf{F} = \nabla f \Rightarrow \nabla \times \mathbf{F} = \nabla \times \nabla f = \mathbf{0}$. $\therefore \oint_C \mathbf{F}\cdot d\mathbf{r} = \iint_S \nabla \times \mathbf{F}\cdot\mathbf{n}\, d\sigma = \iint_S 0\, d\sigma = 0$

15. Let $\mathbf{F} = 2y\,\mathbf{i} + 3z\,\mathbf{j} - x\,\mathbf{k} \Rightarrow \nabla \times \mathbf{F} = -3\,\mathbf{i} + \mathbf{j} - 2\,\mathbf{k}$. $\mathbf{n} = \dfrac{2\,\mathbf{i} + 2\,\mathbf{j} + \mathbf{k}}{3} \Rightarrow \nabla \times \mathbf{F}\cdot\mathbf{n} = -2$.

$\therefore \oint_C 2y\,dx + 3z\,dy - x\,dz = \oint_C \mathbf{F}\cdot d\mathbf{r} = \iint_S \nabla \times \mathbf{F}\cdot\mathbf{n}\, d\sigma = \iint_S -2\, d\sigma = -2\iint_S d\sigma$

where $\iint_S d\sigma$ is the area of the region enclosed by C on the surface S: $2x + 2y + z = 2$.

17. Yes: If $\nabla \times \mathbf{F} = \mathbf{0}$, then the circulation of $\mathbf{F}$ around the boundary C of any oriented surface S in the domain of $\mathbf{F}$ is zero. The reason is: By Stoke's theorem, circulation $= \oint_C \mathbf{F}\cdot d\mathbf{r} = \iint_S \nabla \times \mathbf{F}\cdot\mathbf{n}\, d\sigma = \iint_S \mathbf{0}\cdot\mathbf{n}\, d\sigma = 0$

19. $\dfrac{\partial P}{\partial y} = 0, \dfrac{\partial N}{\partial z} = 0, \dfrac{\partial M}{\partial z} = 0, \dfrac{\partial P}{\partial x} = 0, \dfrac{\partial N}{\partial x} = \dfrac{y^2 - x^2}{(x^2 + y^2)^2}, \dfrac{\partial M}{\partial y} = \dfrac{y^2 - x^2}{(x^2 + y^2)^2} \Rightarrow \text{curl } \mathbf{F} = \left(\dfrac{y^2 - x^2}{(x^2 + y^2)^2} - \dfrac{y^2 - x^2}{(x^2 + y^2)^2}\right)\mathbf{k}$

$= \mathbf{0}$. $x^2 + y^2 = 1 \Rightarrow \mathbf{r} = (a\cos t)\,\mathbf{i} + (a\sin t)\,\mathbf{j} \Rightarrow d\mathbf{r} = (-a\sin t)\,\mathbf{i} + (a\cos t)\,\mathbf{j} \Rightarrow \mathbf{F} = \dfrac{-a\sin t}{a^2}\,\mathbf{i} + \dfrac{a\cos t}{a^2}\,\mathbf{j} +$

$z\,\mathbf{k} \Rightarrow \mathbf{F}\cdot d\mathbf{r} = \dfrac{a^2\sin^2 t}{a^2} + \dfrac{a^2\cos^2 t}{a^2} = 1 \Rightarrow \int_C \mathbf{F}\cdot d\mathbf{r} = \int_0^{2\pi} 1\, dt = 2\pi$

15.8 THE DIVERGENCE THEOREM

1. $\mathbf{F} = \frac{-y\,\mathbf{i} + x\,\mathbf{j}}{\sqrt{x^2 + y^2}} \Rightarrow \text{div}\,\mathbf{F} = \frac{xy - xy}{(x^2 + y^2)^{3/2}} = 0$

3. $\mathbf{F} = -\frac{GM(x\,\mathbf{i} + y\,\mathbf{j} + z\,\mathbf{k})}{(x^2 + y^2 + z^2)^{3/2}} \Rightarrow \text{div}\,\mathbf{F} = -GM\left(\frac{(x^2 + y^2 + z^2)^{3/2} - 3x^2(x^2 + y^2 + z^2)^{1/2}}{(x^2 + y^2 + z^2)^3}\right) -$

$GM\left(\frac{(x^2 + y^2 + z^2)^{3/2} - 3y^2(x^2 + y^2 + z^2)^{1/2}}{(x^2 + y^2 + z^2)^3}\right) - GM\left(\frac{(x^2 + y^2 + z^2)^{3/2} - 3z^2(x^2 + y^2 + z^2)^{1/2}}{(x^2 + y^2 + z^2)^3}\right) =$

$-GM\left(\frac{3(x^2 + y^2 + z^2)^2 - 3(x^2 + y^2 + z^2)(x^2 + y^2 + z^2)}{(x^2 + y^2 + z^2)^{7/2}}\right) = 0$

5. $\frac{\partial}{\partial x}(y - x) = -1, \frac{\partial}{\partial y}(z - y) = -1, \frac{\partial}{\partial z}(y - x) = 0 \Rightarrow \nabla \cdot \mathbf{F} = -2 \Rightarrow \text{Flux} = \int_{-1}^{1}\int_{-1}^{1}\int_{-1}^{1} -2\,dx\,dy\,dz = -16$

7. $\frac{\partial}{\partial x}(y) = 0, \frac{\partial}{\partial y}(xy) = x, \frac{\partial}{\partial z}(-z) = -1 \Rightarrow \nabla \cdot \mathbf{F} = x - 1 \Rightarrow \text{Flux} = \iiint_{\text{solid}} (x - 1)\,dz\,dy\,dx =$

$\int_0^{2\pi}\int_0^{2}\int_0^{r^2} (r\cos\theta - 1)\,dz\,r\,dr\,d\theta = -8\pi$

9. $\frac{\partial}{\partial x}(x^2) = 2x, \frac{\partial}{\partial y}(-2xy) = -2x, \frac{\partial}{\partial z}(3xz) = 3x \Rightarrow \text{Flux} = \iiint_D 3x\,dx\,dy\,dz =$

$\int_0^{\pi/2}\int_0^{\pi/2}\int_0^{2} 3\rho\sin\phi\cos\theta(\rho^2\sin\phi)\,d\rho\,d\phi\,d\theta = 3\pi$

11. $\frac{\partial}{\partial x}(2xz) = 2z, \frac{\partial}{\partial y}(-xy) = -x, \frac{\partial}{\partial z}(-z^2) = -2z \Rightarrow \nabla \cdot \mathbf{F} = -x \Rightarrow \text{Flux} = \iiint_D -x\,dV =$

$\int_0^{2}\int_0^{\sqrt{16-4x^2}}\int_0^{4-y} -x\,dz\,dy\,dx = -\frac{40}{3}$

13. Let $\rho = \sqrt{x^2 + y^2 + z^2}$. Then $\frac{\partial \rho}{\partial x} = \frac{x}{\rho}, \frac{\partial \rho}{\partial y} = \frac{y}{\rho}, \frac{\partial \rho}{\partial z} = \frac{z}{\rho} \Rightarrow \frac{\partial}{\partial x}(\rho x) = \frac{\partial \rho}{\partial x} x + \rho = \frac{x^2}{\rho} = \rho, \frac{\partial}{\partial y}(\rho y) =$ $\frac{\partial \rho}{\partial y} y + \rho = \frac{y^2}{\rho} + \rho, \frac{\partial}{\partial z}(\rho z) = \frac{\partial \rho}{\partial z} z + \rho = \frac{z^2}{\rho} + \rho \Rightarrow \nabla \cdot \mathbf{F} = \frac{x^2 + y^2 + z^2}{\rho} + 3\rho = 4\rho$ since $\rho = \sqrt{x^2 + y^2 + z^2}$

$$\Rightarrow \text{Flux} = \iiint_D 4\rho \, dV = \iiint_D 4\sqrt{x^2 + y^2 + z^2} \, dx\, dy\, dz =$$

$$\int_0^{2\pi} \int_0^{\pi} \int_1^{\sqrt{2}} (4\rho)\rho^2 \sin\phi \, d\rho\, d\phi\, d\theta = 12\pi$$

15. $\frac{\partial}{\partial x}(5x^3 + 12xy^2) = 15x^2 + 12y^2, \frac{\partial}{\partial y}(y^3 + e^y \sin z) = 3y^2 + e^y \sin z, \frac{\partial}{\partial z}(5z^3 + e^y \cos z) = 15z^2 - e^y \sin z \Rightarrow$

$$\nabla \cdot \mathbf{F} = 15x^2 + 15y^2 + 15z^2 \Rightarrow \text{Flux} = \iiint_V (15x^2 + 15y^2 + 15z^2) \, dz\, dy\, dz =$$

$$\int_0^{2\pi} \int_0^{\pi} \int_1^{\sqrt{2}} 15\rho^2(\rho^2) \sin\phi \, d\rho\, d\phi\, d\theta = \int_0^{2\pi} \int_0^{\pi} (12\sqrt{2} - 3)\sin\phi \, d\phi\, d\theta = \int_0^{2\pi} (24\sqrt{2} - 6)d\theta = (48\sqrt{2} - 12)\pi$$

17. $|\mathbf{F} \cdot \mathbf{n}| \le \|\mathbf{F}\|\,\|\mathbf{n}\| \le 1$ since $\|\mathbf{F}\| \le 1$, $\|\mathbf{n}\| = 1$. Then $\iiint_D \nabla \cdot \mathbf{F} \, d\sigma = \iint_S \mathbf{F} \cdot \mathbf{n} \, d\sigma \le$

$$\iint_S |\mathbf{F} \cdot \mathbf{n}| \, d\sigma \le \iint_S 1 \, d\sigma = \text{Area of } S$$

19. Yes: The divergence of a constant field is zero, so the flux of a constant field across any closed orientable surface must be zero as well, by the Divergence Theorem.

21. a) $\mathbf{G} = M\mathbf{i} + N\mathbf{j} + P\mathbf{k} \Rightarrow \text{curl } \mathbf{G} = \left(\frac{\partial P}{\partial y} - \frac{\partial N}{\partial z}\right)\mathbf{i} + \left(\frac{\partial M}{\partial z} - \frac{\partial P}{\partial x}\right)\mathbf{j} + \left(\frac{\partial N}{\partial x} - \frac{\partial M}{\partial y}\right)\mathbf{k}.$

$\nabla \cdot \nabla \times \mathbf{G} = \text{div(curl } \mathbf{G}) = \frac{\partial}{\partial x}\left(\frac{\partial P}{\partial y} - \frac{\partial N}{\partial z}\right) + \frac{\partial}{\partial y}\left(\frac{\partial M}{\partial z} - \frac{\partial P}{\partial x}\right) + \frac{\partial}{\partial z}\left(\frac{\partial N}{\partial x} - \frac{\partial M}{\partial y}\right) = \frac{\partial^2 P}{\partial x \partial y} - \frac{\partial^2 N}{\partial x \partial z} +$ $\frac{\partial^2 M}{\partial y \partial z} - \frac{\partial^2 P}{\partial y \partial x} + \frac{\partial^2 N}{\partial z \partial x} - \frac{\partial^2 M}{\partial z \partial y} = 0$ if the partial derivatives are continuous.

b) $\iint_S \nabla \times \mathbf{G} \cdot \mathbf{n} \, d\sigma = \iiint_D \nabla \cdot \nabla \times \mathbf{G} \, dV = \iiint_D 0 \, dV = 0$ if the divergence theorem applies.

23. a) Let $\mathbf{F}_1 = M_1\,\mathbf{i} + N_1\,\mathbf{j} + P_1\,\mathbf{k}$, $\mathbf{F}_2 = M_2\,\mathbf{i} + N_2\,\mathbf{j} + P_2\,\mathbf{k} \Rightarrow a\mathbf{F}_1 + b\mathbf{F}_2 = (aM_1 + bM_2)\,\mathbf{i} + (aN_1) + bN_2)\,\mathbf{j} +$

$(aP_1 + bP_2)\,\mathbf{k}.$ Then $\nabla \cdot (a\mathbf{F}_1 + b\mathbf{F}_2) = \left(a\,\dfrac{\partial M_1}{\partial x} + b\,\dfrac{\partial M_2}{\partial x}\right) + \left(a\,\dfrac{\partial N_1}{\partial y} + b\,\dfrac{\partial N_2}{\partial y}\right) + \left(a\,\dfrac{\partial P_1}{\partial z} + b\,\dfrac{\partial P_2}{\partial z}\right) =$

$a\left(\dfrac{\partial M_1}{\partial x} + \dfrac{\partial N_1}{\partial y} + \dfrac{\partial P_1}{\partial z}\right) + b\left(\dfrac{\partial M_2}{\partial x} + \dfrac{\partial N_2}{\partial y} + \dfrac{\partial P_2}{\partial z}\right) = a\left(\nabla \cdot \mathbf{F}_1\right) + b\left(\nabla \cdot \mathbf{F}_2\right)$

b) Define $\mathbf{F}_1$ and $\mathbf{F}_2$ as in part a. Then $\nabla \times (a\mathbf{F}_1 + b\mathbf{F}_2) = \left(a\,\dfrac{\partial P_1}{\partial y} + b\,\dfrac{\partial P_2}{\partial y} - \left(a\,\dfrac{\partial N_1}{\partial z} + b\,\dfrac{\partial N_2}{\partial z}\right)\right)\mathbf{i} +$

$\left(a\,\dfrac{\partial M_1}{\partial z} + b\,\dfrac{\partial M_2}{\partial z} - \left(a\,\dfrac{\partial P_1}{\partial x} + b\,\dfrac{\partial P_2}{\partial x}\right)\right)\mathbf{j} + \left(a\,\dfrac{\partial N_1}{\partial x} + b\,\dfrac{\partial N_2}{\partial x} - \left(a\,\dfrac{\partial M_1}{\partial y} + b\,\dfrac{\partial M_2}{\partial y}\right)\right)\mathbf{k} =$

$a\left[\left(\dfrac{\partial P_1}{\partial y} - \dfrac{\partial N_1}{\partial z}\right)\mathbf{i} + \left(\dfrac{\partial M_1}{\partial z} - \dfrac{\partial P_1}{\partial x}\right)\mathbf{j} + \left(\dfrac{\partial N_1}{\partial x} - \dfrac{\partial M_1}{\partial y}\right)\mathbf{k}\right] + b\left[\left(\dfrac{\partial P_2}{\partial y} - \dfrac{\partial N_2}{\partial z}\right)\mathbf{i} + \left(\dfrac{\partial M_2}{\partial z} - \dfrac{\partial P_2}{\partial x}\right)\mathbf{j} + \left(\dfrac{\partial N_2}{\partial x} - \dfrac{\partial N_2}{\partial y}\right)\mathbf{k}\right]$

$= a\nabla \times \mathbf{F}_1 + b\nabla \times \mathbf{F}_2$

c) $\text{div}(g\mathbf{F}) = \nabla \cdot g\mathbf{F} = \dfrac{\partial}{\partial x}(gM) + \dfrac{\partial}{\partial y}(gN) + \dfrac{\partial}{\partial z}(gP) = \left(g\,\dfrac{\partial M}{\partial x} + M\,\dfrac{\partial g}{\partial x}\right) + \left(g\,\dfrac{\partial N}{\partial y} + N\,\dfrac{\partial g}{\partial y}\right) + \left(g\,\dfrac{\partial P}{\partial z} + P\,\dfrac{\partial g}{\partial z}\right) =$

$\left(M\,\dfrac{\partial g}{\partial x} + N\,\dfrac{\partial g}{\partial y} + P\,\dfrac{\partial g}{\partial z}\right) + g\left(\dfrac{\partial M}{\partial x} + \dfrac{\partial N}{\partial y} + \dfrac{\partial P}{\partial z}\right) = \mathbf{F} \cdot \nabla g + g(\nabla \cdot \mathbf{F})$

d) $\nabla \times (g\mathbf{F}) = \left(\dfrac{\partial}{\partial y}(gP) - \dfrac{\partial}{\partial z}(gN)\right)\mathbf{i} + \left(\dfrac{\partial}{\partial z}(gM) - \dfrac{\partial}{\partial x}(gP)\right)\mathbf{j} + \left(\dfrac{\partial}{\partial x}(gN) - \dfrac{\partial}{\partial y}(gM)\right)\mathbf{k} =$

$\left(P\,\dfrac{\partial g}{\partial y} + g\,\dfrac{\partial P}{\partial y} - N\,\dfrac{\partial g}{\partial z} - g\,\dfrac{\partial N}{\partial z}\right)\mathbf{i} + \left(M\,\dfrac{\partial g}{\partial z} + g\,\dfrac{\partial M}{\partial z} - P\,\dfrac{\partial g}{\partial x} - g\,\dfrac{\partial P}{\partial x}\right)\mathbf{j} + \left(N\,\dfrac{\partial g}{\partial x} + g\,\dfrac{\partial N}{\partial x} - M\,\dfrac{\partial g}{\partial y} - g\,\dfrac{\partial M}{\partial y}\right)\mathbf{k}$

$= \left(P\,\dfrac{\partial g}{\partial y} - N\,\dfrac{\partial g}{\partial z}\right)\mathbf{i} + \left(g\,\dfrac{\partial P}{\partial y} - g\,\dfrac{\partial N}{\partial z}\right)\mathbf{i} + \left(M\,\dfrac{\partial g}{\partial z} - P\,\dfrac{\partial g}{\partial x}\right)\mathbf{j} + \left(g\,\dfrac{\partial M}{\partial z} - g\,\dfrac{\partial P}{\partial x}\right)\mathbf{j} + \left(N\,\dfrac{\partial g}{\partial x} - M\,\dfrac{\partial g}{\partial y}\right)\mathbf{k} +$

$\left(g\,\dfrac{\partial N}{\partial x} - g\,\dfrac{\partial M}{\partial y}\right)\mathbf{k} = g(\nabla \times \mathbf{F}) + \nabla g \times \mathbf{F}$

e) $(\mathbf{F}_1 \times \mathbf{F}_2) = \begin{vmatrix} \mathbf{i} & \mathbf{j} & \mathbf{k} \\ M_1 & N_1 & P_1 \\ M_2 & N_2 & P_2 \end{vmatrix} = (N_1 P_2 - P_1 N_2)\,\mathbf{i} - (M_1 P_2 - P_1 M_2)\,\mathbf{j} + (M_1 N_2 - N_1 M_2)\,\mathbf{k} \Rightarrow$

$\nabla \cdot (\mathbf{F}_1 \times \mathbf{F}_2) = \nabla \cdot \left((N_1 P_2 - P_1 N_2)\,\mathbf{i} - (M_1 P_2 - P_1 M_2)\,\mathbf{j} + (M_1 N_2 - N_1 M_2)\,\mathbf{k}\right) =$

$\dfrac{\partial}{\partial x}(N_1 P_2 - P_1 N_2) - \dfrac{\partial}{\partial y}(M_1 P_2 - P_1 M_2) + \dfrac{\partial}{\partial z}(M_1 N_2 - N_1 M_2) = P_2\,\dfrac{\partial N_1}{\partial x} + N_1\,\dfrac{\partial P_2}{\partial x} - N_2\,\dfrac{\partial P_1}{\partial x} - P_1\,\dfrac{\partial N_2}{\partial x}$

$- M_1\,\dfrac{\partial P_2}{\partial y} - P_2\,\dfrac{\partial M_1}{\partial y} + P_1\,\dfrac{\partial M_2}{\partial y} + M_2\,\dfrac{\partial P_1}{\partial y} + M_1\,\dfrac{\partial N_2}{\partial z} + N_2\,\dfrac{\partial M_1}{\partial z} - N_1\,\dfrac{\partial M_2}{\partial z} - M_2\,\dfrac{\partial N_1}{\partial z} =$

$M_2\left(\dfrac{\partial P_1}{\partial y} - \dfrac{\partial N_1}{\partial z}\right) + N_2\left(\dfrac{\partial M_1}{\partial z} - \dfrac{\partial P_1}{\partial x}\right) + P_2\left(\dfrac{\partial N_1}{\partial x} - \dfrac{\partial M_1}{\partial y}\right) + M_1\left(\dfrac{\partial N_2}{\partial z} - \dfrac{\partial P_2}{\partial y}\right) + N_1\left(\dfrac{\partial P_2}{\partial x} - \dfrac{\partial M_2}{\partial z}\right) +$

$P_1\left(\dfrac{\partial M_2}{\partial y} - \dfrac{\partial N_2}{\partial x}\right) = \mathbf{F}_2 \cdot (\nabla \times \mathbf{F}_1) - \mathbf{F}_1 \cdot (\nabla \times \mathbf{F}_2)$

15.P PRACTICE EXERCISES

1. Path 1: $\mathbf{r} = t\mathbf{i} + t\mathbf{j} + t\mathbf{k} \Rightarrow x = t, y = t, z = t, 0 \le t \le 1 \Rightarrow f(g(t),h(t),k(t)) = 3 - 3t^2$ and $\frac{dx}{dt} = 1, \frac{dy}{dt} = 1,$

$$\frac{dz}{dt} = 1 \Rightarrow \sqrt{\left(\frac{dx}{dt}\right)^2 + \left(\frac{dy}{dt}\right)^2 + \left(\frac{dz}{dt}\right)^2}\,dt = \sqrt{3}\,dt \Rightarrow \int_C f(x,y,z)\,ds = \int_0^1 \sqrt{3}\left(3 - 3t^2\right) dt = 2\sqrt{3}$$

Path 2: $\mathbf{r}_1 = t\mathbf{i} + t\mathbf{j}, 0 \le t \le 1 \Rightarrow x = t, y = t, z = 0 \Rightarrow f(g(t),h(t),k(t)) = 2t - 3t^2 + 3$ and $\frac{dx}{dt} = 1, \frac{dy}{dt} = 1,$

$$\frac{dz}{dt} = 0 \Rightarrow \sqrt{\left(\frac{dx}{dt}\right)^2 + \left(\frac{dy}{dt}\right)^2 + \left(\frac{dz}{dt}\right)^2}\,dt = \sqrt{2}\,dt \Rightarrow \int_{C_1} f(x,y,z)\,ds = \int_0^1 \sqrt{2}\left(2t - 3t^2 + 3\right) dt =$$

$3\sqrt{2}$. $\mathbf{r}_2 = \mathbf{i} + \mathbf{j} + t\mathbf{k} \Rightarrow x = 1, y = 1, z = t \Rightarrow f(g(t),h(t),k(t)) = 2 - 2t$ and $\frac{dx}{dt} = 0, \frac{dy}{dt} = 0, \frac{dz}{dt} = 1 \Rightarrow$

$$\sqrt{\left(\frac{dx}{dt}\right)^2 + \left(\frac{dy}{dt}\right)^2 + \left(\frac{dz}{dt}\right)^2}\,dt = dt \Rightarrow \int_{C_2} f(x,y,z)\,ds = \int_0^1 (2 - 2t)\,dt = 1. \therefore \int_C f(x,y,z)\,ds =$$

$$\int_{C_1} f(x,y,z)\,ds + \int_{C_2} f(x,y,z)\,ds = 3\sqrt{2} + 1$$

3. $\mathbf{r} = (a\cos t)\mathbf{j} + (a\sin t)\mathbf{k} \Rightarrow x = 0, y = a\cos t, z = a\sin t \Rightarrow f(g(t),h(t),h(t)) = \sqrt{a^2\sin^2 t} = a\,|\sin t|$ and

$$\frac{dx}{dt} = 0, \frac{dy}{dt} = -a\sin t, \frac{dz}{dt} = a\cos t \Rightarrow \sqrt{\left(\frac{dx}{dt}\right)^2 + \left(\frac{dy}{dt}\right)^2 + \left(\frac{dz}{dt}\right)^2}\,dt = a\,dt \Rightarrow \int_C f(x,y,z)\,ds =$$

$$\int_0^{2\pi} a^2\,|\sin t|\,dt = \int_0^{\pi} a^2 \sin t\,dt + \int_{\pi}^{2\pi} -a^2\sin t\,dt = 4a^2$$

5. a) $\mathbf{r} = \sqrt{2}t\mathbf{i} + \sqrt{2}t\mathbf{j} + (4 - t^2)\mathbf{k}, 0 \le t \le 1 \Rightarrow x = \sqrt{2}t, y = \sqrt{2}t, z = 4 - t^2 \Rightarrow \frac{dx}{dt} = \sqrt{2}, \frac{dy}{dt} = \sqrt{2}, \frac{dz}{dt} = -2t$

$$\Rightarrow \sqrt{\left(\frac{dx}{dt}\right)^2 + \left(\frac{dy}{dt}\right)^2 + \left(\frac{dz}{dt}\right)^2}\,dt = \sqrt{4 + 4t^2}\,dt \Rightarrow M = \int_C \delta(x,y,z)\,ds = \int_0^1 3t\sqrt{4 + 4t^2}\,dt =$$

$4\sqrt{2} - 2$

b) $M = \int_C \delta(x,y,z)\,ds = \int_0^1 \sqrt{4 + 4t^2}\,dt = \sqrt{2} + \ln(1 + \sqrt{2})$

7. $\mathbf{r} = t\mathbf{i} + \frac{2\sqrt{2}}{3}t^{3/2}\mathbf{j} + \frac{t^2}{2}\mathbf{k}, 0 \le t \le 2 \Rightarrow x = t, y = \frac{2\sqrt{2}}{3}t^{3/2}, z = \frac{t^2}{2} \Rightarrow \frac{dx}{dt} = 1, \frac{dy}{dt} = \sqrt{2}\,t^{1/2}, \frac{dz}{dt} = t \Rightarrow$

$\sqrt{\left(\frac{dx}{dt}\right)^2 + \left(\frac{dy}{dt}\right)^2 + \left(\frac{dz}{dt}\right)^2}\,dt = \sqrt{(t + 1)^2}\,dt = |t + 1|\,dt = (t + 1)dt$ on the domain given. Then $M_{yz} =$

$$\int_C x\delta\,ds = \int_0^2 t\left(\frac{1}{t + 1}\right)(t + 1)\,dt = \int_0^2 t\,dt = 2.\ M = \int_C \delta\,ds = \int_0^2 \frac{1}{t + 1}(t + 1)\,dt = \int_0^2 dt = 2$$

$$M_{xz} = \int_C y\delta\,ds = \int_0^2 \frac{2\sqrt{2}}{3}t^{3/2}\left(\frac{1}{t + 1}\right)(t + 1)\,dt = \int_0^2 \frac{2\sqrt{2}}{3}t^{3/2}\,dt = \frac{32}{15}.\ M_{xy} = \int_C z\delta\,ds =$$

7. (Continued)

$\int_0^2 \frac{t^2}{2}\left(\frac{1}{t+1}\right)(t+1)\,dt = \int_0^2 \frac{t^2}{2}\,dt = \frac{4}{3}$. $\therefore\ \bar{x} = M_{yz}/M = \frac{2}{2} = 1$, $\bar{y} = M_{xz}/M = \frac{32/15}{2} = \frac{16}{15}$, $\bar{z} = M_{xy}/M = \frac{4/3}{2} = \frac{2}{3}$. $I_x = \int_C (y^2 + z^2)\delta\,ds = \int_0^2 \left(\frac{8}{9}t^3 + \frac{t^4}{4}\right)dt = \frac{232}{45}$. $I_y = \int_C (x^2 + z^2)\delta\,ds = \int_0^2 \left(t^2 + \frac{t^4}{4}\right)dt = \frac{64}{15}$ $I_z = \int_C (y^2 + x^2)\delta\,ds = \int_0^2 \left(t^2 + \frac{8}{9}t^3\right)dt = \frac{56}{9}$. $R_x = \sqrt{I_x/M} = \sqrt{\frac{232/45}{2}} = \frac{2}{3}\sqrt{\frac{29}{5}}$. $R_y = \sqrt{I_y/M} = \sqrt{\frac{64/15}{2}} = 4\sqrt{\frac{2}{15}}$. $R_z = \sqrt{I_z/M} = \sqrt{\frac{56/9}{2}} = \frac{2}{3}\sqrt{7}$

9. a) $x^2 + y^2 = 1 \Rightarrow \mathbf{r} = (\cos t)\,\mathbf{i} + (\sin t)\,\mathbf{j},\ 0 \le t \le \pi \Rightarrow x = \cos t,\ y = \sin t \Rightarrow \mathbf{F} = (\cos t + \sin t)\,\mathbf{i} - \mathbf{j}$ and $\frac{d\mathbf{r}}{dt} = (-\sin t)\,\mathbf{i} + (\cos t)\,\mathbf{j} \Rightarrow \mathbf{F}\cdot\frac{d\mathbf{r}}{dt} = -\sin t\cos t - \sin^2 t - \cos t \Rightarrow$ Flow $=$ $\int_0^\pi (-\sin t\cos t - \sin^2 t - \cos t)\,dt = -\frac{1}{2}\pi$

b) $\mathbf{r} = -t\,\mathbf{i},\ -1 \le t \le 1 \Rightarrow x = -t,\ y = 0 \Rightarrow \mathbf{F} = -t\,\mathbf{i} - t^2\,\mathbf{j}$ and $\frac{d\mathbf{r}}{dt} = -\mathbf{i} \Rightarrow \mathbf{F}\cdot\frac{d\mathbf{r}}{dt} = t \Rightarrow$ Flow $= \int_{-1}^1 t\,dt = 0$

c) $\mathbf{r}_1 = (1-t)\,\mathbf{i} - t\,\mathbf{j},\ 0 \le t \le 1 \Rightarrow \mathbf{F}_1 = (1-2t)\,\mathbf{i} - (1 - 2t - 2t^2)\,\mathbf{j}$ and $\frac{d\mathbf{r}_1}{dt} = -\mathbf{i} - \mathbf{j} \Rightarrow \mathbf{F}_1\cdot\frac{d\mathbf{r}_1}{dt} = 2t^2 \Rightarrow$ Flow$_1$ $= \int_0^1 2t^2\,dt = \frac{2}{3}$. $\mathbf{r}_2 = -t\,\mathbf{i} + (t-1)\,\mathbf{j},\ 0 \le t \le 1 \Rightarrow \mathbf{F}_2 = -\mathbf{i} - (2t^2 - 2t + 1)\,\mathbf{j}$ and $\frac{d\mathbf{r}_2}{dt} = -\mathbf{i} + \mathbf{j} \Rightarrow \mathbf{F}_2\cdot\frac{d\mathbf{r}_2}{dt} = -2t^2 + 2t \Rightarrow$ Flow$_2$ $= \int_0^1 (-2t^2 + 2t)\,dt = \frac{1}{3}$. $\therefore$ Flow $=$ Flow$_1$ $+$ Flow$_2$ $= 1$

11. $\frac{\partial P}{\partial y} = 0 = \frac{\partial N}{\partial z},\ \frac{\partial M}{\partial z} = 0 = \frac{\partial P}{\partial x},\ \frac{\partial N}{\partial x} = 0 = \frac{\partial M}{\partial y} \Rightarrow$ Conservative

13. $\frac{\partial P}{\partial y} = 0 \ne ye^z = \frac{\partial N}{\partial z} \Rightarrow$ Not Conservative

15. $\frac{\partial f}{\partial x} = 2 \Rightarrow f(x,y,z) = 2x + g(y,z)$. $\frac{\partial f}{\partial y} = \frac{\partial g}{\partial y} = 2y + z \Rightarrow g(y,z) = y^2 + zy + h(z) \Rightarrow f(x,y,z) = 2x + y^2 + zy + h(z)$. $\frac{\partial f}{\partial z} = y + h'(z) = y + 1 \Rightarrow h'(z) = 1 \Rightarrow h(z) = z + C$. $\therefore\ f(x,y,z) = 2x + y^2 + zy + z + C$

17. $\frac{\partial P}{\partial y} = -\frac{1}{2}(x+y+z)^{-3/2} = \frac{\partial N}{\partial z},\ \frac{\partial M}{\partial z} = -\frac{1}{2}(x+y+z)^{-3/2} = \frac{\partial P}{\partial x},\ \frac{\partial N}{\partial x} = -\frac{1}{2}(x+y+z)^{-3/2} = \frac{\partial M}{\partial y} \Rightarrow M\,dx +$ $N\,dy + P\,dz$ is exact. $\frac{\partial f}{\partial x} = \frac{1}{\sqrt{x+y+z}} \Rightarrow f(x,y,z) = 2\sqrt{x+y+z} + g(y,z)$. $\frac{\partial f}{\partial y} = \frac{1}{\sqrt{x+y+z}} + \frac{\partial g}{\partial y} = \frac{1}{\sqrt{x+y+z}}$ $\Rightarrow \frac{\partial g}{\partial y} = 0 \Rightarrow g(y,z) = h(z) \Rightarrow f(x,y,z) = 2\sqrt{x+y+z} + h(z)$. $\frac{\partial f}{\partial z} = \frac{1}{\sqrt{x+y+z}} + h'(z) = \frac{1}{\sqrt{x+y+z}} \Rightarrow h'(z) = 0$

17. (Continued)

$\Rightarrow h(z) = C \Rightarrow f(x,y,z) = 2\sqrt{x+y+z} + C \Rightarrow \int_{(-1,1,1)}^{(4,-3,0)} \frac{dx + dy + dz}{\sqrt{x+y+z}} = f(4,-3,0) - f(-1,1,1) = 0$

19. Over Path 1: $\mathbf{r} = t\mathbf{i} + t\mathbf{j} + t\mathbf{k} \Rightarrow x = t, y = t, z = t$ and $d\mathbf{r} = (\mathbf{i} + \mathbf{j} + \mathbf{k})\,dt \Rightarrow \mathbf{F} = 2t^2\mathbf{i} + \mathbf{j} + t^2\mathbf{k} \Rightarrow$

$\mathbf{F} \cdot d\mathbf{r} = (3t^2 + 1)\,dt \Rightarrow \text{Work} = \int_0^1 (3t^2 + 1)\,dt = 2$

Over Path 2: $\mathbf{r}_1 = t\mathbf{i} + t\mathbf{j}, 0 \le t \le 1 \Rightarrow x = t, y = t, z = 0$ and $d\mathbf{r}_1 = (\mathbf{i} + \mathbf{j})\,dt \Rightarrow \mathbf{F}_1 = 2t^2\mathbf{i} + \mathbf{j} + t^2\mathbf{k} \Rightarrow$

$\mathbf{F}_1 \cdot d\mathbf{r}_1 = (2t^2 + 1)\,dt \Rightarrow \text{Work}_1 = \int_0^1 (2t^2 + 1)\,dt = \frac{5}{3}$. $\mathbf{r}_2 = \mathbf{i} + \mathbf{j} + t\mathbf{k}, 0 \le t \le 1 \Rightarrow x = 1, y = 1, z = t$ and

$d\mathbf{r}_2 = \mathbf{k}\,dt \Rightarrow \mathbf{F}_2 = 2\mathbf{i} + \mathbf{j} + \mathbf{k} \Rightarrow \mathbf{F}_2 \cdot d\mathbf{r}_2 = dt \Rightarrow \text{Work}_2 = \int_0^1 dt = 1.\ \therefore\ \text{Work} = \text{Work}_1 + \text{Work}_2 = \frac{8}{3}$

21. $\mathbf{F} = x\mathbf{i} + y\mathbf{j} + z\mathbf{k} \Rightarrow \frac{\partial P}{\partial y} = 0 = \frac{\partial N}{\partial z}, \frac{\partial M}{\partial z} = 0 = \frac{\partial P}{\partial x}, \frac{\partial N}{\partial x} = 0 = \frac{\partial M}{\partial y} \Rightarrow \mathbf{F}$ is conservative $\Rightarrow$ curl $\mathbf{F} = \mathbf{0} \Rightarrow$

circulation is 0.

23. a) $\mathbf{r} = (e^t \cos t)\mathbf{i} + (e^t \sin t)\mathbf{j} \Rightarrow x = e^t \cos t, y = e^t \sin t$ from (1,0) to $(e^{2\pi}, 0) \Rightarrow 0 \le t \le 2\pi \Rightarrow$

$d\mathbf{r} = (e^t \cos t = e^t \sin t)dt\,\mathbf{i} + (e^t \sin t + e^t \cos t)dt\,\mathbf{j}$. $\mathbf{F} = \frac{x\mathbf{i} + y\mathbf{j}}{(x^2 + y^2)^{3/2}} = \frac{e^t \cos t\,\mathbf{i} + e^t \sin t\,\mathbf{j}}{(e^{2t}\cos^2 t + e^{2t}\sin^2 t)^{3/2}} =$

$\frac{\cos t}{e^{2t}}\mathbf{i} + \frac{\sin t}{e^{2t}}\mathbf{j}$. $\therefore\ \mathbf{F} \cdot d\mathbf{r} = \left(\frac{\cos^2 t}{e^t} - \frac{\sin t \cos t}{e^t} + \frac{\sin^2 t}{e^t} + \frac{\sin t \cos t}{e^t}\right)dt = e^{-t}\,dt$. Thus the work =

$\int_0^{2\pi} e^{-t}\,dt = -e^{-2\pi} + 1.$

b) $\mathbf{F} = \frac{x\mathbf{i} + y\mathbf{j}}{(x^2 + y^2)^{3/2}} \Rightarrow \frac{\partial f}{\partial x} = \frac{x}{(x^2 + y^2)^{3/2}} \Rightarrow f(x,y,z) = -(x^2 + y^2)^{-1/2} + g(y,z)$. $\frac{\partial f}{\partial y} = \frac{y}{(x^2 + y^2)^{3/2}} + \frac{\partial g}{\partial y} =$

$\frac{y}{(x^2 + y^2)^{3/2}} \Rightarrow g(y,z) = C.\ \therefore\ f(x,y,z) = -(x^2 + y^2)^{-1/2} + C$ is a potential function for $\mathbf{F} \Rightarrow$

$\int_C \mathbf{F} \cdot d\mathbf{r} = f(e^{2\pi}, 0) - f(1,0) = 1 - e^{-2\pi}$

25. $M = 2xy + x, N = xy - y \Rightarrow \frac{\partial M}{\partial x} = 2y + 1, \frac{\partial M}{\partial y} = 2x, \frac{\partial N}{\partial x} = y, \frac{\partial N}{\partial y} = x - 1 \Rightarrow$ Flux =

$\iint_R (2y + 1 + x - 1)\,dy\,dx = \int_0^1 \int_0^1 (2y + x)\,dy\,dx = \frac{3}{2}$. Circ = $\iint_R (y - 2x)\,dy\,dx =$

$\int_0^1 \int_0^1 (y - 2x)\,dy\,dx = -\frac{1}{2}$

27. Let $M = \frac{\cos y}{x}$, $N = \ln x \sin y \Rightarrow \frac{\partial M}{\partial y} = -\frac{\sin y}{x}$, $\frac{\partial N}{\partial x} = \frac{\sin y}{x}$. Then $\oint_C \ln x \sin y\, dy - \frac{\cos y}{x} dx =$

$$\iint_R \left(\frac{\sin y}{x} - \left(-\frac{\sin y}{x}\right)\right) dx\, dy = 0$$

29. Let $M = 8x \sin y$, $N = -8y \cos x \Rightarrow \frac{\partial M}{\partial y} = 8x \cos y$, $\frac{\partial N}{\partial x} = 8y \sin x \Rightarrow \int_C 8x \sin y\, dx - 8y \cos x\, dy =$

$$\iint_R (8y \sin x - 8x \cos y)\, dy\, dx = \int_0^{\pi/2}\int_0^{\pi/2} (8y \sin x - 8x \cos y)\, dy\, dx = 0$$

31. A possible parametrization is $\mathbf{r}(\phi,\theta) = 6 \sin\phi \cos\theta\, \mathbf{i} + 6 \sin\phi \sin\theta\, \mathbf{j} + 6 \cos\phi\, \mathbf{k}$ (spherical coordinates) where $\rho = \sqrt{x^2 + y^2 + z^2} = 6$, $\frac{\pi}{6} \le \phi \le \frac{2\pi}{3}$ since $z = -3 \Rightarrow -3 = 6 \cos\phi \Rightarrow \cos\phi = -\frac{1}{2} \Rightarrow \phi = \frac{2\pi}{3}$ and $z = 3\sqrt{3}$ $\Rightarrow 3\sqrt{3} = 6 \cos\phi \Rightarrow \cos\phi = \frac{\sqrt{3}}{2} \Rightarrow \phi = \frac{\pi}{6}$. Also $0 \le \theta \le 2\pi$.

33. A possible parametrization is $\mathbf{r}(r,\theta) = r \cos\theta\, \mathbf{i} + r \sin\theta\, \mathbf{j} + (1 + r)\, \mathbf{k}$ (cylindrical coordinates) where $r = \sqrt{x^2 + y^2}$ $\Rightarrow z = 1 + r$. $1 \le z \le 3 \Rightarrow 1 \le 1 + r \le 3 \Rightarrow 0 \le r \le 2$. Also, $0 \le \theta \le 2\pi$.

35. In a fashion similar to cylindrical coordinates, but working in the xz-plane instead of the xy-plane, let $x = u \cos v$, $z = u \sin v$ where $u = \sqrt{x^2 + z^2}$ and v is the angle formed by (x,y,z), (y,0,0), and (x,y,0) with vertex (y,0,0). Then $\mathbf{r}(u,v) = u \cos v\, \mathbf{i} + 2u^2\, \mathbf{j} + u \sin v\, \mathbf{k}$ is a possible parametrization. $0 \le y \le 2 \Rightarrow 2u^2 \le 2 \Rightarrow u^2 \le 1$ $\Rightarrow 0 \le u \le 1$ since $u \ge 0$. Also, for just the upper half of the paraboloid, $0 \le v \le \pi$.

37. Let $z = 1 - x - y \Rightarrow f_x(x,y) = -1$, $f_y(x,y) = -1 \Rightarrow \sqrt{f_x^2 + f_y^2 + 1} = \sqrt{3} \Rightarrow \text{Area} = \iint_R \sqrt{3}\, dx\, dy =$ $\sqrt{3}$(Area of the circlular region in the xy–plane) $= \pi\sqrt{3}$

39. $\nabla f = 2x\, \mathbf{i} + 2y\, \mathbf{j} + 2z\, \mathbf{k}$, $\mathbf{p} = \mathbf{k} \Rightarrow |\nabla f| = \sqrt{4x^2 + 4y^2 + 4z^2} = 2\sqrt{x^2 + y^2 + z^2} = 2$ and $|\nabla f \cdot \mathbf{p}| = |2z| = 2z$ since $z \ge 0 \Rightarrow S = \iint_R \frac{2}{2z} dA = \iint_R \frac{1}{z} dA = \iint_R \frac{1}{\sqrt{1 - x^2 - y^2}} dx\, dy =$

$$\int_0^{2\pi}\int_0^{1/\sqrt{2}} \frac{1}{\sqrt{1 - r^2}}\, r\, dr\, d\theta = \int_0^{2\pi}\left(1 - \frac{1}{\sqrt{2}}\right) d\theta = 2\pi\left(1 - \frac{1}{\sqrt{2}}\right)$$

41. $\frac{x}{a}+\frac{y}{b}+\frac{z}{c}=1 \Rightarrow$ x-intercept = a, y-intercept = b, z-intercept = c. $F=\frac{x}{a}+\frac{y}{b}+\frac{z}{c} \Rightarrow \nabla F=\frac{1}{a}\mathbf{i}+\frac{1}{b}\mathbf{j}+\frac{1}{c}\mathbf{k} \Rightarrow$

$|\nabla F|=\sqrt{\frac{1}{a^2}+\frac{1}{b^2}+\frac{1}{c^2}}$. $\mathbf{p}=\mathbf{k} \Rightarrow |\nabla F \cdot \mathbf{k}|=\frac{1}{c}$ since $c>0$. Area $= \int_R\int \frac{\sqrt{\frac{1}{a^2}+\frac{1}{b^2}+\frac{1}{c^2}}}{1/c}\,dA =$

$c\sqrt{\frac{1}{a^2}+\frac{1}{b^2}+\frac{1}{c^2}}\int_R\int dA=\frac{1}{2}abc\sqrt{\frac{1}{a^2}+\frac{1}{b^2}+\frac{1}{c^2}}$

43. a) $\nabla f=2y\mathbf{j}-\mathbf{k}$, $\mathbf{p}=\mathbf{k} \Rightarrow |\nabla f|=\sqrt{4y^2+1}$ and $|\nabla f \cdot \mathbf{p}|=1 \Rightarrow d\sigma=\sqrt{4y^2+1}\;dx\,dy \Rightarrow$

$\int_S\int g(x,y,z)\,d\sigma = \int_S\int \frac{yz}{\sqrt{4y^2+1}}\sqrt{4y^2+1}\,dx\,dy = \int_S\int y(y^2-1)\,d\sigma =$

$\int_{-1}^{1}\int_0^3 (y^3-y)\,dx\,dy = 0$

b) $\int_S\int g(x,y,z)\,d\sigma = \int_S\int \frac{z}{\sqrt{4y^2+1}}\sqrt{4y^2+1}\,dx\,dy = \int_{-1}^{1}\int_0^3 (y^2-1)\,dx\,dy = -4$

45. Because of symmetry $\bar{x}=\bar{y}=0$. Let $F(x,y,z)=x^2+y^2+z^2=25 \Rightarrow \nabla F=2x\mathbf{i}+2y\mathbf{j}+2z\mathbf{k} \Rightarrow$

$|\nabla F|=\sqrt{4x^2+4y^2+4z^2}=10$, $\mathbf{p}=\mathbf{k} \Rightarrow |\nabla F \cdot \mathbf{p}|=2z$ since $z \ge 0 \Rightarrow M=\int_R\int \delta(x,y,z)\,d\sigma =$

$\int_R\int z\left(\frac{10}{2z}\right)dA = \int_R\int 5\,dA = 5(\text{Area of the circular region}) = 80\pi$. $M_{xy}=\int_R\int z\delta\,d\sigma =$

$\int_R\int 5z\,dA = \int_R\int 5\sqrt{25-x^2-y^2}\;dx\,dy = \int_0^{2\pi}\int_0^4 5\sqrt{25-r^2}\;r\,dr\,d\theta = \int_0^{2\pi}\frac{490}{3}\,d\theta=\frac{980}{3}\pi$

$\therefore \bar{z}=\frac{\frac{980}{3}\pi}{80\pi}=\frac{49}{12}$. Thus $(\bar{x},\bar{y},\bar{z})=\left(0,0,\frac{49}{12}\right)$. $I_z=\int_R\int (x^2+y^2)\delta\,d\sigma = \int_R\int 5(x^2+y^2)\,dx\,dy =$

$\int_0^{2\pi}\int_0^4 5r^3\,dr\,d\theta = \int_0^{2\pi} 320\,d\theta = 640\pi$. $R_z=\sqrt{I_z/M}=\sqrt{\frac{640\pi}{80\pi}}=2\sqrt{2}$

47. $\frac{\partial}{\partial x}(2xy) = 2y, \frac{\partial}{\partial y}(2yz) = 2z, \frac{\partial}{\partial z}(2xz) = 2x \Rightarrow \nabla \cdot \mathbf{F} = 2x + 2y + 2z \Rightarrow \text{Flux} = \iiint_D (2x + 2y + 2z)\, dV$

$$\int_0^1 \int_0^1 \int_0^1 (2x + 2y + 2z)\, dx\, dy\, dz = \int_0^1 \int_0^1 (1 + 2y + 2z)\, dy\, dz = \int_0^1 (2 + 2z)\, dz = 3$$

49. $\frac{\partial}{\partial x}(-2x) = -2, \frac{\partial}{\partial y}(-3y) = -3, \frac{\partial}{\partial z}(z) = 1 \Rightarrow \nabla \cdot \mathbf{F} = -4 \Rightarrow \text{Flux} = \iiint_D -4\, dV =$

$$-4 \int_0^{2\pi} \int_0^1 \int_{r^2}^{\sqrt{2 - r^2}} dz\, r\, dr\, d\theta = -4 \int_0^{2\pi} \int_0^1 \left(r\sqrt{2 - r^2} - r^3\right) dr\, d\theta = -4 \int_0^{2\pi} \left(-\frac{7}{12} + \frac{2}{3}\sqrt{2}\right) d\theta =$$

$$\frac{2}{3}\pi\left(7 - 8\sqrt{2}\right)$$

51. $\mathbf{F} = 3xz^2\mathbf{i} + y\mathbf{j} - z^3\mathbf{k} \Rightarrow \nabla \cdot \mathbf{F} = 3z^2 + 1 - 3z^2 = 1. \therefore \text{Flux} = \iint_S \mathbf{F} \cdot \mathbf{n}\, d\sigma = \iiint_D \nabla \cdot \mathbf{F}\, dV =$

$$= \int_0^4 \int_0^{\sqrt{16-x^2}/2} \int_0^{y/2} 1\, dz\, dy\, dx = \frac{8}{3}.$$

53. $\nabla f = 2\mathbf{i} + 6\mathbf{j} - 3\mathbf{k} \Rightarrow \nabla \times \mathbf{F} = -2y\,\mathbf{k}.\ \mathbf{n} = \frac{2\mathbf{i} + 6\mathbf{j} - 3\mathbf{k}}{\sqrt{4 + 36 + 9}} = \frac{2\mathbf{i} + 6\mathbf{j} - 3\mathbf{k}}{7} \Rightarrow \nabla \times \mathbf{F} \cdot \mathbf{n} = \frac{6}{7}y.\ \mathbf{p} = \mathbf{k} \Rightarrow$

$$|\nabla f \cdot \mathbf{p}| = 3 \Rightarrow d\sigma = \frac{7}{3}dA \Rightarrow \oint_C \mathbf{F} \cdot d\mathbf{r} = \iint_R \frac{6}{7}y\, d\sigma = \iint_R \frac{6}{7}y\left(\frac{7}{3}dA\right) = \iint_R 2y\, dx\, dy =$$

$$\int_0^{2\pi} \int_0^1 2r \sin\theta\; r\, dr\, d\theta = \int_0^{2\pi} \frac{2}{3}\sin\theta\, d\theta = 0$$

55. $dx = (-2\sin t + 2\sin 2t)\,dt,\ dy = (2\cos t - 2\cos 2t)\,dt.$ Area $= \frac{1}{2}\oint_C x\,dy - y\,dx =$

$$\frac{1}{2}\int_0^{2\pi} ((2\cos t - \cos 2t)(2\cos t - 2\cos 2t) - (2\sin t - \sin 2t)(-2\sin t + 2\sin 2t))\, dt =$$

$$\frac{1}{2}\int_0^{2\pi} (6 - 6\cos t)\, dt = 6\pi$$

57. $dx = \cos 2t\,dt$, $dy = \cos t\,dt$. Area $= \frac{1}{2}\oint_C x\,dy - y\,dx = \frac{1}{2}\int_0^{\pi}\left(\frac{1}{2}\sin 2t\cos t - \sin t\cos 2t\right)dt =$

$\frac{1}{2}\int_0^{\pi}(-\sin t\cos^2 t + \sin t)\,dt = \frac{2}{3}$

APPENDICES

APPENDIX A.2 PROOFS OF THE LIMIT THEOREMS IN CHAPTER 2

1. Let $\lim_{x\to c} f_1(x) = L_1$, $\lim_{x\to c} f_2(x) = L_2$, $\lim_{x\to c} f_3(x) = L_3$. Then $\lim_{x\to c}\left(f_1(x) + f_2(x)\right) = L_1 + L_2$ by Theorem 1. Thus $\lim_{x\to c}\left(f_1(x) + f_2(x) + f_3(x)\right) = \lim_{x\to c}\left(\left(f_1(x) + f_2(x)\right) + f_3(x)\right) = (L_1 + L_2) + L_3 = L_1 + L_2 + L_3$.

   Suppose functions $f_1(x), f_2(x), f_3(x), \ldots, f_n(x)$ have limits $L_1, L_2, L_3, \ldots, L_n$ as $x \to c$. Prove $\lim_{x\to c}\left(f_1(x) + f_2(x) + f_3(x) + \cdots + f_n(x)\right) = L_1 + L_2 + L_3 + \cdots + L_n$, n a positive integer.

   Step 1: For n = 1, we are given that $\lim_{x\to c} f_1(x) = L_1$.

   Step 2: Assume $\lim_{x\to c}\left(f_1(x) + f_2(x) + f_3(x) + \cdots + f_k(x)\right) = L_1 + L_2 + L_3 + \cdots + L_k$ for some k.
   Then $\lim_{x\to c}\left(f_1(x) + f_2(x) + f_3(x) + \cdots + f_k(x) + f_{k+1}(x)\right) =$
   $\lim_{x\to c}\left(\left(f_1(x) + f_2(x) + f_3(x) + \cdots + f_k(x)\right) + f_{k+1}(x)\right) =$
   $\left(L_1 + L_2 + L_3 + \cdots + L_k\right) + L_{k+1} = L_1 + L_2 + L_3 + \cdots + L_k + L_{k+1}$.

   $\therefore$ by Steps 1 and 2 and mathematical induction, $\lim_{x\to c}\left(f_1(x) + f_2(x) + f_3(x) + \cdots + f_n(x)\right)$
   $= L_1 + L_2 + L_3 + \cdots + L_n$

3. Given $\lim_{x\to c} x = c$. Then $\lim_{x\to c} x^n = \lim_{x\to c}\left(x\cdot x\cdot x\cdot \cdots \cdot x\right) = c\cdot c\cdot c \cdots \cdot c = c^n$ by Exercise 2.
   (n factors) (n factors)

5. $\lim_{x\to c}\dfrac{f(x)}{g(x)} = \dfrac{\lim_{x\to c} f(x)}{\lim_{x\to c} g(x)}$ (by Theorem 1) $= \dfrac{f(c)}{g(c)}$ if $g(c) \neq 0$ (by Exercise 4).

APPENDIX A.4 MATHEMATICAL INDUCTION

1. Step 1: For n = 1, $|x_1| = |x_1| \le |x_1|$.
   Step 2: Assume $|x_1 + x_2 + \cdots + x_k| \le |x_1| + |x_2| + \cdots + |x_k|$ for some positive integer k.
   Then $|x_1 + x_2 + \cdots + x_k + x_{k+1}| = |(x_1 + x_2 + \cdots + x_k) + x_{k+1}|$
   $\le |x_1 + x_2 + \cdots + x_k| + |x_{k+1}|$ (by the triangle inequality)
   $\le |x_1| + |x_2| + \cdots + |x_k| + |x_{k+1}|$. $\therefore |x_1 + x_2 + \cdots + x_n| \le |x_1| + |x_2| + \cdots + |x_n|$ for all positive integers n by Steps 1 and 2 and mathematical induction.

3. Step 1: For n = 1, $\frac{d}{dx}(x) = 1 = 1\cdot x^0$.
   Step 2: Assume $\frac{d}{dx}(x^k) = kx^{k-1}$ for some positive integer k.
   Then $\frac{d}{dx}(x^{k+1}) = \frac{d}{dx}(x^k x) = \frac{d}{dx}(x^k)x + x^k\frac{d}{dx}(x) = kx^{k-1}x + 1\cdot x^k = kx^k + x^k = (k+1)x^k = (k+1)x^{(k+1)-1}$.
   $\therefore \frac{d}{dx}(x^n) = nx^{n-1}$ for all positive integers n by Steps 1 and 2 and mathematical induction.

5. Step 1: For $n = 1$, $\frac{2}{3^1} = \frac{2}{3} = 1 - \frac{1}{3^1}$.

Step 2: Assume $\frac{2}{3^1} + \frac{2}{3^2} + \cdots + \frac{2}{3^k} = 1 - \frac{1}{3^k}$ for some positive integer k.

Then $\frac{2}{3^1} + \frac{2}{3^2} + \cdots + \frac{2}{3^k} + \frac{2}{3^{k+1}} = 1 - \frac{1}{3^k} + \frac{2}{3^{k+1}} = 1 - \frac{3^{k+1}}{3^k 3^{k+1}} + \frac{2(3^k)}{3^k 3^{k+1}} = 1 - \left(\frac{3^{k+1} - 2(3^k)}{3^k 3^{k+1}}\right) =$
$1 - \left(\frac{3^k(3-2)}{3^k 3^{k+1}}\right) = 1 - \frac{1}{3^{k+1}}$. $\therefore \frac{2}{3^1} + \frac{2}{3^2} + \cdots + \frac{2}{3^n} = 1 - \frac{1}{3^n}$ for all positive integers n by Steps 1 and 2 and mathematical induction.

7.

n	1	2	3	4	5	6
2^n	2	4	8	16	32	64
n^2	1	4	9	16	25	36

Step 1: For $n = 5$, $2^5 = 32 > 25 = 5^2$.

Step 2: Assume $2^k > k^2$ for some $k \geq 5$, k a positive integer.

Then $2^{k+1} = 2^k(2) > 2k^2$. Now $k \geq 5 \Rightarrow k - 1 \geq 4 \Rightarrow (k-1)^2 \geq 16 \Rightarrow (k-1)^2 - 2 \geq 14 \Rightarrow$
$(k-1)^2 - 2 > 0$. Then $k^2 - 2k + 1 - 2 > 0 \Rightarrow k^2 - 2k - 1 > 0 \Rightarrow k^2 > 2k + 1 \Rightarrow k^2 + k^2 > k^2 + 2k + 1 \Rightarrow 2k^2 > (k+1)^2$.
$\therefore 2^{k+1} > (k+1)^2$. Thus $n^n > n^2$ for positive integers $n \geq 5$ by Steps 1 and 2 and mathematical induction.

9. Step 1: For $n = 1$, $1^2 = \frac{1(1+1)(2(1)+1)}{6}$.

Step 2: Assume $1^2 + 2^2 + 3^2 + \cdots + k^2 = \frac{k(k+1)(2k+1)}{6}$ for some positive integer k.

Then $1^2 + 2^2 + 3^2 + \cdots + k^2 + (k+1)^2 = \frac{k(k+1)(2k+1)}{6} + (k+1)^2 \Rightarrow$

$1^2 + 2^2 + 3^2 + \cdots + k^2 + (k+1)^2 = \frac{k(k+1)(2k+1) + 6(k+1)^2}{6} \Rightarrow$

$1^2 + 2^2 + 3^2 + \cdots + k^2 + (k+1)^2 = \frac{(k+1)(k(2k+1) + 6(k+1))}{6} \Rightarrow$

$1^2 + 2^2 + 3^2 + \cdots + k^2 + (k+1)^2 = \frac{(k+1)(2k^2 + 7k + 6)}{6} \Rightarrow$

$1^2 + 2^2 + 3^2 + \cdots + k^2 + (k+1)^2 = \frac{(k+1)(k+2)(2k+3)}{6} \Rightarrow$

$1^2 + 2^2 + 3^2 + \cdots + k^2 + (k+1)^2 = \frac{(k+1)((k+1)+1)(2(k+1)+1)}{6}$.

$\therefore$ by Steps 1 and 2 and mathematical induction, $1^2 + 2^2 + 3^2 + \cdots + n^2 = \frac{n(n+1)(2n+1)}{6}\left(\frac{1/2}{1/2}\right) =$

$\frac{n\left(n + \frac{1}{2}\right)(n+1)}{3}$ for all positive integers n.

11. a) Step 1: For $n = 1$, $\sum_{k=1}^{1}(a_k + b_k) = a_1 + b_1 = \sum_{k=1}^{1} a_k + \sum_{k=1}^{1} b_k$.

Step 2: Assume $\sum_{k=1}^{j}(a_k + b_k) = \sum_{k=1}^{j} a_k + \sum_{k=1}^{j} b_k$ for some positive integer j.

Then $\sum_{k=1}^{j+1}(a_k + b_k) = \left(\sum_{k=1}^{j}(a_k + b_k)\right) + (a_{j+1} + b_{j+1}) = \sum_{k=1}^{j} a_k + \sum_{k=1}^{j} b_k + a_{j+1} + b_{j+1} =$

11. (Continued)

$$\left(\sum_{k=1}^{j} a_k\right) + a_{j+1} + \left(\sum_{k=1}^{j} b_k\right) + b_{j+1} = \sum_{k=1}^{j+1} a_k + \sum_{k=1}^{j+1} b_k.$$

$\therefore$ by Steps 1 and 2 and mathematical induction, $\sum_{k=1}^{n} (a_k + b_k) = \sum_{k=1}^{n} a_k + \sum_{k=1}^{n} b_k$ for all positive integers n.

b) Step 1: For n = 1, $\sum_{k=1}^{1} (a_k - b_k) = a_1 - b_1 = \sum_{k=1}^{1} a_k - \sum_{k=1}^{1} b_k.$

Step 2: Assume $\sum_{k=1}^{j} (a_k - b_k) = \sum_{k=1}^{j} a_k - \sum_{k=1}^{j} b_k$ for some positive integer j.

Then $\sum_{k=1}^{j+1} (a_k - b_k) = \left(\sum_{k=1}^{j} (a_k - b_k)\right) + (a_{j+1} - b_{j+1}) = \sum_{k=1}^{j} a_k - \sum_{k=1}^{j} b_k + a_{j+1} - b_{j+1} =$

$$\left(\sum_{k=1}^{j} a_k\right) + a_{j+1} - \left(\left(\sum_{k=1}^{j} b_k\right) + b_{j+1}\right) = \sum_{k=1}^{j+1} a_k - \sum_{k=1}^{j+1} b_k.$$

$\therefore$ by Steps 1 and 2 and mathematical induction, $\sum_{k=1}^{n} (a_k - b_k) = \sum_{k=1}^{n} a_k - \sum_{k=1}^{n} b_k$ for all positive integers n.

c) Step 1: For n = 1, $\sum_{k=1}^{1} ca_k = ca_1 = c \sum_{k=1}^{1} a_k.$

Step 2: Assume $\sum_{k=1}^{j} ca_k = c \sum_{k=1}^{j} a_k$ for some positive integer j.

Then $\sum_{k=1}^{j+1} ca_k = \sum_{k=1}^{j} ca_k + ca_{j+1} = c\left(\sum_{k=1}^{j} a_k\right) + ca_{j+1} = c\left(\left(\sum_{k=1}^{j} a_k\right) + a_{j+1}\right) = c \sum_{k=1}^{j+1} a_k.$

$\therefore$ by Steps 1 and 2 and mathematical induction, $\sum_{k=1}^{n} ca_k = c \sum_{k=1}^{n} a_k$ for all positive integers n.

d) Step 1: For n = 1, $\sum_{k=1}^{1} a_k = a_1 = c = 1 \cdot c.$

Step 2: Assume $\sum_{k=1}^{j} a_k = jc$ for some positive integer j.

Then $\sum_{k=1}^{j+1} a_k = \sum_{k=1}^{j} a_k + a_{j+1} = jc + c = (j + 1)c.$

$\therefore$ by Steps 1 and 2 and mathematical induction, $\sum_{k=1}^{n} a_k = nc$ for all positive integers n and

where $a_k = c$, a constant.

APPENDIX A.7 DETERMINANTS AND CRAMER'S RULE

1. $\begin{vmatrix} 2 & 3 & 1 \\ 4 & 5 & 2 \\ 1 & 2 & 3 \end{vmatrix}\begin{matrix} 2 & 3 \\ 4 & 5 \\ 1 & 2 \end{matrix} = 30 + 6 + 8 - 5 - 8 - 36 = -5$

3. $\begin{vmatrix} 1 & 2 & 3 & 4 \\ 0 & 1 & 2 & 3 \\ 0 & 0 & 2 & 1 \\ 0 & 0 & 3 & 2 \end{vmatrix} = 1\begin{vmatrix} 1 & 2 & 3 \\ 0 & 2 & 1 \\ 0 & 3 & 2 \end{vmatrix} = 1\begin{vmatrix} 2 & 1 \\ 3 & 2 \end{vmatrix} = 1$

5. a) $\begin{vmatrix} 2 & -1 & 2 \\ 1 & 0 & 3 \\ 0 & 2 & 1 \end{vmatrix} = \begin{vmatrix} 2 & -5 & 2 \\ 1 & -6 & 3 \\ 0 & 0 & 1 \end{vmatrix} = 1\begin{vmatrix} 2 & -5 \\ 1 & -6 \end{vmatrix} = -7$

b) $\begin{vmatrix} 2 & -1 & 2 \\ 1 & 0 & 3 \\ 0 & 2 & 1 \end{vmatrix} = \begin{vmatrix} 2 & -1 & 2 \\ 1 & 0 & 3 \\ 4 & 0 & 5 \end{vmatrix} = -(-1)\begin{vmatrix} 1 & 3 \\ 4 & 5 \end{vmatrix} = -7$

7. a) $\begin{vmatrix} 1 & 1 & 0 & 0 \\ 0 & 0 & -2 & 1 \\ 0 & -1 & 0 & 7 \\ 3 & 0 & 2 & 1 \end{vmatrix} = \begin{vmatrix} 1 & 1 & 0 & 7 \\ 0 & 0 & -2 & 1 \\ 0 & -1 & 0 & 0 \\ 3 & 0 & 2 & 1 \end{vmatrix} = -(-1)\begin{vmatrix} 1 & 0 & 7 \\ 0 & -2 & 1 \\ 3 & 2 & 1 \end{vmatrix} = \begin{vmatrix} 1 & 0 & 7 \\ 0 & -2 & 1 \\ 0 & 2 & -20 \end{vmatrix} =$

$1\begin{vmatrix} -2 & 1 \\ 2 & -20 \end{vmatrix} = 38$

b) $\begin{vmatrix} 1 & 1 & 0 & 0 \\ 0 & 0 & -2 & 1 \\ 0 & -1 & 0 & 7 \\ 3 & 0 & 2 & 1 \end{vmatrix} = \begin{vmatrix} 1 & 1 & 0 & 0 \\ 0 & 0 & -2 & 1 \\ 1 & 0 & 0 & 7 \\ 3 & 0 & 2 & 1 \end{vmatrix} = -1\begin{vmatrix} 0 & -2 & 1 \\ 1 & 0 & 7 \\ 3 & 2 & 1 \end{vmatrix} = -1\begin{vmatrix} 0 & -2 & 1 \\ 1 & 0 & 7 \\ 0 & 2 & -20 \end{vmatrix} = -1(-1)$

$\begin{vmatrix} -2 & 1 \\ 2 & -20 \end{vmatrix} = 38$

9. $D = \begin{vmatrix} 1 & 8 \\ 3 & -1 \end{vmatrix} = -25.\ x = \dfrac{\begin{vmatrix} 4 & 8 \\ -13 & -1 \end{vmatrix}}{-25} = \dfrac{100}{-25} = -4,\ y = \dfrac{\begin{vmatrix} 1 & 4 \\ 3 & -13 \end{vmatrix}}{-25} = \dfrac{-25}{-25} = 1$

11. $D = \begin{vmatrix} 4 & -3 \\ 3 & -2 \end{vmatrix} = 1.\ x = \dfrac{\begin{vmatrix} 6 & -3 \\ 5 & -2 \end{vmatrix}}{1} = 3,\ y = \dfrac{\begin{vmatrix} 4 & 6 \\ 3 & 5 \end{vmatrix}}{1} = 2$

13. $D = \begin{vmatrix} 2 & 1 & -1 \\ 1 & -1 & 1 \\ 2 & 2 & 1 \end{vmatrix} = \begin{vmatrix} 2 & 1 & -1 \\ 3 & 0 & 0 \\ 4 & 3 & 0 \end{vmatrix} = -1\begin{vmatrix} 3 & 0 \\ 4 & 3 \end{vmatrix} = -9.\ x = \dfrac{\begin{vmatrix} 2 & 1 & -1 \\ 7 & -1 & 1 \\ 4 & 2 & 1 \end{vmatrix}}{-9} = \dfrac{\begin{vmatrix} 2 & 1 & -1 \\ 9 & 0 & 0 \\ 6 & 3 & 0 \end{vmatrix}}{-9} =$

$\dfrac{-1\begin{vmatrix} 9 & 0 \\ 6 & 3 \end{vmatrix}}{-9} = 3,\ y = \dfrac{\begin{vmatrix} 2 & 2 & -1 \\ 1 & 7 & 1 \\ 2 & 4 & 1 \end{vmatrix}}{-9} = \dfrac{\begin{vmatrix} 2 & 2 & -1 \\ 3 & 9 & 0 \\ 4 & 6 & 0 \end{vmatrix}}{-9} = \dfrac{-1\begin{vmatrix} 3 & 9 \\ 4 & 6 \end{vmatrix}}{-9} = -2,\ z = \dfrac{\begin{vmatrix} 2 & 1 & 2 \\ 1 & -1 & 7 \\ 2 & 2 & 4 \end{vmatrix}}{-9} =$

$\dfrac{\begin{vmatrix} 3 & 0 & 9 \\ 1 & -1 & 7 \\ 4 & 0 & 18 \end{vmatrix}}{-9} = \dfrac{-1\begin{vmatrix} 3 & 9 \\ 4 & 18 \end{vmatrix}}{-9} = 2$

15. $D = \begin{vmatrix} 1 & 0 & -1 \\ 0 & 2 & -2 \\ 2 & 0 & 1 \end{vmatrix} = 2\begin{vmatrix} 1 & -1 \\ 2 & 1 \end{vmatrix} = 6.\ x = \dfrac{\begin{vmatrix} 3 & 0 & -1 \\ 2 & 2 & -2 \\ 3 & 0 & 1 \end{vmatrix}}{6} = \dfrac{2\begin{vmatrix} 3 & -1 \\ 3 & 1 \end{vmatrix}}{6} = 2,\ y = \dfrac{\begin{vmatrix} 1 & 3 & -1 \\ 0 & 2 & -2 \\ 2 & 3 & 1 \end{vmatrix}}{6} =$

$\dfrac{\begin{vmatrix} 1 & 3 & -1 \\ 0 & 2 & -2 \\ 0 & -3 & 3 \end{vmatrix}}{6} = \dfrac{1\begin{vmatrix} 2 & -2 \\ -3 & 3 \end{vmatrix}}{6} = 0,\ z = \dfrac{\begin{vmatrix} 1 & 0 & 3 \\ 0 & 2 & 2 \\ 2 & 0 & 3 \end{vmatrix}}{6} = \dfrac{2\begin{vmatrix} 1 & 3 \\ 2 & 3 \end{vmatrix}}{6} = -1$

17. $D = \begin{vmatrix} 2 & h \\ 1 & 3 \end{vmatrix} = 6 - h = 0 \Rightarrow h = 6.\ x: \begin{vmatrix} 8 & h \\ k & 3 \end{vmatrix} = 24 - hk = 24 - 6k = 0 \Rightarrow k = 4$

a) When $h = 6$, $k = 4$, there are infinitely many solutions.

b) When $h = 6$, $k \neq 4$, there are no solutions.

19. $au + bv + cw = 0 \Rightarrow v = \dfrac{-au - cw}{b},\ au' + bv' + cw' = 0 \Rightarrow v' = \dfrac{-au' - cw'}{b},\ au'' + bv'' + cw'' = 0 \Rightarrow$

$v'' = \dfrac{-au'' - cw''}{b} \Rightarrow \begin{vmatrix} u & v & w \\ u' & v' & w' \\ u'' & v'' & w'' \end{vmatrix} = \begin{vmatrix} u & \dfrac{-au - cw}{b} & w \\ u' & \dfrac{-au' - cw'}{b} & w' \\ u'' & \dfrac{-au'' - cw''}{b} & w'' \end{vmatrix} = \dfrac{1}{b}\begin{vmatrix} u & -au - cw & w \\ u' & -au' - cw' & w' \\ u'' & -au'' - cw'' & w'' \end{vmatrix} =$

19. (Continued)

$$\frac{1}{b}(u(-au' - cw')w'' + u''(-au - cw)w' + w(-au'' - cw'')u' - u''(-au' - cw')w - u'(-au - cw)w'' - u(-au'' - cw'')w') = \frac{1}{b}(0) = 0$$